MUSCLE DEVELOPMENT
Molecular and Cellular Control

MUSCLE DEVELOPMENT
Molecular and Cellular Control

EDITED BY

Mark L. Pearson
Frederick Cancer Research Facility

Henry F. Epstein
Baylor College of Medicine

COLD SPRING HARBOR LABORATORY / 1982

MUSCLE DEVELOPMENT: Molecular and Cellular Control

Cover and book design by Emily Harste
Printed in United States of America

Front cover: Photos show cultures of the L6 line of skeletal muscle cells before (top) and after (bottom) differentiation. Residual myoblasts in the differentiated culture have been removed to show only long multinucleated myotubes. (Photos by J. Garrels)

Back cover: (Top) R-loop analysis of a representative hybrid molecule of genomic clone 287 A-3 with RNA from L6E9 myotubes. (Bottom) Summary diagram of exon and intron locations in clone 287 A-3. (From Nadal-Ginard et al., this volume)

Library of Congress Cataloging in Publication Data

Main entry under title:

Muscle development--molecular and cellular control.

Includes bibliographical references and index.
1. Muscle--Congresses. 2. Embryology--Congresses.
3. Cellular control mechanisms--Congresses.
I. Pearson, Mark L. II. Epstein, Henry F., 1944- .
III. Cold Spring Harbor Laboratory. [DNLM: 1. Muscles--Embryology--Congresses. 2. Muscles--Cytology--Congresses.
WE 500 M9865 1981]
QL831.M87 1982 591.1'852 82-72381
ISBN 0-87969-154-9

All Cold Spring Harbor Laboratory publications are available through booksellers or may be ordered directly from Cold Spring Harbor Laboratory, Box 100, Cold Spring Harbor, New York 11724.

SAN 203-6185

First Row: R. Firtel/J.-C. Perriard, M. Pearson
Second Row: D. Fischman, S. Lowey/B. Storti, S. Kaufman
Third Row: F. Stockdale/M. Buckingham/J. Garrels
Fourth Row: Cocktail Party at Airslie

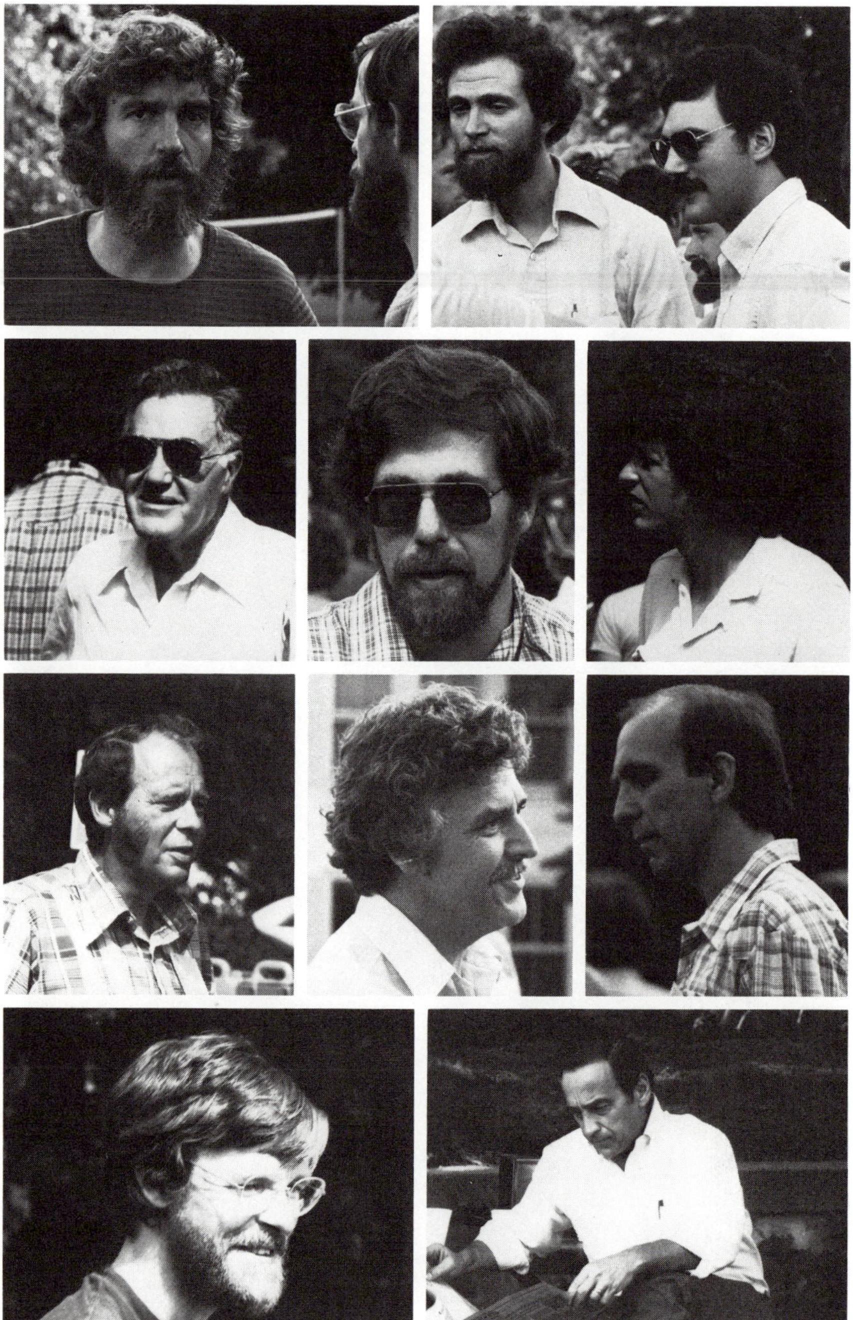

First Row: Z. Hall/G. Crouse, R. Zivin
Second Row: I.R. Konigsberg/H. Epstein/C. Clegg
Third Row: J. Colemen/R. Przybylski/S. Hauschka
Fourth Row: C. Slayter/S. Appel

Meeting Participants

Allen, Ronald, *Dept. of Nutrition and Food Science, University of Arizona, Tucson*
Anderson, M. John, *Dept. of Embryology, Carnegie Institution of Washington, Baltimore, Maryland*
Appel, Stanley, *Dept. of Neurology, Baylor College of Medicine, Houston, Texas*
Appel, Vicki, *Dept. of Neurology, Baylor College of Medicine, Houston, Texas*
Armant, Rany, *Dept. of Cell Biology, Worcester Foundation, Shrewsbury, Massachusetts*
Baldwin, Albert, Jr., *Dept. of Biology, University of Virginia, Charlottesville*
Bandman, Everett, *Dept. of Zoology, University of California, Berkeley*
Bautch, Victoria, *Dept. of Biological Chemistry, University of Illinois Medical Center, Chicago*
Bayne, Ellen, *Carnegie Institution of Washington, Baltimore, Maryland*
Benfield, Pamela, *Cancer Biology Program, Frederick Cancer Research Facility, Frederick, Maryland*
Benoff-Rind, Susan, *Dept. of Anatomy and Cell Biology, State University of New York, Downstate Medical Center, Brooklyn*
Benyajati, Cheeptip, *Cancer Biology Program, Frederick Cancer Research Facility, Frederick, Maryland*
Bernstein, Sanford, *Dept. of Biology, University of Virginia, Charlottesville*
Bixby, John, *Dept. of Biology, University of California, San Diego, La Jolla*
Blau, Helen, *Dept. of Pharmacology, Stanford University School of Medicine, California*
Bleisch, W.V., *Rockefeller University, New York, New York*
Bonner, Philip, *Dept. of Biological Sciences, University of Kentucky, Lexington*
Bonner, Susan, *Dept. of Pathology, University of Kentucky, Lexington*
Bragg, P.W., *Dept. of Genetics, University of Connecticut, Storrs*
Buckingham, Margaret, *Dept. of Molecular Biology, Pasteur Institute, Paris, France*
Butler-Browne, Gillian, *Dept. of Molecular Biology, Pasteur Institute, Paris, France*
Caplan, Arnold, *Dept. of Biology, Case Western Reserve University, Cleveland, Ohio*
Cates, George, *Dept. of Biochemistry, University of Western Ontario, London, Canada*
Cherney, Barry, *Dept. of Biology, Case Western Reserve University, Cleveland, Ohio*
Chiquet, Matthias, *Dept. of Embryology, Carnegie Institute of Washington, Baltimore, Maryland*
Chizzonite, Richard, *Dept. of Medicine, University of Chicago, Illinois*
Clark, William, *Dept. of Medicine, University of Chicago, Illinois*
Claycomb, William, *Louisiana State University School of Medicine, New Orleans*
Clegg, Chris, *Dept. of Biochemistry, University of Washington, Seattle*
Cohen, M.W., *Dept. of Physiology, McGill University, Montreal, Canada*
Colella, Rita, *Bureau of Biological Research, Rutgers University, Piscataway, New Jersey*
Coleman, John, *Dept. of Biology and Medicine, Brown University, Providence, Rhode Island*
Cossu, Giulio, *Wistar Institute, Philadelphia, Pennsylvania*
Craig, Susan, *Dept. of Physiological Chemistry, Johns Hopkins University, Baltimore, Maryland*
Crain, William, *Worcester Foundation, Shrewsbury, Massachusetts*
Crouse, Gray, *Cancer Biology Program, Frederick Cancer Research Facility, Frederick, Maryland*

Davey, David, *McGill University, Montreal, Canada*
Davidson, E.H., *Dept. of Biology, California Institute of Technology, Pasadena*
Davidson, Norman, *Dept. of Biology, California Institute of Technology, Pasadena*
Delaporte, Chris, *Laboratory of Biology and Pathology, Pasteur Institute, Paris, France*
Dhillon, Daljit Singh, *University of Strathclyde, Glasgow, Scotland*
Dhoot, G.K., *Dept. of Immunology, University of Birmingham, England*
Di Mauro, Salvatore, *Research Center for Muscular Dystrophy, Columbia University, New York, New York*
Donady, J., *Dept. of Biology, Wesleyan University, Middletown, Connecticut*
Dottin, Robert, *Dept. of Biology, Johns Hopkins University, Baltimore, Maryland*
Ecob, Marion, *Muscular Dystrophy Laboratory, University of Newcastle, England*
Ehrismann, Ruth, *State University of New York Upstate Medical Center, Syracuse*
Elson, Hannah, *Dept. of Pathology, University of California, San Diego, La Jolla*
Emerson, Charles, *Dept. of Biology, University of Virginia, Charlottesville*
Emmerling, Mark, *Carnegie Institution of Washington, Baltimore, Maryland*
Engel, Joanne, *Veterans Administration Hospital, Stanford Universty, Palo Alto, California*
Eppenberger, Hans, *Dept. of Cell Biology, Swiss Federal Institute of Technology, Zurich, Switzerland*
Epstein, Henry, *Dept. of Neurology, Baylor College of Medicine, Houston, Texas*
Everett, Alan, *Dept. of Medicine, University of Chicago, Illinois*
Falkenthal, Scott, *Dept. of Chemistry, California Institute of Technology, Pasadena*
Fambrough, D.M., *Dept. of Embryology, Carnegie Institution of Washington, Baltimore, Maryland*
Fernandez, Salvador, *Dept. of Physiology, University of Connecticut Health Center, Farmington*
Firtel, Richard, *Dept. of Biology, University of California, San Diego, La Jolla*
Fischman, Donald, *State University of New York Downstate Medical Center, Brooklyn*
Fiszman, Marc Y., *Dept. of Molecular Biology, Pasteur Institute, Paris, France*
Frair, Patricia, *McGill University, Montreal, Canada*
Fuchs, Sara, *Dept. of Chemical Immunology, Weizmann Institute of Science, Rehovot, Israel*
Gambke, Brigette, *Dept. of Anatomy, University of Pennsylvania School of Medicine, Philadelphia*
Gardner, John, *Dept. of Embryology, Carnegie Institution of Washington, Baltimore, Maryland*
Garfinkel, Leonard, *Dept. of Pediatrics, Harvard Medical School, Boston, Massachusetts*
Garrels, James I., *Cold Spring Harbor Laboratory, New York*
Gauthier, Geraldine, *Dept. of Anatomy, University of Massachusetts Medical School, Worcester*
Geisler, Norbert, *Dept. of Biophysical Chemistry, Max-Planck Institute, Goettingen, Federal Republic of Germany*
Gilfix, Brian, *Dept. of Biochemistry, University of Western Ontario, London, Canada*
Giometti, Carol, *Dept. of Molecular Anatomy, Argonne National Laboratory, Illinois*
Glass, Charles, *Dept. of Zoology, University of California, Berkeley*
Gordon, Joel, *State University of New York, Stony Brook*
Grebenau, Ruth, *Dept. of Cell Biology, Albert Einstein College of Medicine, Bronx, New York*
Gubits, David, *Albert Einstein College of Medicine, Bronx, New York*
Gunning, Peter, *Veterans Administration Hospital, Stanford University, Palo Alto, California*
Guttman, Susan, Dept. of Pharmacology, Stanford University, California
Hall, Zach, *Dept. of Physiology, University of California, San Francisco*
Hance, Allan, *Division of Respiratory Medicine, Stanford University, California*
Hastings, Ken, *Dept. of Biology, University of Virginia, Charlottesville*
Hauschka, Stephen, *Dept. of Biochemistry, University of Washington, Seattle*
Heinemann, Stephen, *Dept. of Neurobiology, Salk Institute, San Diego, California*
Heywood, S.M., *Dept. of Genetics, University of Connecticut, Storrs*
Hirsh, David, *University of Colorado, Boulder*
Holland, Paul, *Montreal Neurological Institute, Canada*
Holtzer, H., *Dept. of Anatomy, University of Pennsylvania School of Medicine, Philadelphia*
Hornig, Danuta, *Dept. of Pediatrics, Harvard Medical School, Boston, Massachusetts*
Horwitz, A.F., *University of Pennsylvania School of Medicine, Philadelphia*
Hudecki, Michael, *Dept. of Cell and Molecular Biology, State University of New York, Buffalo*

Huiatt, Ted, *Brandeis University, Waltham, Massachusetts*
Hurko, Orest, *National Hospital, London, England*
Hutchison, C.J., *Institute of Neurology, London, England*
Inestrosa, Nibaldo, *Dept. of Physiology, University of California, San Francisco*
Jacob, Michele, *Dept. of Biology, University of California, San Diego, La Jolla*
Kalderon, Nurit, *Rockefeller University, New York, New York*
Karn, J., *Medical Research Council Laboratory, London, England*
Kaufman, Stephen, *University of Illinois, Urbana*
Kavinsky, Clifford, *Chicago, Illinois*
Kedes, Laurence, *Veterans Administration Hospital, Stanford University School of Medicine, Palo Alto, California*
Kidokoro, Yoshi, *Salk Institute, San Diego, California*
Kittler, Ellen, *Dept. of Biology, University of Virginia, Charlottesville*
Konieczny, Stephen, *Biomedical Division, Brown University, Providence, Rhode Island*
Konigsberg, Irwin, *Dept. of Biology, University of Virginia, Charlottesville*
Kucera, Jan, *Veterans Administration Hospital, Stanford University, California*
Kuncio, Gerald, *Dept. of Anatomy, Temple University Medical School, Philadelphia, Pennsylvania*
Lawrence, Jeanne, *Brown University, Providence, Rhode Island*
Lazarides, Elias, *Dept. of Biology, California Institute of Technology, Pasadena*
Liew, C., *Banting Institute, University of Toronto, Canada*
Lin, James, *Cold Spring Harbor Laboratory, New York*
Linden, Diana, *Dept. of Physiology, University of California School of Medicine, Los Angeles*
Linkhart, Thomas, *Dept. of Biochemistry, University of Washington, Seattle*
Lowey, Susan, *Rosenstiel Center, Brandeis University, Waltham, Massachusetts*
MacBride, Robert, *Dept. of Anatomy, Oral Roberts University, Tulsa, Oklahoma*
Mahdavi, Vijale, *Dept. of Pediatrics, Harvard Medical School, Boston, Massachusetts*
Maisonpierre, Peter, *Dept. of Biology, University of Virginia, Charlottesville*
Masaki, Tomoh, *State University of New York Downstate Medical Center, Brooklyn*
Matsuda, Ryoichi, *Dept. of Zoology, University of California, Berkeley*
Matsumura, Fumio, *Cold Spring Harbor Laboratory, New York*
McCarthy, T.L., *Dept. of Genetics, University of Connecticut, Storrs*
McMahan, U. Jack, *Dept. of Neurobiology, Stanford University School of Medicine, California*
Merlie, John, *Dept. of Biological Sciences, University of Pittsburgh, Pennsylvania*
Merrifield, Peter, *Dept. of Biology, University of Virginia, Charlottesville*
Miller, David, *Dept. of Neurology, Baylor College of Medicine, Houston, Texas*
Minty, Adrian, *Dept. of Molecular Biology, Pasteur Institute, Paris, France*
Molinaro, Mario, *Institute of Istologia, Rome, Italy*
Montarras, Didier, *Dept. of Molecular Biology, Pasteur Institute, Paris, France*
Moody-Corbett, F., *Dept. of Physiology, McGill University, Montreal, Canada*
Murray, Susan, *Dept. of Biology, Wesleyan University, Middletown, Connecticut*
Nadal-Ginard, Bernardo, *Dept. of Pediatrics, Harvard Medical School, Boston, Massachusetts*
Nameroff, Mark, *University of Washington School of Medicine, Seattle*
Nayaki, R.C., *Institute of Neurology, London, England*
Nguyen, Hanh Thi, *Dept. of Pediatrics, Harvard Medical School, Boston, Massachusetts*
Nigg, Erich, *Dept. of Biology, University of California, San Diego, La Jolla*
Noonan, Douglas, *Dept. of Anatomy, Case Western Reserve University, Cleveland, Ohio*
Nudel, Uri, *Dept. of Cell Biology, Weizmann Institute of Science, Rehovot, Israel*
O'Neill, Michael, *Dept. of Biological Sciences, University of Maryland, Baltimore*
Obinata, Takashi, *Dept. of Biology, Chiba University, Japan*
Ordahl, Charles, *Dept. of Anatomy, Temple University Medical School, Philadelphia, Pennsylvania*
Parent, James, *Howard University Cancer Center, Washington, D.C.*
Paterson, Bruce, *NCI, National Institutes of Health, Bethesda, Maryland*
Patrick, James, *Dept. of Neurobiology, Salk Institute, San Diego, California*
Pauw, Peter, *Dept. of Genetics, Iowa State University, Ames*
Payne, Michael, *Dept. of Biology, University of Virginia, Charlottesville*
Pearson, Mark L., *Cancer Biology Program, Frederick Cancer Research Facility, Frederick, Maryland*
Periasamy, M., *Dept. of Pediatrics, Harvard Medical School, Boston, Massachusetts*

Perriard, J.-C., *Dept. of Cell Biology, Swiss Federal Institute for Technology, Zurich, Switzerland*
Podleski, T.R., *Langmuir Laboratory, Cornell University, Ithaca, New York*
Prives, Joav, *Dept. of Anatomical Science, State University of New York, Stony Brook*
Przybylski, Ronald, *Dept. of Anatomy, Case Western Reserve University, Cleveland, Ohio*
Quinn, Chris, *Institute of Neurology, London, England*
Quinn, Lebris, *University of Washington School of Medicine, Seattle*
Reinach, Fernando, *Dept. of Anatomy and Cell Biology, State University of New York Downstate Medical Center, Brooklyn*
Richer, Leroy, *U.S. Dept. of Agriculture, U.S. Meat Animal Research Center, Clay Center Nebraska*
Robbins, Jeffrey, *Dept. of Biochemistry, University of Missouri, Columbia*
Rotundo, Richard, *Dept. of Embryology, Carnegie Institution of Washington, Baltimore, Maryland*
Rozek, Charles, *Dept. of Chemistry, California Institute of Technology, Pasadena*
Rubin, Lee, *Dept. of Neurobiology, Rockefeller University, New York, New York*
Rubinstein, Neal, *Dept. of Anatomy, University of Pennsylvania School of Medicine, Philadelphia*
Salpeter, Miriam, *Cornell University, Ithaca, New York*
Scarpa, Sigfrido, *Instituto Superiore di Sanita, Rome, Italy*
Schwartz, J.L., *Dept. of Embryology, Carnegie Institution of Washington, Baltimore, Maryland*
Schwartz, Robert, *Dept. of Cell Biology, Baylor College of Medicine, Houston, Texas*
Sebbane, R., *Dept. of Biological Science, University of Pittsburgh, Pennsylvania*
Shani, Moshe, *Weizmann Institute of Science, Rehovot, Israel*
Shanske, Sara, *Columbia University College of Physicians and Surgeons, New York, New York*
Siddiqui, M.A.Q., *Dept. of Biochemistry, Roche Institute of Molecular Biology, Nutley, New Jersey*
Silberstein, Laura, *Dept. of Physiology, University of California, San Francisco*
Silver, Geri, *Dept. of Anatomy and Cell Biology, State University of New York Downstate Medical Center, Brooklyn*
Sinha, A.M., *Dept. of Medicine, University of Chicago, Illinois*
Slater, C.R., *Muscular Dystrophy Laboratory, University of Newcastle, England*
Smith, Martin, *Dept. of Biology, University of California, San Diego, La Jolla*
Stein, Joseph, *Dept. of Endocrinology, University of Texas Medical School, Houston, Texas*
Stockdale, Frank, *Dept. of Oncology, Stanford University Medical Center, California*
Storti, Robert, *Dept. of Biochemistry, University of Illinois Medical Center, Chicago*
Strittmatter, Warren, *Dept. of Neurology, Baylor College of Medicine, Houston, Texas*
Strohman, R.C., *Dept. of Zoology, University of California, Berkeley*
Sutherland, William, *Dept. of Biology, University of Virginia, Charlottesville*
Turner, David, *Dept. of Biochemistry, State University of New York Upstate Medical Center, Syracuse*
Umeda, Patrick, *Dept. of Medicine, University of Chicago, Illinois*
Vandenburgh, Herman, *Dept. of Pathology and Laboratory Medicine, Miriam Hospital, Providence, Rhode Island*
Wakelam, M.J.O., *Universität Konstanz, Federal Republic of Germany*
Waller, G., *Brandeis University, Waltham, Massachusetts*
Walsh, Frank, *Institute of Neurology, London, England*
Wang, Kuan, *Dept. of Chemistry, University of Texas, Austin*
Waterston, R.H., *Dept. of Genetics, Washington University, St. Louis, Missouri*
Webster, Cecelia, *Dept. of Pharmacology, Stanford Medical Center, California*
Weldon, Peter, *Dept. of Physiology, McGill University, Montreal, Canada*
Whalen, R.G., *Dept. of Molecular Biology, Pasteur Institute, Paris, France*
Winkelmann, Donald, *Brandeis University, Waltham, Massachusetts*
Wolitzky, Barry, *Dept. of Cell Biology, State University of New York, Buffalo*
Wright, Woodring, *University of Texas Health Sciences Center, Dallas*
Wydro, R., *Dept. of Pediatrics, Harvard Medical School, Boston, Massachusetts*
Yamashiro-Matsumura, Shigeko, *Cold Spring Harbor Laboratory, New York*
Yaross, Marcia, *Dept. of Oncology, Stanford University Medical Center, California*

Yasin, Rose, *Institute of Neurology, London, England*
Zak, Radovan, *Dept. of Medicine, University of Chicago, Illinois*
Zalin, Rosalind, *Dept. of Anatomy, University College, London, England*
Zehner, Zenra, *NCI, National Institutes of Health, Bethesda, Maryland*
Ziskind-Conhaim, Lea, *Dept. of Physiology, University of California, San Francisco*
Zivin, Robert, *Cancer Biology Program, Frederick Cancer Research Facility, Frederick, Maryland*
Zubrzycka, E., *Best Institute, University of Toronto, Canada*

Preface

Research in muscle development has attracted scientists with a wide spectrum of interests ranging from protein structure and gene expression to human disease. As a biological system, muscle is complex enough to be interesting and simple enough to permit detailed physical and genetic analysis. In the past, many of the questions being studied in this area appeared to be unrelated. What, for example, were the common elements relating the formation of neuromuscular junctions and the structural complexities of various forms of myosin? However, muscle research, like many other fields of biological science, has undergone a profound transformation in the past few years, fueled by spectacular advances in cloning genes, antibodies, and muscle cells. That we were entering a transition phase in this field became clear at the European Molecular Biology Organization (EMBO) workshop, organized by David Yaffe and his colleagues at Shoresh, Israel, in March, 1980. It seemed then that a meeting encompassing both traditional and modern approaches to understanding muscle structure, function, and development might serve a catalytic function, not only for integrating what was known, but also for synthesizing new experimental strategies. The result was the meeting on "Molecular and Cellular Control of Muscle Development" (held September, 1981, at Cold Spring Harbor Laboratory) which provided the impetus for the eventual compilation of this book.

Many people contributed significant amounts of time and effort to make the meeting a truly remarkable scientific event. Steve Kaufman and Jim Garrels were invaluable colleagues on the organizing committee. Stan Appel, Charlie Emerson, Rick Firtel, Don Fischman, Zach Hall, and Frank Stockdale helped by organizing and chairing sessions. The Muscular Dystrophy Association (MDA) was especially generous, both in support of the meeting and in encouraging publication of this book by a significant advance purchase. We are especially indebted to Dr. Mel Moss, MDA's Director of Research, for his enthusiastic support. Essential financial support was also provided by the National Cancer Institute, Frederick Cancer Research Facility; the National Science Foundation; and Merck, Sharp, and Dohme Research Laboratories.

The meeting would not have been possible without the advice, encouragement, and support of Jim Watson, Director of Cold Spring Harbor

Laboratory, who, fortunately, had already been thinking of such a meeting when he was first approached. We are also grateful to Gladys Kist and the staff in the Meetings Office for their help with the plethora of arrangements necessary for such a meeting, and to Nancy Ford, Nadine Dumser, Michaela Cooney, Mary Cozza, and Gail Anderson of the staff in the Publications Department for patiently guiding this book to completion with no (known) loss of life along the way. Finally, this meeting would not have been nearly so exciting without the energy and enthusiasm of the participants, many of them not represented directly in this volume, although their thoughts and criticisms have done much to shape what is recorded here.

Mark L. Pearson
Henry F. Epstein

Contents

Membrane Events

Embryogenesis

REGULATION OF MUSCLE PROTEINS

Introduction: Regulation of Muscle Proteins

HENRY F. EPSTEIN
Program in Neurosciences
Departments of Neurology, Biochemistry, and Medicine
Baylor College of Medicine
Houston, Texas 77030

Changes in the expression of proteins follow two major patterns during muscle development. One pattern is the end result of muscle differentiation. Physiologically and morphologically distinct adult muscle types may exhibit different forms of a specific protein. For example, fast and slow skeletal muscles produce characteristic forms of heavy and light myosin polypeptide chains (Frank and Weeds 1974; Weeds 1976; Whalen et al., this volume). The other pattern occurs during development. Certain muscles may sequentially produce different forms of a muscle-characteristic protein. The switch in isoactins from the β and γ forms of myoblasts to α-actin as the major protein subunit of thin filaments in embryonic myotubes is now a classic example (Garrels and Gibson 1976).

Protein chemistry and immunology have been key methods in the study of these transformations of specific proteins during muscle development. Different forms of a particular muscle protein have been characterized by their unique peptide maps and, in a limited number of cases, unique amino acid sequences. These differences in covalent structure have been associated with changes in the antigenic determinants of the particular protein. These protein forms can react with specific antibodies in sufficiently different manners, both quantitatively and qualitatively, to permit the isolation of antibody fractions specific to one isoform. These highly specific antibodies can be used to locate isoforms within muscle and to trace the production of isoforms during development.

Research discussed in the papers that follow focuses upon myosin as the molecule of interest. Myosin was the first contractile protein to be identified; its role as the enzyme that catalyzes chemical energy in the form of

ATP into the mechanical work of muscle contraction is well established. Accordingly, myosin has been the subject of both chemical and immunological studies for over 20 years. These methodological approaches have been exploited in studying the presence of different myosins during muscle development.

Myosin multiplicity appears to be both phylogenetically and histologically general, in that individual muscles from nematodes to birds and mammals contain more than one myosin type. Different kinds of muscles, such as cardiac and skeletal muscles in mammals, produce distinct sets of myosins. Even distinguishable parts of the same muscles may be differentiated at the molecular level. For example, in the mammalian heart, the atria and ventricles produce different myosins. As a particular muscle goes through development, the kinds of myosins and the amount of each myosin may change. In avian and mammalian skeletal muscles, at least three forms of myosin heavy chains are expressed sequentially: embryonic, neonatal, and adult. These homologous forms of myosin subunit contribute to the sequential appearance of myosin isozymes during development.

The patterns of developmental changes themselves may be further modified. The hormone thyroxine appears to regulate the forms of myosin in the cardiac ventricles of the rabbit (Everett et al., this volume). Certain mutants in the nematode *Caenorhabditis elegans* can affect the production of specific myosin isoforms. Some of these mutants are in the structural gene coding for only one of several kinds of myosin heavy chains (Epstein et al., Karn et al; both this volume). These results provide explicit evidence that more than one gene exists to code for these multiple forms of myosin subunits.

The hypothesis that families of genes are required to code for the multiple forms of myosin and other muscle proteins was also strongly suggested by the chemical and immunological analyses of specific proteins. Direct confirmation of this hypothesis at the gene level has now been accomplished for the actin and myosin heavy-chain genes of certain species by the chemical analysis of the DNA base sequences in such genes. The mechanisms of how such multigene families arose and how specific family members are differentially expressed are still not understood.

Several paradoxes require explanation if we are to understand the functional significance of the different myosin isozymes. Specific genes code for the distinct myosin heavy chains in both fast and slow adult embryonic muscles of the chicken. Specialized mechanisms must control the differential synthesis of these chains. Despite this high degree of differentiation at a genetic level, the structurally distinct myosins assembled from these different subunits when purified and studied biochemically exhibit indistinguishable enzymatic properties (Lowey et al., this volume). Are there other functions served by these molecules, of which we are unaware, that require the different isoforms? Another question concerns the two classes of

vertebrate myosin light chains that constitute at least two gene families requiring elaborate genetic regulation for the differential expression of their members. Recent biochemical experiments suggest that the enzymatic properties of the rabbit skeletal-muscle myosin head do not require any light chains (Wagner and Giniger 1981). What then is the physiological function of the light chains?

A plausible answer to these apparent paradoxes of genes and functions may be found in considering evolution and speciation. The multiplicity of muscle proteins and even the very presence of certain proteins may reflect not only the present physiological demands of the particular biological system under study but also may be the consequence of past events in evolution. Thus, myosin light chains may have regulatory significance in certain muscles, such as in many invertebrates (Lehman et al. 1973) or in vertebrate smooth muscle, but not in skeletal muscle (Adelstein and Eisenberg 1980). Their continued presence in skeletal muscle may reflect some secondary difficulty in deleting the set of light-chain genes that is unrelated to the primary function of the light chains themselves. The multiplicity of isoforms in different stages of development may in part also derive from events in the evolutionary past, such as a set of genes being duplicated and being linked with positively selected characters in the origin of the species. In this regard, biochemists who study myosin isozymes may be contributing to evolutionary biology as well as to the understanding of development.

REFERENCES

Adelstein, R.S. and E. Eisenberg. 1980. Regulation and kinetics of the actin-myosin-ATP interaction. *Annu. Rev. Biochem.* **49:** 921.

Frank, G. and A.G. Weeds. 1974. The amino acid sequence of the alkali light chains of rabbit skeletal muscle myosin. *Eur. J. Biochem.* **44:** 317.

Garrels, J.I. and W. Gibson. 1976. Identification and characterization of multiple forms of actin. *Cell* **9:** 793.

Lehman, W., J. Kendrick-Jones, and A.G. Szent-Gyorgyi. 1973. Myosin-linked regulatory systems: Comparative studies. *Cold Spring Harbor Symp. Quant. Biol.* **37:** 319.

Wagner, P.D. and E. Giniger. 1981. Hydrolysis of ATP and reversible binding to F-actin by myosin heavy chains free of all light chains. *Nature* **292:** 560.

Weeds, A.G. 1976. Light chains from slow-twitch muscle myosin. *Eur. J. Biochem.* **66:** 157.

Immunological Studies of Myosin Isoforms in Nematode Embryos

HENRY F. EPSTEIN
DAVID M. MILLER III
Program in Neurosciences
Departments of Neurology, Biochemistry, and Medicine
Baylor College of Medicine
Houston, Texas 77030

LANI A. GOSSETT
RALPH M. HECHT
Department of Biochemical and Biophysical Sciences
University of Houston, Central Campus
Houston, Texas 77004

Structural analysis of myosin and other muscle proteins by peptide mapping and amino acid sequencing first suggested that these isoforms are the products of different genes. The use of antibodies specific to distinct structural gene products has permitted detailed investigations of myosin structure, biosynthesis and degradation, and cellular location as muscle development proceeds. We have been interested in studying the assembly and function of myosin using such techniques in the small nematode *Caenorhabditis elegans*, a laboratory animal that offers genetic dissection and manipulation as tools in the deciphering of gene regulation in terms of specific protein synthesis during muscle development.

The first application of this combined genetic and immunochemical approach in the nematode provided evidence for the multiplicity of myosin heavy chains and its relation to different muscle types (Epstein et al. 1974). In this work, three isoforms of myosin heavy chain were identified: A, B, and C. The B isoform appeared specific to body-wall muscle; the C isoform, specific to pharyngeal muscle; and the A isoform, shared by both muscles.

The B isoform was clearly specified by a gene not encoding either the A or C isoforms; several *unc-54* mutants had distinct effects upon the B isoform alone. Peptide mapping of the isolated myosin heavy chains or of the homodimeric myosin molecules verified that the A and B myosin heavy chains have multiple differences in amino acid sequence; these experiments coupled with the genetic results implied that the major isoforms are encoded by separate genes (Epstein et al. 1977; MacLeod et al. 1977a,b, 1981; Schachat et al. 1977, 1978). Immunocytochemical studies indicated that myosins A and B (homodimers of the heavy-chain isoforms) coexist within the same body-wall muscle cells and sarcomeres (Mackenzie et al. 1978). Developmental studies showed that both myosins are synthesized at a constant ratio in late embryos, larvae, and adults (Garcea et al. 1978).

RESULTS AND DISCUSSION

Reactivity of Monoclonal Antibodies with Myosin Isoforms and Paramyosin

Purified nematode myosin (Harris and Epstein 1977) was used as the immunogen in sensitizing BALB/c mice. Paramyosin, a protein closely associated with myosin in nematode thick filaments, is a variable contaminant. The spleens of these mice were hybridized with the nonsecreting S194/5.XXO.BU.1 myeloma cell line (Trowbridge 1978), viable hybridomas were cloned, and proliferating clones were screened for reactivity with either nematode myosin or paramyosin by solid-phase assay (Sogin and Hinkle 1980). Some 24 clones were selected on this basis.

Several considerations must guide the characterization of antibody reactivity. In general, the reactivity of the immunoglobulin must be tested not only against antigens that might have contaminated the immunogen preparation but also antigens that may have some structural homology to the major antigen. The actual reaction product should be analyzed by a technique that permits resolution and detection of the other antigens at concentrations significantly lower than that of the major antigen. Reactions at a wide range of antibody-to-antigen ratios should be performed in order to detect lower affinity interactions. We have reacted antibodies with nematode proteins labeled in vivo with $^{35}SO_4^{=}$ (Garcea et al. 1978), separated the immunoprecipitates by SDS-polyacrylamide gel electrophoresis, and detected the antigens by autoradiography of the dried gels (Schachat et al. 1978). Alternatively, the proteins were separated electrophoretically and transferred to nitrocellulose sheets (Towbin et al. 1979). The positions of antigens on these sheets were determined by reaction with the specific antibody and then with secondary antibody linked to peroxidase. We have used both methods to characterize the immunoglobulin products of our mouse hybridomas.

Figure 1 shows the analysis of such a reaction. Immunoglobulin was

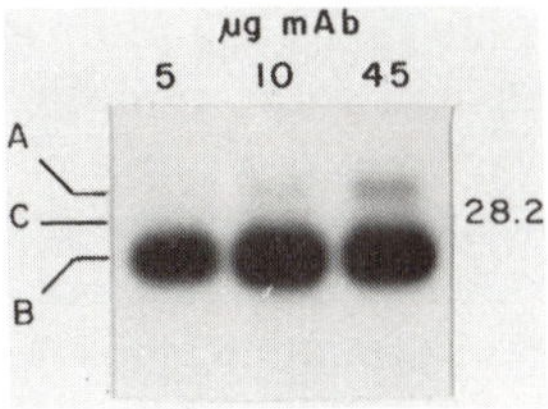

FIGURE 1
Polyacrylamide gel analysis of the nematode myosin isoforms that react with different concentrations of monoclonal antibody from hybridoma clone 28.2. The reaction volume in each case was 0.4 ml. (*A,B,C*) The three established isoforms of nematode myosin heavy chains. ^{35}S-labeled heavy chains were detected by autoradiography of the 4.5% polyacrylamide-SDS gel slab.

purified from the ascites fluid of a mouse tumor derived from clone 28.2. ^{35}S-labeled nematode myosins were reacted overnight at 4°C with the immunoglobulin immobilized on Sepharose beads. After the reaction, the myosins were removed from the beads and electrophoresed on a 4.5% polyacrylamide-SDS slab gel. In molar terms, the ratios of immunoglobulin to total nematode myosin were 15:1, 30:1, and 135:1. Before reaction with antibody, the proportions of A, B, and C isoforms were about 0.28, 0.64, and 0.08, respectively (Schachat et al. 1977). At the lowest ratio of antibody to myosin, the fluorograph in Figure 1 detects only myosin heavy-chain B. At higher ratios, however, a second reaction of considerably weaker affinity is detected. This analysis suggests that experiments utilizing concentrations of this antibody of 12.5 μg/ml or less would involve interactions specific to myosin B. Experiments requiring higher concentrations would be potentially complicated by the additional, weaker interactions with myosin A.

Table 1 shows the summary of results of immunoreaction product

TABLE 1
Reactivity of Monoclonal Antibodies with Nematode Myosin Isoforms

Clone	Myosin heavy-chain A	Myosin heavy-chain C	Myosin heavy-chain B	Paramyosin
5-19	+	+	+	−
25.1.1	+	+	+	−
10.2.1	+	(+)	(+)	−
5-14	+	−	(+)	−
5-17	+	−	(+)	−
9.2.1	−	+	(+)	−
5-11	−	+	(+)	−
5-4	(+)	−	+	−
5-8	(+)	−	+	−
5-12	(+)	−	+	−
5-13	(+)	−	+	−
28.2	(+)	−	+	−
5-9	−	−	−	+

+, strong reaction; (+), weak reaction; −, no detectable reaction.

analysis by the nitrocellulose transfer method with 14 antimyosin or antiparamyosin monoclonal antibodies. The antibody concentrations were varied over a range of about 1 ng/ml to 1 mg/ml. The reaction products were assayed by the peroxidase method (D.M. Miller and H.F. Epstein, unpubl.). As in the experiment described in the legend to Figure 1, most of the anti-nematode myosin monoclonal antibodies show preferential affinity to one isoform with some level of cross-reactivity with at least one other isoform. None of the antimyosin antibodies reacts with nematode paramyosin; the anti-nematode paramyosin antibody does not react with nematode myosin. It is clear that significant structural homologies exist between the members of the nematode myosin heavy-chain family; in most cases, the immunological reactivities based upon these homologies are weak. Most of the myosin determinants, although homologous, are structurally distinct between isoforms.

Immunological Dissection of Nematode Embryonic Development

In the mature *C. elegans* animal, there are 95 body-wall muscle cells and 20 pharyngeal muscle cells (J.E. Sulston and E. Schierenberg, pers. comm.). About 85% of nematode myosin is in the body wall; almost all of the rest is in the pharynx. At hatching, the pharynx is complete, but there are only 81 body-wall muscle cells (Sulston and Horvitz 1977). At this stage, the body and pharynx contribute about equally to the myosin content of the nematode (Garcea et al. 1978).

Specific rabbit antimyosin and antiparamyosin polyclonal antibodies (Gossett et al. 1982) can react with intact nematode embryos of various stages and with squashes of these structures (Gossett and Hecht 1980). As followed by indirect immunofluorescence, myosin and paramyosin first become detectable in a small number of cells in embryos between 400 and 450 cells. A maximum of 81 cells react in 550-cell embryos. The same number of cells react with either antimyosin or antiparamyosin antibodies; these cells are primitive body-wall muscle cells, which then align by pairs into the characteristic four columns. It is important to note that myosin and paramyosin are produced before the embryonic cells construct sarcomeres or become spindle-shaped. The embryonic cleavages in temperature-sensitive lethal mutants, such as B1, B126, and B209, arrest prematurely. Despite this disruption of cell proliferation in which DNA synthesis, nuclear division, and cell division may become uncoupled from one another (Hecht et al. 1982), myosin and paramyosin can be detected in a subset of cells. These results suggest that early muscle development in nematode embryos may not require either the normal number of nuclear or cytoplasmic divisions nor the apposition of specific types of neighboring cells (Gossett et al. 1982).

The availability of monoclonal antibodies with preferential reactivities to

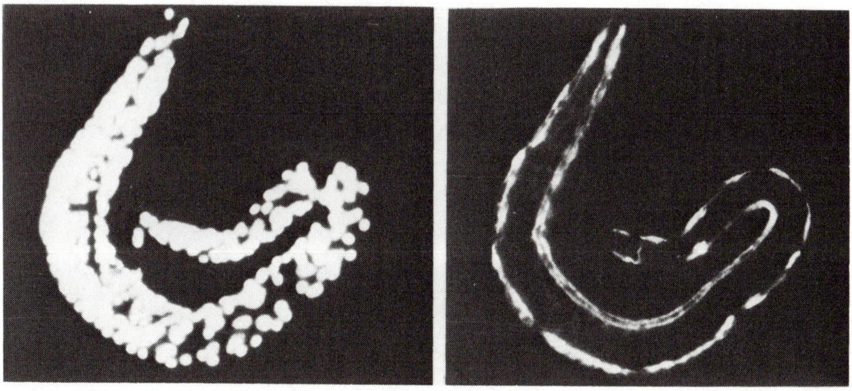

FIGURE 2
Reaction of clone 28.2 antibody with vermiform embryo. (*Left*) Hoechst 33258 reaction with nuclei. (*Right*) Immunofluorescent localization of antibody reaction.

specific myosin isoforms offers exciting possibilities for embryonic studies. Two of the monoclonal antibodies have been reacted with squashes of normal embryos. The monoclonal antibody from clone 28.2 that preferentially reacts with myosin B and the monoclonal antibody from clone 9.2.1 that preferentially reacts with myosin C were used. Intense, specific immunofluorescence with these monoclonal antibodies was not detected until the early vermiform stage. Further experiments are required to determine whether these results arise from a detectability problem intrinsic to the monoclonal antibodies or whether the specific isoforms reach significant concentrations only in late embryos.

Figure 2 shows a vermiform embryo that has been reacted with Hoechst 33258, a fluorescent dye that binds DNA, and with clone 28.2 immunoglobulin, followed by rhodamine-labeled goat anti-mouse antibody. The photomicrograph on the left shows the nuclei of the organism with the head at the upper left. The photomicrograph on the right shows the presence of myosin B in the embryonic body wall. Careful examination reveals the double column of body-wall muscle cells. A very slight reactivity with the pharynx may be seen near the head. Since the concentration of antibody in these experiments is 163 μg/ml, this reactivity may be due to the weak affinity of clone 28.2 antibody with myosin A, which is shared between the body-wall and pharyngeal muscles.

Figure 3 shows a similar comparison of nuclei of all cells and the location of reaction with the myosin-C-specific clone 9.2.1 antibody. The pharyngeal muscle of this vermiform embryo reacts strongly with the antibody, whereas there is no detectable reaction with the body wall. The anatomic specificity of the reactions of these monoclonal antibodies with similarly prepared embryos of the same stage serves as a control for the penetrability of the specific antibodies and of the fluorescent goat anti-mouse antibody.

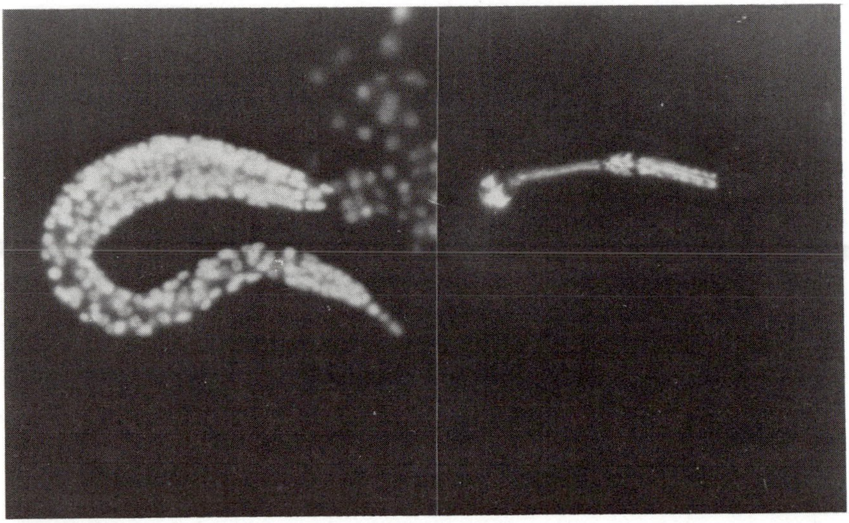

FIGURE 3
Reaction of clone 9.2.1 antibody with vermiform embryo. (*Left*) Hoechst 33258 reaction with nuclei. (*Right*) Immunofluorescent localization of antibody reaction.

The results indicate that in a simple organism such as the nematode, monoclonal antibodies with defined reactivities to myosin isoforms can serve as markers of specific types of muscle cells.

SUMMARY

Monoclonal antibodies have been selected that react preferentially with specific myosin isoforms and with paramyosin of the nematode *C. elegans*. These reactivities were established by polyacrylamide gel analysis of the immunological reaction products obtained over a wide range of antibody-to-myosin ratios. Most of the antimyosin antibodies show strong affinity for one of the three nematode myosin isoforms and much weaker affinity for one of the other isoforms. None of the antimyosin antibodies reacts with nematode paramyosin, and the antiparamyosin antibody does not react with nematode myosin. These results suggest that significant structural homologies exist between the three myosin isoforms despite their established differences in peptide structure and genetic specification. Most of these structurally homologous sites are nevertheless distinct from one another as indicated by the markedly different affinities of the specific monoclonal antibodies. The nematode myosin heavy-chain family bears no detectable antigenic relationship to nematode paramyosin, although both kinds of proteins have many grossly similar structural properties.

The antimyosin antibodies can serve as useful markers of specific differentiated states of muscle in nematode embryos. Antibodies from clone

28.2 show preferential reactivity to myosin B and stain the body-wall muscle cells by immunofluorescence. A markedly lower affinity of this antibody for myosin A correlates with a slight staining of the pharynx. Clone 9.2.1 produces antibody with strong affinity for myosin C and stains the pharyngeal muscle specifically. These and other monoclonal antibodies are being used to establish the pattern of expression of specific myosin isoforms during embryonic muscle differentiation in the nematode. The availability of temperature-sensitive mutants that arrest embryonic development and these specific immunological reagents will be useful in establishing early embryonic events required for muscle differentiation in the nematode.

ACKNOWLEDGMENTS

We thank Gary Berliner, Lois Bolton, and Janet Jackson for excellent research assistance. D.M.M. is a fellow of the Muscular Dystrophy Association of America, and H.F.E. holds a Research Career Development Award from the National Institute of Child Health and Human Development. The research was supported by grants from the National Institute of Aging, the National Institute of Child Health and Human Development, and the Jerry Lewis Neuromuscular Disease Research Center of the Muscular Dystrophy Association of America (to H.F.E.) and from the Muscular Dystrophy Association of America (to R.M.H).

REFERENCES

Epstein, H.F., F.H. Schachat, and J.A. Wolff. 1977. Molecular genetics of nematode myosin. In *Pathogenesis of human muscular dystrophies* (ed. L.P. Rowland), p. 460. Excerpta Medica, Amsterdam.

Epstein, H.F., R.H. Waterston, and S. Brenner. 1974. A mutant affecting the heavy chain of myosin in *Caenorhabditis elegans. J. Mol. Biol.* **90:** 291.

Garcea, R.L., F. Schachat, and H.F. Epstein. 1978. Coordinate synthesis of two myosins in wild-type and mutant nematode muscle during larval development. *Cell* **15:** 421.

Gossett, L.A. and R.M. Hecht. 1980. A squash technique demonstrating embryonic nuclear cleavage of the nematode *Caenorhabditis elegans. J. Histochem. Cytochem.* **28:** 507.

Gossett, L.A., R.M. Hecht, and H.F. Epstein. 1982. Muscle differentiation in normal and cleavage-arrested mutant embryos of *Caenorhabditis elegans. Cell* **30:** 193.

Harris, H.E. and H.F. Epstein. 1977. Myosin and paramyosin of *Caenorhabditis elegans:* Biochemical and structural properties of wild-type and mutant proteins. *Cell* **10:** 709.

Hecht, R.M., S.M. Wall, D.F. Schomer, J.A. Oró, and A.H. Bartel. 1982. DNA replication may be uncoupled from nuclear and cellular division in temperature-sensitive embryonic lethal mutants of *Caenorhabditis elegans. Dev. Biol.* (in press).

Mackenzie, J.M., Jr., F. Schachat, and H.F. Epstein. 1978. Immunocytochemical

localization of two myosins within the same muscle cells in *Caenorhabditis elegans. Cell* **15:** 413.

MacLeod, A.R., J. Karn, and S. Brenner. 1981. Molecular analysis of the *unc*-54 myosin heavy-chain gene of *Caenorhabditis elegans. Nature* **291:** 386.

MacLeod, A.R., R.H. Waterston, and S. Brenner. 1977a. An internal deletion mutant of a myosin heavy chain in *Caenorhabditis elegans. Proc. Natl. Acad. Sci.* **74:** 5336.

MacLeod, A.R., R.H. Waterston, R.M. Fishpool, and S. Brenner. 1977b. Identification of the structural gene for a myosin heavy chain in *Caenorhabditis elegans. J. Mol. Biol.* **114:** 133.

Schachat, F.H., R.L. Garcea, and H.F. Epstein. 1978. Myosins exist as homodimers of heavy chains: Demonstration with specific antibody purified by nematode mutant myosin affinity chromatography. *Cell* **15:** 405.

Schachat, F.H., H.E. Harris, and H.F. Epstein. 1977. Two homogeneous myosins in body-wall muscle of *Caenorhabditis elegans. Cell* **10:** 721.

Sogin, D.C. and P.C. Hinkle. 1980. Immunological identification of the human erythrocyte glucose transporter. *Proc. Natl. Acad. Sci.* **77:** 5725.

Sulston, J.E. and H.R. Horvitz. 1977. Post-embryonic cell lineages of the nematode *Caenorhabditis elegans. Dev. Biol.* **56:** 110.

Towbin, H., T. Staehelin, and J. Gordon. 1979. Electrophoretic transfer of proteins from polyacrylamide gels to nitrocellulose sheets: Procedure and some applications. *Proc. Natl. Acad. Sci.* **76:** 4350.

Trowbridge, I.S. 1978. Interspecies spleen-myeloma hybrid producing monoclonal antibodies against mouse lymphocyte surface glycoprotein, T200. *J. Exp. Med.* **148:** 313.

Characterization of Myosins from Embryonic and Developing Chicken Pectoralis Muscle

SUSAN LOWEY
*PAMELA A. BENFIELD**
DENISE D. LeBLANC
GUILLERMINA WALLER
DONALD A. WINKELMANN
Rosenstiel Basic Medical Sciences Research Center
Brandeis University
Waltham, Massachusetts 02254

GERALDINE F. GAUTHIER
Department of Anatomy
University of Massachusetts Medical School
Worcester, Massachusetts 01605

As part of a continuing effort to understand the elusive nature of the small subunits in adult skeletal-muscle myosin, closely related but distinct myosins were sought whose structural features could be related to their functional properties. A comparison between the myosins from adult and embryonic chicken pectoralis muscles seemed especially advantageous, since these myosins might be expected to differ in their enzymic activity. Chicken pectoralis myosin from the adult muscle has a high ATPase activity, consistent with its derivation from a fast-twitch muscle. According to Bárány's axiom (1967), the enzymic activity of a myosin is directly related to the speed of contraction of the parent muscle. Since developing muscles are generally slow in their contractile properties (Buller et al. 1960), the myosin from embryonic pectoralis might be expected to have a low rate of ATP hydrolysis.

The prediction of a "slow-type" myosin in embryonic muscle appeared to be supported by immunocytochemical studies. The fibers of developing

*Present address: Basic Research Program-LBI, Cancer Biology Program, Frederick Cancer Research Facility, Frederick, Maryland 21701.

chicken (Masaki and Yoshizaki 1974) and rat muscles (Gauthier et al. 1978) reacted with antibodies specific for both fast and slow myosin. An example of such a reaction is shown in Figure 1. All of the embryonic muscle fibers fluoresce brightly, which indicates that they contain proteins that cross-react with antibodies prepared against adult fast (pectoralis) myosin and adult slow (anterior latissmus dorsi [ALD]) myosin.

These observations can be explained in one of several ways. (1) Developing muscles contain a mixture of adult slow and fast myosins at a time when each muscle fiber receives axons from more than one motoneuron. As "polyneuronal innervation" is withdrawn, the adult fiber-type pattern is established in which each fiber has predominantly one type of myosin. For example, in the case of the posterior latissimus dorsi (PLD) muscle, the adult fibers react almost exclusively with antibodies specific for fast myosin, and no reaction is visible with antibodies to ALD myosin (Fig. 1, bottom). (2) Developing muscles contain an embryonic myosin that is distinct from adult slow or fast myosin, but that cross-reacts immunochemically with the adult forms. Since all myosins will undoubtedly prove to have extensive sequence homology (Karn et al., this volume), it

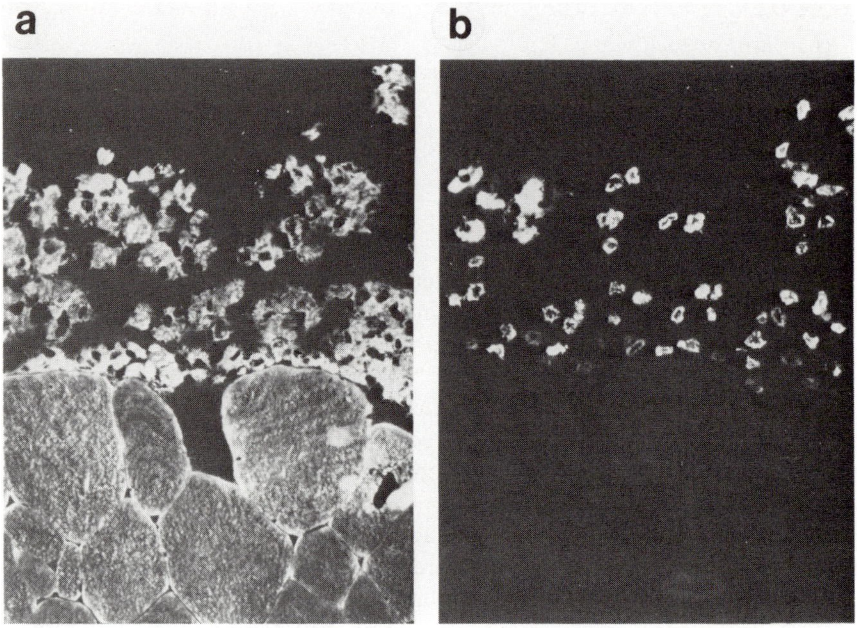

FIGURE 1
Fluorescent transverse frozen sections of chicken PLD. An 11-day embryo (*top*) was sectioned with adult PLD (*bottom*) and reacted with anti-Δ1 (fast myosin) (*a*) and anti-ALD (slow myosin) (*b*). All fibers in the developing muscle (*top*) react with both antibodies, whereas the adult muscle (*bottom*) reacts only with anti-Δ1. (Reprinted, with permission, from Gauthier 1980.)

would not be surprising to find cross-reacting determinants shared between embryonic and adult myosins. (3) There is more than one embryonic myosin in the early stages of development; in fact, several slow- and fast-type isozymes may exist. These embryonic myosins may be preferentially distributed among different cell lineages.

The discussion that follows is an attempt to distinguish among these several possibilities on the basis of recent biochemical and immunological data obtained in our laboratory.

RESULTS AND DISCUSSION

The most direct approach to resolving the question of myosin type is to isolate and characterize the myosin from an embryonic muscle of known physiological properties. The best described muscle from a physiological point of view is the chicken PLD (Gordon and Vrbova 1975); this muscle contracts slowly up to about day 18, at which time its speed of contraction increases rapidly to adult levels. Unfortunately, the small size of the PLD precludes it from extensive biochemical investigation, and we chose instead the more prominent chicken pectoralis muscle. Although no physiological data are available for this muscle, immunocytochemical data suggest that the pectoralis is probably slower in its development than the PLD (Gauthier et al. 1982) and therefore is likely to contract slowly at 11–12 days of incubation. Another advantage of the chicken pectoralis muscle is its relative homogeneity in fiber type in the adult (Gauthier and Lowey 1977). Mammalian muscles for the most part tend to be more heterogeneous, thereby complicating the interpretation of myosin polymorphism in embryonic muscles.

Characterization of Embryonic Chicken Pectoralis Myosin

Myosin was isolated from pectoralis muscle at various stages of development, the earliest being about 11 days of incubation. The myosin was chromatographically purified by the standard procedures in our laboratory and judged to be reasonably homogeneous by the criteria of extinction measurements and electrophoretic mobility on SDS-polyacrylamide gels. Unexpectedly, the enzymic activity of embryonic chicken myosin at 11–12 days of incubation was very similar to that obtained for the adult myosin (Table 1). These values represent a high level of ATPase activity and not the low rates usually associated with myosin from a slow adult muscle (Bárány 1967). Earlier work had shown a change in the Ca^{++}-ATPase activity of the myosin from several animals during development, including chickens (Trayer and Perry 1966). We have no explanation for this discrepancy, except to note that techniques to detect proteolysis were not as advanced then as now, and some protein modification may have occurred. A more recent study found only a slightly lower Ca^{++}- and K^{+}-ATPase activity for the

TABLE 1
ATPase Activities of Myosin Isozymes

	K^+(EDTA) (sec^{-1})	Ca^{++} (sec^{-1})	Actin-activated V_{max}(sec^{-1})	Actin-activated K_m (μM)
Adult	10.8	1.5	2.8	31.0
Embryonic	12.3	1.8	3.1	23.0

The activity is expressed in (mole of inorganic phosphate [P_i])/(mole myosin site).(sec). Kinetic constants for the actin-activated ATPase were obtained by the use of double-reciprocal plots; the measurements were made in the pH stat at pH 8.0, 25°C, with actin concentrations in the range 15–110 μM.

myosin of rabbit embryos, and adult values were attained immediately after birth (Sreter et al. 1975). One should not assume, however, that results obtained for one species will apply to another. The reason for the slow speed of contraction of chick embryos remains unclear; it has been suggested that an immature membrane system may lead to a reduction in the free Ca^{++} concentration available to the cross bridges during contraction.

Despite the identical steady-state kinetics for embryonic and adult chicken myosin, one should not infer that adult fast myosin is synthesized in the embryo. Embryonic myosin, solubilized in pyrophosphate buffer, migrates as a single band in nondenaturing polyacrylamide gels (data not shown). This is in contrast to adult pectoralis myosin, which has been found to separate into three bands. The least mobile band contains only alkali 1 light chain ($LC1_f$), the most mobile band has alkali 2 light chain ($LC3_f$), and the middle band contains both light chains (Hoh 1978; D'Albis et al. 1979; Lowey et al. 1979). Embryonic myosin was found to comigrate with the slowest band, corresponding to a homodimeric molecule containing an alkali 1 light chain on each head. This result is not too surprising in view of the observation that the mobility of chicken pectoralis myosin in pyrophosphate, nondenaturing gels is largely determined by the type of light chain present (Lowey et al. 1979). Since chicken pectoralis myosin contains primarily alkali 1 light chain before hatching, one would not expect to see any alkali 2 homodimers (nor alkali 1/alkali 2 heterodimers) until about 1 week posthatch, when alkali 2 is present in appreciable amounts (see Fig. 3b).

Do these results mean that embryonic myosin is identical to the homodimeric alkali 1 light-chain isozyme of adult chicken myosin? The answer is no. Peptide mapping of embryonic myosin by chymotryptic digestion of denatured myosin shows a distinctive heavy-chain fingerprint (Fig. 2). The two-dimensional electrophoretic pattern of myosin from pectoralis muscle at 11–12 days of incubation is clearly different from that ob-

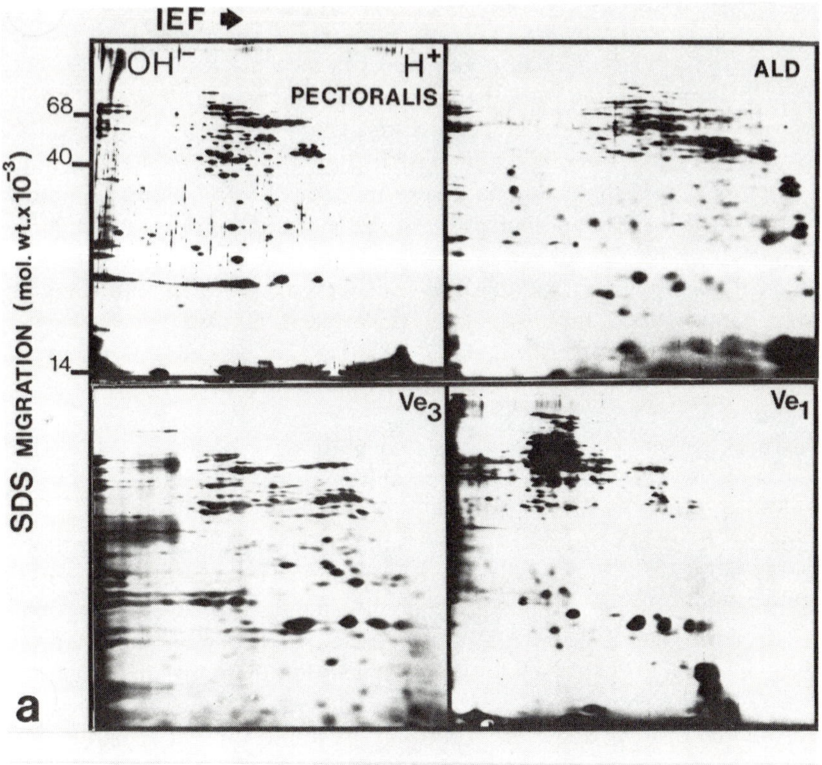

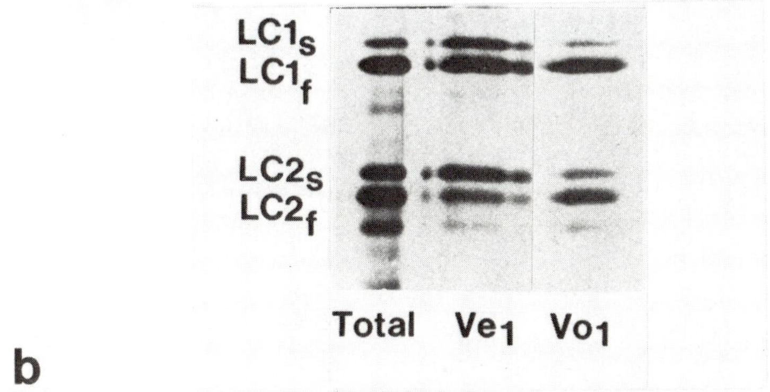

FIGURE 2

Electrophoretic gel patterns of embryonic and adult chicken myosins. (*a*) Two-dimensional O'Farrell gels of chymotryptic cleavage products of adult pectoralis, adult ALD, and two fractions (Ve_1 and Ve_3) of 11-day embryonic pectoralis myosin. Ve_1 is eluted from an anti-slow (ALD) heavy-chain column, and Ve_3 is eluted from an antipectoralis myosin column. (*b*) One-dimensional SDS gels of the light chains associated with the retained fraction Ve_1 and the void volume fraction Vo_1. Note the depletion of the slow light chains after passage through the immunoadsorbent.

tained for adult pectoralis and ALD myosin (Benfield et al. 1981). It is also different from the peptide map obtained for homodimeric alkali 1 myosin, which is indistinguishable from the peptide pattern for unfractionated adult chicken myosin (P.A. Benfield, pers. comm.).

From these results we can conclude that the bulk of chicken embryonic myosin is different from adult fast myosin, although they have common properties, such as immunological cross-reactivity and a fast rate of ATP hydrolysis. But what is the basis for the cross-reactivity of embryonic myosin with antibodies to adult slow (ALD) myosin? Nondenaturing gels show no band with the mobility of an ALD myosin. Two-dimensional gels (data not shown) and one-dimensional gels (Fig. 2b) do show, however, a small amount ($<10\%$) of slow light chains that comigrate with adult slow light chains.

Are these slow light chains associated with slow heavy chains? It is unlikely that 10% slow heavy chains would be detected in a digest of total embryonic myosin. To answer this question, we prepared an immunoadsorbent specific for ALD myosin heavy chain to see if we could isolate a slow-type myosin from the pool of embryonic myosins. This proved to be possible, although not simple. Starting with about 6 mg of chromatographed embryonic myosin, less than 0.5 mg of cross-reactive protein was retained by the anti-slow (ALD) heavy-chain column. Peptide mapping of the eluted fraction (Ve_1) by two-dimensional gel electrophoresis was a formidable task due to the low protein concentration. Only by application of a highly sensitive silver-staining technique (Switzer et al. 1979) was it possible to detect the faint peptide spots. The remainder of the embryonic myosin ($\sim 90\%$) was bound by an immunoadsorbent specific for adult pectoralis myosin (Ve_3). A comparison of the electrophoretic maps of the slow-type embryonic myosin (Ve_1) to those of the fast-type embryonic myosin (Ve_3) shows that they are distinct from each other and are also different from adult pectoralis and ALD myosins (Fig. 2a). Moreover, one can also see from the one-dimensional electrophoretic pattern of the light chains (Fig. 2b) that the fraction retained by the anti-slow heavy-chain column is enriched in slow light chains. This result is particularly significant in that it implies a specific association between the slow light chains and the cross-reactive slow heavy chains.

By affinity chromatography, we have selected a small population of molecules (Ve_1) that cross-react with adult slow myosin. Since this fraction bears no resemblance to adult ALD myosin, it is highly unlikely that any adult slow myosin is present in the total embryonic myosin pool. However, it remains to be shown that there is no adult fast myosin present. Although two-dimensional gels can probably set an upper limit of about 20% adult pectoralis myosin in the embryonic myosin, they are too insensitive to exclude the possibility of 10% or less adult isozyme. An alternative and highly sensitive method for detecting the presence of adult chicken myosin during development has been the use of monoclonal antibodies. By this technique

it can be shown that adult pectoralis myosin is not present prior to hatching (see below).

Characterization of Posthatch Myosins by Chemical and Immunological Methods

Having established the absence of adult fast and slow myosin in embryonic chicken pectoralis muscle, the next question is when do the adult isozmes appear? It was originally thought that adult myosin would gradually replace embryonic myosin following hatching. Immunofluorescence studies had shown a decrease in reactivity of the pectoralis fibers with anti-ALD antibody by 1 day posthatch (Gauthier et al. 1982), which suggests a decline in slow-type embryonic myosin. However, the appearance of adult fast myosin could not be determined with anti-pectoralis myosin antibody, since the latter cross-reacts with embryonic fast-type myosin. It was therefore necessary to isolate the myosin at various times after hatching and characterize the myosin by two-dimensional peptide mapping, as described above for embryonic myosin.

Unexpectedly, the peptide map of posthatch myosin did not resemble that of embryonic myosin, nor could it be explained by a mixture of embryonic and adult myosins (data not shown). This pattern persisted through 25 days posthatch and must be considered to represent yet another myosin isozyme, a "neonatal" form (Fig. 3a). After 25 days, one can begin to see trace amounts of the spots characteristic of adult myosin; however, the fully adult cleavage pattern is not attained until several months after hatching.

An alternative approach to the detection of adult isozymes became available with the preparation of monoclonal antibodies specific for adult chicken pectoralis myosin. One antibody in particular proved to be especially valuable in that it did not cross-react with embryonic myosin. This antibody, specific for the light meromyosin (LMM) region of myosin, could therefore be used to monitor the appearance of the adult isozyme during development in a modified radioimmunoassay (Fig. 4). The amount of posthatch myosin needed to inhibit the binding of anti-LMM antibody to a fixed concentration of adult myosin can be used as a measure of the fraction of adult myosin in the competing myosin preparation. The lack of reactivity of the monoclonal antibody with myosin samples prior to 2 weeks posthatch provides strong independent evidence in favor of the existence of a neonatal myosin. The possibility that the unique two-dimensional peptide map seen after hatching could be due to a mixture of embryonic and adult myosins is rendered highly unlikely by these immunological data.

SUMMARY

The following conclusions about the development of chicken pectoralis muscle can be drawn from the work described here. (1) Embryonic chicken

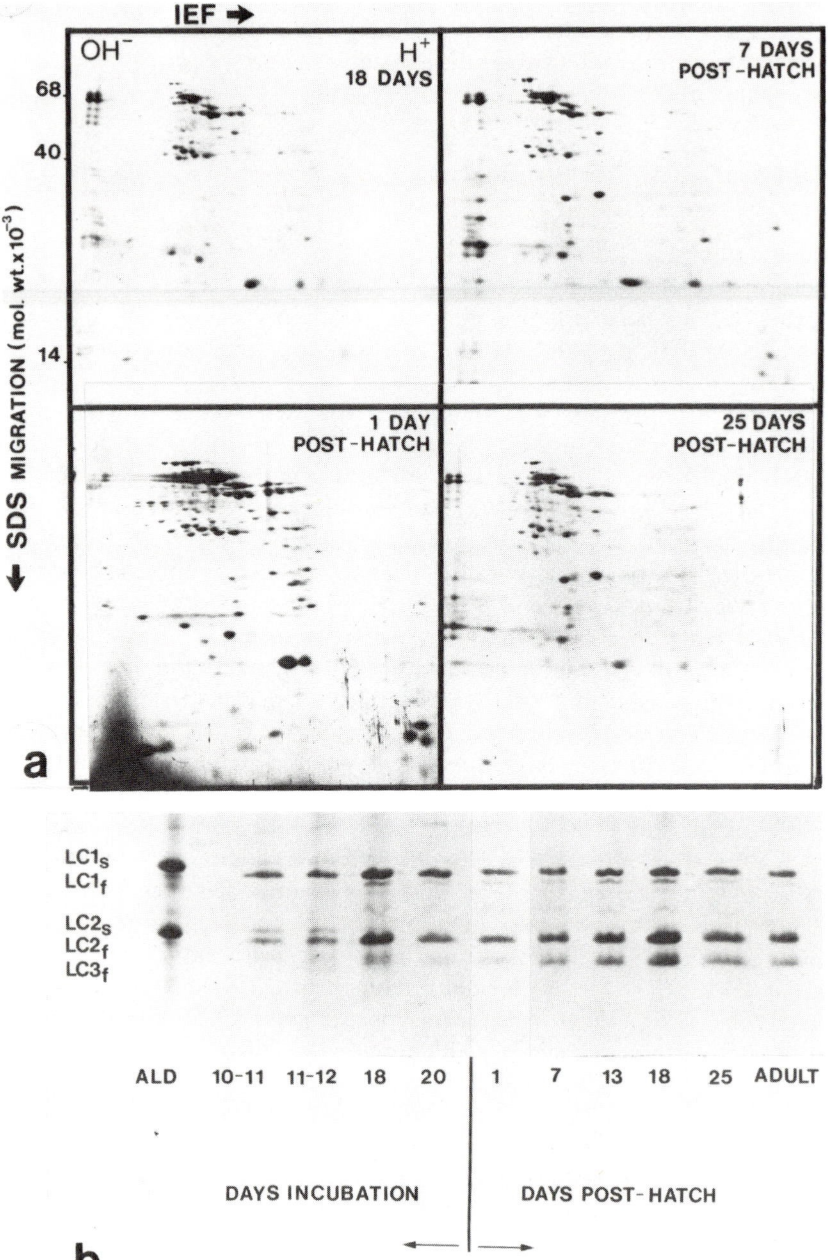

FIGURE 3

Electrophoretic gel patterns of embryonic and neonatal chicken myosins. (*a*) Two-dimensional gels of chymotryptic peptides from myosins isolated at various stages of development. (*b*) One-dimensional gels of the light chains associated with the corresponding heavy chains depicted in *a*. Note the disappearance of $LC1_s$ and $LC2_s$ after 12 days of incubation and the increase in $LC3_f$ following hatching.

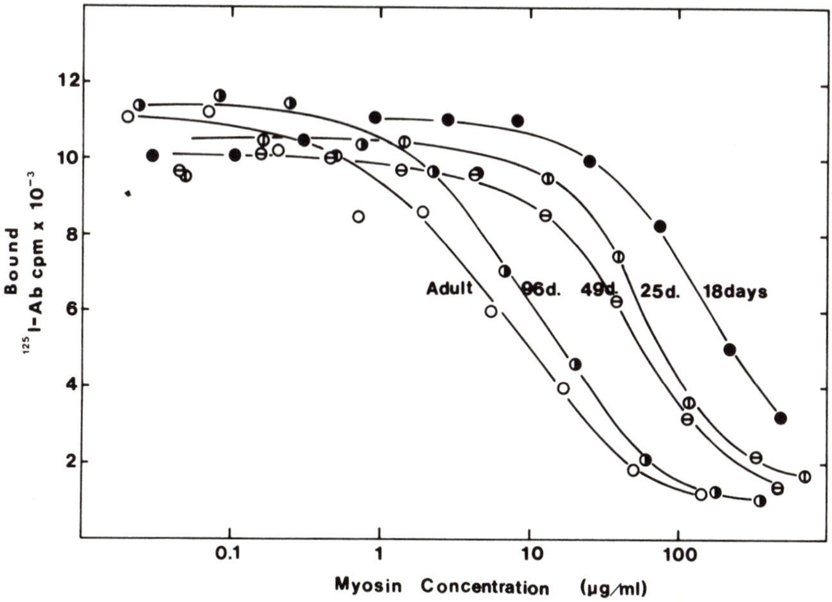

FIGURE 4

Solid-phase radioimmunoassay competition curves for the detection of adult chicken myosin in developing muscles. Serial dilutions of myosin prepared from chicken pectoralis muscle at 18, 25, 49, and 96 days posthatch and adult muscle (290 days) were used to inhibit the binding of adult-specific monoclonal anti-LMM antibody to adult myosin adsorbed onto a plastic microtiter plate. ^{125}I-labeled rabbit anti-mouse Fab was used at a fixed concentration as the detecting agent. The shift in the competition curves toward the adult curve with increasing age reflects the increase in the amount of adult myosin in the samples.

pectoralis myosin consists of at least two isozymes, a slow type and a fast type, both of which are distinct from adult slow (ALD) and adult fast (pectoralis) myosin. (2) Adult fast myosin does not appear in measurable amounts until about 2 weeks posthatch and reaches adult levels later than 14 weeks after hatching. (3) The transition from embryonic to adult isozymes is marked by the appearance of a neonatal myosin. This new heavy-chain isozyme is accompanied by an increased synthesis of the alkali 2 light chain, which reaches adult levels by about 1 week posthatch.

In summary, three classes of heavy-chain isozymes appear sequentially during chicken muscle development: $\text{myosin}_{emb} \rightarrow \text{myosin}_{neo} \rightarrow \text{myosin}_{adult}$. A class may consist of several related proteins, as shown here for embryonic myosin and in the literature for adult myosin. This pattern of development is qualitatively similar to that observed for rat muscles (Whalen et al. 1981). The functional significance of these many variants remains obscure, but their similarity in ATPase activity appears to rule out any direct relationship to the speed of contraction in developing muscles.

ACKNOWLEDGMENTS

This work was supported by grants from the National Institutes of Health and the Muscular Dystrophy Association of America (S.L. and G.F.G.), and the National Science Foundation (S.L.). P.A.B. and D.A.W. acknowledge receipt of fellowships from the Muscular Dystrophy Association of America.

REFERENCES

Bárány, M. 1967. ATPase activity of myosin correlated with speed of muscle shortening. *J. Gen. Physiol.* **50:** 197.

Benfield, P.A., S. Lowey, and D.D. LeBlanc. 1981. Fractionation and characterization of myosins from embryonic chicken pectoralis muscle. *Biophys. J.* **33:** 243a.

Buller, A.J., J.C. Eccles, and R.M. Eccles. 1960. Differentiation of fast and slow muscles in the cat hind limb. *J. Physiol.* **150:** 399.

D'Albis, A., C. Pantaloni, and J.-J. Bechet. 1979. An electrophoretic study of native myosin isozymes and of their subunit content. *Eur. J. Biochem.* **99:** 261.

Gauthier, G.F. 1980. Distribution of myosin isoenzymes in adult and developing muscle fibers. In *Plasticity of Muscle* (ed. D. Pette), p. 83. de Gruyter, Berlin.

Gauthier, G.F. and S. Lowey. 1977. Polymorphism of myosin among skeletal muscle fiber types. *J. Cell Biol.* **74:** 760.

Gauthier, G.F., S. Lowey, and A.W. Hobbs. 1978. Fast and slow myosin in developing muscle fibres. *Nature* **274:** 25.

Gauthier, G.F., S. Lowey, P.A. Benfield, and A.W. Hobbs. 1982. Distribution and properties of myosin isozymes in developing avian and mammalian skeletal muscle fibers. *J. Cell Biol.* **92:** 471.

Gordon, T. and G. Vrbová. 1975. The influence of innervation on the differentiation of contractile speeds of developing chick muscles. *Pflügers Arch. Gesamte Physiol.* **360:** 199.

Hoh, J.F.Y. 1978. Light chain distribution of chicken skeletal muscle myosin isoenzymes. *FEBS Lett.* **90:** 297.

Lowey, S., P.A. Benfield, L. Silberstein, and L.M. Lang. 1979. Distribution of light chains in fast skeletal myosin. *Nature* **282:** 522.

Masaki, T. and C. Yoshizaki. 1974. Differentiation of myosin in chick embryos. *J. Biochem.* **76:** 123.

Sreter, F.A., M. Bálint, and J. Gergely. 1975. Structural and functional changes of myosin during development. Comparison with adult fast, slow, and cardiac myosin. *Dev. Biol.* **46:** 317.

Switzer, R.C., III, C.R. Merril, and S. Shifrin. 1979. A highly sensitive silver stain for detecting proteins and peptides in polyacrylamide gels. *Anal. Biochem.* **98:** 231.

Trayer, I.P. and S.V. Perry. 1966. The myosin of developing skeletal muscle. *Biochem. J.* **345:** 87.

Whalen, R.G., S.M. Sell, G.S. Butler-Browne, K. Schwartz, P. Bouveret, and I. Pinset-Härström. 1981. Three myosin heavy-chain isozymes appear sequentially in rat muscle development. *Nature* **292:** 805.

Characterization of Myosin Isozymes Appearing during Rat Muscle Development

ROBERT G. WHALEN
LAWRENCE B. BUGAISKY
GILLIAN S. BUTLER-BROWNE
SUSAN M. SELL
Départment de Biologie Moléculaire
Institut Pasteur
75724 Paris, France

KETTY SCHWARTZ
Institut National pour la Santé
et la Recherche Médicale
(INSERM) Unité 127
Hôpital Lariboisière
75010 Paris, France

INGRID PINSET-HÄRSTRÖM
Départment de Biologie
Centre d'Etudes Nucléaires de Saclay
91190 Gif sur Yvette, France

Myosin is a multisubunit protein, composed of two heavy chains (200,000 M_r), two potentially phosphorylatable light chains (18,000–20,000 M_r), and two nonphosphorylatable light chains (16,000–27,000 M_r). Different isozymic forms of myosin are characteristic of different cell types. Adult striated skeletal muscle contains two major types of myosin; these two types differ significantly in the level of actin-activated ATPase activity, a property that is considered to account for the physiological differences between fast- and slow-contracting muscle fibers (Jolesz and Streter 1981; Salmons and Henricksson 1981).

Specific myosin isozymes are also present at different stages of skeletal-muscle development. Although it was previously thought that fetal muscle contained one or several adult myosin types (reviewed in Whalen 1981), current evidence suggests that distinct forms of myosin heavy and light chains are found. In developing rat leg muscle, we have recently shown that two developmental forms of myosin, embryonic and neonatal, appear and disappear sequentially before adult fast myosin becomes predominant (Whalen et al. 1981).

At present, 11–12 myosin heavy-chain (MHC) polypeptides and 12 myosin light-chain (MLC) subunits have been identified in different vertebrate muscle and nonmuscle tissues (Table 1). In Table 1, consideration is restricted to results of studies using protein-chemical techniques (sequencing, peptide and polypeptide mapping, and gel electrophoresis). Isozymes of native myosin can be formed from the different heavy and light chains. The various combinations depend on which subunits are present at the same time in a given cell and on which interactions are allowed and which are forbidden. Little is known about these possibilities, although in vitro recombination experiments have shown, e.g., that cardiac MLCs can interact with skeletal fast-twitch MHCs (Wagner and Weeds 1977).

A number of methodological approaches can be used to study myosin polymorphism, to analyze both the individual subunits and the native isozymes. Several approaches have been applied to the study of myosin forms appearing in early muscle development, and our data concerning rat muscle development are briefly reviewed.

RESULTS AND DISCUSSION

Distinct MHCs in Developing Muscle

One of the major biochemical approaches that we have used is that of polypeptide mapping of MHCs. This approach is based on the following considerations. Cleveland et al. (1977) observed that a number of proteases retain sufficient activity in the presence of SDS to allow partial proteolysis. Under these denaturing conditions, the presence of the light chains does not seem to influence the cleavage pattern of the heavy chain (Whalen et al. 1979). Moreover, because the MHC is so large, it can be cleaved into many fragments of sizes that are conveniently analyzed by standard SDS-polyacrylamide gel electrophoresis. The use of two-dimensional gel electrophoresis provides an additional high-resolution analytical tool with which to study the partial proteolytic products of the MHC, which is particularly useful when the myosin preparations are mixtures of known isozymes.

To investigate the nature of the MHC present at early stages of muscle development, myosins from cultured myotubes of the myogenic cell line L6 and of primary cultures and from fetal muscle tissue were analyzed by

TABLE 1
Distinct Vertebrate Myosin Subunits Identified by Protein-chemical Methods

MHC			MLC		
source	no.	reference[a]	source	no.	reference[a]
Skeletal muscle			skeletal muscle		
fast-twitch	2	Starr and Offer (1973)	fast-twitch	3	Weeds (1976)
slow-twitch	1	Weeds and Burridge (1975)	slow-twitch	3[b]	Weeds (1976)
masseter	1	Rowlerson et al. (1981)	masseter	2	Rowlerson et al. (1981)
embryonic	1	Whalen et al. (1979)	embryonic	(1)[c]	Whalen et al. (1978)
neonatal	1	Whalen et al. (1981)	cardiac muscle		
Cardiac muscle			ventricular	(2)[b]	Whalen et al. (1978)
ventricular	2	Hoh et al. (1979)	atrial	2[c]	Whalen et al. (1982)
atrial	1[d]	Flink et al. (1978)	smooth muscle	2[e]	Burridge (1974)
Smooth muscle	1	Burridge and Bray (1975)	nonmuscle		
Nonmuscle	2	Burridge and Bray (1975)			

[a] For brevity, only one reference is given for each case.
[b] The two ventricular MLCs are indistinguishable by two-dimensional gel electrophoresis from two of the slow-twitch proteins.
[c] The embryonic MLC is indistinguishable by two-dimensional gel electrophoresis from the nonphosphorylatable atrial MLC (Whalen et al. 1982).
[d] A recent preliminary report has suggested that the atrial MHC may be identical to one of the ventricular types (Chizzonite et al. 1981).
[e] To date, no results distinguish between the smooth-muscle MLC and the nonmuscle MLC.

polypeptide mapping (Whalen et al. 1979). These three MHCs had maps different from those of the various adult myosins examined (fast, slow, and cardiac). The maps of the MHC from the two cultured-cell myosins were very similar. Fetal myosin gave a map in which most of the polypeptides corresponded to those found in the cultured myosin, although a small amount of another MHC isozyme was seen (see below). We concluded from the above results that developing muscle, either in the animal or in culture, contained a MHC different from the adult types. We refer to this distinct form as embryonic heavy chain or MHC_{emb}.

The disappearance of the MHC_{emb} was followed to determine at what point in muscle development the adult heavy chain appears. Myosin isolated from muscle of the hind leg at various times after birth was examined by polypeptide mapping. Although the MHC_{emb} is the predominant form seen in the late stages of gestation, another heavy-chain cleavage pattern is observed in the neonatal stages (up to at least 2 weeks after birth) that is also distinct from the adult form (Whalen et al. 1981). This neonatal heavy chain (MHC_{neo}) is found in small amounts in the polypeptide map of myosin from fetal tissue, as noted above. The isozyme transitions occurring during rat muscle development are summarized schematically in Figure 1.

In the MHC polypeptide mapping experiments described above, the position of the various cleavage products within the MHC polypeptide was not determined. Recently, we have prepared the subfragment 1 (aminoterminal half) and the rod fragment (carboxyterminal half) from the different native myosin isozymes. Partial proteolytic cleavage in the presence of SDS was then carried out. The results show that both of these fragments differ among the embryonic, neonatal, fast-twitch, and slow-twitch myosins. These results argue against the possibility that one of these two structurally and

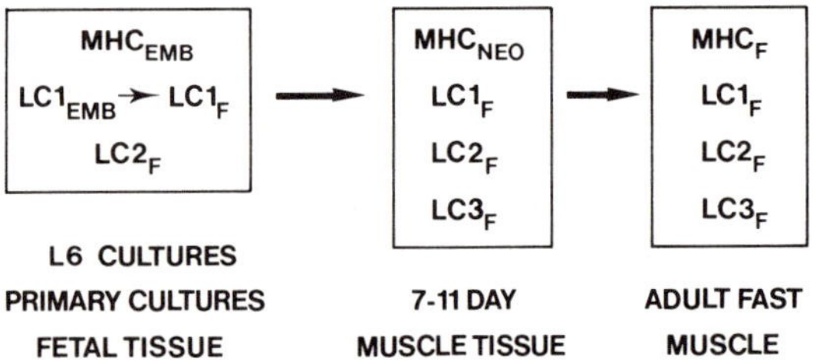

FIGURE 1
Schematic illustration of the various myosin subunit combinations present at different times during rat muscle development. This scheme is meant to show the major isozymes present and does not give other, possibly minor, combinations present during the various transitions. (Adapted from Whalen et al. 1981.)

functionally distinct "domains" of the MHC might be highly conserved among the various isozymes.

To investigate the structural differences between the various MHCs using an independent technique, antibodies to myosin were prepared to test the cross-reactivity of several heterologous myosins, using complement fixation as an assay. Antibodies were obtained using SDS-denatured myosins as antigens. These antibodies to denatured myosin were found to cross-react only poorly with the native protein (Schwartz et al. 1980). Thus, before testing in the complement-fixation assays, the myosins were first denatured with SDS, and the excess SDS was removed (Lompré et al. 1979). In this way it was shown that the predominant MHCs of developing muscles (MHC_{emb} and MHC_{neo}) are immunologically distinct from either adult fast or adult slow skeletal-muscle myosin (Whalen et al. 1979, 1981).

Since the leg muscle from which these myosin preparations were obtained is mostly (>90%) of the fast-contracting type in the adult, these results would seem to suggest that three myosin isozymes characterize the development of fast muscles, as illustrated schematically in Figure 1. However, the implication that the muscle fibers found even at the fetal stages are destined to become of the fast type should be regarded as tentative. It has been suggested that most fast and slow fibers evolve from two separate fiber populations, a "slow" population being predominant at some fetal stages (Kelly and Rubinstein 1981).

The isozymes MHC_{emb} and MHC_{neo} are the major forms found in developing muscle. It is possible that these forms occur in small amounts in adult muscle, either at a low level in many fibers or as the major form in a small percentage of fibers. An example of this type of developmental distribution is the case of the embryonic light chain $LC1_{emb}$ (Whalen et al. 1978). This subunit is found in developing skeletal and ventricular heart muscle but not in these adult muscles. However, it is the major adult form found in the heart atria and Purkinje fibers (Whalen et al. 1982).

Native Myosin Isozymes

Electrophoresis of myosin in nondenaturing conditions (pyrophosphate buffer) (D'Albis et al. 1979; Hoh et al. 1979) allows analysis of native myosin isozymes. This approach is especially valuable since it can be applied to small amounts of crude extracts of muscle tissue or cultured cells. The results described above give indications of the MHCs found at given times in development and allow an interpretation of the patterns obtained by electrophoresis of the corresponding native myosins. The patterns observed for rat myosins (Whalen et al. 1981) are shown schematically in Figure 2. In some situations, several native isozymes can be resolved. For example, adult fast muscle gives three bands, due to the formation of isozymes based on light-chain content. Two types of nonphosphorylatable

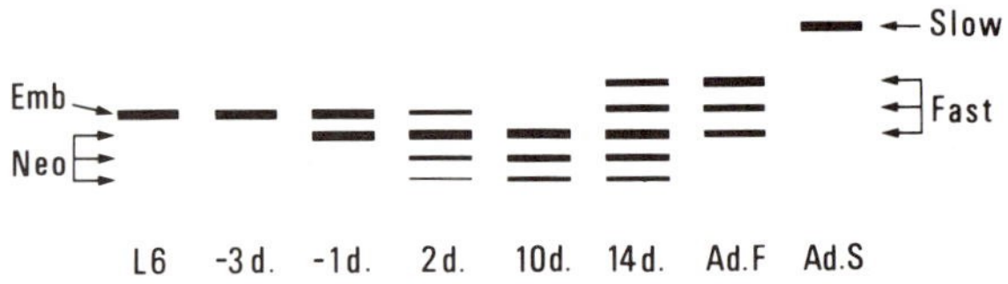

FIGURE 2
Schematic representation of the separation of native myosin isozymes obtained after electrophoresis in pyrophosphate buffer. Migration is from top (cathode) to bottom (anode). The separation of any two bands is on the order of 1 mm on the original gels. The different thicknesses of the bands indicate the relative amounts present. The times indicated are days relative to birth in the rat (−3 days=18 days of gestation). (Results are from Whalen et al. 1981.)

light chains are present and can form two homodimer and one heterodimer species in native myosin (see Lowey et al. 1980).

In cultured L6 cells and in fetal muscle tissue taken several days before birth, a single band is found corresponding to myosin containing MHC_{emb}. At the end of gestation, a second band appears due to the presence of MHC_{neo}. Up to this point in development, the combinations of MLCs present apparently do not give rise to further bands such as are found in adult fast myosin. Several days after birth, the embryonic myosin band disappears and neonatal myosin persists through the second postnatal week. By 10–11 days after birth, three native isozymes are seen; since the light-chain composition at this time is the same as in the adult (see Fig. 1), this result is presumed to be due to isozymes formed by the light chains. At about 14 days after birth, adult fast myosin begins to appear in these muscles. The band corresponding to adult slow myosin is separated from all of the other isozymes found.

Electrophoresis of native myosin will probably be the method of choice to screen various interesting developmental, physiological, and pathological situations, using as a guide the interpretation of the patterns given above. Clearly, any interesting results require confirmation by the use of at least one other independent method.

Myosin Function

Very little is currently known about the function of the embryonic and neonatal isozymes. Previous work has suggested that the Ca^{++}-activated or the K^+/EDTA-activated ATPase activities of these myosins might be lower than those of adult fast myosin and thus account for the slow contraction speed of newborn mammalian muscles (for references, see Whalen et al.

1981). Resolution of this particular point will require careful measurements of the actin-activated ATPase, which is the most physiological of the myosin ATPase activities. However, there is no reason to invoke the myosin ATPase as the sole rate-limiting step for contraction speed in developing muscles.

Another major function of myosin is to form filaments. In our studies, synthetic filaments of myosin were made by controlled dilution and then examined in the electron microscope. It has been shown (Pinset-Härström and Truffy 1979) that the relative concentrations of Mg^{++} and ATP play a significant role in determining the morphology of the filaments observed. The effect of these parameters has been studied on synthetic filaments made from embryonic, neonatal, and adult fast and slow myosins (Whalen et al. 1981). The formation of bipolar filaments with physiological diameters occurs over a narrower range of Mg^{++} concentrations for the non-adult forms of myosin than for the adult forms. It can be speculated that these results reflect a less stable filament structure for these developmental forms or that they indicate conformational differences between the various myosins. Further study using this approach might provide some insight into this problem.

Relationship of Isozyme Transitions to Innervation

The type of nervous innervation in the adult animal determines whether a muscle will be of the fast- or slow-contracting type (Jolesz and Sreter 1981; Salmons and Henriksson 1981). It is well established that developing muscle fibers are contacted by several nerves early in development and that this polyneuronal innervation regresses to a monosynaptic situation by about the end of the second week after birth in the rat. A correlation thus exists between the time when adult fast myosin begins to appear in newborn muscle and the time that the adult configuration of the neuromuscular junction is established. It is therefore of interest to examine the role of innervation in general and the role of polyneuronal innervation specifically in controlling isozyme transitions.

Two observations suggest that the relationship between innervation and isozyme type during development is not strict. First, we find that cutting the sciatic nerve at 7 days postnatal does not prevent the appearance of adult fast myosin in the hind-leg muscles, although neonatal myosin may persist for a longer time as a consequence of the denervation. Second, we have found, using immunocytochemistry, that the developing soleus muscle (mostly slow type in the adult) at 3 weeks after birth is composed of two fiber populations. About half the fibers contain adult slow myosin, whereas the other half contain neonatal myosin; no significant amount of adult fast myosin is observed. However, it is known that polyneuronal innervation has essentially disappeared from this muscle by this time (Brown et al. 1976). This latter observation would seem to disprove the notion that the

disappearance of neonatal myosin and the corresponding appearance of an adult form is tightly linked to the regression of polyneuronal innervation.

CONCLUSION

A large number of polypeptides account for the various nonmuscle, smooth-, cardiac-, and skeletal-muscle myosins identified to date (Table 1). The cloning of genes for these proteins will likely reveal a further multiplicity of subunits. Several approaches are currently available for the study of myosin isozymes, and others remain to be exploited. It is now clear that during muscle development, several myosin types can be found that do not occur in large amounts in the corresponding adult muscles (Fig. 1). A number of problems remain concerning (among others) (1) the function of the various myosin isozymes, (2) their precise distribution during development and in the adult, (3) the role of the nerve in controlling the phenotype of the developing muscle, and (4) the extent to which regenerating muscle mimics the events occurring in development. The biochemical, immunological, and molecular techniques alluded to above should help answer some of these questions.

ACKNOWLEDGMENTS

The experimental work reported in this paper was supported by the Pasteur Institute, the Centre National de la Recherche Scientifique, the Institut National pour la Santé et de la Recherche Médicale, the Centre d'Energie Atomique, and the Muscular Dystrophy Association of America.

REFERENCES

Brown, M.C., J.K.S. Jansen, and D. Van Essen. 1976. Polyneuronal innervation of skeletal muscle in new-born rats and its elimination during maturation. *J. Physiol.* **261:** 387.

Burridge, K. 1974. A comparison of fibroblast and smooth muscle myosins. *FEBS Lett.* **45:** 14.

Burridge, K. and D. Bray. 1975. Purification and structural analysis of myosins from brain and other non-muscle tissues. *J. Mol. Biol.* **99:** 1.

Chizzonite, R.A., A.W. Everett, M. Rabinowitz, and R. Zak. 1981. Myosin heavy chains from normal rabbit atria and thyrotoxic rabbit ventricle: Are they identical? *J. Cell Biol.* **91:** 346a.

Cleveland, D.W., S.G. Fischer, M.W. Kirschner, and U.K. Laemmli. 1977. Peptide mapping by limited proteolysis in sodium dodecyl sulfate and analysis by gel electrophoresis. *J. Biol. Chem.* **252:** 1102.

D'Albis, A., C. Pantaloni, and J.-J. Bechet. 1979. An electrophoretic study of native myosin isozymes and of their subunit content. *Eur. J. Biochem.* **99:** 261.

Flink, I.L., J.H. Rader, S.K. Banerjee, and E. Morkin. 1978. Atrial and ventricular cardiac myosins contain different heavy chain species. *FEBS Lett.* **94:** 125.

Hoh, J.F.Y., G.P.S. Yeoh, M.A.W. Thomas, and L. Higginbottom. 1979. Structural differences in the heavy chains of rat ventricular myosin isoenzymes. *FEBS Lett.* **97:** 330.

Jolesz, F. and F.A. Sreter. 1981. Development, innervation, and activity-pattern induced changes in skeletal muscle. *Annu. Rev. Physiol.* **43:** 531.

Kelly, A.M. and N.A. Rubinstein. 1981. Why are fetal muscles slow? *Nature* **288:** 266.

Lompré, A.M., P. Bouveret, J. Léger, and K. Schwartz. 1979. Detection of antibodies specific to sodium dodecyl sulfate treated proteins. *J. Immunol. Methods* **28:** 143.

Lowey, S., P.A. Benfield, L. Silberstein, and L.M. Lang. 1980. Distribution of light chains in fast skeletal myosin. *Nature* **282:** 522.

Pinset-Härström, I. and J. Truffy. 1979. Effect of adenosine triphosphate, inorganic phosphate and divalent cations on the size and structure of synthetic myosin filaments. *J. Mol. Biol.* **134:** 173.

Rowlerson, A., B. Pope, J. Murray, R.G. Whalen, and A.G. Weeds. 1981. A novel myosin present in cat jaw-closing muscle. *J. Muscle Res. Cell Motil.* **2:** 415.

Salmons, S. and J. Henriksson. 1981. The adaptive response of skeletal muscle to increased use. *Muscle and Nerve* **4:** 94.

Schwartz, K., A.M. Lompré, P. Bouveret, C. Wisnewsky, and B. Swynghedauw. 1980. Use of antibodies against dodecyl-sulfate-denatured heavy meromyosins to probe structural differences between muscular myosin isoenzymes. *Eur. J. Biochem.* **104:** 341.

Starr, R. and G. Offer. 1973. Polarity of the myosin molecule. *J. Mol. Biol.* **81:** 17.

Wagner, P.D. and A.G. Weeds. 1977. Studies on the role of myosin alkali light chains. Recombination and hybridization of light chains and heavy chains in subfragment-1 preparations. *J. Mol. Biol.* **109:** 455.

Weeds, A.G. 1976. Light chains from slow-twitch muscle myosin. *Eur. J. Biochem.* **66:** 157.

Weeds, A.G. and K. Burridge. 1975. Myosin from cross-reinnervated cat muscles. Evidence for reciprocal transformations of heavy chains. *FEBS Lett.* **57:** 203.

Whalen, R.G. 1981. Contractile protein isoforms in developing muscle: The nature of the myosin isozymes. *Ad. Physiol. Sci.* **5:** 63.

Whalen, R.G., G.S. Butler-Browne, and F. Gros. 1978. Identification of a novel form of myosin light chain present in embryonic muscle tissue and cultured muscle cells. *J. Mol. Biol.* **126:** 415.

Whalen, R.G., S.M. Sell, A. Eriksson, and L.-E. Thornell. 1982. Myosin subunit types in skeletal and cardiac tissues and their developmental distribution. *Dev. Biol.* (in press).

Whalen, R.G., K. Schwartz, P. Bouveret, S.M. Sell, and F. Gros. 1979. Contractile protein isozymes in muscle development: Identification of an embryonic form of myosin heavy chain. *Proc. Natl. Acad. Sci.* **76:** 5197.

Whalen, R.G., S.M. Sell, G.S. Butler-Browne, K. Schwartz, P. Bouveret, and I. Pinset-Härström. 1981. Three myosin heavy-chain isozymes appear sequentially in rat muscle development. *Nature* **292:** 805.

Cardiac Isomyosins: Influence of Thyroid Hormone on Their Distribution and Rate of Synthesis

*ALAN W. EVERETT**
*RICHARD A. CHIZZONITE**
*WILLIAM A. CLARK**
RADOVAN ZAK†*
*Departments of *Medicine and*
†Pharmacological and Physiological Sciences
The University of Chicago
Chicago, Illinois 60637

The heart in a number of animal species has been shown to contain several isomyosins whose content in this tissue changes depending on the developmental (Hoh et al. 1978), hormonal (Banerjee and Morkin 1977; Sartore et al. 1981), or physiologic (Lompré et al. 1979) state of the animal. The relative amounts of these isomyosins in the heart have important physiologic implications because these proteins differ in their ATPase activities (Banerjee and Morkin 1977; Hoh et al. 1978). The expression of one or another isomyosin would be expected to have a significant effect on the contractile properties of the tissue, because a correlation between the maximal speed of muscle shortening and the myosin ATPase activity appears to hold not only in skeletal muscle but also in heart muscle (Hamrell and Low 1978).

Recently, evidence has been presented for a developmental change in the relative amounts of the isomyosins of the heart (Lompré et al. 1981; Chizzonite et al. 1982). The findings indicated that, in species such as the rabbit, three isomyosins are expressed during embryonic and neonatal life, but only one of these forms (the low ATPase or V_3 form) is present in the adult. What is responsible for this developmental isomyosin switch? One of the

factors that may regulate the expression of isomyosins in the heart is thyroid hormone. Administration of the hormone to rabbits quickly changes the myosin type from a low to a high ATPase form (Banerjee and Morkin 1977). Thyroid hormone also reverses the effects of hypophysectomy on the cardiac isomyosins in the rat (Hoh et al. 1978). Induction of thyrotoxicosis may therefore be useful in the investigation of processes that regulate myosin expression.

In this study we first compared developmental changes in cardiac isomyosins with serum levels of thyroid hormone in the rabbit. Next, to evaluate the mechanisms that control the expression of isomyosin genes, we measured the rates of synthesis of the cardiac isomyosin heavy chains (HC_α and HC_β) after administering thyroxine (T_4) to rabbits. Here, we report the results of our investigations.

RESULTS AND DISCUSSION

Cardiac Isomyosins and Thyroid Hormone Levels in Rabbits

The cardiac isomyosin composition and serum levels of thyroid hormone (T_3 and T_4) at various stages of development are given in Figure 1. Native isomyosins (V_1, V_2, and V_3) were separated by polyacrylamide gel electrophoresis, and their relative amounts were determined by densitometry after staining with Coomassie blue. The thyroid hormone levels were determined by radioimmunoassay.

Isomyosins with both high (V_1) and low (V_3) ATPase activity were evident in the heart as early as 17 days of gestation; the amount of V_3 was two to three times greater than that of V_1. The relative proportions of these isomyosins remained constant during fetal development. The V_2 isomyosin was not observed in the fetus. After birth, there is a marked accumulation of V_1 and V_2 relative to V_3 until, by 14 days postpartum, the amount of V_1 exceeds that of V_3 in some rabbits. Thereafter, V_1 and V_2 gradually diminish, whereas V_3 accumulates; by 4 months of age, V_1 represents less than 10% of the total myosin and in some animals is undetectable.

The time course for the change in the isomyosin content of the rabbit heart that we have observed differs in some respects from that reported by Lompre et al. (1981). These investigators found that the concentration of V_1 in the heart reached a maximum by 1 week postpartum in rabbits, whereas, in our experience, this occurs at 2 weeks. Moreover, Lompre and co-workers observed a more rapid elimination of V_1 than that reported here. These findings illustrate the variability of the phenomenon, which may be related to strain differences or other unknown factors.

Figure 1 shows that a significant concentration of free T_4 is detectable in fetal serum from at least 23 days of gestation. The less sensitive procedure used for measurement of total T_3 and T_4 did not indicate the presence of thyroid hormone at this time. Serum levels of T_3 and T_4 began to increase

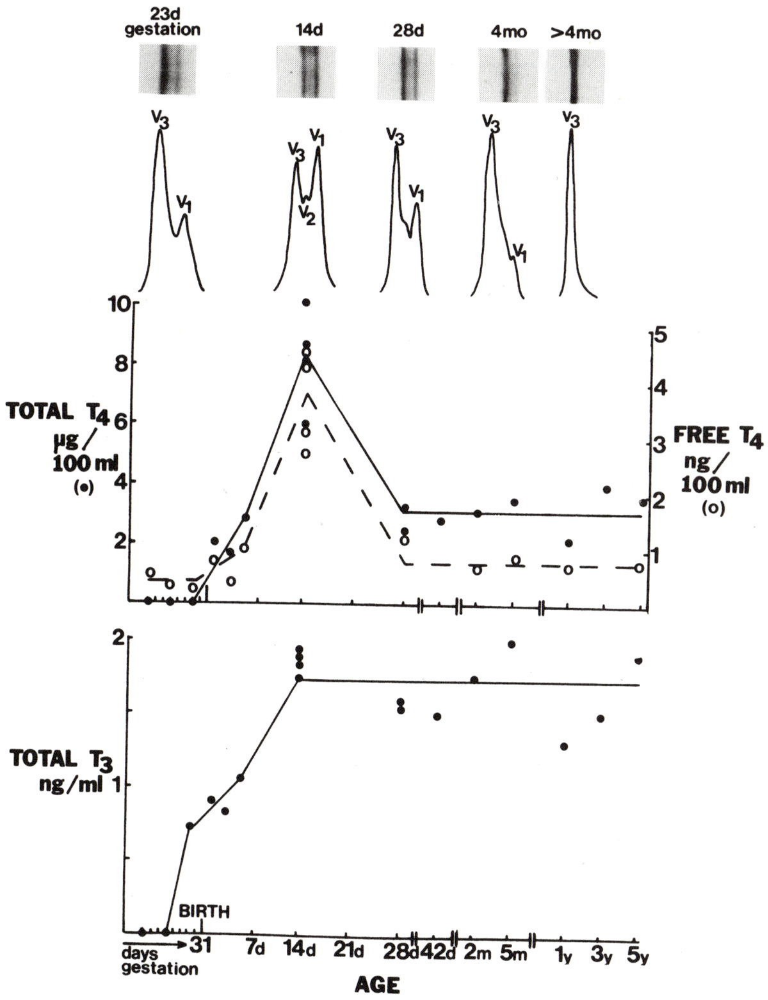

FIGURE 1
Cardiac isomyosin composition and serum levels of thyroid hormone (T_3 and T_4) in the rabbit. The relative amounts of the three ventricular isomyosins (V_1, V_2, and V_3) were determined by densitometric scans of Coomassie-blue-stained polyacrylamide gels after electrophoresis of myosin under non-dissociating conditions (see legend to Fig. 2). Levels of thyroid hormone were measured with a standard clinical radioimmunoassay kit (Clinical Assays, Cambridge, Massachusetts).

dramatically around the time of birth, reaching a maximum at about 2 weeks of age; thereafter, the concentration of T_3 remained fairly constant, whereas that of T_4 fell over the next 2 weeks to a concentration that remained stable throughout the life of the animal.

A relationship between serum hormone levels and isomyosin V_1 expression in the heart is most apparent during neonatal life, when the concentrations of V_1 in the heart and of T_4 in the serum both increase to a maximum at 2 weeks and then both decline thereafter. A relationship is not so obvious, however, after 4 weeks when T_4 levels stabilize and V_1 is continually lost from the heart. Perhaps the levels of T_4 are too low to influence expression of V_1 in the heart after this time. Unlike T_4, T_3 does not decrease in concentration in serum after 2 weeks, and yet it is the major mediator of thyroid hormone action. Thus, either there are other factors that play a role in myosin expression in the heart during development, or serum levels of T_3 and T_4 are inadequate indicators of thyroid hormone action at the cellular level. For example, monodeiodination of T_4 in the heart may contribute significantly to the intracellular concentration of T_3. Thus, a fall in both serum and intracellular levels of T_4 would result in a decline in the production of T_3 in the heart, independent of serum. Monodeiodination of T_4 is a significant source of T_3 in the brain and pituitary (Crantz and Larsen 1980) and, perhaps, also in the heart during early development.

Clearly, the effect of T_3 and T_4 on myosin expression in the heart deserves further investigation. A better understanding will require measurements of intracellular or even intranuclear levels of T_3 and T_4 and also of the binding capacity of thyroid hormone receptors in the heart throughout development.

Rates of Synthesis of Cardiac Isomyosins

Since cardiac myosin normally has a half-life of 5–6 days (Everett et al. 1981), a change in the proportions of the isomyosins in the heart can be determined by the rates of synthesis and/or degradation of each protein. A greater understanding of these processes will yield insight into the mechanisms by which the concentrations of specific isomyosins are regulated in the heart.

As a first step toward this objective, we compared the fractional rates of synthesis of the heavy chains of the V_1 (HC_α) and V_3 (HC_β) isomyosins in normal and T_4-injected rabbits. Animals 4–5 weeks of age were selected, since both isomyosins are normally present in significant amounts, with the ratio of V_1 to V_3 averaging 40:60 (S.D.=11, n=13).

To determine the rates of synthesis, we infused rabbits intravenously with [^{3}H]leucine for 1 hour. Myosin was extracted from the hearts and passed over affinity columns of monoclonal antibodies (McAbs) specific for the heavy chains of either the V_1 or the V_3 isomyosin (Chizzonite et al. 1982). Figure 2 shows the results of the chromatographic separation of these isomyosins from a myosin preparation containing V_1, V_2, and V_3 (gel 3). When a V_1-specific antibody (McAb 37) is used, both V_1 and V_2 are adsorbed onto the column,

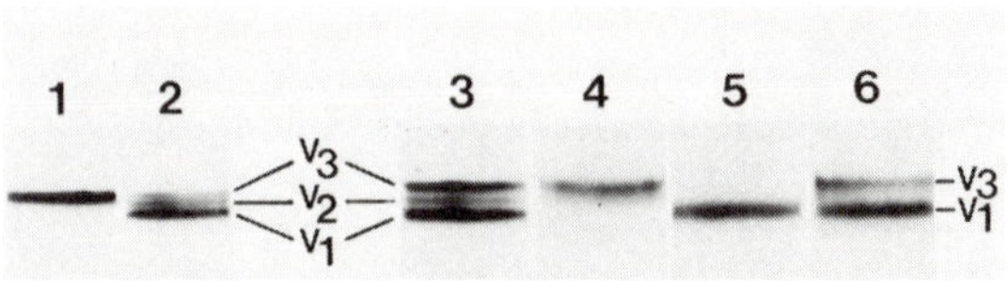

FIGURE 2
Electrophoresis of native cardiac myosins under nondissociating conditions. Cardiac tissue was minced and extracted with ten volumes of Guba-Straub buffer (pH 6.5) containing 1 mM EDTA. Polyacrylamide tube gels (4% total acrylamide, 3% bis-acrylamide) were prepared in 30 mM sodium pyrophosphate buffer (pH 8.8), 1 mM EDTA, 2 mM cysteine, and 10% glycerol (Hoh et al. 1978), and electrophoresis of the crude myosin extract was carried out for 24 hr at a constant voltage gradient of 11 V/cm. Migration is from the top (cathode). (*Gel 1*) Normal rabbit (>4 months old); (*gel 2*) 4-kg rabbit (>4 months old) injected i.m. with 100 μg of L-T_4/kg/day; (*gel 3*) original myosin preparation before passage over an affinity column of McAb 37 (*gel 4*) and McAb 52 (*gel 5*). The excluded fractions from these columns were mixed in *gel 6*.

whereas the V_3 isomyosin is excluded in pure form. With a V_3-specific antibody (McAb 52), both V_2 and V_3 are adsorbed onto the column, whereas V_1 is excluded. Since V_2 was undetectable in the excluded fractions from these affinity columns, both antigenic determinants recognized by the antibodies must be present on this isomyosin. This may indicate that V_2 in rabbits consists of one each of the V_1 (HC_α) and V_3 (HC_β) heavy chains, similar to the V_2 isomyosin described in the rat myocardium (Hoh et al. 1979).

The heavy chains from these affinity-purified isomyosins were isolated by polyacrylamide gel electrophoresis in the presence of SDS. The fractional rates of synthesis of the heavy chains of the V_1 and V_3 isomyosins in two normal and two T_4-injected rabbits are given in Table 1. In one normal rabbit (1), the rates of synthesis of HC_α and HC_β were very similar; in the other (rabbit 2), HC_β was synthesized at a rate 70% greater than that of HC_α. This may indicate that the two rabbits were in different developmental stages. The depressed synthesis rate of HC_α relative to HC_β in rabbit 2 is probably related to the elimination of V_1 from the myocardium. Administration of T_4 for 2 days resulted in about a fourfold increase in the fractional rate of synthesis of HC_α and about a threefold decrease in the fractional rate of synthesis of HC_β, when compared to the rates for the respective heavy chains in normal animals. The animal treated with T_4 for 4 days also showed elevation and depression in the fractional rates of synthesis of HC_α and HC_β, respectively.

TABLE 1
Fractional Rates of Synthesis of the Heavy Chains of Isomyosins V_1 and V_3 in Rabbit before and after Treatment with L-T_4

Rabbit no.	Weight (g)	Condition	K_s (day^{-1})	
			HC_α	HC_β
1	800	untreated	0.13	0.10
2	735	untreated	0.07	0.12
3	585	2 days T_4	0.41	0.04
4	740	4 days T_4	0.24	0.05

The rabbits were infused via an ear vein with 1.5 mCi of [^{3}H]leucine for 1 hr. Rabbits 3 and 4 received 200 μg of L-T_4/kg/day. After extraction of myosin from the ventricles, isomyosins V_1 and V_3 were isolated by affinity chromatography (Fig. 2). The fractional synthesis rate (K_s) was calculated from the specific radioactivities of leucine both in the heavy chains of V_1 (HC_α) and V_3 (HC_β) and free in the serum (Everett et al. 1981).

Such a combined and opposing effect of thyroid hormone on the synthesis of the two heavy chains in the heart would result in the very rapid redistribution of the isomyosin composition to predominantly V_1 (Fig. 2, gels 1 and 2), independent of possible changes in the degradation rates of these heavy chains.

Since the stimulatory effect of thyroid hormone on protein synthesis is at the transcriptional level (Seo et al. 1981), accumulation in the heart of the mRNA for HC_α would be expected. It would be of great interest to know the fate of the mRNA for HC_β after administration of thyroid hormone and to determine whether the decrease in the fractional rate of synthesis of this heavy chain results from a decrease in the transcriptional rate of the mRNA or from an increase in the degradation rate of the mRNA, or both.

SUMMARY

The ventricles of the rabbit heart contain three isomyosins (V_1, V_2, and V_3), which vary in their relative content in the myocardium depending on the developmental stage of the animal. One of these isomyosins (V_1) is inducible with thyroid hormone. During normal neonatal development, the change in the amount of V_1 (the high ATPase isomyosin) in the heart is better correlated with changes in serum levels of T_4 than those of T_3. The concentration of V_1 is at a maximum in the myocardium at the age of about 2 weeks, when the serum T_4 level is at its peak. After this time, the concentrations of both T_4 in the serum and V_1 in the heart decrease. In mature animals (3 kg and more), V_1 is present in very small and sometimes undetectable amounts but can be induced to reappear by injection of L-T_4. Administra-

tion of T_4 induces a large increase in the fractional rate of synthesis of the V_1 heavy chain (HC_α) and a marked reduction in the fractional rate of synthesis of the V_3 heavy chain (HC_β). These combined effects would lead to a rapid redistribution of the isomyosin profile in the heart.

ACKNOWLEDGMENTS

This work was supported in part by U.S. Public Health Service grants (HL-20592, HL-16627, and HL-09172) from the National Heart, Lung, and Blood Institute, and a grant from the Muscular Dystrophy Association of America. A.W.E. and R.A.C. are fellows of the Chicago Heart Association.

REFERENCES

Banerjee, S.K. and E. Morkin. 1977. Actin-activated adenosine triphosphatase activity of native and *N*-ethylmaleimide-modified cardiac myosin from normal and thyrotoxic rabbits. *Circ. Res.* **41:** 630.

Chizzonite, R.A., A.W. Everett, W.A. Clark, S. Jakovcic, M. Rabinowitz, and R. Zak. 1982. Isolation and characterization of two molecular variants of myosin heavy chain from rabbit ventricle: Change in their content during growth and after treatment with thyroid hormone. *J. Biol. Chem.* **257:** 2056.

Crantz, F.R. and P.R. Larsen. 1980. Rapid thyroxine to 3,5,3′-triiodothyronine conversion and nuclear 3,5,3′-triiodothyronine binding in rat cerebral cortex and cerebellum. *J. Clin. Invest.* **65:** 935.

Everett, A.W., G. Prior, and R. Zak. 1981. Equilibration of leucine between the plasma compartment and leucyl-tRNA in the heart, and turnover of cardiac myosin heavy chain. *Biochem. J.* **194:** 365.

Hamrell, B.B. and R.B. Low. 1978. The relationship of mechanical V_{max} to myosin ATPase activity in rabbit and marmot ventricular muscle. *Pflügers Arch.* **377:** 119.

Hoh, J.F.Y., P.A. McGrath, and P.T. Hale. 1978. Electrophoretic analysis of multiple forms of rat cardiac myosin: Effect of hypophysectomy and thyroxine replacement. *J. Mol. Cell. Cardiol.* **10:** 1053.

Hoh, J.F.Y., G.P.S. Yeoh, A.W. Thomas, and L. Higginbottom. 1979. Structural differences in the heavy chains of rat ventricular myosin isoenzymes. *FEBS Lett.* **97:** 330.

Lompré, A.M., K. Schwartz, A. d'Albis, G. Lacombe, N. van Thiem, and B. Swyneghedauw. 1979. Myosin isoenzyme redistribution in chronic heart overload. *Nature* **282:** 105

Lompré, A.M., J.J. Mercadier, C. Wisnewski, P. Bouveret, C. Pantaloni, A. d'Albis, and K. Schwartz. 1981. Species and age dependent changes in the relative amounts of cardiac myosin isoenzymes in mammals. *Dev. Biol.* **84:** 286.

Sartore, S., L. Gorza, S. Pierobon-Bormioli, L. Dalla-Libera, and S. Schiaffino. 1981. Myosin types and fiber types in cardiac muscle. I. Ventricular myocardium. *J. Cell Biol.* **88:** 226.

Seo, H., C. Wunderlich, G. Vassart, and S. Refetoff. 1981. Growth hormone responses to thyroid hormone in the neonatal rat. *J. Clin. Invest.* **67:** 569.

Major Phosphoproteins of Cultured Muscle Cells: Kinetics of Phosphorylation and Induction by Heat Shock

JAMES I. GARRELS
Cold Spring Harbor Laboratory
Cold Spring Harbor, New York 11724

Many intracellular responses to extracellular stimuli lead ultimately to protein phosphorylation reactions, which modify the functions of specific proteins. In muscle cells, much is known about the control of glycogen metabolism and the modulation of contractile activity by cAMP-dependent and Ca^{++}-dependent protein kinases (for reviews, see Stull et al. 1980; Cohen 1982). In these systems the protein kinases and the substrate proteins have been purified and characterized, and for glycogen metabolism the complex regulatory pathway has been elucidated in some detail. However, for other cellular responses, such as differentiation or growth control, no detailed information is available regarding the role of protein phosphorylation.

All animal cells contain numerous phosphoproteins and a variety of protein kinases. Selective stimulation of cAMP-dependent protein kinases (Steinberg and Coffino 1979) or tyrosine-specific protein kinases (Cooper and Hunter 1981; Hunter and Cooper 1981) results in the altered phosphorylation of only a small fraction of the phosphoproteins in the cell. If most protein phosphorylation serves a regulatory function, then many different regulatory pathways and many kinase systems must be present. Alternatively, many protein phosphorylation events might not regulate function but might instead be important for transport, assembly, or stability of the protein. Some knowledge of the time course and reversibility of individual protein phosphorylations can help to distinguish between these possibilities.

We have used two-dimensional gel electrophoresis to detect and characterize the major phosphoproteins of the L6 skeletal-muscle cell line (Yaffe 1968). The major phosphoproteins we observe in L6 cells do not correspond to the major contractile proteins or to known enzymes, but some of them are induced by heat shock. The major heat-shock phosphoprotein and several other major phosphoproteins are phosphorylated slowly, and these phosphorylation reactions proceed to completion, suggesting that they do not dynamically regulate protein function.

RESULTS

The proteins and phosphoproteins of L6 skeletal-muscle cell cultures were examined by two-dimensional gel electrophoresis (Garrels 1979a). Cultures of partially differentiated L6 cells (myoblasts and myotubes) were radiolabeled with [^{35}S]methionine for 2 hours or with [^{33}P]phosphate for 24 hours, and the proteins were resolved on two-dimensional gels containing either pH 5–7 or pH 6–8 ampholytes in the first dimension. Over 1000 ^{35}S-labeled proteins and over 300 ^{33}P-labeled phosphoproteins were detected on these gels, as shown in Figure 1.

We have matched the patterns shown in Figure 1 by careful superposition of the films to determine which of the spots in the total protein pattern represent phosphoproteins. Most of the identifications have been confirmed on two or more gels representing different pH ranges or slab gel concentrations, and most of the matches have been confirmed in other experiments with radiolabeled L6 proteins and phosphoproteins. Some of the corresponding spot positions are indicated by the arrowheads in Figure 1. Visual analysis of these patterns shows that the major ^{33}P-labeled proteins are not the same as the major ^{35}S-labeled proteins. The abundant contractile and structural proteins such as actin, tubulin, tropomyosin, and α-actinin contain no detectable phosphate. Vimentin, myosin light chain 2, and vinculin contain minor phosphorylated forms to the acidic side of the major ^{35}S-labeled spots (Fig. 1). Some of the acidic proteins and phosphoproteins (from a different labeling experiment of partially fused L6 cells) are shown at a higher level of sensitivity in Figure 2. This figure clearly shows the phosphorylated form of vimentin and the absence of phosphorylation over the tubulins and tropomyosins. In addition, it also shows many of the minor phosphoproteins that can be detected on gels of higher sensitivity.

Phosphoproteins Involved in the Heat-shock Response

Some of the phosphoproteins, including the major phosphoprotein at M_r 90,000, appeared to be coincident with proteins recently studied in HeLa cells and chick cells as part of the heat-shock response (Thomas et al. 1982). To determine the response of L6 cells to elevated temperatures, we subjected

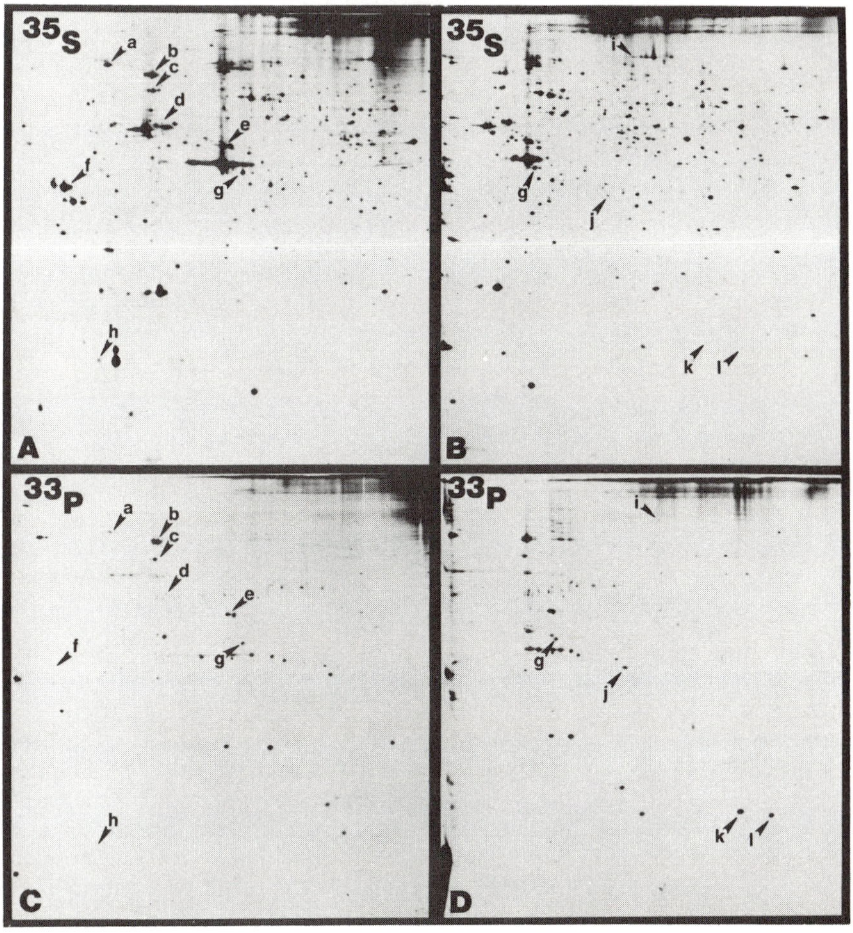

FIGURE 1
Two-dimensional resolution of proteins (*A,B*) and phosphoproteins (*C,D*). Samples were prepared from radiolabeled proteins of partially differentiated L6 muscle cell cultures and electrophoresed as described in Garrels (1979a). The gels in *A* and *C* were run using pH 5–7 ampholytes in the first dimension; those in *B* and *D* were run with pH 6–8 ampholytes. All second-dimension gels contained 12.5% acrylamide. Approximately 1,300,000 dpm of the [^{33}P]phosphate-labeled sample and 300,000 dpm of the [^{35}S]methionine-labeled sample were applied to each gel. The gels were prepared for fluorography and exposed to film for 1 week. The indicated proteins are 100K heat-shock protein (*a*), 90K heat-shock protein (*b*), 80K heat-shock protein (*c*), vimentin, phosphorylated form (*d*), unknown phosphoprotein (*e*), β-tropomyosin, unphosphorylated form (*f*), unknown muscle-specific protein (*g*), myosin light chain 2, phosphorylated form (*h*), vinculin, phosphorylated form, and unknown phosphoproteins (*j–l*). The phosphorylated proteins *h* and *i* are visible on the original films, though not on the photograph. Protein *f* has no phosphorylated form, as confirmed in Fig. 2.

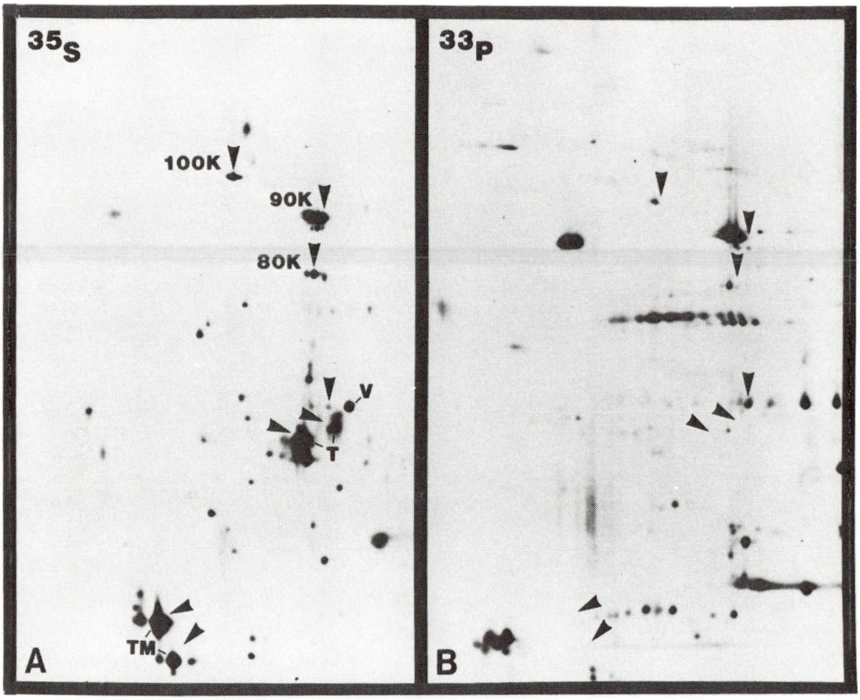

FIGURE 2
Acidic proteins and phosphoproteins of partially differentiated L6 muscle cells. Samples labeled with [^{35}S]methionine (560,000 dpm [*A*]) and [^{33}P]phosphate (1,700,000 dpm [*B*]) were electrophoresed on two-dimensional gels containing pH 5–7 ampholytes in the first dimension and 7.5% acrylamide in the second dimension. The gels were prepared for fluorography and exposed to film for 1 week. The major heat-shock proteins are indicated, as well as the forms of tubulin (T), vimentin (V), and tropomyosin (TM). The arrowheads point accurately to corresponding positions in the two images for alignment.

mixed myoblast and myotube cultures to a temperature of 42°C for 1–4 hours. The cells were labeled with [^{35}S]methionine during the last hour, and the proteins were resolved by two-dimensional gel electrophoresis (Fig. 3). Within 2 hours of shock, dramatic inductions in rate of synthesis were observed for two proteins, labeled 80K and 100K. In contrast, only the most acidic forms of the 90K complex were induced.

Quantitative computerized densitometry (Garrels 1979a) was carried out to determine the relative amounts of induction of the phosphorylated and unphosphorylated forms of the major heat-shock proteins in L6 cells. For the 100K protein, 4 hours of heat shock resulted in 7- and 8-fold inductions of the phosphorylated and unphosphorylated forms, respectively, whereas the phosphorylated and unphosphorylated forms of the 80K protein were

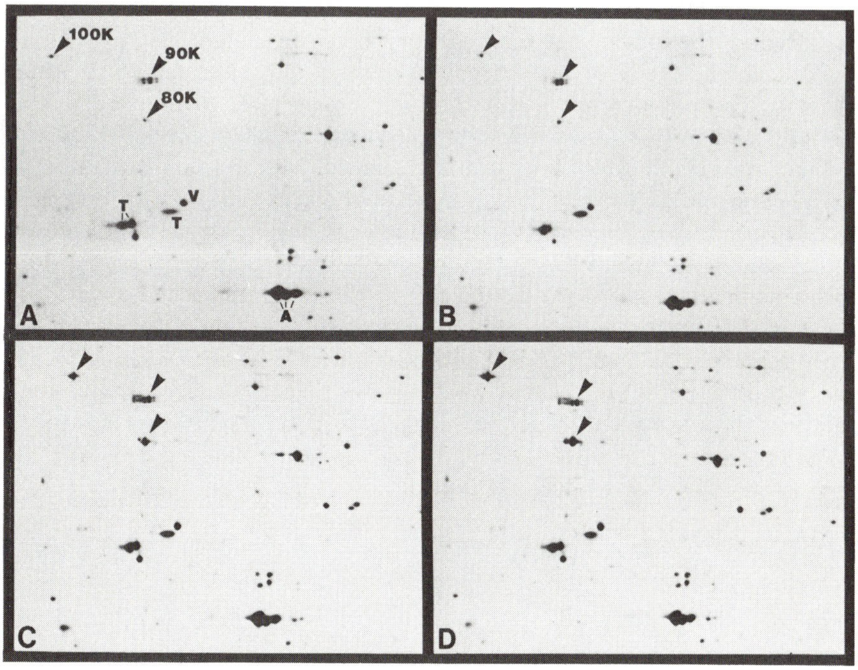

FIGURE 3
Heat-shock response of cultured L6 cells. Cells were exposed to a heat shock of 42°C for periods of 0 hr (*A*), 1 hr (*B*), 2 hr (*C*), and 4 hr (*D*). During the last hour of treatment, cells were labeled with [^{35}S]methionine without a change of media. The control was labeled for 1 hr at 37°C. Shown are the major heat-shock proteins, vimentin (V), the forms of tubulin (T), and the forms of actin (A). The gels contained pH 5–7 ampholytes in the first dimension and 10% acrylamide in the second dimension.

both induced by 13-fold. The induction of the 90K protein is twofold for the complex as a whole, but the pattern for each component is much more complex. In the control cells, there are predominantly two electrophoretic mobilities, with at least three differently charged components at each mobility. These appear not to be induced by 2–4 hours of heat shock, but two related series of spots, more acidic and of slightly higher apparent molecular weight, were induced.

90K Protein Complex

Because the components of the 90K complex are not induced in unison by heat shock in L6, we carried out experiments to learn more about the kinetic relationships between each of these components. First, it was determined, using highly exposed gels of ^{33}P-labeled proteins, that phosphate incorpora-

tion can be detected in each of the components except the most basic (Fig. 2). Second, the time course of phosphorylation was determined by labeling L6 cells (unshocked) for various lengths of time from 30 minutes to 16 hours (Fig. 4). The major component of the 90K complex after 16 hours of labeling is indicated by the small arrow in each panel. Detectable labeling with [^{35}S]methionine does not appear at this position within the first 30 minutes of labeling and only begins to appear after 1 hour. At early times, two more acidic forms are prominent. These are the same three charged forms seen in unshocked cells labeled for 1 hour (see Fig. 3A). The "maturation" of these proteins did not lead to forms as acidic as those induced by the heat shock (Fig. 3C,D). The heat-shock-induced components of the 90K complex probably represent two independent series of multiply phosphorylated forms, since their apparent molecular weights differ slightly from the major components of unshocked cells (see Fig. 2) and since they appear to be simultaneously converted from acidic to basic forms during a pulse-chase experiment (J.I. Garrels, unpubl.).

Kinetics of Phosphorylation for Other Proteins

The kinetics of phosphorylation for several other proteins can be easily observed among the proteins seen in Figure 4. In the series just above actin and in the series just below actin (see arrowheads), a conversion of basic to acidic forms takes place over time. In each series the most basic form is unphosphorylated, whereas the other more acidic forms are phosphorylated. This is the region of the gel where others have found the regulatory subunit of type-I cAMP-dependent protein kinase (Steinberg et al. 1977). The lower series is an unknown muscle-specific protein (see Garrels 1979b). The interesting feature in each case is that more than 4 hours are required to reach the steady-state level of phosphorylation, and the ultimate conversion to the phosphorylated forms seems to be essentially complete. Additional experiments (J.I. Garrels, unpubl.) have shown that [^{35}S]methionine incorporated during a brief pulse label disappears from the basic forms and appears in the more acidic forms during a "chase" period in unlabeled medium. This rules out differential stabilities as an explanation of the changing intensities of the phosphorylated and unphosphorylated forms. Because these phosphorylation reactions are slow and appear to progress to completion, they may represent modifications necessary for transport, assembly, or structural maturation, rather than playing a role in the regulation of enzymatic activity.

DISCUSSION

Quantitative two-dimensional gel electrophoresis methods can be used to detect phosphoproteins in cultured L6 muscle cells, regardless of their functional activity. Such analysis can help us to determine how many of the pro-

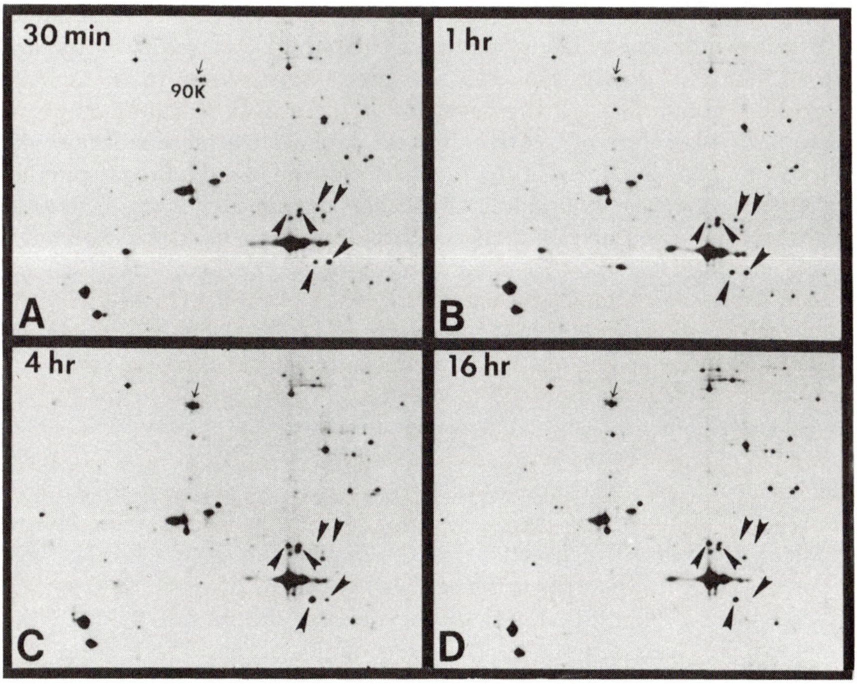

FIGURE 4
Kinetics of phosphorylation observed by differential periods of [^{35}S]methionine incorporation. L6 cells were labeled continuously at 37°C for different lengths of time, as indicated, before harvest, to observe the conversion of unphosphorylated forms into phosphorylated derivatives. All arrowheads point to corresponding positions in the four panels. The small arrow points to a form of the 90K protein that is not observed at early times of labeling. The arrowheads indicate different series of apparently related spots that shift in charge during continued labeling. The gels contained pH 5–7 ampholytes in the first dimension and 12.5% acrylamide in the second dimension.

teins are modified by phosphorylation and, ultimately, to classify the phosphoproteins by tissue-specificity, by subcellular localization, by kinetics of phosphorylation and dephosphorylation, and by their responses to physiological stimuli.

We have examined some of the major phosphoproteins of the L6 muscle cell line to determine how they are altered in response to external stimuli and to determine the kinetics of their phosphorylation. Myogenic differentiation, which was previously examined in L6 cells (Garrels 1979b), did not lead to major changes in the state of phosphorylation for any of the most intensely labeled phosphoproteins identified here, although the synthesis of many of these proteins is induced or repressed during differentiation. Likewise, in the heat-shock response, several phosphoproteins were induced dramatically; however, the relative degree of phosphophorylation was un-

changed (Fig. 3). Kinetically, several of the major acidic phosphoproteins were shown to accumulate phosphate steadily over a period of hours, and by 16 hours of labeling, the phosphorylated forms were virtually the only forms detectable (Fig. 4). We have not yet rigorously classified all of the detectable phosphoproteins by kinetics of phosphate accumulation and turnover, but it is clear that many of the most intensely labeled phosphoproteins do not exist in a rapidly established, dynamic equilibrium between phosphophorylated and unphosphorylated forms, as might be expected if the phosphorylation reaction were controlling an enzymatic activity.

The heat-shock proteins are among the major phosphoproteins in L6 and other cell types. The heat-shock protein at M_r 90,000 (90K) is especially interesting because it is abundant in many different cell types of vertebrate species from chick to man and is induced by many physiological stresses besides heat, including noxious agents such as arsenite, heavy metals, or amino acid analogs (Kelley and Schlesinger 1978; Johnston et al. 1980; Levinson et al. 1980; Thomas et al. 1982). It has been reported to be associated with the $pp60^{src}$ transforming protein in Rous-sarcoma-virus-transformed cells (Brugge et al. 1981; Oppermann et al. 1981). Phosphoserine, but not phosphotyrosine, has been detected in the 90K protein (Oppermann et al. 1981). We find four different electrophoretic mobilities within this 90K complex, and at each mobility at least three differently charged forms can be found. A brief heat shock of L6 cells leads to the induction of only two of the four mobility groups. The induced forms are somewhat more acidic and have slightly higher apparent molecular weights than the major uninduced forms. All four forms of the 90K protein can be synthesized by in vitro translation of poly(A)$^+$ mRNA from HeLa or chick embryo cells (G.P. Thomas et al., in prep.), indicating that the apparent size heterogeneity may be due to multiple gene products or to some unknown modification other than glycosylation.

Many hundreds of phosphoproteins are detectable by two-dimensional gel electrophoresis. We have found phosphate labeling coincident with approximately 13% of the 500 most intense spots of the [^{35}S]methionine-labeled pattern; however, we have not yet determined how many of these represent multiple phosphorylations of the same protein. Surely, many of the more minor proteins also contain phosphate; in the gels shown in Figure 1, over 200 minor phosphoproteins did not coincide with any detectable [^{35}S]methionine-labeled protein.

The further characterization of the phosphoproteins detectable on two-dimensional gels, in different cell types and during cellular responses to various environmental stimuli, will be of great interest and should help us to determine which of the minor phosphoproteins act as key regulators of cellular functions. Whether any of the minor phosphoproteins play critical regulatory or structural roles in myogenic differentiation remains to be determined.

ACKNOWLEDGMENTS

I thank Dr. J.D. Watson for enthusiastic support in the development of our system for quantitative two-dimensional gel electrophoresis. I also thank J. Leibold, T. Kelly, and P. Renna for technical assistance. This work was supported by grants from the Muscular Dystrophy Association of America and from the National Institutes of Health (GM-26298).

REFERENCES

Brugge, J.S., E. Erikson, and R.L. Erikson. 1981. The specific interaction of the Rous sarcoma virus transforming protein, $pp60^{src}$, with two cellular proteins. *Cell* **25:** 363.

Cohen, P. 1982. The role of protein phosphorylation in neural and hormonal control of cellular activity. *Nature* **296:** 613.

Cooper, J.A. and T. Hunter. 1981. Changes in protein phosphorylation in Rous sarcoma virus-transformed chicken embryo cells. *Mol. Cell. Biol.* **1:** 165.

Garrels, J.I. 1979a. Two-dimensional gel electrophoresis and computer analysis of proteins synthesized by clonal cell lines. *J. Biol. Chem.* **254:** 7961.

Garrels, J.I. 1979b. Changes in protein synthesis during myogenesis in a clonal cell line. *Dev. Biol.* **73:** 134.

Hunter, T. and J.A. Cooper. 1981. Epidermal growth factor induces rapid tyrosine phosphorylation of proteins in A431 human tumor cells. *Cell* **24:** 741.

Johnston, D., H. Oppermann, J. Jackson, and W. Levinson. 1980. Induction of four proteins in chick embryo cells by sodium arsenite. *J. Biol. Chem.* **255:** 6975.

Kelley, P.M. and M.J. Schlesinger. 1978. The effect of amino acid analogs and heat shock on gene expression in chicken embryo fibroblasts. *Cell* **15:** 1277.

Levinson, W., H. Oppermann, and J. Jackson. 1980. Transition series metals and sulfhydryl reagents induce the synthesis of four proteins in eukaryotic cells. *Biochim. Biophys. Acta* **606:** 170.

Oppermann, J., A.D. Levinson, L. Levintow, H.E. Varmus, J.M. Bishop, and S. Kawai. 1981. Two cellular proteins that immunoprecipitate with the transforming protein of Rous sarcoma virus. *Virology* **113:** 736.

Steinberg, R.A. and P. Coffino. 1979. Two-dimensional gel analysis of cyclic AMP effects in cultured S49 lymphoma cells: Protein modification, induction, and repressions. *Cell* **18:** 719.

Steinberg, R.A., P.H. O'Farrell, U. Friedrich, and P. Coffino. 1977. Mutations causing charge alterations in regulatory subunits of the cAMP-dependent protein kinases of cultured S49 lymphoma cells. *Cell* **10:** 381.

Stull, J.T., D.R. Manning, C.W. High, and D.K. Blumenthal. 1980. Phosphorylation of contractile proteins in heart and skeletal muscle. *Fed. Proc.* **39:** 1552.

Thomas, G.P., W.J. Welch, M.B. Mathews, and J.R. Feramisco. 1982. Molecular and cellular effects of heat shock and related treatments of mammalian tissue culture cells. *Cold Spring Harbor Symp. Quant. Biol.* **46:** 985.

Yaffe, D. 1968. Retention of differentiation potentialities during prolonged cultivation of myogenic cells. *Proc. Natl. Acad. Sci.* **61:** 477.

GENE STRUCTURE AND EXPRESSION

Introduction: Muscle Gene Structure and Expression

MARK L. PEARSON
PAMELA A. BENFIELD
ROBERT A. ZIVIN
Basic Research Program-LBI
Cancer Biology Program
Frederick Cancer Research Facility
Frederick, Maryland 21701

Recent advances in molecular and cell biology have made eukaryotic organisms directly accessible to genetic analysis at the nucleotide level. Understanding the molecular mechanisms that regulate gene expression during tissue differentiation in complex organisms is a long-standing challenge in biology that is likely to benefit from the application of this new methodology. Muscle development offers a particularly attractive system for study in this regard. The formation of mature muscle requires the coordinate expression of many different genes coding for the specialized proteins that are abundant in muscle. These proteins have been subject to extensive biochemical and biophysical analysis over the past 20 years. Through such analysis it has become clear that many "muscle-specific" proteins are actually isoforms of proteins that are commonly found in much lower amounts in all eukaryotic cells. Here they form part of the cytoskeleton and are involved in cell motility. Although much is known about these gene products, little is known about the genetic mechanisms that control their specific activation and inactivation during development. Fortunately, several different biological systems used for the biochemical and physiological analysis of muscle development are amenable to genetic studies as well. Several myogenic cell lines exist that exhibit many of the early stages of muscle-specific gene induction and offer opportunities for applying somatic cell genetic methods to the analysis of muscle development. In even more complex systems, such as the soil roundworm *Caenorhabditis* and the fruit fly *Drosophila,* remarkable advances have been made in the isolation and characterization of mutant organisms with altered mobility phenotypes. The application of recombinant DNA methods to questions concerning

gene structure and function in these and other systems is discussed in the research papers that follow.

STRUCTURAL ORGANIZATION OF MUSCLE GENES

The genes for several major muscle contractile proteins have been found to belong to extensive multigene families. For example, the measured number of actin genes varies from 4 in the nematode *Caenorhabditis elegans* to as many as 30 in man (Engel et al.)[1]. Similar findings are reported for the myosin heavy-chain genes (Buckingham et al.; Nadal-Ginard et al; Umeda et al.). Whether all of these genes detected by DNA hybridization are actually transcribed and expressed is unclear. As far as is known at present, genes within such families demonstrate only limited genetic linkage. Of the four actin genes in *C. elegans,* three are linked (within 12 kb) and one is separate (Hirsh et al.), whereas in *Drosophila* the six actin genes reside in six separate locations (Fyrberg et al.). Consequently, the regulation of the expression of individual members of such gene families must involve *trans*-acting elements. Examples of such elements may be the *unc-52* and *sup-3* functions in *C. elegans* that interact with specific myosin heavy-chain genes (Zengel and Epstein 1980; Waterston et al. 1982).

To date the most extensive structural information available for a particular gene is that for the members of the actin gene family. These genes have been counted and their structures partially determined in a variety of phylogenetically diverse organisms. Even a simple organism like the slime mold *Dictyostelium discoideum* has as many as 17 such genes, which in some cases are developmentally regulated and in others (pseudogenes) are not expressed at all (McKeown et al.). The actin genes in vertebrates code for multiple actin isoforms. As expected, these different actin genes show extensive nucleotide sequence homology within their protein-coding regions. Even the aminoterminal sequence Met-Cys, which is not retained in the mature protein, is remarkably conserved. In contrast, sequence divergence of the 3′-noncoding regions has allowed the identification and quantitation of some of the individual members of this gene family and their transcripts by cDNA hybridization to nitrocellulose blots.

Further exploiting sequence differences in regions flanking the actin genes proper, Hirsh et al. have used DNA sequence polymorphism as a marker in genetic crosses to map the actin genes in *C. elegans.* Attention has also been paid to the position of the intervening sequences (introns) that often interrupt the coding sequences of eukaryotic genes. This interest was prompted by the suggestion that introns might separate functional domains in proteins and by theories concerning intron conservation during evolution (Gilbert 1979). In the actin genes, the total number of intervening sequences

[1]The references listed herein refer to chapters in this volume unless otherwise indicated.

varies, but their positions are partially conserved, at least within the deuterostomes. This leads to the conclusion (Crain et al.; Fyrberg et al.; Nudel et al.) that intervening sequences in this gene family have been both gained and lost at a limited number of locations during evolution.

MECHANISMS CONTROLLING MUSCLE GENE EXPRESSION

The availability of genomic and cDNA probes specific for different members within a gene family allows analysis of the transcription of these genes during muscle development. Fortunately, probes covering nontranslated regions are often characteristic of a particular gene and can therefore be used to examine the expression of that member of the gene family. In this way, it has been shown that in multigene families that code for muscle-specific proteins, different genes show their own characteristic pattern of expression.

The bulk of the evidence now available indicates that the major control of gene expression is experienced at the level of mRNA transcription. For example, the changing expression of actin genes in the sea urchin (Crain et al.) and in developing chick muscle (Ordahl et al.) is clearly shown to be accompanied by changes in the steady-state levels of actin mRNA. Mechanisms that might regulate transcription during muscle development include (1) regulatory signals within the 5′- and 3′-flanking sequences that could interact with RNA polymerase to control transcription initiation and termination specificity, (2) structural modification of RNA polymerase via accessory proteins, (3) conformational changes in the nucleosome structure of chromatin that lead to the selective accessibility of muscle genes, (4) effects of DNA methylation, (5) gene amplification, and (6) differential processing and modification of mRNA transcripts.

Many of these possibilities have been explored and are described in this volume. Using cDNA clones from quail breast myoblasts, Hastings and Emerson examine a "coordinately regulated gene set" of muscle-specific genes and have begun looking for common regulatory elements in flanking sequences. In addition, Hastings and Emerson and Ordahl et al. point out the importance of studying not only genes whose expression is turned *on* during myogenesis but also those genes whose transcription is turned *off*.

To assay the accessibility of muscle genes for transcription, Shani et al. have examined the DNase-I sensitivity of the genes for myosin light chain 2 and actin in L8 rat myoblasts. They have shown that these genes become sensitive to nuclease digestion at the time their expression is turned on during myogenesis. However, whether this increased DNase-I sensitivity is the result or the cause of increased transcription remains unclear. Buckingham and co-workers present interesting data from several rodent tissues and cell lines on the expression of the gene for actin and the two myosin light chains, LC_1 and LC_3. On the basis of mRNA blotting experiments, they conclude

that the two myosin light-chain transcripts share 3′-noncoding sequences. Whether these two proteins are the products of different genes or are the result of differential processing of the same transcript remains to be determined.

A particularly provocative finding is reported by Schwartz and Zimmer, who have obtained evidence for a transient amplification of the α-actin gene in chicken at the time the gene is maximally induced. Subsequently, the amplified sequences appear to be degraded quickly. Gene amplification has been demonstrated previously for rRNA and insect chorion genes, although in these cases the amplified genes are not lost as precipitously as in the actin case. It is also curious that the amplified chicken actin genes do not appear to have the same restriction patterns as the unamplified gene. This observation raises questions concerning the functional activities of the amplified copies. Nevertheless, amplification of the muscle genes during myogenesis could provide another general mechanism for the quantitative induction of muscle-specific proteins.

Somatic cell genetic methods have led to the isolation of mutant myoblast cell lines, which should facilitate the genetic and biochemical dissection of early steps in the pathway regulating myogenesis. In L6 rat myoblasts, some α-amanitin-resistant mutations in RNA polymerase II perturb muscle differentiation, presumably by altering the normal transcriptional program activated during myogenesis (Pearson and Crerar). In the same cell line, Nadal-Ginard et al. describe a temperature-sensitive mutant that exhibits a commitment-defective phenotype at high temperature.

The evidence presented in the research papers that follow indicates that expression of the muscle genes is regulated primarily at the level of mRNA synthesis; however, modulation of gene expression via secondary posttranscriptional control mechanisms may also be of physiological importance. For instance, it has been reported that a unique low-molecular-weight inhibitory "translational control" RNA is present in myosin mRNA-containing messenger ribonucleoproteins (mRNPs) and may determine subsequent myosin translation efficiencies (Bester et al. 1975; Pluskal and Sarkar 1981). The role of such mRNPs in both actin and myosin mRNA sequestration and turnover also remains unclear (Buckingham et al. 1974). In view of the possibility of artifacts arising in translation systems used to quantitate mRNA levels via their translational activity in vitro (see Benoff and Nadal-Ginard 1979), these early reports may have attached too much significance to posttranslational control mechanisms as they affect the expression of muscle genes.

ROLE OF ISOFORM HETEROGENEITY

The functional significance of the extensive heterogeneity of muscle protein isoforms remains obscure. Different isoforms could possess hitherto unrecognized secondary activities of physiological importance in addition

to their commonly known functions. For example, Perriard et al. point out that muscle creatine kinase is localized in the M line, where it may play a structural role as well as serving as an ATP-generating enzyme. Perhaps the ubiquitous brain creatine kinase isozyme serves an analogous structural role in cell motility and cytokinesis. The protein isoform heterogeneity could also reflect a need for differential induction in specific muscle cell types (Fyrberg et al.).

Information defining the role of protein structural heterogeneity will come from the study of mutants that are defective in the expression of certain members of a particular gene family. This type of study has already been used to good effect in the genetic analysis of the nematode *C. elegans.* This organism has a relatively simple genome and a convenient life cycle, both of which facilitate its genetic and biochemical analysis. In this volume, Waterston et al. and Karn et al. report the identification of mutations in the myosin heavy-chain gene *unc-54* of the nonsense, missense, and deletion types. The phenotypic consequence of these mutations for the organism (*unc*oordinated or slow movement) also permits second-site suppressor mutations to be selected by screening for normal movement. Both the myosin heavy-chain mutations and their suppressors will be useful in relating structural changes in myosin to its activity and its assembly.

SUMMARY

Significant advances have been made in the isolation and characterization of genomic DNA clones that code for several of the major muscle proteins. Nucleotide sequence analysis of such cloned genes now provides the most direct route to the determination of protein primary structure, as demonstrated so elegantly by Karn et al. for the myosin heavy chain of the body wall of *C. elegans.* The problem of analyzing the mechanisms that regulate the expression of muscle genes is complicated by the number and sequence similarities of individual members of these contractile protein gene families. This structural complexity at the genetic level plus the universal importance of contractile proteins in the cytoskeleton of all eukaryotic cells make such an analysis a challenge to the molecular biologist.

The availability of well-characterized cloned genes should ultimately permit us to use accurate RNA polymerase-II transcription systems to study the factors that regulate the expression of these genes in vitro. We hope that this approach, coupled with the use of appropriate mutants, will lead to the eventual reconstruction of control systems that allow the selective and coordinate expression of muscle genes. In addition, the development of gene transfer techniques in cells of the muscle lineage may make possible the reintroduction of cloned genes carrying mutations engineered in vitro for subsequent studies of the effects of these genes on muscle development and muscle function.

ACKNOWLEDGMENTS

Research was sponsored by the National Cancer Institute, Department of Health and Human Services, under contract no. N01-C0-23909 with Litton Bionetics. The contents of this publication do not necessarily reflect the views or policies of the Department of Health and Human Services, nor does mention of trade names, commercial products, or organizations imply endorsement by the U.S. Government.

REFERENCES

Benoff, S. and B. Nadal-Ginard. 1979. Cell-free translation of mammalian myosin heavy-chain messenger ribonucleic acid from growing and fused-L_6E_9 myoblasts. *Biochemistry* **18:** 494.

Bester, A.J., D.S. Kennedy, and S.M. Heywood. 1975. Two classes of translational control RNA: Their role in the regulation of protein synthesis. *Proc. Natl. Acad. Sci.* **72:** 1523.

Buckingham, M.E., D. Caput, A. Cohen, R.G. Whalen, and F. Gros. 1974. The synthesis and stability of cytoplasmic messenger RNA during myoblast differentiation in culture. *Proc. Natl. Acad. Sci.* **71:** 1466.

Gilbert, W. 1979. Introns and exons: Playgrounds of evolution. *ICN-UCLA Symp. Mol. Cell. Biol.* **14:** 1.

Pluskal, M.G. and S. Sarkar. 1981. Cytoplasmic low molecular weight ribonucleic acid species of chick embryonic muscles, a potent inhibitor of messenger ribonucleic acid translation *in vitro. Biochemistry.* **20:** 2048.

Waterston, R.H., D.G. Moerman, D. Baillie, and T.R. Lane. 1982. Mutations affecting myosin heavy chain accumulation and function in the nematode *Caenorhabditis elegans.* In *Disorders of the Motor Unit.* (ed. D.L. Schotland), p. 747. Wiley, New York.

Zengel, J.M. and H.F. Epstein. 1980. Mutants altering coordinate synthesis of specific myosins during nematode muscle development. *Proc. Natl. Acad. Sci.* **77:** 852.

Actin Genes in Dictyostelium

MICHAEL McKEOWN
*CAROL MacLEOD**
RICHARD FIRTEL
Department of Biology
University of California, San Diego
La Jolla, California 92093

Dictyostelium discoideum is a simple eukaryote that undergoes multicellular development and differentiation. Vegetative amoebae emerge from spores and feed on bacteria or grow in axenic medium. Upon removal of the food source, the amoebae initiate a 24–26-hour developmental program leading to the production of multicellular fruiting bodies with two cell types: (1) stalks, which are highly vacuolated and incapable of growth, and (2) spores, which can repeat the life cycle when favorable conditions return.

Actin is a major developmentally regulated protein. Little or no actin is synthesized in freshly activated spores, but the level of actin synthesis increases rapidly during the first few hours after activation such that actin represents approximately 1% of the newly synthesized protein during vegetative growth. Removal of the food source results in a threefold increase in the rate of actin synthesis in the first 3 hours of development. During later development, the rate of actin synthesis drops below that in vegetative amoebae (Tuchman et al. 1974; Alton and Lodish 1977; Geri and Ennis 1977; MacLeod 1979; MacLeod et al. 1980). Actin synthesis also varies from cell type to cell type; prespore cells synthesize little actin and contain little actin mRNA, whereas prestalk cells continue to synthesize substantial amounts of actin and have high levels of actin mRNA (Alton and Brenner 1979; Coloma and Lodish 1981; M. Mehdy et al., in prep.). The changes in the rate of actin synthesis are found to nearly parallel changes in the level of actin RNA (Tuchman et al. 1974; Altoń and Lodish 1977; Kindle 1978; Kindle and Firtel 1978; MacLeod et al. 1980; Margolskee and Lodish 1980a,b; M. Mehdy et al., in prep.).

*Present address: Salk Institute for Biological Studies, La Jolla, California 92138.

RESULTS

Isoelectric Forms of Actin

There are two size classes of actin mRNA with lengths of about 1.25 kb and 1.35 kb (Kindle and Firtel 1978; McKeown et al. 1978). When translated in vivo or in vitro, these give rise to four separate protein forms on two-dimensional gels (Kindle and Firtel 1978; MacLeod et al. 1980). That these isoelectric forms are actin has been confirmed by affinity chromatography of in-vitro- and in-vivo-synthesized proteins on immobilized DNase I, followed by two-dimensional gel electrophoreses. Using the O'Farrell (1975) system of two-dimensional gels versus that described by Garrels (see Garrels and Gibson 1976), only two actin forms are seen, one of which appears to be an unacetylated form of the other (Rubenstein and Deuchler 1979). When actin purified by polymerization is sequenced, there appears to be only a single protein with less than 5% of the actin having a different sequence (Vandekerckhove and Weber 1980). Since only 30% of the input actin was recovered after purification, it is possible that minor forms were lost.

Proteins synthesized in vivo from pulse-labeled cells and actins synthesized in vitro from isolated RNA have been examined through development. As shown in Figures 1 and 2, the relative intensities of the two major isoelectric forms remain constant during development.

Actin-gene Arrangement

We have isolated 10 of the 17 *Dictyostelium* actin genes and cDNAs for at least two more genes. Figure 3 shows the restriction maps of the cloned genes and adjacent regions. The maps are all aligned by the actin genes. The sequences outside the genes are not identical. This implies that the genes are not part of a large tandem array with homogeneous spacers, although they may be linked to each other in a manner similar to globin genes (Efstratiadis et al. 1980). In fact there is some linkage; pDd *actin 2* contains two genes, *sub 1* and *sub 2*, and pDd *actin M6* is near another actin gene in the genome (McKeown et al. 1978). The actin genes have many restriction sites in common and are homologous in the coding region as has been shown by electron microscope heteroduplex mapping (McKeown et al. 1978). Especially notable are the *Hin*dIII site at the 5′ end of the genes and the *Ava*II/*Sau*96I site at the 3′ end. These are found in every gene and cDNA clone yet examined. Variation in restriction sites indicates that the genes vary slightly in primary sequence. DNA sequencing and thermal denaturation studies on heteroduplexes indicate a 2–10% mismatch between genes (see below).

Restriction maps of the cDNA clones lead to a similar conclusion (McKeown and Firtel 1982). Especially noteworthy is that the distance between the *Hin*dIII and *Ava*II/*Sau*96I sites is the same in the genes and cDNAs, implying that there are no intervening sequences, or very small ones, in the central regions of the genes.

5′ Region of the Actin Genes

We have partially sequenced 12 different genes or cDNAs (Firtel et al. 1979; McKeown and Firtel 1981a,b, 1982; McKeown et al. 1981). There is only 2–10% sequence divergence between regions sequenced in different pairs of genes. Almost all of this difference is confined to the third positions of codons and does not lead to amino acid substitutions. Two notable exceptions to this are the genes *actin 2-sub 2* and *actin 3*, each of which have multiple amino acid substitutions and appear not to be transcribed during early development (McKeown and Firtel 1981a; McKeown et al. 1982).

The sequence of the 5′-untranslated and flanking region and the amino acid sequence in the aminoterminal region for a number of genes are shown in Figure 4. The similarity within the coding region is obvious. It is also obvious that the region immediately 5′ to the coding region is greater than 90% A + T and is not conserved from gene to gene. This difference within the 5′-untranslated regions of the genes has allowed us to use nuclease S1 mapping to determine the 5′ ends of individual mRNAs and to quantitate individual gene expression. A 5′-end fragment from the *Hin*dIII site lying within codons 7–9 is labeled with ^{32}P using polynucleotide kinase and is then hybridized in DNA excess to poly(A)$^{+}$ RNA. Hybrids form between this probe and the cellular actin RNA. Actin RNA derived from the gene used as probe will hybridize up to the 5′ end of the RNA, whereas RNAs derived from other actin genes will hybridize only as far as the end of the common protein-coding region near the AUG. When these hybrids are treated with single-strand-specific nuclease, longer DNA fragments will result from protection by gene-specific RNA, and shorter fragments will result from protection by RNA from other actin genes. The fragments are easily separated on a DNA sequencing gel, and individual bands can be cut out and counted. The ratio of radioactivity in these two sizes of fragments can be used to calculate the fraction of actin RNA derived from a particular gene.

Figure 5 shows such a gel for pDd *actin 8*. The longer fragments correspond to the region of the 5′ end of the mRNA; the shorter fragments correspond to the region around the AUG initiation codon. Figure 4 shows the approximate 5′ ends of mRNA derived from a number of genes as determined in this way. All of the transcribed genes initiate transcription about 40 nucleotides upstream from the AUG initiation codon, giving rise to mRNAs with similar-length 5′-untranslated regions. Although the 5′ sequences of the genes are different, all of the transcribed genes share certain similar regions. They all have a TATA box (Gannon et al. 1979; Goldberg 1979) (labeled A in Fig. 4) about 30 nucleotides upstream from the start of transcription and a T-rich region (labeled B in Fig. 4) between the TATA box and the transcription initiation site. These are found at the 5′ end of all other expressed *Dictyostelium* genes studied to date (McKeown et al. 1981; A.K. Kimmel and R.A. Firtel, in prep.). Two genes, *actin 2-sub 2* and *actin 3* have not been detected as being transcribed. If they are expressed, they

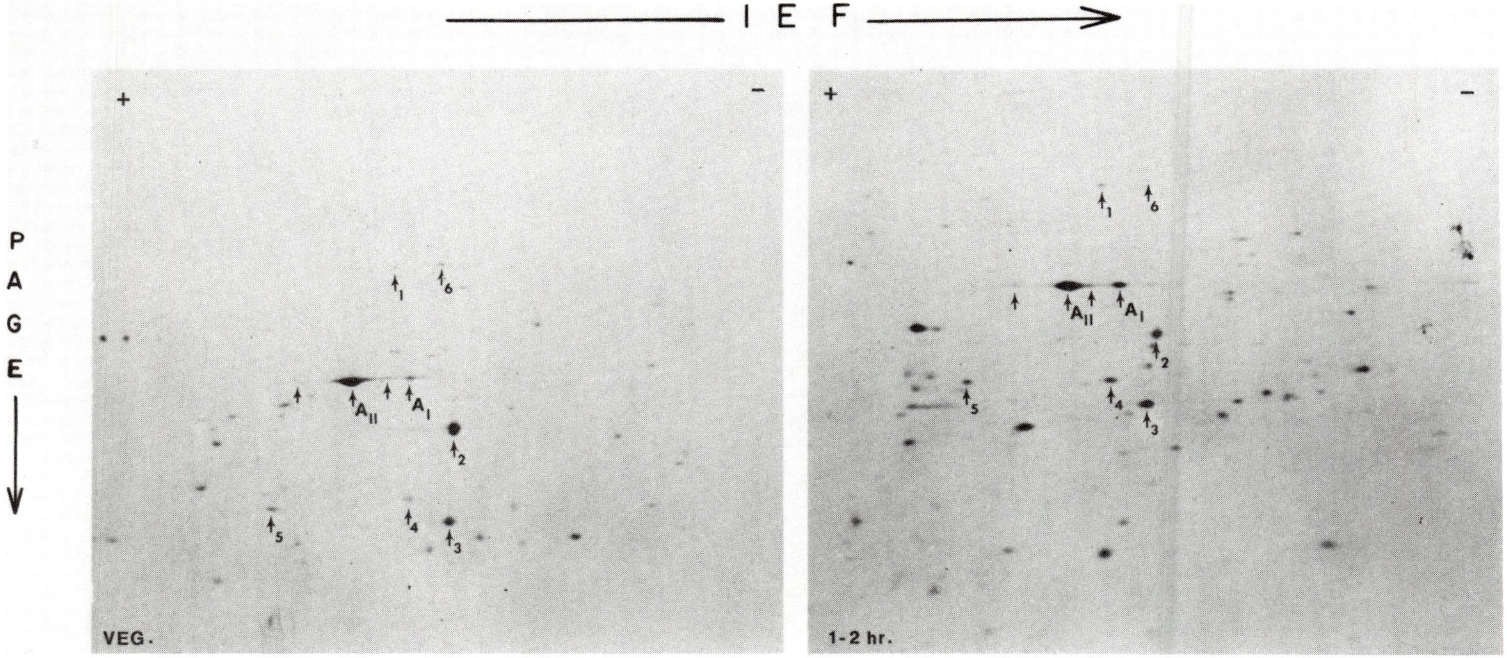
I E F
P A G E
+
−
1
6
A
II
A
I
2
3
4
5
VEG.
1-2 hr.

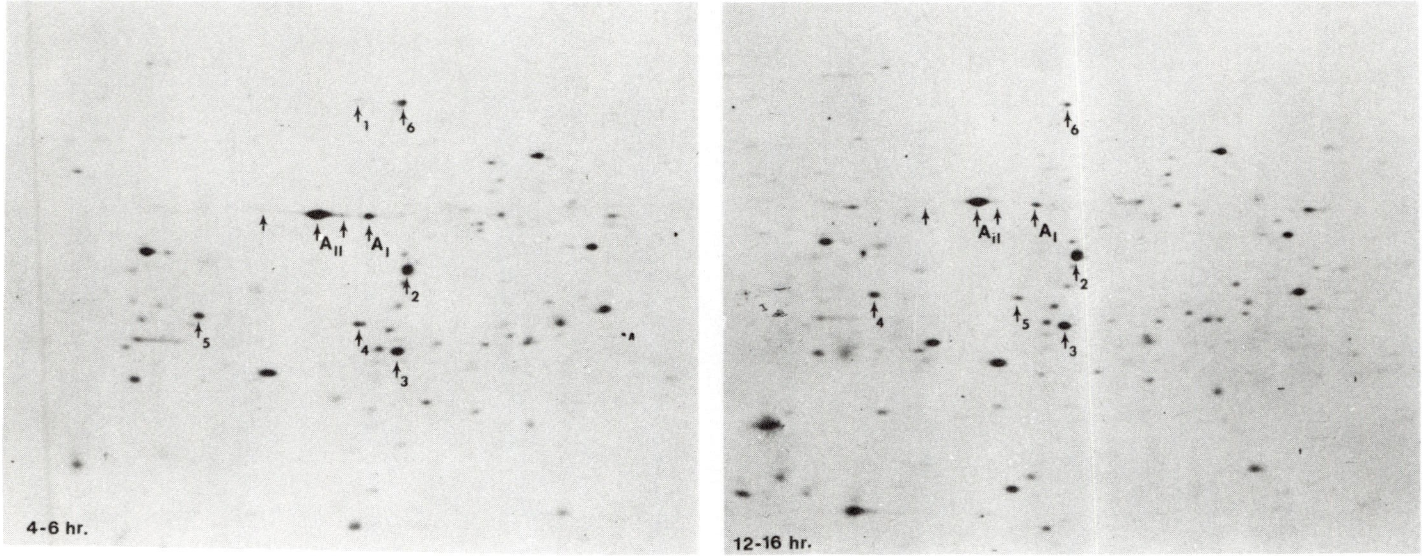

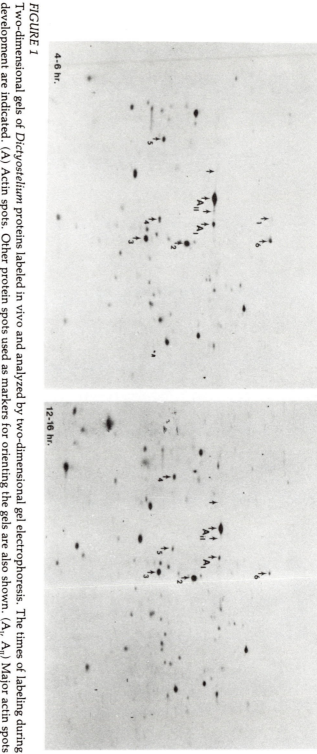

FIGURE 1
Two-dimensional gels of *Dictyostelium* proteins labeled in vivo and analyzed by two-dimensional gel electrophoresis. The times of labeling during development are indicated. (A) Actin spots. Other protein spots used as markers for orienting the gels are also shown. (A_I, A_{II}) Major actin spots (MacLeod 1979).

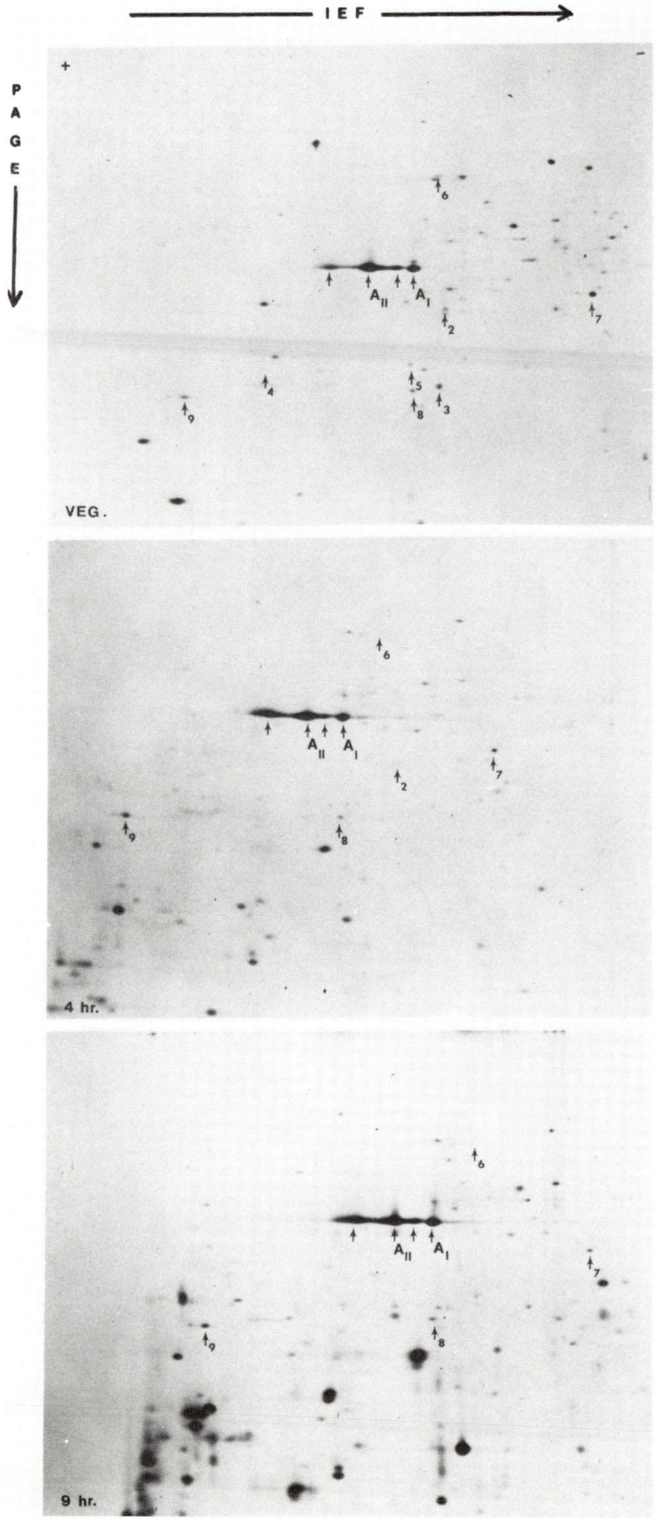

IEF
P A G E
+
−
VEG.
4 hr.
9 hr.
A_II
A_I
2
3
4
5
6
7
8
9

give rise to less than 1 copy of RNA per cell. These genes have aberrant 5′ sequences. *sub 2* lacks a TATA box and *actin 3* lacks the oligo(T) region.

It has also been possible to quantitate gene expression by this method. Table 1 summarizes all of the data gathered on specific gene expression. At least 9 of the 17 actin genes are expressed (cDNAs B1 and ITL-1 may have been derived from the same gene). These are not expressed at similar levels. Some genes, such as *M6*, give rise to less than 5% of the actin RNA, whereas other genes, such as *actin 8*, give rise to up to 30% of the actin RNA. The genes are not coordinately expressed. The level of expression of *actin 8* and *actin 6* appear to change relative to the expression of other actin genes during development, indicating that some members of the actin-gene family are differentially regulated. From analysis of total actin mRNA levels during development, we feel that *actin 6*, expressed at a high level in vegetative cells, is turned off during development. In contrast, the relative expression of *actin 8* increases during a time when total actin mRNA increases, indicating that the rate during expression of *actin 8* preferentially increases during early development.

3′ Region of the Actin Genes

Figure 6 shows the sequences of the 3′-untranslated regions of many of the actin genes. The three cDNAs, pcDd *actin 8*, pcDd actin III-12/A1, and pcDd actin ITL-1 all contain poly(A) at their 3′ ends. The *actin 8* cDNA has a 3′-untranslated region of 43 nucleotides, whereas the other two cDNAs have 3′-untranslated regions about 120 nucleotides or longer. This difference in the length of the 3′-untranslated regions is sufficient to account for the difference in length between the two size classes of actin mRNA. The genomic clones *actin 2-sub 1*, *actin 5*, and *actin 6* contain an oligo(A) sequence in a position similar to the poly(A) region of the longer cDNAs. In pcDd actin A1, the poly(A) tail actually consists of an oligo(A) region separated from a longer poly(A) by a single base. This is the structure suggested for *Dictyostelium* genes and mRNAs by Firtel and Jacobson (1977).

In contrast to the 5′ end of the genes, there is substantial sequence homology within the 3′ ends of the actin genes. Figure 7 shows this homology. The five genes shown all code for the longer mRNA class and share homology for about the first 40 bases of the 3′-untranslated region. Beyond this region there are two families of sequences. Within a family the sequence homology extends to the poly(A). Beyond the oligo(A) region there is no homology evident between any of the actin genes. If the

FIGURE 2 (see facing page)
Two-dimensional gels of actin proteins labeled in vitro. RNA was isolated at the time points shown, translated in vitro, and analyzed by two-dimensional gel electrophoresis. Protein labeling patterns are the same as those indicated in Fig. 1. (For details, see MacLeod 1979.)

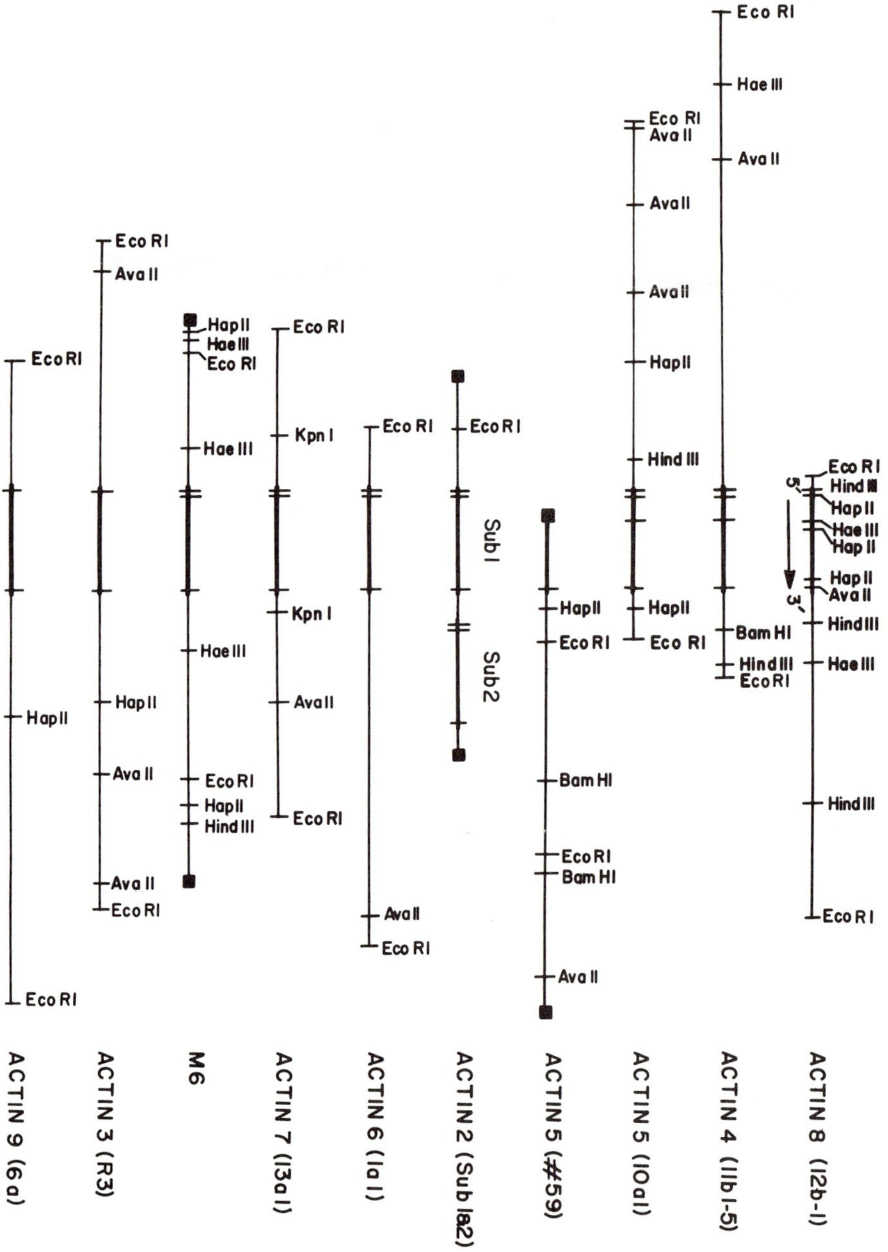

FIGURE 3

Restriction maps of *Dictyostelium* actin genes. All of the plasmids shown were mapped by standard techniques. The maps are all aligned by the actin-gene-coding regions, shown as a thick, solid bar, with the direction of transcription from left to right. (■) A:T tails of clones generated by random shear (pDd *actin 2*, pDd *actin M6)* or by partial *Hae*III digestion (pDd *actin 5* [#59]).

3′-untranslated region of the cDNA pcDd actin ITL-1 is used to probe a DNA blot of *Eco*RI-digested *Dictyostelium* DNA, five bands of hybridization are seen. All of these correspond to actin genes. As expected, one of these bands corresponds to the band containing *actin 5*, another member of the same 3′-end family (McKeown and Firtel 1981b). These results indicate that this 3′-end sequence is only found associated with actin genes and that there are three other actin genes in the *actin 5*/actin ITL-1 3′-end family. We believe the 3′-end similarities indicate more closely related genes and will make it possible to infer the evolution of the *Dictyostelium* actin genes.

SUMMARY

The 17 actin genes of *Dictyostelium* constitute a disperse multigene family with some clustering of genes. This arrangement of genes is similar to that found for the actin genes of other organisms, e.g., *Drosophila* (Fyrberg et al. 1980, 1981; Tobin et al. 1980) and sea urchin (Durica et al. 1980; Scheller et al. 1981).

Similar to other multigene families, most of the actin genes are expressed, but there appear to be at least two pseudogenes, *actin 2-sub 2* and *actin 3*. These have multiple amino acid differences from the major sequence and have grossly altered 5′ sequences in the expected promoter regions. No RNA has been detected from these genes. *actin 2-sub 2* is closely linked to the expressed gene *actin 2-sub 1*. *sub 2* is more closely related to *sub 1* than to any other actin gene we have examined and shares some sequence homology with *sub 1* in the 3′-untranslated region (McKeown and Firtel 1981b). It is conceivable that *sub 2* resulted as a duplication of *sub 1* and evolved by nucleotide-sequence drift into a pseudogene. The relationship of *actin 3* to other genes is not known.

The actin genes are quantitatively and qualitatively regulated during development. Tissue-specific differences in expression of individual genes have not been examined. The lack of coordinate control of the actin genes is consistent with the substantial differences in their 5′-noncoding sequences. The development of a *Dictyostelium* transformation system (K.P. Hirth et al., in prep.) should make it possible to determine which regions around the genes are important for gene control in a homologous system.

The similarity of the proteins produced by the different genes indicates that substantial selective pressure is being exerted to maintain the actin protein sequence in the face of a moderate level of nucleotide substitution. The large variation in sequence among the 5′ regions may indicate that these regions undergo much different kinds or levels of selection than the coding regions. The 3′-untranslated regions continue to show sequence similarity, probably indicating both the relationship between genes and some selective pressure. There is no sequence homology 3′ to the region coding for the mRNA.

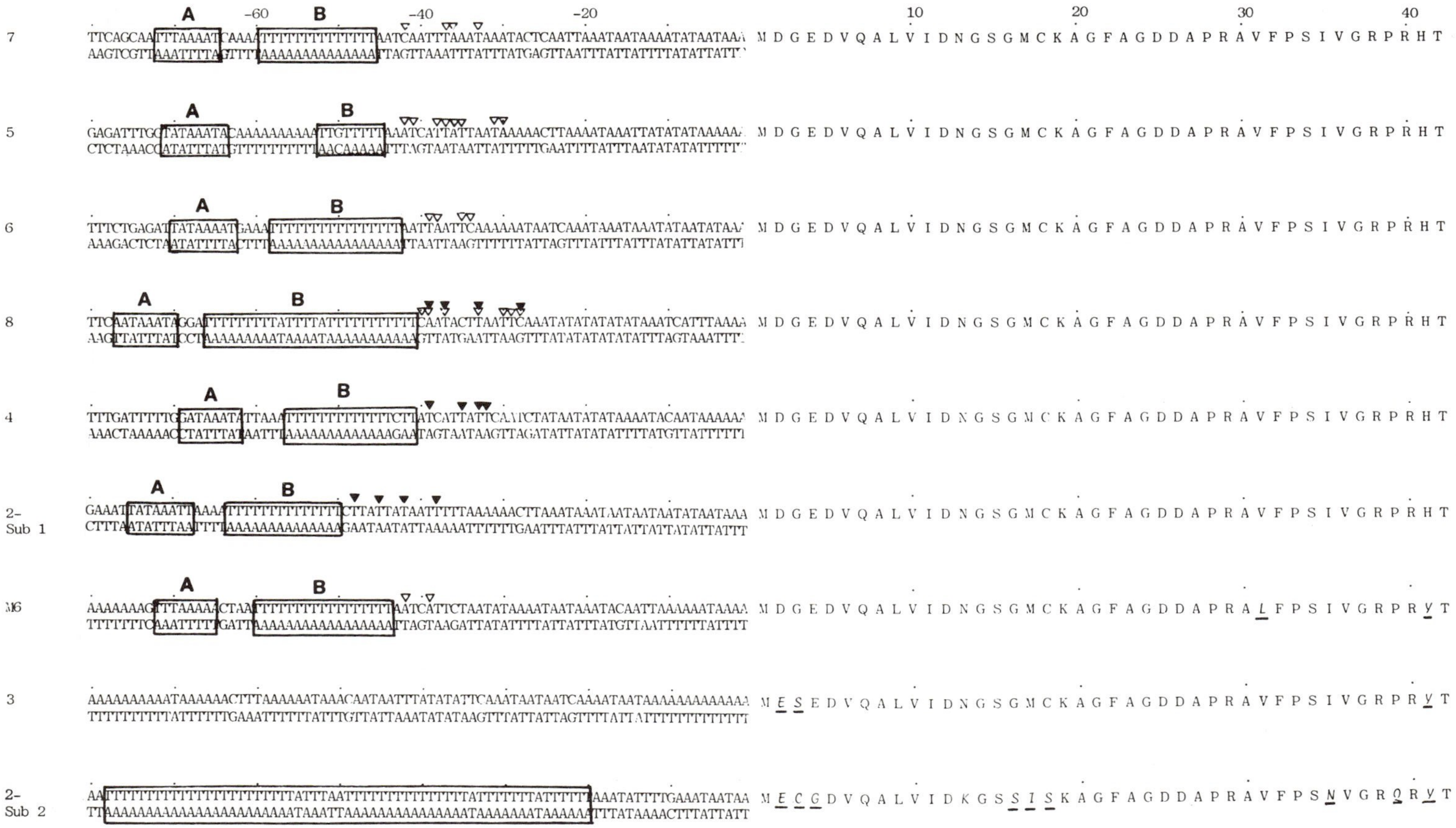
A
B
-60
-40
-20
10
20
30
40
7
5
6
8
4
2-
Sub 1
M6
3
2-
Sub 2
M D G E D V Q A L V I D N G S G M C K A G F A G D D A P R A V F P S I V G R P R H T
M D G E D V Q A L V I D N G S G M C K A G F A G D D A P R A L F P S I V G R P R Y T
M E S E D V Q A L V I D N G S G M C K A G F A G D D A P R A V F P S I V G R P R Y T
M E C G D V Q A L V I D K G S S I S K A G F A G D D A P R A V F P S N V G R Q R Y T

FIGURE 5
5′-end mapping of *actin 8*. Poly(A)$^+$ RNA from cells starved for 3 hr was hybridized to an excess of the 5′ *Hind*III fragment from pDd *actin 8*. This DNA was labeled with ^{32}P at its 5′ end, and only the mRNA cDNA strand was used. Hybrids were treated with nuclease S1 at either 0.1 units/μl (*lane a*) or 0.33 units/μl (*lane b*) for 45 min at 37°C. S1-resistant fragments were separated by electrophoresis adjacent to a sequencing ladder of the same DNA. A 1.5-nucleotide correction for the difference between the ends of chemically cleaved and S1-cleaved DNA is necessary to determine the bases marked in Fig. 4.

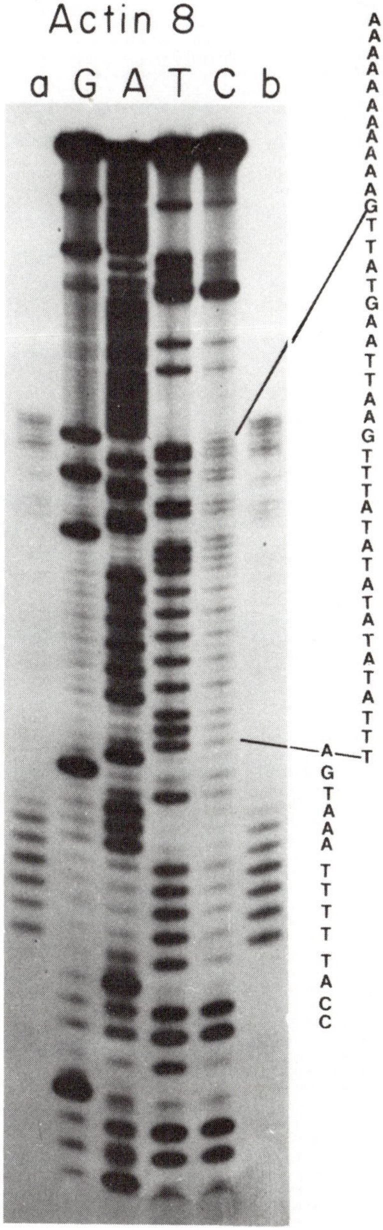

FIGURE 4 (see facing page)
5′-Untranslated and flanking sequences, and aminoterminal amino acid sequences of actin genes. The nucleotide sequence of the regions 5′ to the AUG initiation codon is shown. One-letter abbreviations for the aminoterminal amino acid sequences are used within the coding region. Altered amino acids are shown in italics and underlined. The TATA box and oligo(dT) sequences of the proposed *Dictyostelium* promoters are boxed and labeled A and B, respectively. (∇, ▼) Potential transcription initiation sites determined using nuclease S1 and mung bean nuclease, respectively. (For details, see McKeown et al. [1981, 1982].)

```
                         10        20        30        40        50        60        70        80        90        100       110
                  Ter     .         .         .         .         .         .         .         .         .         .         .
pcDd actin 8-1    TAAATTATTTAATAAATAATAAAAAAACAAATTGTTGTAATAATCT(A45)

pDd actin 8       TAAATTATTTAATAAATAATAAAAAAACAAATTGTTGTAATAATCTAATATTTTCTTTTTTTTTTAATTTTTTTTTTTTAAATCTTAATAATTATTAAGTTATTTTAATTTTT

pcDd actin        TAAATCATGATGAAAGTGCTTCACATAAAAATAATAATAATAATATAACAATAATAATATTTAAATGTATAATAAAATTTAATTACTTTTTTTTTAATGGTTGTTGATCTTTA
III-12/Al
pcDd actin        TAAACAAATAATTAAAACTAGTGATGAAAGTGCTTCTCACAAACAATTATGTAAAATATATAATAAAATACATTATTTAATCATTTTTATTTTTGTTTTAGTTGTTGATCTTT
ITL-1

pDd actin 5       TAAACAAAAAAAAAAAACCGAGTGATGAAAGCGCTTCTCACAAAATTATGAAAAATATTTAATAGTATAATAATTTTAAATCTTTTTTATTTTTTTTAGTTGTTGTCTTTATC

pDd actin 6       TAAACTAAACAATTAAAACCAGTGATGAAATGTCTTCTCACACTTAACAATATAATATTTATATGTATAATAATAAAATCTCAATAAAATATAATTCTTATTTTTATTTTTGA

pDd actin 2-      TAAATTAATTAAAAAAATTTAGTGATGAAAGTGCTTCTCACACAAAAATTATTATATATGTACAATAATAACAATAAAAACCCAATAAAATATAAACTTTTTTCTTTGATAGT
sub 1
pDd actin 2-      TAAATTAATTAATTAAAAATTTTAATGATGAAATTGTTTCTCACAAACAAACAATTTAATAAATGCACAAATAAAAAAAAAATAATAATAATTATTTTTTTTTGAATAAT
sub 2

pDd actin 4       TAAACAATTAATAAATATGTTGTATTTTATTTAAACTTAAATTATAAAAAAAAACCATTTTTAATTTTTCNATTTTTAATTTATTTTTTTTTTAGTTTAATATTGTATTTTT

                                 120       130       140       150       160       170       180       190       200       210       220
                                  .         .         .         .         .         .         .         .         .         .         .
pDd actin 8 cont.  TTTTTTTTTTTTTTTTTTTTTTTTTTTTTTTTCTATCAAAAAAATCAAATATATTTAAAAAATTTATTATTTACAGTACATTTTGAATGGTGAAGATAAATATATGCATT

pcDd actin            TCCGACCTT(AAAAAAAAAAAAAAAAAAAA)T(poly A)
III-12/Al
pcDd actin            ATAAGACTATTTAAAATTAATTGT(poly A)
ITL-1

pDd actin 5           CGACTTTAAAAATAAAAAAATTGTAAAAAAAAAAAGTTTATTTGTTTAATTTTATTGTTGTTTTTAATTTTTTGCAACCATTAATTTATAAAGCCAGCACCTAGCTATAA

pDd actin 6           ATCGGTTGTTGTCTTTATCCAGCCATCAAAAAAAAAAAAAAAAAAAAAAAAAAAAAAAAAGCAAATGAAACTATTAATTTTTTTTTTTTTAACCCAAAATT

pDd actin 2-          CGTTGATCTTTATCCGACCTTTAAAAAAAAAAAAAAATTTCGTTTACTTTTATTTTATTTATTATTTTATTATTAAAAAAAAAGGTTGTTTTAAATTTGGAGA
sub 1
pDd actin 2-          AATTGACTTTAATATAAAAAAAGATATAGAAATTTTTTTATTTTAATTGATTTTAAATTAAAAACTTAGCCAAAAAAAAAAAAAAAAAAAAAAAAAAAATCCGAGTTTTGA
sub 2

pDd actin 4           AATTTATATTTTTATTTTTCAAATTATCTAAAATAAACAATTCAAAAAAAAAAATATATAAAAAAGTGTTTTAAAAATAATCATGTATTAATTTTGTTTTTATTATTTAT
```

TABLE 1
Actin-gene Expression during Development

Gene	Hours after initiation of development					
	0	3	6	8	13	20
actin 8	18	27	c	30	19	22
actin 6	23	9		6	10	2
actin 5	5	4		5	3	4
actin M6	1	3		2	2	3
actin 2-sub 1		+				
actin 4		+				
actin 7		+				
Actin ITL-1			c			
Actin III-12/A1			c			
Actin B1			c			
actin 2-sub 2	—	—		—		
actin 3	—	—				

Numbers represent the percentages of actin RNA derived from a specific gene at the times shown. This number was determined from experiments such as that shown in Fig. 5. The upper and lower sets of nuclease-S1-protected fragments were cut from the gel and counted. Percentages are determined as counts in the upper bands divided by counts in the upper plus lower bands × 100. Each number is the average of two experiments. +, RNA has been detected from a given gene by the 5′-end mapping procedure; −, no RNA has been detected; c, a cDNA for that gene was isolated from the mRNA population present in cells 6 hr into development.

Our lab is currently isolating the remaining actin genes. We are preparing to use the *Dictyostelium* transformation system to examine the regions necessary for proper expression and regulation of the actin genes.

ACKNOWLEDGMENTS

We thank Alan Kimmel, Stephen Poole, and John Brandis for helpful discussion and Elise Lamar, Robin Timm, Karen Farrington, Drina White, and Colleen Silan for technical assistance. M.M. was a National Science Foundation predoctoral fellow and was also supported by National Institutes of Health predoctoral traineeships. C.M. was supported by a National Institutes of Health predoctoral traineeship. R.A.F. is the recipient of an American Cancer Society Faculty research award. This work was par-

FIGURE 6 (see facing page)
3′-Noncoding regions of the actin genes. The sequences shown have the same polarity as mRNA and begin with the TAA termination codon. Parentheses indicate that the exact number of bases in a homopolymer run was not determined. The sequence AATAAA, thought to be important for poly(A) addition, has been underlined.

	Ter
pcDd actin ITL-1	TAAAC AAATAATTAAAACTAGTGATGAAAGTGCTTCTCACAAAcAATTATGtAAAATATaTAATAaaATAc AtTaTTtAAT
pDd actin 5	TAAACaAAAaAAAAAAAcCgAGTGATGAAAGcGCTTCTCACAAA ATTATGaAAAATATtTAATAgtATAatAaAtTTaAAT
pcDd actin III-12/A1	TAAAtcA TGATGAAAGTGCTTCaCAtAAAAAtAATAATAataATaTAaCAATAATAATAtTtAAATgT
pDd actin 2-sub 1	TAAAtTAAtTAAAAAAAtTTAGTGATGAAAGTGCTTCTCACAcAAAAAtTAtTAtATATgTA CAATAATAAcAATAAAAaCc
pDd actin 6	TAAACTAAAcAATTAAAATcAGTGATGAAAtgtCTTCTCACActtAAcAatATA ATATtTA tAtgtATAATAATAAAAaCc

pcDd actin ITL-1 cont.	CaTTTTTATTTTTgtTTTAGTTGTTGaTCTTTATCCGACTaTttAAA AttAATTGT poly(A)
pDd actin 5	CtTTTTTATTTTT TTTAGTTGTTG TCTTTATCCGACTtTaaAAAtaaAaaAATTGT A_{11}
pcDd actin III-12/A1	atAATAAAATATAATT ACTTTTTT TTTaATgGTtGTTGAT CTTTATCCGACCTT(A_{20})T poly(A)
pDd actin 2-sub 1	C AATAAAATATAA ACTTTTTTcTTTgATAGTCGTTGAT CTTTATCCGACCTTtA_{14}
pDd actin 6	C AATAAAATATAATTcttaTTTTTatTTTttgAaTCGgTtgTtgtCTTTATCCagCCaTcA_{33}

FIGURE 7
Families of genes as determined by 3′-end sequence homologies. The sequences of genes and cDNAs for the longer mRNA are aligned so as to maximize homology within a family of sequences (McKeown and Firtel 1981b). Capital letters indicate a base found in 50% or more of the compared sequences. All five genes are compared for the first 44 nucleotides. Beyond nucleotide 44, there are two groups of sequences as shown by the horizontal line. AATAAA is underlined.

tially supported by grants from the National Institutes of Health and the National Science Foundation to R.A.F.

REFERENCES

Alton, T.A. and M. Brenner. 1979. Comparison of proteins synthesized by anterior and posterior regions of *Dictyostelium discoideum* pseudoplasmodia. *Dev. Biol.* **71:** 1.

Alton, T.A. and H.F. Lodish. 1977. Developmental changes in messenger RNAs and protein synthesis in *Dictyostelium discoideum. Dev. Biol.* **60:** 180.

Coloma, A. and H.F. Lodish. 1981. Synthesis of spore- and stalk-specific proteins during differentiation of *Dictyostelium discoideum. Dev. Biol.* **81:** 238.

Durica, D., J.A. Schloss, and W.R. Crain. 1980. Organization of actin gene sequences in the sea urchin: Molecular cloning of an intron-containing DNA sequence for a cytoplasmic actin. *Proc. Natl. Acad. Sci.* **77:** 5683.

Efstratiadis, A., J.W. Posakony, T. Maniatis, R. Lawn, C. O'Connel, R.A. Spritz, J.K. DeRiel, B.G. Forget, S.M. Weissman, J.L. Slighton, A.E. Blechl, O. Smithies, F.E. Barelle, C.C. Shoulders, and N.J. Proudfoot. 1980. The structure and evolution of the β-globin gene family. *Cell* **21:** 653.

Firtel, R.A. and A. Jacobson. 1977. Structural organization and transcription of the genome of *Dictyostelium discoideum. Int. Rev. Biochem.* **15:** 377.

Firtel, R.A., R. Timm, A. Kimmel, and M. McKeown. 1979. Unusual nucleotide sequence at the 5′ end of actin genes in *Dictyostelium discoideum. Proc. Natl. Acad. Sci.* **76:** 6206.

Fyrberg, E.A., K.L. Kindle, N. Davidson, and A. Sodja. 1980. The actin genes of *Drosophila:* A dispersed multigene family. *Cell* **19:** 365.

Fyrberg, E.A., B.J. Bond, N.D. Hershey, K.S. Mixter, and N. Davidson. 1981. The actin genes of *Drosophila:* Protein coding regions are highly conserved but intron positions are not. *Cell* **24:** 107.

Gannon, F., K. O'Hare, F. Perrin, J. Le Pennee, C. Benoist, M. Cocket, R. Breathnach, A. Royal, A. Garapin, B. Cami, and P. Chambon. 1979. Organisation and sequences at the 5′ end of a cloned complete ovalbumin gene. *Nature* **278:** 428.

Garrels, J.I. and W. Gibson. 1976. Identification and characterization of multiple forms of actin. *Cell* **9:** 793.

Geri, J. and H.L. Ennis. 1977. Developmental changes in RNA and protein synthesis during germination in *D. discoideum* spores. *Dev. Biol.* **67:** 189.

Goldberg, M. 1979. "Sequence analysis of *Drosophila* histone genes." Ph.D. thesis, Stanford University, California.

Kindle, K.L. 1978. "The analysis of genome structure and transcription in *Dictyostelium discoideum* using three recombinant plasmids." Ph.D. thesis, University of California, San Diego.

Kindle, K.L. and R.A. Firtel. 1978. Identification and analysis of *Dictyostelium* actin genes, a family of moderately repeated genes. *Cell* **15:** 763.

MacLeod, C. 1979. "Gene expression in *Dictyostelium discoideum:* A genetic and biochemical analysis." Ph.D. thesis, University of California, San Diego.

MacLeod, C., R.A. Firtel, and J. Papkoff. 1980. Regulation of actin gene expression during spore germination in *Dictyostelium discoideum. Dev. Biol.* **76:** 263.

Margolskee, J.P. and H.F. Lodish. 1980a. Half lives of messenger RNA species during growth and differentiation of *Dictyostelium discoideum*. *Dev. Biol.* **74:** 37.

———. 1980b. The regulation of the synthesis of actin and two other proteins induced early in *Dictyostelium discoideum* development. *Dev. Biol.* **74:** 50.

McKeown, M. and R.A. Firtel. 1981a. Differential expression and 5′ end mapping of actin genes in *Dictyostelium*. *Cell* **24:** 799.

———. 1981b. Evidence for sub-families of actin genes in *Dictyostelium* as determined by comparison of 3′ end sequences. *J. Mol. Biol.* **151:** 593.

———. 1982. Actin multigene family of *Dictyostelium*. *Cold Spring Harbor Symp. Quant. Biol.* **46:** 495.

McKeown, M., A.K. Kimmel, and R.A. Firtel. 1981. Organization and expression of the actin multi-gene family in *Dictyostelium*. *ICN-UCLA Symp. Mol. Cell. Biol.* **23:** 107.

McKeown, M., K.-P. Hirth, C. Edwards, and R.A. Firtel. 1982. Examination of the regulation of the actin multigene family in *Dictyostelium discoideum*. In *Embryonic development part A: Genetic aspects. Progress in clinical and biological research* (ed. M. Burger and R. Weber), vol. 85A, p. 51. Alan R. Liss, New York.

McKeown, M., W.C. Taylor, K.L. Kindle, R.A. Firtel, W. Bender, and N. Davidson. 1978. Multiple, heterogeneous actin genes in *Dictyostelium*. *Cell* **15:** 789.

O'Farrell, P.H. 1975. High resolution two-dimensional electrophoresis of proteins. *J. Biol. Chem.* **250:** 4007.

Rubenstein, P. and J. Deuchler. 1979. Acetylated and nonacetylated actins in *Dictyostelium discoideum*. *J. Biol. Chem.* **254:** 11142.

Scheller, R.H., L.B. McAllister, W.R. Crain, D.S. Durica, S.W. Posakony, R.J. Britten, and E.H. Davidson. 1981. Organization and expression of multiple actin genes in the sea urchin. *Mol. Cell. Biol.* **1:** 609.

Tobin, S.L., E. Zulauf, F. Sanchez. E.A. Craig, and B.J. McCarthy. 1980. Multiple actin related sequences in the *Drosophila melanogaster* genome. *Cell* **19:** 21.

Tuckman, J., T.A. Alton, and H.F. Lodish. 1974. Preferential synthesis of actin during early development of the slime mold *Dictyostelium discoideum*. *Dev. Biol.* **40:** 116.

Vandekerckhove, J. and K. Weber. 1980. Vegetative *Dictyostelium* cells containing 17 actin genes express a single major actin. *Nature* **284:** 475.

Isolation and Genetic Mapping of the Actin Genes of Caenorhabditis elegans

DAVID HIRSH
JAMES G. FILES
STEPHEN H. CARR
Department of Molecular, Cellular, and Developmental Biology
University of Colorado
Boulder, Colorado 80309

The nematode *Caenorhabditis elegans* contains a small genome of 8×10^7 bp of DNA (Sulston and Brenner 1974). The adult animal has 953 somatic cells whose complete embryonic and postembryonic cell lineages are known (Sulston and Horvitz 1977; Deppe et al. 1978; Kimble and Hirsh 1979; J.E. Sulston and E. Schierenberg, pers. comm.). Many are muscle cells or cells containing thin filaments (Zengel and Epstein 1980). *C. elegans* is a self-fertilizing hermaphrodite that is convenient for genetic analyses, and a detailed genetic map has been constructed (Brenner 1974; Herman and Horvitz 1980). Approximately 20 muscle genes have been identified on the basis of mutant phenotypes that include uncoordinated behavior and defective muscles (Waterston et al. 1980). Among these 20 muscle genes, only 2 have been connected with functional gene products; *unc-54* codes for a major myosin, and *unc-15* codes for paramyosin (Epstein et al. 1974; Waterston et al. 1977). No actin gene has been identified in *C. elegans.* It is possible that one of the known muscle genes codes for actin, but actin loci may be difficult to detect genetically. For example, mutations in actin genes might be lethal or cryptic if multiple, identical genes are present. Nevertheless, the actin genes of *C. elegans* are important objects of study because actin is an abundant protein in *C. elegans,* because its synthesis is regulated developmentally in specific cell lineages, and because the genetic manipulation of actin genes seems feasible once these genes are recognized and mapped. We therefore began a search for the actin genes of *C. elegans* using nongenetic methods.

RESULTS AND DISCUSSION

Isolation and Characterization of Actin Genes in C. elegans

We isolated the *C. elegans* actin genes using a cDNA actin clone from *Dictyostelium discoideum* as a hybridization probe (Kindle and Firtel 1978). Restriction endonuclease digests of *C. elegans* DNA transferred to Southern filters and hybridized to the *Dictyostelium* clone yielded three to five bands, depending on the endonuclease used, indicating that the number of actin genes in this organism is small. We also screened a λ Charon-10–*C. elegans* recombinant DNA library using the *Dictyostelium* DNA as a hybridization probe. Phage, which hybridized to the cDNA, were analyzed by restriction-endonuclease mapping and hybridization to the cDNA clone to distinguish which of the fragments contained the actin sequences (Fig. 1). Several recombinant phages carrying *C. elegans* DNA inserts indicated that there are four actin genes: Three are clustered within a 12-kb region, and a fourth seems to be located elsewhere. The orientations of the genes were established by hybridization to the 5′ and 3′ halves of the *Dictyostelium* cDNA clone.

We carried out DNA sequence analysis on the 5′ ends of these four actin genes to distinguish how many different genes were present. Gene I is identical to gene III in the first 166 nucleotides of the coding region and in the 20 nucleotides upstream from the AUG initiation codon, with the exception of the nucleotide at the −1 position. Gene II differs from gene I in nine posi-

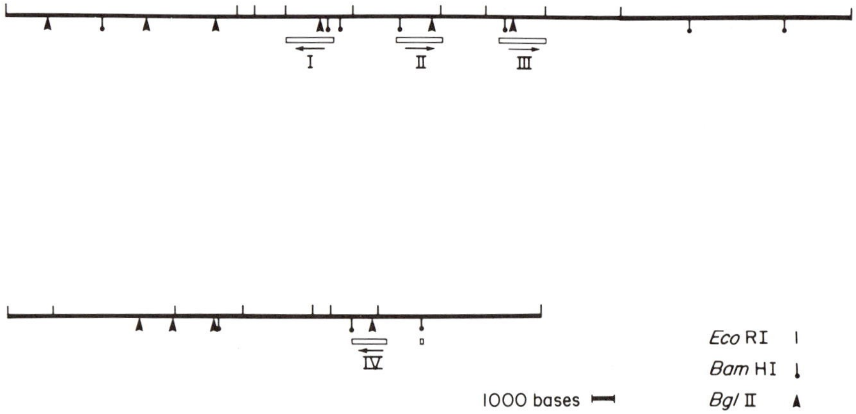

FIGURE 1
Restriction endonuclease cleavage maps of the regions containing actin genes. The two regions of DNA were isolated on λ Charon-10–*C. elegans* recombinant DNA phages. (▭) Regions that hybridize to the cloned *Dictyostelium* actin cDNA (pDdBl) and to one another. (→) Directions of potential transcription determined by hybridization to the 5′ or the 3′ fragments of the *Dictyostelium* actin cDNA clone.

tions within the first 158 nucleotides of the coding region. All nine differences are in wobble positions, and only the difference at nucleotide 15 results in an amino acid change. Gene II differs from gene I extensively in the region upstream from the AUG codon. The sequence of gene IV differs from genes I and II, but none of the nucleotide changes results in an amino acid difference in the region between codons 20 and 35. The analysis of the 5′ end of gene IV has been slower, and the 5′ end is not yet known because a 2000-bp intron is present within codon 20. Gene III contains a 300-bp intron between codons 63 and 64. The 3′ ends of the introns contain the consensus junction sequence TTCAGG, and the 5′ ends contain the consensus junction sequence GTAAGTT. Sequence analysis of gene I has not been carried far enough to determine whether it contains an intron between codons 63 and 64, but from the sizes of the restriction fragments, it appears to contain an intron of the same size and location as that in gene III.

The amino acid sequences derived from these DNA sequences show that *C. elegans* actins are similar to other actins (Fig. 2). Genes I and III have glutamic acid at the fifth codon and gene II has aspartic acid. It is not known whether gene IV differs from these other two sequences, because the sequencing of the aminoterminal region has not been finished. These amino acid sequences of the aminoterminal ends of *C. elegans* actins resemble the sequences found in cytoplasmic actins of vertebrates and not the sequences found in muscle actins of vertebrates, even though *C. elegans* has distinct muscle cells containing paracrystalline arrays of thick and thin filaments. It appears that these four genes are the only actin genes in *C. elegans* because when either the *Dictyostelium* clone or genes cloned from *C. elegans* are hybridized under less stringent conditions to the whole genome, we see no more bands than were seen originally with the *Dictyostelium* probe. Consequently, we conclude that these *C. elegans* actin sequences resemble those of cytoplasmic actins of vertebrates, not their muscle actins. Similar findings have been reported for the six *Drosophila* actin sequences by Fyrberg et al. (1981).

In summary, *C. elegans* has a small family of four actin genes. Three of the four genes are clustered within 12 kb of DNA. All of the genes are very similar to one another and to the amino acid sequences of other actin genes. The genes have sequences that thus far resemble cytoplasmic, not muscle, actins of vertebrates.

Locating Actin Genes on the Genetic Map

Although we had cloned the actin genes of *C. elegans,* none had been located on the genetic map. Yet genetic mapping is essential in order to determine the functional and physiological roles of the different actin genes

C. elegans	(1)	(5)	(10)	(17) (18)	(20)	(25)	(77)
II	(Met)-Cys-Asp-Asp-Asp-Val-Ala-	[Ala-Leu-Val]-Val-	[Asp-Asn-Gly-Ser-Gly]-	Met-Cys-	[Lys-Ala-Gly-Phe-Ala-Gly-Asp-Asp-Ala-Pro-Arg]	. . .	—
I	(Met)-Cys-Asp-Asp-Glu-Val-Ala-	-Val-		-Met-Cys-			—
III	(Met)-Cys-Asp-Asp-Glu-Val-Ala	-Val-		-Met-Cys-			-Val
IV				-Met-Cys			-Val

FIGURE 2

Amino acid sequences derived from DNA sequences of *C. elegans* actins. The amino acids enclosed in the horizontal boxes are constant for all of the *C. elegans* actins as well as for other known actin sequences. Amino acids at positions 7, 11, 17, 18, and 77 distinguish muscle actin from nonmuscle actin in various species. These *C. elegans* sequences resemble nonmuscle actin at these positions. It is not known whether the terminal methionine is retained.

and to study their developmental regulation. We therefore used DNA polymorphisms as phenotypic markers in standard genetic crosses to locate the actin DNA on the genetic map of *C. elegans*. We used two strains of *C. elegans*, the Bristol and Bergerac strains, which interbreed and also carry extensive DNA polymorphisms (Emmons et al. 1980). One polymorphism is adjacent to the cluster of three actin genes and can be seen when comparing the restriction endonuclease digests of the DNAs from the two strains. A 1.7-kb DNA insert is adjacent to actin gene III in the Bergerac strain but not in the Bristol strain. Consequently, most of the restriction fragments carrying actin gene III are 1.7 kb larger than the corresponding Bristol DNA fragments.

Linkage of the polymorphism and, hence, the adjacent actin cluster are determined in the following manner. Separate stocks of Bristol worms were prepared, each homozygous recessive for a known morphological marker on each of the six linkage groups. (In the case of linkage group III, which is longer than the other linkage groups, two morphological markers were used.) Each of these Bristol stocks was used to prepare heterozygous Bristol males, which were crossed to Bergerac worms. (Heterozygous Bristol males were used because the phenotypes often interfere with mating, and the Bergerac males mate poorly.) The F1 individuals were isolated, and F2 worms homozygous for the Bristol markers were picked and raised. Picking the homozygous F2 worms ensured recovering the Bristol chromosome, because each marker entered the cross on a specific Bristol chromosome. The other five unmarked chromosomes presumably assorted randomly. Approximately 60 F2 animals homozygous for each marked chromosome were picked and grown to approximately 10^7 worms. The DNA from these worms was isolated, digested with *Hin*dIII, size-separated, transferred to a Southern blot, and hybridized to a ^{32}P-labeled cloned actin DNA. In addition, the F2 animals from each cross, neither of which carried the Bristol marker nor subsequently segregated the Bristol marker, were also picked and their DNAs analyzed. These latter worms were homozygous for each of the respective Bergerac chromosomes. In the parental DNAs, a *Hin*dIII fragment from the Bergerac strain was 1.7 kb larger than the corresponding fragment from the Bristol strain and clearly distinguished the actin cluster between the two strains (Fig. 3). When the Bristol chromosome V was homozygous, only the Bristol actin pattern appeared, and when the Bergerac chromosome V was homozygous, only the Bergerac actin pattern appeared. Conversely, when chromosomes I–IV and X were homozygous, both the Bristol and Bergerac actin patterns were present, indicating random assortment of the actin genes on chromosome V. Therefore, this polymorphism and the adjacent actin gene cluster are both on linkage group V.

The location of the actin gene cluster within linkage group V was determined by three-factor crosses that also used the polymorphism as one of the phenotypic markers. The genetic map of linkage group V is shown in Figure

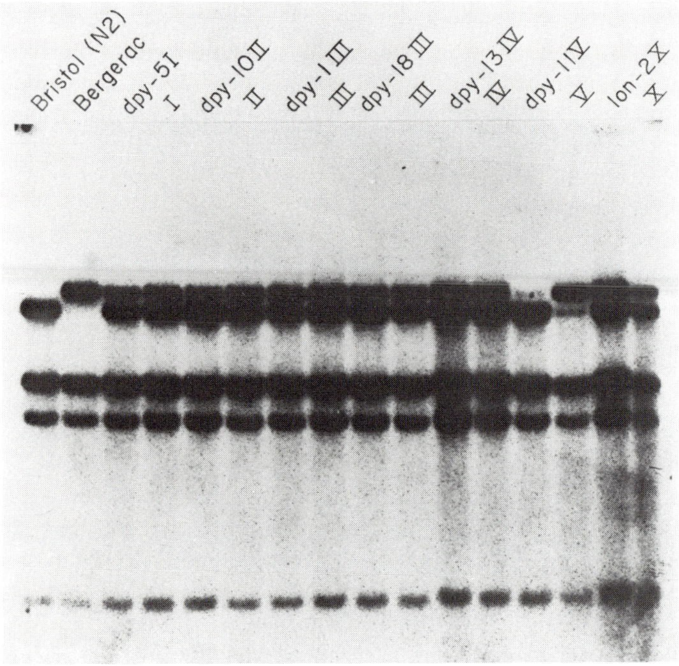

FIGURE 3
Linkage group determination. DNAs were isolated from the parental Bristol and Bergerac strains and from each of the hybrid strains in which one of the chromosomes was kept homozygous Bristol or Bergerac. DNAs were cut with *Hin*dIII, which revealed the 1700-bp interstrain difference in one of the fragments containing an actin sequence, when the Southern blots were hybridized with a ^{32}P-labeled *Dictyostelium* actin cDNA clone.

4. The Bristol strain *dpy-11 unc-76* was crossed with a Bergerac strain, and eight dumpy recombinants were picked that were not uncoordinated. Each animal was made homozygous for the *dpy-11* recombinant chromosome, grown up, and its DNA isolated. Similarly, two uncoordinated recombinants that were not dumpy were isolated and made homozygous for the *unc-76* recombinant chromosome, grown up, and their DNAs isolated. The DNAs were cut with *Hin*dIII, size-separated, transferred to Southern filters, and hybridized to a cloned actin DNA. Seven of the eight *dpy-11* recombinants and one *unc-76* recombinant showed the Bristol pattern. One *dpy-11* and one *unc-76* recombinant showed the Bergerac pattern. This indicates that the polymorphism and therefore the actin gene cluster reside near *dpy-11*, roughly 2% of the distance between *dpy-11* and *unc-76* (Fig. 4). These data could also be interpreted to mean that the actin cluster is at

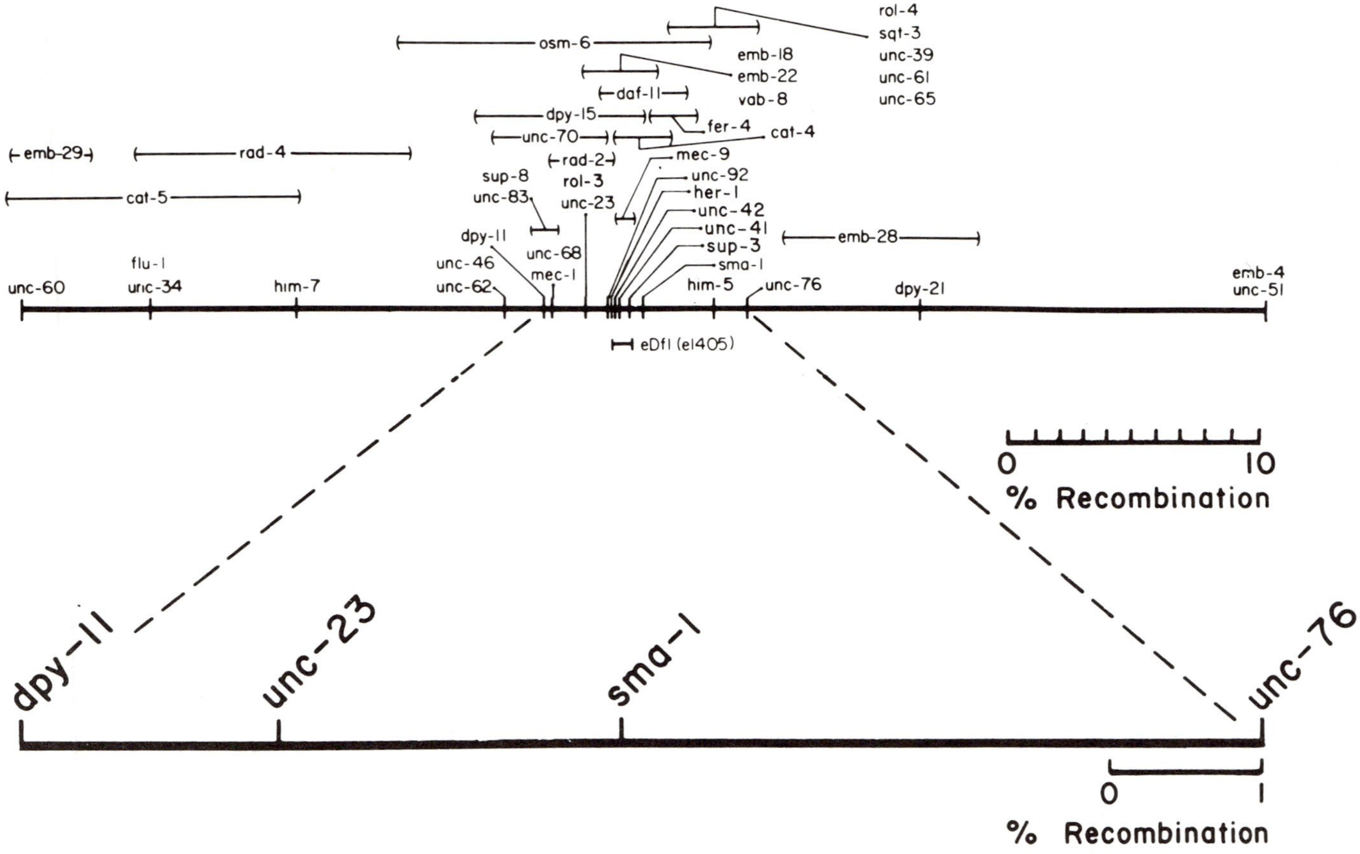

83 *FIGURE 4*
Linkage group V of *C. elegans*. The region in which three-factor crosses were carried out for the actin genes is expanded.

the left end of linkage group V and that the two Bergerac actin patterns were obtained by double crossovers. This is unlikely because the actin cluster is tightly linked to the *dpy-11* marker on chromosome V.

We used two more three-factor crosses to determine where within the *dpy-11 unc-76* interval the actin gene cluster resides. The Bristol strain *dpy-11 unc-23* was crossed to the Bergerac strain, recombinant progeny were picked, individuals were made homozygous for their recombinant chromosomes, and the DNAs were analyzed as before. All seven *dpy-11* recombinants showed the Bergerac pattern, and all eight *unc-23* recombinants showed the Bristol pattern, indicating that the actin cluster resides to the right of the *dpy-11 unc-23* interval.

A three-factor cross was also done between the Bristol strain *sma-1 unc-76* and the Bergerac strain. All seven *sma-1* recombinants showed the Bristol pattern and all eight *unc-76* recombinants showed the Bergerac pattern, indicating that the actin gene cluster resides to the left of the *sma-1 unc-76* interval. Therefore, the actin gene cluster is located between *unc-23* and *sma-1*. This distance represents 2% genetic recombination, which is roughly equivalent to 400 kb of DNA. Clearly, finer mapping must be carried out, using the polymorphism as a marker, and chromosomal rearrangements in this region must be examined to resolve which genetic locus within this region is an actin gene or whether the actin gene is a locus that has not yet been identified genetically.

General Applications of DNA Polymorphisms to Genetic Mapping

This work has shown that DNA polymorphisms can be used to establish the linkage group and map position of a DNA fragment that has been cloned. Often, as in the case of the actin genes of *C. elegans*, the function of the gene is known because the DNA fragment of interest has been identified either directly or indirectly with a cDNA clone. Wyman and White (1980) and Botstein et al. (1980) have shown that DNA polymorphisms can also be used along with pedigrees to establish human linkage maps where intentional crosses are not feasible. Human DNA polymorphisms can also serve as prenatal diagnostic indicators of closely linked genetic disorders (Kan and Dozy 1978).

DNA polymorphisms can be used for genetic mapping provided the differences between the strains or individuals are not so great as to distort either recombination or random segregation. Linkage disequilibrium and nonrandom segregation in hybrid strains have been described (Sturtevant 1936; Schafer 1978). Initially, we too observed linkage disequilibrium among the segregants of Bergerac/Bristol hybrids (Hirsh et al. 1979). This disequilibrium was probably due to more rapid growth rates of animals that carried certain combinations of Bristol or Bergerac chromosomes. The disequilibrium was overcome in the current studies by picking each of the in-

dividual F2 animals from an interstrain cross and rearing it and its progeny under noncompeting growth conditions until a saturated culture was obtained, however long it took. In this way, we obtained straightforward linkage and recombination data.

By continuing to use the *C. elegans* DNA polymorphism adjacent to the actin cluster, we should be able to identify the actin locus within the 2% recombination interval defined thus far. Identifying the actin locus will enable us to analyze the actin genes genetically, which should be extremely valuable in conjunction with DNA sequence analysis. Such studies will also shed light on the tissue-specific expression of actin genes during development and on the interrelationships of the four members of the gene family. Gene IV remains unmapped at this time. When a neighboring DNA polymorphism is located, it should also be possible to determine the map position of gene IV.

ACKNOWLEDGMENTS

We thank Kimberly Johnson, Michael Krause, and George Cox for their help with some of these experiments. This work was supported by grants GM-26515 and GM-1985 from the National Institutes of Health and grant 1-472 from the National Foundation/March of Dimes.

REFERENCES

Botstein, D., R.L. White, M. Skolnick, and R.W. Davis. 1980. Construction of a genetic linkage map in man using restriction fragment length polymorphisms. *Am. J. Hum. Genet.* **32:** 314.

Brenner, S. 1974. The genetics of *Caenorhabditis elegans. Genetics* **77:** 71.

Deppe, U., E. Schierenberg, T. Cole, C. Krieg, D. Schmitt, B. Yoder, and G. Ehrenstein. 1978. Cell lineages of the embryo of the nematode *Caenorhabditis elegans. Proc. Natl. Acad. Sci.* **75:** 376.

Emmons, S.W., B. Rosenzweig, and D. Hirsh. 1980. The arrangement of repeated sequences in the DNA of the nematode *Caenorhabditis elegans. J. Mol. Biol.* **144:** 481.

Epstein, H.F., R.H. Waterston, and S. Brenner. 1974. A mutant affecting the heavy chain of myosin in *Caenorhabditis elegans. J. Mol. Biol.* **90:** 291.

Fyrberg, E.A., B.J. Bond, N.D. Hershey, K.S. Mixter, and N. Davidson. 1981. The actin genes of *Drosophila:* Protein regions are highly conserved but intron positions are not. *Cell* **24:** 107.

Herman, R. and H.R. Horvitz. 1980. Genetic analysis of *Caenorhabditis elegans.* In *Nematodes as model biological systems* (ed. B. Zuckerman), vol. 1, p. 227. Academic Press, New York.

Hirsh, D., S.W. Emmons, J.G. Files, and M.R. Klass. 1979. Stability of the *C. elegans* genome during development and evolution. *ICN-UCLA Symp. Mol. Cell. Biol.* **14:** 205.

Kan, Y.W. and A.M. Dozy. 1978. Polymorphism of DNA sequence adjacent to human beta-globin structural gene: Relationship to sickle mutation. *Proc. Natl. Acad. Sci.* **75:** 5631.

Kimble, J. and D. Hirsh. 1979. The post-embryonic cell lineages of the hermaphrodite and male gonads in *Caenorhabditis elegans. Dev. Biol.* **70:** 396.

Kindle, K.L. and R.A. Firtel. 1978. Identification and analysis of *Dictyostelium* actin genes, a family of moderately repeated genes. *Cell* **15:** 763.

Schafer, U. 1978. Sterility in *Drosophila hydei* × *D. neohydei* hybrids. *Genetica* **49:** 205.

Sturtevant, A.H. 1936. Preferential segregation in triplo-IV females of *Drosophila melanogaster. Genetics* **21:** 444.

Sulston, J.E. and S. Brenner. 1974. The DNA of *Caenorhabditis elegans. Genetics* **77:** 95.

Sulston, J.E. and H.R. Horvitz. 1977. Post-embryonic cell lineages of the nematode *Caenorhabditis elegans. Dev. Biol.* **56:** 110.

Waterston, R.H., R.M. Fishpool, and S. Brenner. 1977. Mutants affecting paramyosin in *Caenorhabditis elegans. J. Mol. Biol.* **117:** 679.

Waterson, R.H., J.N. Thomson, and S. Brenner. 1980. Mutants with altered muscle structure in *Caenorhabditis elegans. Dev. Biol.* **77:** 271.

Wyman, A.R. and R. White. 1980. A highly polymorphic locus in human DNA. *Proc. Natl. Acad. Sci.* **77:** 6754.

Zengel, J.M. and H.F. Epstein. 1980. Identification of genetic elements associated with muscle structure in the nematode *Caenorhabditis elegans. Cell Motil.* **1:** 73.

Structural Studies of the Drosophila melanogaster *Actin Genes*

ERIC A. FYRBERG
Department of Biology
The Johns Hopkins University
Baltimore, Maryland 21218

*BEVERLEY J. BOND**
N. DAVIS HERSHEY†
*KATHARINE S. MIXTER**
NORMAN DAVIDSON†
**Division of Biology and †Department of Chemistry*
California Institute of Technology
Pasadena, California 91125

Biochemical investigations during the last three decades have established that actin is a major structural component of all eukaryotic cells (reviewed by Pollard and Weihing 1974). Actin has been isolated from several lower eukaryotes, as well as from a variety of invertebrate and vertebrate tissues, including nonmuscle and muscle cell lineages and a number of established cell lines. In nonmuscle cells, actin filaments mediate a wide spectrum of contractile processes, including cell motility, cytokinesis, exocytosis, endocytosis, and several types of internal movements, e.g., the separation of paired chromosomes during mitosis (Stossel 1978). Additionally, in nonmuscle cells, actin filaments form a cytoskeletal network that is believed to integrate the activities of many formed elements in the cytoplasm (Wolosewick and Porter 1979; Batten et al. 1980). In striated muscle, actin filaments form a precisely ordered interdigitating array that slides past the thick myosin filaments to generate contractile force (Huxley 1969).

The actin genes are particularly interesting subjects for studies of evolutionary and developmental aspects of eukaryotic gene regulation. The

primary amino acid sequence of actin is highly conserved, which allows comparison of actin gene structure across enormous evolutionary distances. Despite this high degree of conservation, multiple actin proteins (termed isoforms) have been documented in many metazoans (Garrels and Gibson 1976; Storti et al. 1976; Whalen et al. 1976; Fyrberg and Donady 1979). In several cases, these isoforms are known to be the products of small gene families (Cleveland et al. 1980; Fyrberg et al. 1980; Tobin et al. 1980; Scheller et al. 1981; Hirsch et al., this volume). Synthesis of these actin isoforms is precisely controlled during metazoan ontogeny, which indicates that the activities of particular actin genes are regulated by distinct mechanisms.

During the last few years, we have investigated the structure and expression of the actin genes of *Drosophila melanogaster. Drosophila* is a convenient organism for such studies and is suitable for classical genetic and cytogenic analyses as well. Ultimately, it should be possible to isolate mutants having lesions in particular actin genes and to use these strains to examine the assembly and function of particular actin isoforms. In the following sections, we summarize our progress to date.

RESULTS

Isolation of All Members of the Drosophila *Actin Gene Family*

To isolate all members of the actin gene family, we screened a recombinant bacteriophage library of *Drosophila* genomic DNA with a probe derived from λDmA2, our original actin gene isolate (Fyrberg et al. 1980). From a screen of 40,000 phages, 30, which hybridized strongly, were selected. Restriction maps of these phages indicated that they fell into six distinct classes. To demonstrate that these six classes of phages accounted for all six members of the *Drosophila* actin gene family, we digested a representative of each class with *Eco*RI and separated the digestion fragments on an agarose gel. In an adjacent lane of the same gel, an *Eco*RI digest of genomic *Drosophila* DNA was electrophoresed. DNA fragments were then blotted to nitrocellulose and hybridized with a labeled actin gene probe. As shown in Figure 1, the seven prominent bands of the genomic *Eco*RI digest can all be accounted for by the six representative phages. We have designated these phages λDmA1–λDmA6, on the basis of the decreasing sizes of the *Eco*RI fragments that hybridize to the actin probe. Each of the phages contains a single hybridizing *Eco*RI fragment, except λDmA6, which contains two.

Chromosomal Localization of Drosophila *Actin Genes*

By separately hybridizing the actin-gene-containing phage DNAs in situ to larval polytene chromosomes, we have determined the chromosomal location of each: λDmA1, 88F; λDmA2, 5C; λDmA3, 42A; λDmA4, 57A; λDmA5, 87E; and λDmA6, 79B (data not shown). These six regions are widely separated on the polytene chromosomes (in fact, they are located on

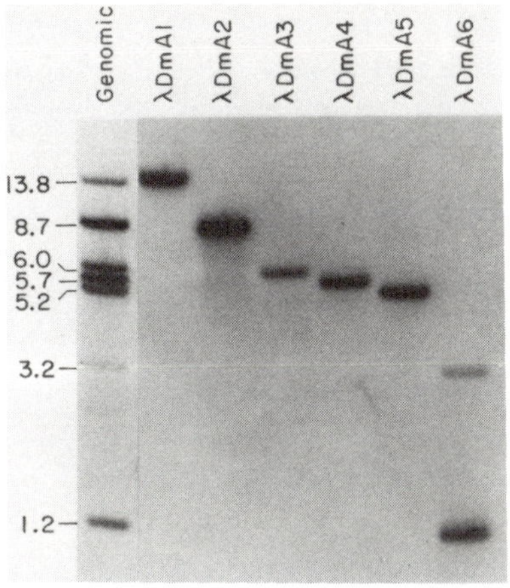

FIGURE 1
Blot-hybridization analysis of λDmA1–λDmA6. An *Eco*RI digest of genomic *Drosophila* DNA (*left lane*) was electrophoresed in parallel with *Eco*RI digests of λDmA1–λDmA6. Hybridization to the λDmA2 actin gene probe revealed that these six phages contain all of the *Drosophila* actin genes.

three different chromosomes). This experiment therefore provides direct proof that the *Drosophila* actin genes are widely dispersed, rather than closely linked or clustered, and confirms the result of Tobin et al. (1980) and Fyrberg et al. (1980) who reported that actin gene cDNA probes hybridized to these six locations.

Restriction Mapping of Chimeric Phages Containing Genomic Actin Sequences

Restriction maps prepared from λDmA1–λDmA6 (Fig. 2) failed to reveal a particular arrangement of restriction sites that could be used to position the actin structural genes. To localize the actin-coding regions, digested phage DNAs were blotted to nitrocellulose and hybridized to the λDmA2 structural gene probe. These experiments demonstrated that each of the six phages contained a single actin structural gene. The genes were localized within reasonably small (2–6 kb) DNA fragments (see Fig. 2) that were subcloned in pBR322 and analyzed further.

Visualization of Intervening Sequences within the DmA4 and DmA6 Structural Genes

To map precisely the protein-coding region of each *Drosophila* actin gene, we formed heteroduplexes between various pairs of genes and examined the resultant molecules in the electron microscope. In each case, a duplex segment of 1.1 kb, corresponding to the minimum size of the protein-coding region, terminates in single-strand forks (data not shown). We interpret this to mean that although the protein-coding regions of these genes are homologous, untranslated and flanking sequences are not. In two cases, single-strand loops, tentatively identified as intervening sequences, were seen: one of 630 nucleotides near the 5′ end of DmA4 and one of 360 nucleotides near codon 300 of DmA6. Introns were not seen in other *Drosophila* actin genes;

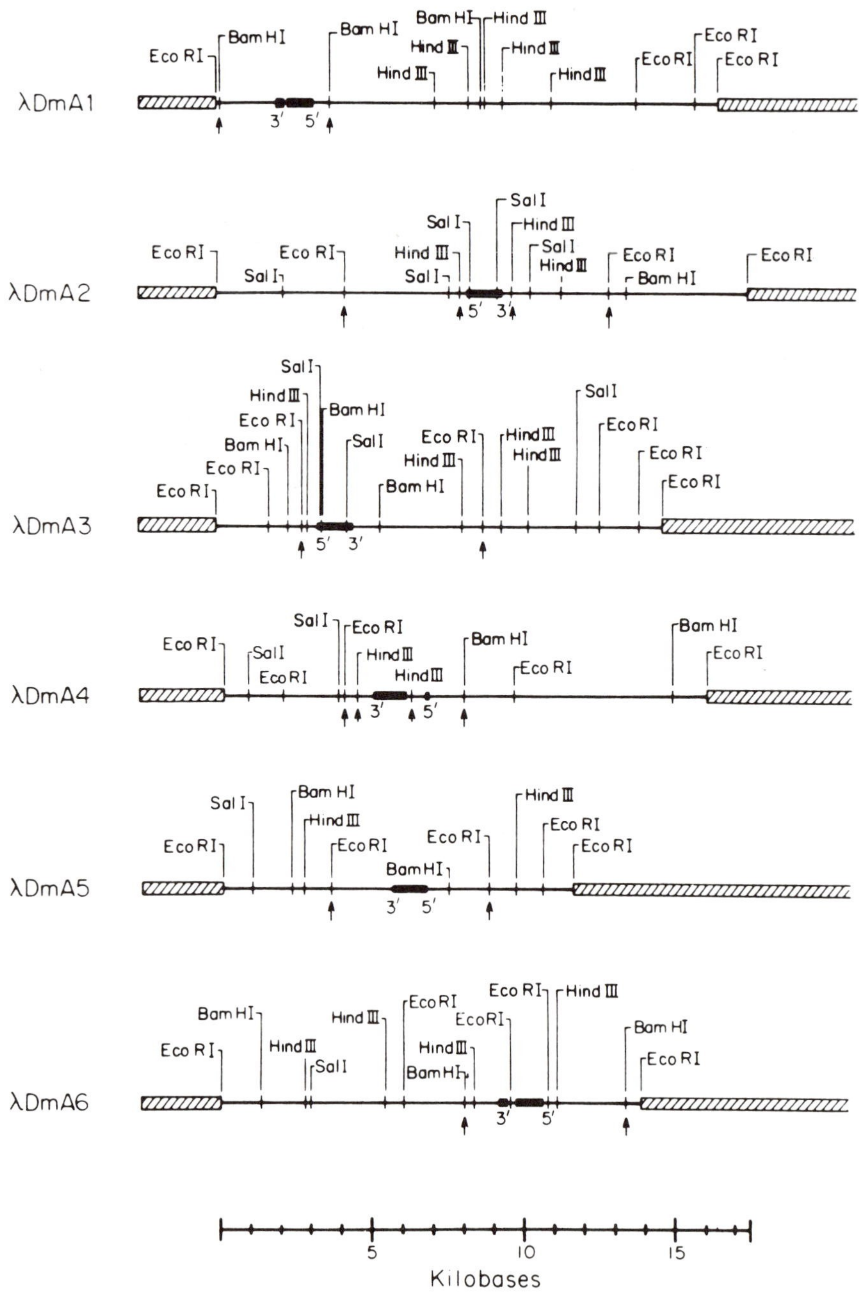

FIGURE 2
Restriction maps of λDmA1–λDmA6. (■) The positions of protein-coding regions, with interruptions apparent in DmA1, DmA4, and DmA6. (↑) Fragments subcloned in pBR322. (▨) Vector sequences.

however, these experiments would not detect very small (<100 nucleotides) introns within protein-coding regions or introns in untranslated regions.

The nucleotide sequence information presented in Figure 3 confirms that the DmA4 and DmA6 introns are not present in other *Drosophila* actin genes and, additionally, localizes an intron within DmA2. The sequence near the 5′ end of the protein-coding region of DmA2 is compared to that of DmA4 (Fig. 3A). As can be seen, DmA4 is split within the glycine codon at position 13, whereas DmA2 is not. Instead, DmA2 appears to be split 8 nucleotides upstream from the start codon, since a eukaryotic intron-exon junction sequence appears at this point. This finding agrees well with our previous R-loop analysis of DmA2 (Fyrberg et al. 1980).

Nonconservation of the DmA6 intervening sequence is also illustrated (Fig. 3B). The DmA6 interruption is within the glycine codon at position 307. From the DmA2 and DmA3 sequence data, it is clear that these genes are not interrupted at this position. However, F. Sánchez et al. (pers. comm.) have recently found that DmA1 is split by a 60-nucleotide intron within codon 307, demonstrating that at least one other *Drosophila* actin gene is split in the identical position.

All Drosophila *Actins Are Similar in Sequence to Vertebrate Cytoplasmic Actins*

A comparison of the aminoterminal tryptic peptides of each *Drosophila* actin to those of vertebrate skeletal muscle and cytoplasmic actin isoforms is shown in Figure 4. Each of the derived actin amino acid sequences closely resembles the vertebrate cytoplasmic sequences, partial exceptions being DmA1 and DmA6. Therefore, *Drosophila* apparently does not synthesize actins comparable in amino acid sequence to that of vertebrate skeletal muscle. As described in the following section, four of these actin isoforms appear to be expressed predominantly in *Drosophila* muscle tissue.

Figure 4 reveals another surprising result, namely the presence of a cysteine codon following the initiator methionines of each of the *Drosophila* actin genes. All vertebrate actins begin with aspartic or glutamic acid and, recently, Vandekerckhove and Weber (1981) have demonstrated that *Drosophila* actin present in Schneider L-2 cells also begins with a string of acidic amino acids (Asp-Glu-Glu). Therefore, the cysteine residues at the amino termini of *Drosophila* actins must be removed by in vivo processing. Recent structural studies of the chicken α-actin gene by Ordahl et al. (this volume) have shown that it also contains a cysteine codon following that for the initiator methionine. The role of aminoterminal cysteine in actin metabolism remains to be elucidated.

Expression of Drosophila *Actin Genes Is Temporally Regulated*

In preliminary experiments, we have examined the expression of *Drosophila* actin genes as a function of development. We constructed hybridization

```
          ovalbumin G··ATTTCC-TTGCAG|CTT
eucaryotic consensus····TYTYYYTXCAG|G
              ····ATTCTTTCCATTGCAG|

                            5                  10               15
                   Bl-Asp-Asp-Asp-Ile-Ala-Ala-Leu-Val-Val-Asp-Asn-Gly-Ser-Gly-Met-Cys-Lys
                      Glu-Glu-Glu                      Ile

DmA2  |CTTACAAA Met Cys Asp Glu Glu Val Ala Ala Leu Val Val Asp Asn Gly Ser Gly Met Cys Lys
                ATG TGT GAC GAA GAA GTT GCT GCT CTG GTT GTC GAC AAC GGC TCT GGC ATG TGC AAG

DmA4 TAAAACAAA Met Cys Asp Asp Glu Val Ala Ala Leu Val Val Asp Asn Gly Ser Gly Met Cys Lys
                ATG TGT GAC GAT GAA GTT GCT GCT CTG GTC GTT GAC AAT G|GC TCC GGC ATG TGC AAG

                               |GTGCGTGG····~600 nucleotides····TGTTATCCTGCAG|
                               G|GTAAGT····eucaryotic consensus····TYTYYYTXCAG|G
```

A

```
                 300             305             310
Rabbit Skeletal  Gly-Gly-Thr-Thr-Met-Tyr-Pro-Gly-Ile-Ala-Asp-Arg-Met-Gln-Lys

DmA2             Gly Gly Thr Thr Met Tyr Pro Ser Ile Ala Asp Arg Met Gln Lys
                 GGT GGC ACC ACC ATG TAC CCT AGC ATC GCC GAC CGT ATG CAG AAG
                                              *
DmA3             Gly Gly Thr Thr Met Tyr Pro Gly Ile Ala Asp Arg Met Gln Lys
                 GGC GGC ACC ACC ATG TAC CCG GGA ATC GCT GAC CGC ATG CAA AAG

DmA6             Gly Gly Thr Thr Met Tyr Pro Gly Ile Ala Asp Arg Met Gln Lys
                 GGC GGC ACT ACC ATG TAT CCA G|GT ATC GCT GAC CGT ATG CAA AAG

                      |GTGCGTAG···336 nucleotides····CTGTCCTGTTCAG|
                      G|GTAAGT···eucaryotic consensus···TYTYYYTXCAG|G
```

B

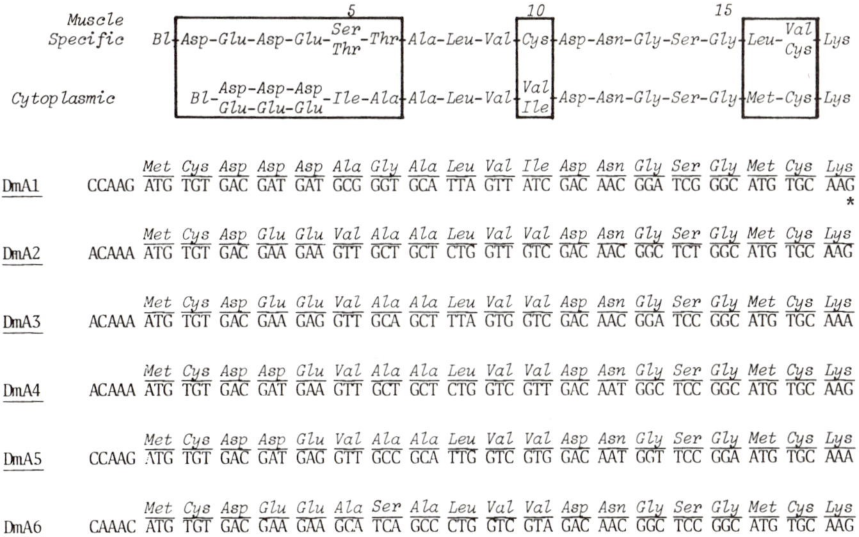

FIGURE 4
Nucleotide sequences of the 5′ ends of the *Drosophila* actin genes. These data reveal that the sequence of each actin aminoterminal tryptic peptide resembles those of vertebrate cytoplasmic actins. Thus, *Drosophila* does not encode an actin whose primary sequence is comparable to that of vertebrate skeletal muscle.

probes for transcripts of individual actin genes by subcloning the region of each that is complementary to the 3′-untranslated region of the homologous mRNA. RNA blotting experiments utilizing these hybridization probes have revealed that individual *Drosophila* actin genes have distinct patterns of expression. Two of the genes, DmA2 and DmA3, are expressed during all stages of development, as well as in several *Drosophila* cell lines. Both of these genes likely encode cytoplasmic or cytoskeletal actins. The remaining four actin genes are expressed predominantly during the differentiation of particular muscle cell lineages. DmA4 and DmA5 are expressed most intensively during the differentiation of either larval intersegmental muscle or the cephalic and abdominal musculature of the adult. DmA1 and DmA6 are expressed only during the differentiation of adult musculature, which is derived from cells of the imaginal disks. These three pairs of genes therefore

FIGURE 3 (see facing page)
Nucleotide sequences surrounding introns within the *Drosophila* actin genes. (*A*) DmA4 is interrupted within the glycine codon at position 13. A comparable intervening sequence is not seen in DmA2, which instead appears to be interrupted 8 nucleotides upstream from the ATG start codon. (*B*) DmA6 is interrupted within the glycine codon at position 307, whereas at least two other genes, DmA2 and DmA3, are not.

define three patterns of actin gene expression, which can be correlated to morphological changes in the *Drosophila* body plan and, in particular, to the differentiation and reorganization of muscle cells during *Drosophila* development.

DISCUSSION

Our characterization of the *Drosophila* actin genes has revealed several unexpected findings. Most striking is the nonconservation of positions of introns. A total of 11 different intron positions have been observed within the protein-coding regions of the actin genes of yeast, *Caenorhabditis*, *Drosophila*, sea urchin, chicken, and rat (Gallwitz and Sures 1980; Fyrberg et al. 1981; Scheller et al. 1981; C. Ordahl, pers. comm.; Hirsh et al., this volume). One generalization that can be drawn from these composite data is that the intron positions within actin genes of closely related species tend to be more similar than those within actin genes of distantly related species. The meaning of this observation is not yet clear, due to the paucity of information regarding the evolutionary origin and functional significance of introns. Gilbert (1979) suggested that introns facilitate gene evolution by shuffling exons into novel combinations and proposed that "primordial" genes would contain the largest number of introns, many of which would subsequently be eliminated by precise excision events. From the results of Lomedico et al. (1979), it is clear that introns can occasionally be removed by excision events. However, we believe that comparison of the structures of actin genes of the above-mentioned species strongly suggests that introns have both inserted into and excised from actin genes during the course of evolution.

The finding that the nucleotide sequences of the *Drosophila* actin genes are conserved to such a high degree naturally leads one to question the selective value of six gene copies. Our preliminary data indicate that each gene may be expressed only in a particular epigenetic situation. This result can be explained most readily by assuming that each structural gene has become associated with a regulatory "element" (either a localized nucleotide sequence or a much larger chromosomal region) that facilitates its expression in particular cell lineages while precluding it in others. If such regulatory elements do exist, then we would propose that the selective value of multiple actin genes is primarily to provide regulatory flexibility for the developmental expression of actin. We hope that when the mechanisms of eukaryotic gene regulation are understood more fully we will have sufficient evidence to support or refute this supposition.

ACKNOWLEDGMENTS

We thank Drs. Charles Ordahl and Sara Tobin for communicating results prior to publication. This work was supported by research grants from the

National Institutes of Health and the Muscular Dystrophy Association of America. B.J.B. and K.S.M. were supported by a National Institutes of Health predoctoral training grant.

REFERENCES

Batten, B.E., J.L. Aalberg, and E. Anderson. 1980. The cytoplasmic filamentous network in cultured ovarian granulosa cells. *Cell* **21:** 885.

Cleveland, D.W., M.A. Lopata, R.J. MacDonald, N.J. Cowan, W.J. Rutter, and M.W. Kirschner. 1980. Number and evolutionary conservation of α- and β-tubulin and cytoplasmic β- and γ-actin genes using specific cloned cDNA probes. *Cell* **20:** 95.

Fyrberg, E.A. and J.J. Donady. 1979. Actin heterogeneity in primary embryonic culture cells from *Drosophila melanogaster. Dev. Biol.* **68:** 487.

Fyrberg, E.A., K.L. Kindle, N. Davidson, and A. Sodja. 1980. The actin genes of *Drosophila:* A dispersed multigene family. *Cell* **19:** 365.

Fyrberg, E.A., B.J. Bond, N.D. Hershey, K.S. Mixter, and N. Davidson. 1981. The actin genes of *Drosophila*: Protein coding regions are highly conserved but intron positions are not. *Cell* **24:** 107.

Gallwitz, D. and I. Sures. 1980. Structure of a split yeast gene: Complete nucleotide sequence of the actin gene in *Saccharomyces cerevisiae. Proc. Natl. Acad. Sci.* **77:** 2546.

Garrels, J.I. and W. Gibson. 1976. Identification and characterization of multiple forms of actin. *Cell* **9:** 793.

Gilbert, W. 1979. Introns and exons: Playgrounds of evolution. *ICN-UCLA Symp. Mol. Cell. Biol.* **14:** 1.

Huxley, H.E. 1969. The mechanism of muscular contraction. *Science* **164:** 1356.

Lomedico, P., N. Rosenthal, A. Efstratiadis, W. Gilbert, R. Kolodner, and R. Tizard. 1979. The structure and evolution of the two nonallelic rat preproinsulin genes. *Cell* **18:** 545.

Pollard, T.D. and R.R. Weihing. 1974. Actin and myosin in cell movement. *Crit. Rev. Biochem.* **2:** 1.

Scheller, R.H., L.B. McAllister, W.R. Crain, D.S. Durica, J.W. Posakony, T.L. Thomas, R.J. Britten, and E.H. Davidson. 1981. Organization and expression of multiple actin genes in the sea urchin. *Mol. Cell. Biol.* **1:** 609.

Storti, R.V., D.M. Coen, and A. Rich. 1976. Tissue specific forms of actin in the developing chick. *Cell* **8:** 521.

Stossel, T.P. 1978. Contractile proteins in cell structure and function. *Annu. Rev. Med.* **29:** 427.

Tobin, S.L., E. Zulauf, F. Sánchez, E.A. Craig, and B.J. McCarthy. 1980. Multiple actin-related sequences in the *Drosophila melanogaster* genome. *Cell* **19:** 121.

Vandekerckhove, J. and K. Weber. 1981. Actin typing on total cellular extracts. *Eur. J. Biochem.* **113:** 595.

Whalen, R.G., G.S. Butler-Browne, and F. Gros. 1976. Protein synthesis and actin heterogeneity in calf muscle cells in culture. *Proc. Natl. Acad. Sci.* **73:** 2018.

Wolosewick, J.J. and K.R. Porter. 1979. Microtrabecular lattice of the cytoplasmic ground substance: Artifact or reality. *J. Cell. Biol.* **82:** 114.

Structure and Developmental Expression of Actin Genes in the Sea Urchin

WILLIAM R. CRAIN, JR.
DAVID S. DURICA *
ALAN D. COOPER†
KEVIN VAN DOREN‡
FREDERIC D. BUSHMAN
Cell Biology Group
Worcester Foundation for Experimental Biology
Shrewsbury, Massachusetts 01545

The actin gene family is proving to be an especially interesting set of genes in which to examine the relationship between gene structure and evolution and gene expression. Actin is found in all eukaryotic cells, is highly conserved throughout the phylogenetic tree, and performs multiple important and universal functions (Pollard and Weihing 1974; Goldman et al. 1976). Since the actins from organisms as phylogenetically diverse as slime molds and mammals are nearly identical, the genes must have been formed quite early in evolution and be subject to considerable selective pressure to maintain the amino acid sequence. In fact, actin genes are among the most ancient eukaryotic protein-coding genes that are currently under investigation. In addition, it has been demonstrated that members of the actin gene family are differentially expressed in various tissues and stages of development from a variety of organisms (Garrels and Gibson 1976; Storti and Rich 1976; Whalen et al. 1976; Devlin and Emerson 1978; Storti et al. 1978;

Present addresses: *Department of Medical Biochemistry, College of Medicine, Texas A & M University, College Station, Texas 77843; ‡Cold Spring Harbor Laboratory, Cold Spring Harbor, New York 11724.

†Permanent address: Department of Chemistry, Worcester State College, Worcester, Massachusetts 10602.

Crain et al. 1981; McKeown and Firtel 1981; Scheller et al. 1981; Zulauf et al. 1981). This gene family thus consists of members whose protein-coding sequences are closely related but are differentially expressed and therefore may be associated with different regulatory regions or sequences in the genome. A thorough understanding of the structure and expression of these genes may thus make it possible to establish relationships between certain features of the genes and the spatial and temporal regulation of their expression.

In this paper we review the currently delineated features of the structure and expression of the actin gene family in the sea urchin *Strongylocentrotus purpuratus*. The sea urchin is a particularly useful experimental animal in which to examine these genes, since it is easy to obtain several differentiated adult tissues and large quantities of embryos from early developmental stages. It is thus possible to perform a reasonably comprehensive analysis of the differential expression of these genes during differentiation and development.

RESULTS AND DISCUSSION

Actin Gene Structure and Organization

In our examination of the actin genes in the sea urchin, we initially isolated and identified two different recombinant DNA clones containing sea urchin actin genes, pSpG17 and pSpG28 (in the plasmid vector pBR322), by hybridization with a *Drosophila* actin-coding DNA fragment (Durica et al. 1980). In addition, a set of sea urchin actin-coding genomic clones in λ Charon 4 and several cDNA clones have also been isolated (Scheller et al. 1981). Southern blot analysis, solution hybridization data, and comparison of the restriction maps of all the available clones indicate that there are 11–20 actin-coding sequences in the sea urchin genome and that we have cloned representatives of at least 11 nonallelic genes (Durica et al. 1980; Scheller et al. 1981). At least four of the λ clones contain two actin genes located 5–9 kb apart, demonstrating linkage of some members of this family. Alignment of the restriction maps of these clones further suggests that at least three actin genes are linked in the genome (Scheller et al. 1981).

To explore the relationship between the fine structure, evolution, and expression of these genes, we are determining the nucleotide sequence of two members of the family, those present on plasmids pSpG17 and pSpG28. Figure 1 shows a map of these two clones indicating the sequenced regions. The nucleotide sequence of the entire gene in plasmid pSpG17 has been determined, including introns and about 500 nucleotides adjacent to coding regions in both the 5′ and 3′ directions. In pSpG28, we have determined about 40% of the protein-coding sequence and a small portion of four introns (Cooper and Crain 1982, and unpubl.).

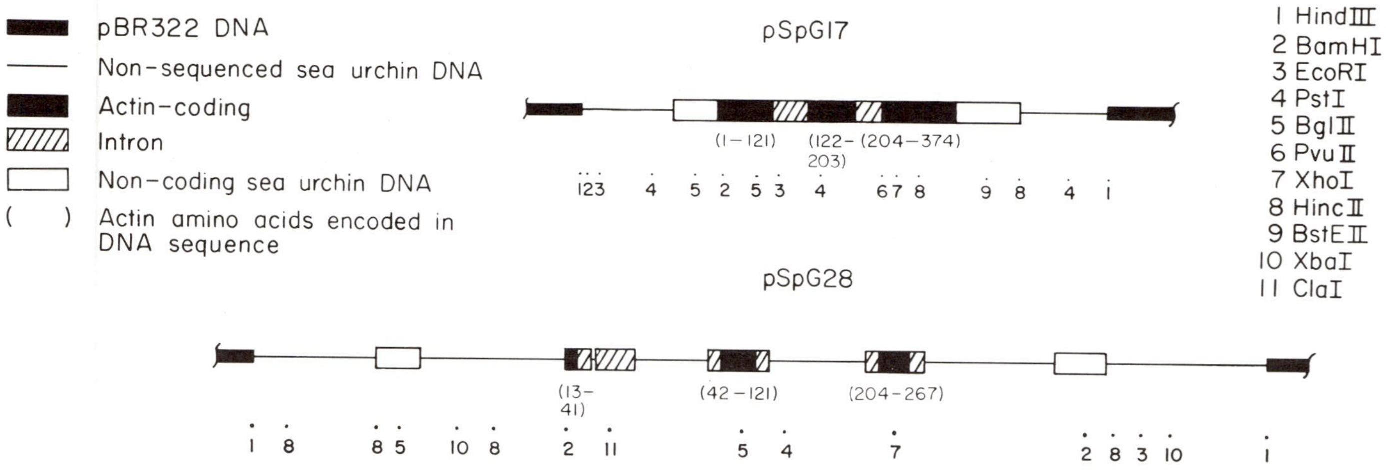

FIGURE 1
Map of sequenced regions of two actin genes in the sea urchin. The positions of the indicated restriction enzyme sites were mapped, and regions of the DNA sequence were determined using the procedure of Maxam and Gilbert (1980).

A major feature of the actin genes in the sea urchin, which has been revealed by this DNA sequence analysis, is the presence and precise location of introns in each of these genes. The presence of these introns has been determined by comparing the encoded amino acid sequence within each gene to the known amino acid sequence for rabbit skeletal-muscle actin. pSpG17 is seen to have its actin-coding sequence split within amino acid coding positions 121-122 and 203-204. Sequence analysis of pSpG28 indicates that, like pSpG17, it is also split at amino acid coding positions 121-122 and 203-204. The two genes therefore have two intron locations in common. However, pSpG28 also contains at least two additional introns at positions 41-42 and 267-268 that are not present in pSpG17. (Intron locations are designated here with a two-amino-acid notation. This convention is used since sequence redundancy at the coding-intron borders makes it difficult to determine whether the split lies between or within the codons.) Analysis of heteroduplexes formed between various cloned members of the actin gene family in the sea urchin indicates that all genes seem to contain either two or four introns at the same approximate locations found by DNA sequence analysis. It is interesting to note that in every case where linkage of actin genes can be demonstrated, the linked genes have a two-intron structure. In addition, those genes examined with the two-intron structure appear to be closely related, with at most 2% sequence divergence (Scheller et al. 1981). In contrast, DNA sequence analysis of several noncontiguous protein-coding regions demonstrates that the four-intron gene, pSpG28, differs by 10% from the two-intron gene (pSpG17). The data therefore suggest that an evolutionarily recent gene duplication may have given rise to multiple copies of the two-intron actin genes.

In addition to defining the structure of these genes, DNA sequence analysis yields information about the encoded proteins. This type of information may make it possible to define whether or not there are actin variants with unique functions. Considerable information on the amino acid sequence of several vertebrate actins indicates that although the amino acid sequence of actins from different species is extremely similar, there are substitutions at specific sites within the polypeptide chain that appear to be characteristic of muscle versus nonmuscle actins (Vandekerckhove and Weber 1978, 1979). The sequence of the sea urchin actin gene in pSpG17 encodes a nonmuscle amino acid at 16 out of 22 of these sites, a musclelike amino acid at 3 out of 22 sites, one previously unreported substitution seen earlier only in *Drosophila,* and one substitution found in *Physarum* and yeast and not in mammalian actins. The currently available sequence of pSpG28 extends across seven of these positions and encodes a musclelike amino acid at three sites, a nonmusclelike amino acid at three sites, and one previously unreported substitution (A.D. Cooper and W.R. Crain, in prep.). According to the available data on the relationship between actin variants and amino acid sequence, pSpG17 encodes a mostly nonmusclelike

actin, and pSpG28 encodes a more musclelike protein. The protein encoded by each sequence shares features of both muscle and nonmuscle invertebrate actins.

Actin Gene Expression in Early Development of the Sea Urchin

The synthesis of actin, which is low in unfertilized eggs, increases strikingly between 8 hours (late cleavage) and 24 hours (mesenchyme blastula) of development in the sea urchin embryo (Crain et al. 1981; Durica and Crain 1982) The expression of some members of this gene family are therefore temporally regulated during early development of this species. To begin an investigation on the basis of this differential gene expression during sea urchin embryogenesis, we have determined the size and relative abundance of actin mRNA from selected stages of development. In these experiments, total polysomal RNA, polysomal poly(A)$^+$ RNA, total cytoplasmic RNA, and cytoplasmic poly(A)$^+$ RNA have been examined by RNA blot analysis using cloned actin-coding DNA fragments as hybridization probes. Several interesting points are evident from the results shown in Figure 2. First, there are two actin mRNA size classes on polysomes, 2.2 kb and 1.8 kb in length (A and B), and there are no additional detectable actin-coding RNAs in cytoplasmic RNA that are not present on polysomes (C and D). Second, the relative abundance of each of these RNA classes increases sharply between 8 hours and 18 hours of development. Since the timing of this increase in actin mRNA abundance closely parallels that of the increase in actin synthesis, it is clear that, to a large extent, the synthesis of actin is a reflection of actin mRNA abundance in early development in this animal. Furthermore, the kinetics of accumulation of actin-coding RNA in the cytoplasm and on polysomes are essentially indistinguishable (Fig. 3), suggesting that regulation of translation of the newly accumulated cytoplasmic actin message does not play a major role in determining the level of actin synthesis.

The presence of relatively low levels of actin mRNA between 2 hours and 8 hours of development raised the question of whether this RNA is a result of new synthesis after fertilization or whether it was present in the unfertilized egg (i.e., maternal actin mRNA). RNA blot analysis of polysomal and nonpolysomal RNA fractions from unfertilized egg and 1-hour and 2-hour embryos has demonstrated that both actin mRNA size classes are present in egg, both on and off polysomes, and that their relative abundance remains constant at these stages (Crain et al. 1981). It seems likely then that the maternal actin mRNA directs the low levels of actin synthesis during approximately the first 8 hours of development. After 8 hours the abundance of actin message increases in the cytoplasm, and the level of actin synthesis rises correspondingly.

To quantitate the relative increase in the abundance of the actin mRNAs, the hybridization bands were cut out of the filters and counted by liquid

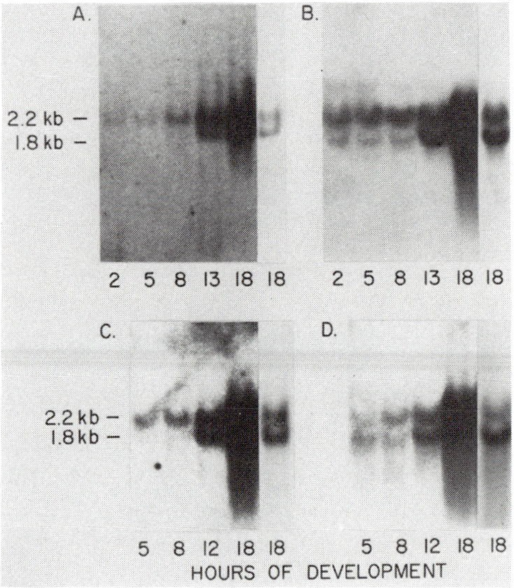

FIGURE 2
RNA blot analysis of actin-coding RNAs in early development of the sea urchin. In the experiment in each panel, the RNA was denatured in glyoxal and formamide, electrophoresed on 1% agarose gels, and transferred to DBM paper. (*A*) A 30-μg sample of total polysomal RNA from each developmental stage was passed once over an oligo(dT)-cellulose column. All of the RNA that bound to the column (poly[A]$^+$-enriched) was then loaded into one lane of the gel (the time of isolation is indicated). The specific activity of the probe was 5.2×10^7 cpm/μg. The first five lanes were exposed for 8 days with preflashed Kodak XR-5 film with an intensifying screen. The sixth lane shows a 20-hr exposure of the fifth lane under similar conditions. (*B*) A 30-μg sample of total polysomal RNA from each developmental stage was loaded into one lane of the gel (the time of isolation is indicated). The specific activity of the probe was 3.8×10^8 cpm/μg. The first five lanes were exposed for 44 hr, as described above. The sixth lane shows a 40-hr exposure of the fifth lane, in which Kodak No-Screen film was used without an intensifying screen. (Reprinted, with permission, from Crain et al. 1981.) (*C*) A 30-μg sample of cytoplasmic RNA from each stage was passed once over an oligo(dT)-cellulose column. All of the RNA from each stage that bound to the column (poly[A]$^+$-enriched) was loaded into a single lane of the gel. The first four lanes were exposed for 5 days with preflashed Kodak XR-5 film and an intensifying screen. The fifth lane shows a 40-hr exposure of the fourth lane on Kodak No-Screen film without an intensifying screen. (*D*) A 30-μg sample of total cytoplasmic RNA from each developmental stage was loaded into one lane of the gel. The exposure of the lanes was as described in *C*.

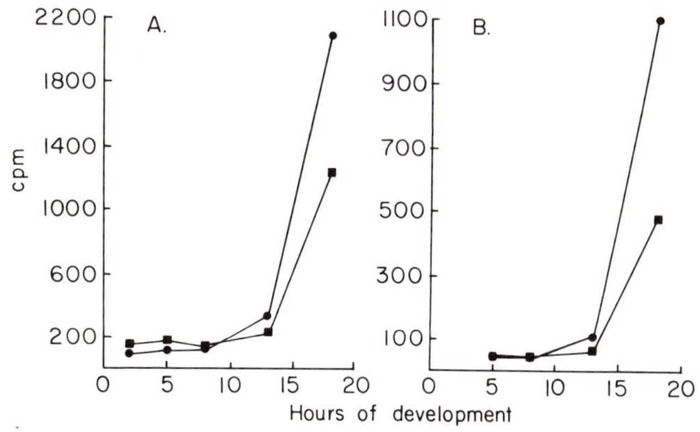

FIGURE 3
Kinetics of accumulation of actin mRNA on polysomes and in the cytoplasm during early development. The regions of the RNA filters shown in Fig. 2B and D containing the indicated autoradiographic bands, which had been visualized on film, were cut out and counted in a liquid scintillation counter. Three separate areas that did not contain bands were also cut out and counted to determine the background level. These regions were similar in size to the regions containing bands. The counts determined in this manner were then plotted as shown. (*A*) Polysomal RNA. (*B*) Cytoplasmic RNA. (●) 1.8-kb actin mRNA. (■) 2.2-kb actin mRNA.

scintillation counting. The data from such an analysis of polysomal and cytoplasmic RNA are plotted in Figure 3. From this analysis, it can be seen that the 2.2-kb RNA increases approximately 8–9-fold in both polysomal and cytoplasmic RNA between 2–5 hours and 18 hours and that the 1.8-kb RNA increases 22–24-fold in both polysomal and cytoplasmic RNA during the same time interval (for calculation, see Crain et al. 1981). It is interesting that, although the two actin RNA classes accumulate over the same time interval, they accumulate to significantly different extents. That is, the ratio of the smaller (1.8-kb) actin RNA to the larger (2.2-kb) RNA increases 2.5–3-fold during these stages of development, suggesting the possibility of noncoordinate expression of these two actin mRNAs, which are known to be derived from different genes (see below).

Different Actin Genes Code for Developmentally Expressed mRNAs

The initial approach to identifying the genes from which the embryonic messages are derived has been to perform RNA blot analysis with RNA from blastula-stage embryos using hybridization probes that show specificity for different actin genes. This has been done by selecting DNA fragments as probes that are complementary to nonprotein-coding regions on the 3′

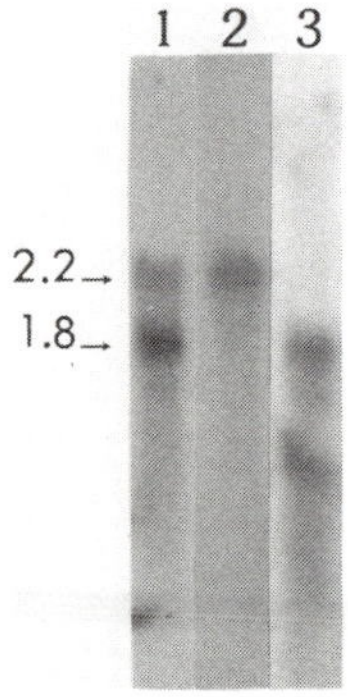

FIGURE 4
RNA blot analysis of blastula RNA with hybridization probes for different actin genes. RNA from blastula-stage embryos was denatured in glyoxal and formamide, electrophoresed on 1% agarose gels, transferred to nitrocellulose paper, and hybridized with the designated probes using a modification of the procedure of Thomas (1980). Each lane contained 10 μg of blastula RNA. (*1*) Hybridized with ^{32}P-labeled DNA from the entire plasmid pSpG17. (*2*) Hybridized with a fragment from pSpG17, which does not contain protein-coding sequence and is bounded by the *Bst*EII and rightward *Hin*dIII sites shown in Fig. 1. (*3*) Hybridized with the actin cDNA clone pSpEC4.

end of the mRNAs and that are unique to a gene or a subset of the genes. Figure 4 displays the result of such an experiment using two actin-gene-specific probes. The probe used in lane 2 is a fragment from pSpG17 that lies between the *Bst*EII site (Fig. 1) and the rightward *Hin*dIII site (Fig. 1). This fragment, which does not contain protein-coding sequence and lies to the 3′ side of coding sequence, does not cross-hybridize with sequences on pSpG28 (F.D. Bushman and W.R. Crain, unpubl.) or with at least four other cloned actin genes under standard hybridization criterion (Scheller et al. 1981). This probe shows a strong preference for the larger (2.2-kb) RNA, as compared to a probe containing protein-coding sequence (lane 1), indicating that this message is transcribed either from the gene in pSpG17 or from one very closely related to it. A second specific probe (lane 3) is an actin cDNA clone, pSpEC4 (Brushkin et al. 1981). This DNA, which consists primarily of the 3′-untranslated region from an actin gene other than that present on pSpG17, shows a strong preference for a 1.8-kb actin mRNA. These data therefore clearly indicate that at least two actin mRNAs from different genes are expressed in early sea urchin embryos. The positive identification of two actin mRNAs of different sizes does not eliminate the possibility that additional mRNAs from other genes are also present within the 2.2-kb and 1.8-kb size classes. Further analysis with additional gene-specific probes will allow detection of any additional embryonic actin mRNAs.

ACKNOWLEDGMENTS

We thank Ms. Agneta Brown for excellent technical assistance. This research was supported by U.S. Public Health Service grants GM-24620, GM-25492, P30-CA-12708, and RR-05528 from the National Institutes of Health and by funds from the Mimi Aaron Greenberg Memorial Institute. A.D.C. was supported by a National Science Foundation Faculty Development Award, and D.S.D., by a postdoctoral fellowship from the Muscular Dystrophy Association of America.

REFERENCES

Bruskin, A.M., A.L. Tyner, D.E. Wells, R.M. Showman, and W.H. Klein. 1981. Accumulation in embryogenesis of five mRNAs enriched in the ectoderm of the sea urchin pluteus. *Dev. Biol.* **87:** 308.

Cooper, A.D. and W.R. Crain. 1982. Complete nucleotide sequence of a sea urchin actin gene. *Nucleic Acids Res.* (in press).

Crain, W.R., D.S. Durica, and K. Van Doren. 1981. Actin gene expression in developing sea urchin embryos. *Mol. Cell Biol.* **1:** 711.

Devlin, R.B. and C.P. Emerson. 1978. Coordinate regulation of contractile protein synthesis during myoblast differentiation. *Cell* **13:** 599.

Durica, D.S. and W.R. Crain. 1982. Analysis of actin synthesis in early sea urchin development. *Dev. Biol.* (in press).

Durica, D.S., J.A. Schloss, and W.R. Crain. 1980. Organization of actin gene sequences in the sea urchin: Molecular cloning of an intron-containing DNA sequence coding for a cytoplasmic actin. *Proc. Natl. Acad. Sci.* **77:** 5683.

Garrels, J.I. and W. Gibson. 1976. Identification and characterization of multiple forms of actin. *Cell* **9:** 793.

Goldman, R., T. Pollard, and J. Rosenbaum, eds. 1976. "Cell motility." *Cold Spring Harbor Conf. Cell Proliferation* **3.**

Maxam, A.M. and W. Gilbert. 1980. Sequencing end-labeled DNA with base-specific chemical cleavage. *Methods Enzymol.* **65:** 499.

McKeown, M. and R.A. Firtel. 1981. Differential expression and 5′ end mapping of actin genes in *Dictyostelium*. *Cell* **24:** 799.

Pollard, T.D. and R.R. Weihing. 1974. Actin and myosin and cell movement. *Crit. Rev. Biochem.* **2:** 1.

Scheller, R.H., L.B. McAllister, W.R. Crain, D.S. Durica, J.W. Posakony, T.L. Thomas, R.J. Britten, and E.H. Davidson. 1981. Organization and expression of multiple actin genes in the sea urchin. *Mol. Cell Biol.* **1:** 609.

Storti, R. and A. Rich. 1976. Chick cytoplasmic actin and muscle actin have different structural genes. *Proc. Natl. Acad. Sci.* **73:** 2346.

Storti, R., S. Horovitch, M. Scott, A. Rich, and M. Pardue. 1978. Myogenesis in primary cell cultures from *Drosophila melanogaster*: Protein synthesis and actin heterogeneity during development. *Cell* **13:** 589.

Thomas, P.S. 1980. Hybridization of denatured RNA and small DNA fragments transferred to nitrocellulose. *Proc. Natl. Acad. Sci.* **77:** 5201.

Vandekerckhove, J. and K. Weber. 1978. Mammalian cytoplasmic actins are the products of at least two genes and differ in primary structure in at least 25 identified positions from skeletal muscle actins. *Proc. Natl. Acad. Sci.* **75:** 1106.

——. 1979. The complete amino acid sequence of actins from bovine aorta, bovine heart, bovine fast skeletal muscle, and rabbit slow skeletal muscle. *Differentiation* **14:** 123.

Whalen, R.F., G.S. Butler-Browne, and F. Gross. 1976. Protein synthesis and actin heterogeneity in calf muscle cells in culture. *Proc. Natl. Acad. Sci.* **73:** 2018.

Zulauf, E., F. Sánchez, S.L. Tobin, U. Rdest, and B.J. McCarthy. 1981. Developmental expression of a *Drosophila melanogaster* actin gene encoding actin I. *Nature* **292:** 556.

Human Actin Proteins Are Encoded by a Multigene Family

JOANNE ENGEL
PETER GUNNING
LARRY KEDES
Howard Hughes Medical Institute and Department of Medicine
Stanford University School of Medicine and
Veterans Administration Medical Center
Palo Alto, California 94305

Actin is a highly conserved protein that is found in abundance in all eukaryotic cells. It is functionally involved in cell motility, mitosis, muscle contraction, and maintenance of cytoskeletal structure (Pollard and Weihing 1974; Goldman et al. 1976). This diversity of roles within a cell is reflected by actin isoforms that differ in their isoelectric point. Whereas yeast makes use of only a single actin protein, mammalian cells utilize at least six different actin variants (Vandekerckhove and Weber 1979). The two mammalian cytoplasmic actins β and γ are found in most cells as components of the cytoskeleton (Clarke and Spudich 1977). The amino acid sequence of these proteins is well conserved in all eukaryotes studied thus far. The other four mammalian actins, which include the skeletal-muscle α-actin, cardiac-muscle actin, and two smooth-muscle actins, are tissue specific and are more closely related to one another than they are to cytoplasmic actins (Vandekerckhove and Weber 1979). This conservation of primary structure may reflect their common role in muscle contraction.

Studies utilizing recombinant DNA technology have shown that the organization and structure of the genes encoding actin proteins are highly variable among different eukaryotes. Yeast encodes only a single actin gene, which contains an intervening sequence between the codons encoding amino acids 3 and 4 (Gallwitz and Sures 1980; Ng and Abelson 1980). In contrast, the slime mold contains 17 actin genes, none of which are inter-

rupted by intervening sequences (Kindle and Firtel 1978). Some of these 17 genes appear to be linked in the genome, and not all of the genes are functional (McKeown et al. 1978; McKeown and Firtel 1981) The six actin genes of *Drosophila* are dispersed to different chromosomes, and the location of their intervening sequences is not conserved (Fyrberg et al. 1980; Tobin et al. 1980). Among the deuterostomes, sea urchins encode 11 actin genes, some of which are linked on a chromosome. Again, the locations of intervening sequences that interrupt these genes are variable (Scheller et al. 1981).

We have chosen to examine the organization and structure of actin genes in the human genome as a prelude to investigating the regulation of this multigene family in normal and diseased cells. We have utilized cloned actin genes from other species to show that humans encode 25–30 actin genes. By examining restriction enzyme sites for DNA polymorphism, we find that these loci are highly conserved among different individuals. Characterization of 12 actin-containing fragments from a library of cloned human DNA demonstrates that most of these sequences are not closely linked in the genome. Furthermore, at least ten of these cloned sequences contain all or most of an actin gene and represent ten nonallelic loci.

RESULTS AND DISCUSSION

The Human Genome Encodes 25–30 Actin Genes

To determine the number of actin gene fragments contained in the human genome, we hybridized a cloned chicken β-actin cDNA probe to nitrocellulose blots of human nuclear DNA that had been digested with restriction enzymes that cleave infrequently. The data in Figure 1 show that 10–30 bands of hybridization can be distinguished when human DNA from a single individual is cleaved with *Eco*RI, *Pvu*II, *Hin*dIII, or *Bam*HI. A similar spectrum of hybridization is observed when identical blots are hybridized to probes derived from two different *Drosophila* actin gene clones or from a human γ-actin cDNA clone (data not shown). Interestingly, several of the fragments (the 2.3-kb and 3.7-kb *Eco*RI fragments, the 8-kb *Bam*HI fragment, the 6-kb and 8-kb *Hin*dIII fragments, and the 8-kb *Pvu*II fragments) exhibit more intense hybridization to all four of these different probes.

To assess whether each of these actin-encoding restriction fragments contains at least one actin gene, we made use of 5′-end and 3′-end coding-region probes derived from the chicken β-actin cDNA clone. This heterologous probe was digested with *Bgl*I plus *Pst*I to yield a 0.5-kb fragment containing 300 bases from the 5′-untranslated region of the mRNA and the first 200 bases of the coding region. A 3′-end coding-region probe was isolated by digesting the chicken β-actin cDNA clone with *Kpn*I to generate a fragment that included 300 bases of the coding region plus 200 bases from the 3′-untranslated region of the mRNA (J. Engel, unpubl.). We reasoned that

any restriction fragment that hybridized to both probes must contain all or most of at least one actin gene. It should be noted that this approach assumes that the entire sequence coding for chicken β-actin cDNA is homologous to the sequences coding for the human actin gene and that its specificity derives from the short length of the coding region contained in the 5′ and 3′ probes.

The appropriate *Bgl*I, *Pst*I, and *Kpn*I fragments of the chicken β-actin cDNA clone were eluted from a gel and labeled by nick translation. No cross-hybridization of the fragments to one another or to any of the DNA fragments located between the two end fragments was observed (data not shown). Figure 2 shows an autoradiograph of a nitrocellulose blot of human nuclear DNA from a single individual that had been digested with *Eco*RI, *Hin*dIII, *Pvu*II, or *Kpn*I and hybridized to these probes. It can be seen that most of the 25–30 fragments generated by digestion with *Eco*RI hybridize to both the 5′-end and 3′-end coding-region probes. Similar results are seen with the other three restriction enzymes that we tested. Thus, the human genome encodes approximately 25–30 actin genes. Since the DNA used in these blots was isolated from a single individual and these bands show very few polymorphisms (see below), they probably represent at least 25 different loci.

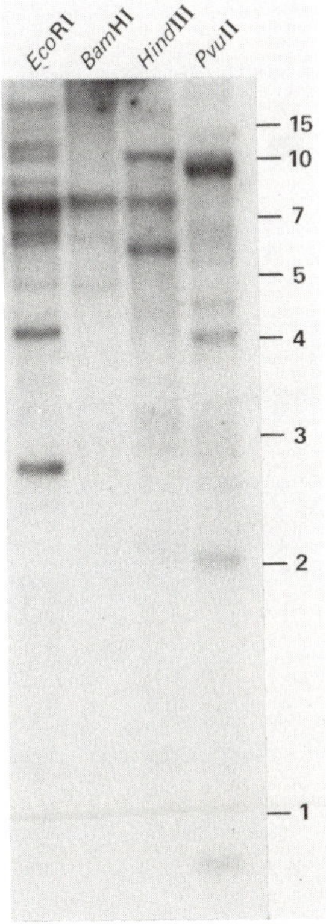

FIGURE 1
Determination of the number of restriction enzyme fragments in the human genome homologous to actin genes. DNA (2 μg) from an individual was digested with *Eco*RI, *Bam*HI, *Hin*dIII, or *Kpn*I (as indicated), electrophoresed through a 0.6% agarose gel, transferred to a nitrocellulose filter, and hybridized with a nick-translated probe prepared from the chicken β-actin cDNA clone. (*Right*) Sizes of the fragments (in kilobases).

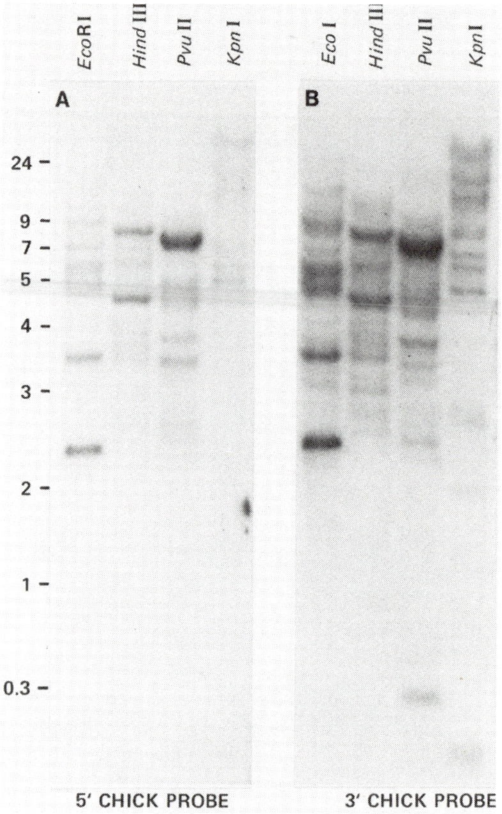

FIGURE 2
Hybridization of 5′-end or 3′-end coding-region probes to Southern blots of human nuclear DNA cleaved with restriction enzymes. DNA (2 μg) from an individual was digested with *Eco*RI, *Hin*dIII, *Kpn*I, or *Pvu*II (as indicated), electrophoresed through a 0.6% agarose gel, transferred to a nitrocellulose filter, and hybridized with a 5′-end coding-region probe (*A*) or with a 3′-end coding-region probe (*B*) as described in the text. (*Left*) Sizes of the fragments (in kilobases).

Actin Gene Loci Exhibit Little Population Polymorphism

To search for population polymorphism in actin genes, we examined DNA isolated from different individuals for changes in or near actin gene sequences that would result in alterations of restriction fragment sizes. Figure 3 shows the hybridization spectrum from 13 nonconsanguineous individuals of three different races and from one human primary cell culture grown in vitro. The DNA was cleaved with *Eco*RI, transferred to nitrocellulose paper, and hybridized to a chicken β-actin cDNA probe. Four bands of hybridization (indicated by arrowheads) are unique to some of the individuals and thus represent either DNA sequence changes at the *Eco*RI restriction enzyme recognition site at these loci or DNA insertions or deletions between adjacent *Eco*RI sites that bound actin-coding regions. (The intense 4.6-kb band of hybridization seen in lane 6 is an artifact resulting from plasmid contamination in this DNA sample; when a similar blot is probed with nick-translated pBR322, only this band of hybridization is observed.)

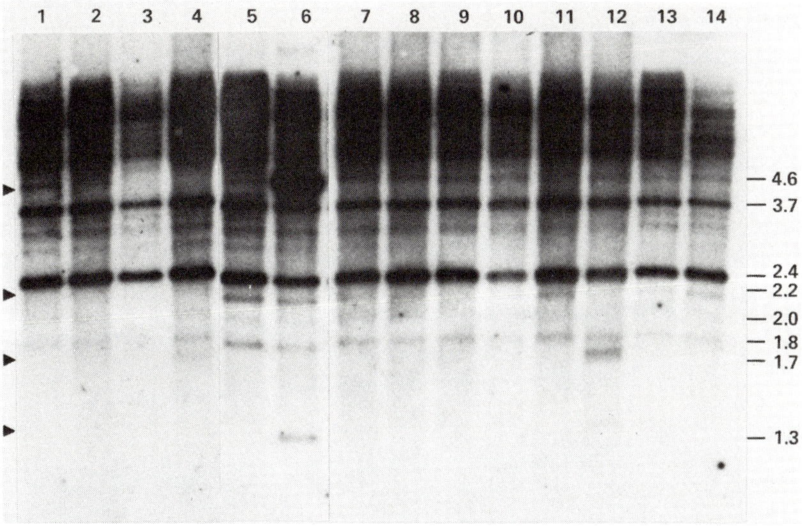

FIGURE 3
Search for polymorphism in human DNA restriction fragments containing actin genes. DNA (1 μg) from 13 different individuals and from one primary cell culture line derived from a single individual was digested with *Eco*RI, electrophoresed through a 0.6% agarose gel, transferred to a nitrocellulose filter, and hybridized to a nick-translated probe prepared from the chicken β-actin cDNA clone. (*Right*) Sizes (in kilobases). (►) Four new *Eco*RI fragments detected in some of the individuals. (*1–4*) DNA electrophoresed from black individuals; (*5–8*) DNA from Asian individuals; (*9–12, 14*) DNA from Caucasian individuals; (*13*) DNA from human myoblasts cultured in vitro.

At least 25 bands of hybridization can be discerned that remain invariant in these 14 different samples. The failure to find extensive polymorphism strongly supports the notion that the majority of bands that hybridize to the coding-region probe represent different actin gene loci.

Isolation of Human Genomic Clones Containing Actin-coding Sequences

To further characterize human actin genes, we screened a cloned library of human nuclear DNA sequences for recombinants that contain actin-coding regions. Using a combination of heterologous actin gene probes derived from a chicken β-actin cDNA clone and two *Drosophila* genomic actin gene clones, we screened approximately 50% of a bacteriophage λ library constructed from a partial *Eco*RI digestion of DNA of a single human fetus. Following two rounds of plaque purification, 14 clones continued to give positive signals when hybridized against these heterologous actin gene probes. These clones were further studied.

Recombinant Clones Are Specifically Homologous to Human Actin mRNA

To demonstrate that the recombinant clones that we isolated indeed contain actin-coding regions, we utilized a positive mRNA selection assay. Figure 4 shows that the RNA that hybridized to seven of the human actin clones is specifically enriched for sequences directing the synthesis of a 42,000-dalton protein that comigrates with authentic rabbit α-actin. Similar results were observed with the other seven clones isolated; however, three of these clones (λHRL51, λHRL54, and λHRL65) hybridize very poorly to RNA isolated from HeLa or human muscle cells (J. Engel, unpubl.). In each case, the protein products synthesized in vitro were further identified as actin by coelectrophoresis on two-dimensional gels with authentic actin standards and by partial proteolysis mapping (data not shown).

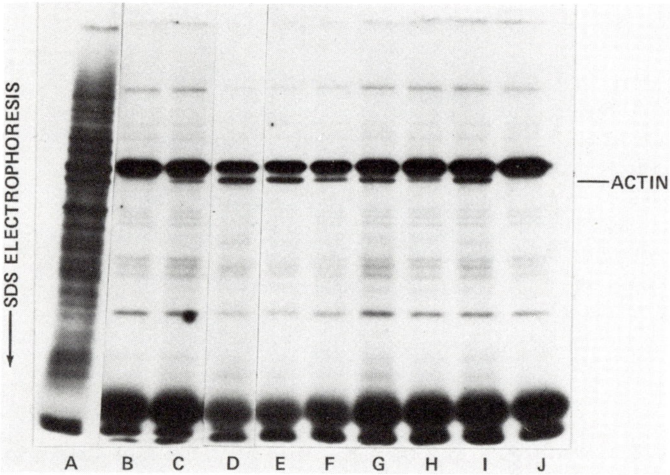

FIGURE 4

Fluorograph of radiolabeled proteins synthesized in vitro by mRNA that hybridized to seven human genomic actin clones. Total cellular RNA was hybridized to nitrocellulose filters to which DNA from each of the actin clones had been bound. The specifically hybridizing mRNA was translated in reticulocyte lysate (Amersham), and the [^{35}S]methionine-labeled products were analyzed by electrophoresis on a SDS-12.5% polyacrylamide gel, followed by fluorography. (*A*) HeLa cell RNA; (*B*) mRNA selected by a blank filter; (*C*) λHRL84 mRNA; (*D*) λHRL45 mRNA; (*E*) λHRL35 mRNA; (*F*) λHRL34 mRNA; (*G*) λHRL25 mRNA; (*H*) λHRL21 mRNA; (*I*) λHRL24 mRNA; (*J*) no exogenous mRNA added. (*Right*) The position of the unlabeled α-actin marker detected by Coomassie-blue staining. (The major band of ^{35}S-labeled material migrating above actin is an artifact of the reticulocyte lysate translation system.)

Physical Characterization of Human Actin Gene Clones

We next compared the DNA structure of the individual clones to assess their genetic identity. DNA from each of the 14 actin-encoding clones was isolated, digested to completion with *Eco*RI, transferred to nitrocellulose paper, and hybridized to each of the three heterologous actin gene probes. Two pairs of the 14 clones have identical *Eco*RI fragments and patterns of hybridization. Each of the other 12 clones contains nonoverlapping sets of *Eco*RI fragments (data not shown). Furthermore, the size of the *Eco*RI fragment in each of the 12 clones that hybridizes to any of the three heterologous actin gene probes is unique to that clone (data not shown). Figure 5 illustrates the arrangement and orientation of the *Eco*RI fragments within

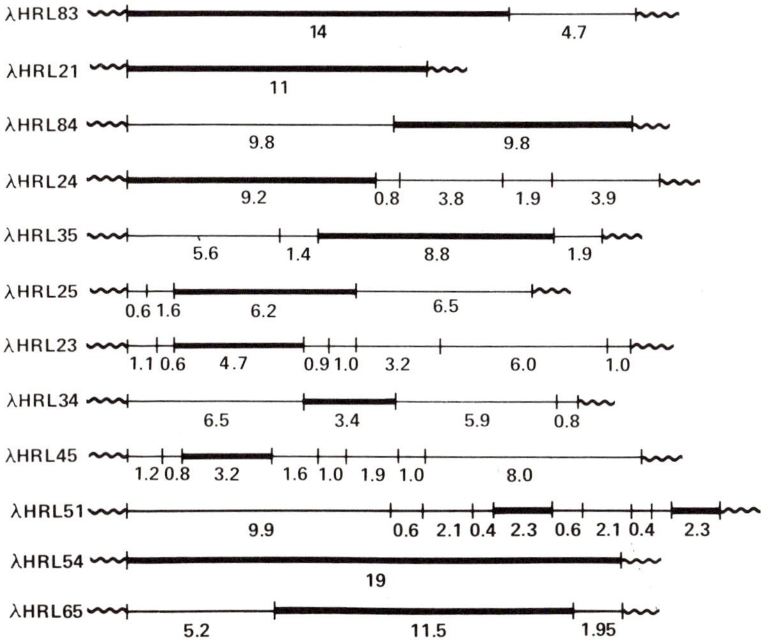

FIGURE 5

*Eco*RI restriction endonuclease maps of recombinant phages containing actin-gene-coding regions. The *Eco*RI fragments are delineated by vertical lines, and the fragments encoding actin sequences are shown by heavy black lines. λ Charon 4A arms are indicated by wavy lines, and the sizes of the *Eco*RI fragments are shown beneath each fragment. The maps were deduced by partial and double restriction enzyme digests in conjunction with hybridization of appropriate nick-translated subcloned fragments to nitrocellulose blots of these digests. λHRL65 contains actin-coding sequences on both the 11.5-kb and the 1.95-kb *Eco*RI fragments. The order of the terminal four *Eco*RI fragments in λHRL51 that comprise the tandem duplication is tentative.

each of the 12 different clones and indicates the *Eco*RI fragment to which the actin-coding region has been localized. Most of the actin-coding fragments are bound by large flanking regions of non-actin-coding sequence, and none of the clones are derived from overlapping regions of the genome. Interestingly, λHRL51 contains a series of contiguous *Eco*RI segments that are repeated in this clone at least twice. The 2.3-kb *Eco*RI fragments that have been identified as containing actin-coding region within this repeat unit comigrate with the more intensely hybridizing 2.3-kb *Eco*RI fragment observed in the genomic Southern blots (see Fig. 1). The structure of this clone thus suggests that at least some of the actin genes may be tandemly linked in the human genome.

To further investigate whether each copy of the 2.3-kb *Eco*RI fragment in λHRL51 contains actin-coding sequences, the 2.3-kb *Eco*RI fragment was gel purified and subjected to restriction enzyme digestion. The multiple copies of this fragment are not all identical. *Taq*I cleavage yields fragments 2.3 kb, 2.0 kb, and 0.3 kb in length. Likewise, *Msp*I digestion generates four restriction fragments that total approximately 4.6 kb in length (data not shown). Hybridization of these blots to an actin-coding-region probe derived from the chicken β-actin cDNA clone demonstrates that the 2.3-kb and 2.0-kb *Taq*I fragments contain actin gene sequences (data not shown). Thus, although the tandem copies of the 2.3-kb *Eco*RI fragment in λHRL51 are not identical and can be assigned to two classes, both classes encode actin genes. The presence of this gene arrangement in the genome may account for the strong hybridization to four different actin probes of the 2.3-kb *Eco*RI fragment observed in the digests of total genomic DNA.

Each Clone Contains All or Most of an Actin Gene

There are two possible explanations of our isolation of a larger number of different actin-encoding clones from the human DNA library. These 12 actin-coding-region fragments could represent different parts of a few large genes. Alternatively, we may have isolated 12 of the approximately 25 genes that we have demonstrated to be present in the human genome. The second hypothesis would be especially interesting in light of the fact that only six different actin proteins have been identified previously in mammals.

To distinguish between these two possibilities, we made use of the 5′-end and 3′-end coding-region probes derived from the chicken β-actin cDNA clone. Any clone that hybridizes to both of these probes must contain all or most of an actin gene. Furthermore, all such clones hybridizing to both 5′ and 3′ probes must be distinct from one another; they could not be derived from different parts of the same gene because they would contain overlapping coding regions.

*Eco*RI digests of each of the λHRL actin clones were blotted into nitrocellulose paper and hybridized to either the chicken 5′-end or 3′-end

coding-region probe. All of the clones, except for λHRL54 and λHRL65, hybridize to both probes (data not shown). Thus, the human genome contains at least ten different actin genes. That these ten different actin genes do not each represent alleles of five different loci is suggested by the lack of conservation of *Eco*RI fragments in each of the isolated clones. This conclusion is further born out by the dissimilar restriction enzyme maps of each of the *Eco*RI coding-region fragments from each of the clones (data not shown).

SUMMARY

We have described the isolation of 12 different nonoverlapping recombinant phages containing sequences coding for human actin genes. These recombinants were isolated from a library of cloned human DNA fragments derived from a single individual. Nine of these clones hybridize well to human α-, β- and/or γ-actin mRNA, as tested by a positive mRNA selection assay. Three of the clones (λHRL51, λHRL54, and λHRL65), originally selected by hybridization to the heterologous actin gene probes, hybridized poorly to mixtures containing these three species of human actin mRNA. These three human clones may contain actin gene sequences that have diverged from the major classes of transcribed human actin genes. On the basis of the following considerations, however, they could encode smooth-muscle actin. Shani and co-workers (Shani et al. 1981) have found that rat smooth-muscle actin mRNA has diverged significantly from that of both the rat skeletal muscle and cytoplasmic actin genes. In contrast, the rat cardiac-muscle mRNA is still quite homologous to the rat skeletal-muscle actin gene, as assayed by thermal stability or S1 protection experiments. Similar results have been reported in studies of the chicken actin gene (Schwartz and Rothblum 1981). Thus, we speculate that a human smooth-muscle actin gene would likewise not hybridize well to human α-, β-, or γ-actin mRNA and that λHRL51, λHRL54, and λHRL65 encode smooth-muscle actin proteins.

Several pieces of evidence support the conclusion that these 12 different actin gene isolates represent at least 10 nonallelic loci that are not tightly linked in the human genome. Each clone contains a unique set of *Eco*RI fragments, a result that suggests that they do not derive from similar regions of allelic genes. The finding that none of the clones contains terminal *Eco*RI fragments of similar size demonstrates that these loci are not overlapping. Finally, 10 of the 12 isolated genes contain sequences corresponding to both the 5′ end and the 3′ end of the actin-coding region and thus represent at least 11 distinct genes. It is interesting to note that λHRL51 contains actin-coding region on tandemly repeated 2.3-kb *Eco*RI fragments. These fragments may correspond to the 2.3-kb *Eco*RI fragment on genomic Southern blots observed to hybridize more intensely to heterologous or homologous actin gene probes.

We have shown elsewhere (J. Engel et al., in prep.) that the majority of these genes that we have isolated encode human cytoplasmic actin proteins. Whether these genes are functional is unknown. Three of the isolates (λHRL34, λHRL45, and λHRL84) hybridize to a probe derived from the 3′-untranslated region of human γ-actin cDNA clone (P. Gunning et al., unpubl.), a finding that suggests that they may in fact be transcribed. In the case of other multigene families such as the globin genes, the majority of these multiple gene copies are at least capable of being expressed (Proudfoot 1980).

Precedence for the existence of multiple cytoplasmic actin genes derives from studies of the slime mold. *Dictyostelium* encodes 17 cytoplasmic actin genes (Kindle and Firtel 1978), and at least 6 of these genes are transcribed into mRNAs specifying different actin proteins. Some of these genes code for an identical protein but are transcribed at different times during development (McKeown and Firtel 1981). These data indicate that tissue and developmental state-specific expression of actin genes arose during evolution long before the advent of mammalian muscle-specific actin genes. Our finding of multiple human actin genes raises the possibility that similar tissue-specific or developmental stage-specific expression of actin genes may also occur in man.

ACKNOWLEDGMENTS

We acknowledge the advice and encouragement of Drs. Rob Maxson, Tim Mohun, Glen Gormezano, Phyllis Ponte, and Alex Mauron. We thank Drs. Don Cleveland, Sally Tobin, and Eric Fyrberg for kindly donating actin gene probes, and Drs. Steven Embury and Susan Guttman for their gracious gifts of human DNA.

J.E. was supported by the Medical Scientist Training Program, P.G. is a research associate, and L.K. is an investigator of the Howard Hughes Medical Institute.

REFERENCES

Clarke, M. and J. Spudich. 1977. Non-muscle contractile proteins. The role of actin and myosin in cell motility and shape determination. *Annu. Rev. Biochem.* **46:** 497.

Fyrberg, E.A., K.L. Kindle, N. Davidson, and A. Sodja. 1980. The actin genes of *Drosophila,* a dispersed multigene family. *Cell* **19:** 365.

Gallwitz, D. and I. Sures. 1980. Structure of split yeast gene: Complete nucleotide sequence of the actin gene in *Saccharomyces cerevisiae. Proc. Natl. Acad. Sci.* **77:** 2546.

Goldman, R., T. Pollard, and J. Rosenbaum, eds. 1976. "Cell Motility." *Cold Spring Harbor Conf. Cell Proliferation* **3.**

Kindle, K.L. and R.A. Firtel. 1978. Identification and analysis of Dictyostelium actin genes, a family of moderately repeated genes. *Cell* **15:** 763.

McKeown, M. and R.A. Firtel. 1981. Differential expression and 5′ end mapping of actin genes in Dictyostelium. *Cell* **24:** 799.

McKeown, M., W. Taylor, K. Kindle, R. Firtel, W. Bender, and N. Davidson. 1978. Multiple heterogeneous actin genes in Dictyostelium. *Cell* **15:** 789.

Ng, R. and J. Abelson. 1980. Isolation and sequence of the gene for actin in *Saccharomyces cerevisiae*. *Proc. Natl. Acad. Sci.* **77:** 3912.

Pollard, T.D. and R.R. Weihing. 1974. Actin and myosin and cell movement. *Crit. Rev. Biochem.* **2:** 1.

Proudfoot, N. 1980. Pseudogenes. *Nature* **286:** 840.

Scheller, R.H., L.B. McAllister, W.R. Crain, D.S. Durica, J. Posakony, T.L. Thomas, R.J. Britten, and E.H. Davidson. 1981. Organization and expression of multiple actin genes in the sea urchin. *Mol. Cell. Biol.* **1:** 609.

Schwartz, R.J. and K.N. Rothblum. 1981. Gene switching in myogenesis; differential expression of the actin multigene family. *Biochem.* **20:** 4122.

Shani, M., U. Nudel, D. Zevin-Sonkin, and R. Zakut. 1981. Skeletal muscle actin mRNA. Characterization of the 3′ untranslated region. *Nucleic Acids Res.* **9:** 579.

Tobin, S.L., E. Zulauf, F. Sánchez, E.A. Craig, and B.J. McCarthy. 1980. Multiple actin-related sequences in the *Drosophila melanogaster* genome. *Cell* **19:** 121.

Vandekerckhove, J. and K. Weber. 1979. The amino acid sequence of actin from chicken skeletal muscle actin and chicken gizzard smooth muscle actin. *FEBS Lett.* **102:** 219.

Mutationally Altered Myosins in Caenorhabditis elegans

ROBERT H. WATERSTON
SUZANNE BOLTEN
HAZEL L. SIVE
DONALD G. MOERMAN
Department of Genetics
Washington University School of Medicine
St. Louis, Missouri 63110

In studies of myogenesis and muscle contraction in *Caenorhabditis elegans*, more than 20 genes have been found that affect muscle structure (Waterston et al. 1977, 1980; Zengel and Epstein 1980). Of these, one gene, *unc-54 I* (Brenner 1974), has proved to be especially well suited for a detailed molecular genetic analysis. The *unc-54* gene codes for the major myosin heavy chain of the body-wall musculature (Epstein et al. 1974; MacLeod et al. 1977a,b), with one or more unidentified genes coding for a minor body-wall myosin as well as pharyngeal and other myosins (Waterston et al. 1982b). More than 100 independently isolated mutations have been found for the *unc-54* gene, and these have served as the basis for an increasingly sophisticated analysis of the gene and its products. Whereas others have focused on the cloning and sequencing of the gene (Karn et al. 1980, and this volume; MacLeod et al. 1981) or the effects of mutants on thick filament formation (Mackenzie and Epstein 1980), our efforts have concentrated on a thorough genetic analysis of the gene. These studies include the isolation of new alleles of *unc-54*, the construction of a fine-structure map of the gene, and the interpretation of this genetic map in physical terms.

RESULTS AND DISCUSSION

Four Functional Classes of unc-54 *Mutations Can Be Distinguished*

As might be expected with a gene coding for a protein as complex as a myosin heavy chain, the effects of the different *unc-54* mutations vary. Based on their

effects on the amount of *unc-54* protein accumulated and on the structure of the body-wall musculature, the mutations can be conveniently subdivided into four groups at present. The largest group consists of mutations that result in the absence or near absence of *unc-54* protein product, e.g., *unc-54 (e190)* and *unc-54 (e1300)* (MacLeod et al. 1977b). These null mutations, when homozygous, result in a marked reduction of the thick filament number and a disorganization of the remaining thick filaments in the body-wall musculature. These mutants still contain myosin heavy chains that are the products of other genes.

The remaining three classes of *unc-54* mutations all contain normal or near normal levels of the *unc-54* myosin heavy chain. In one class with two members, *e675* and *s291*, the *unc-54* polypeptide is smaller as a result of an internal deficiency in the light meromyosin region of the protein (MacLeod et al. 1977a; D.G. Moerman et al., in prep.). Animals homozygous for this class of mutations have a muscle structure resembling that of null mutations, indicating that the myosin, although present, may have altered assembly into thick filaments.

Two different classes of putative missense mutations are now known. Both have normal-molecular-weight *unc-54* myosin but differ in their effect on muscle structure. In one type, e.g., *e1152* and *e1301*, the defective myosin is associated with a decreased number of thick filaments (Epstein et al. 1974; MacLeod et al. 1977b), and in the other type, e.g., *s74* and *s95*, the myosin is incorporated into a nearly normal muscle structure (Moerman et al. 1982). Both classes of mutations result in slow animals, but those with abnormal assembly when homozygous may result in animals so paralyzed that they are inviable.

Genetic Fine-structure Analysis of the unc-54 *Mutations*

These four classes of mutations could provide useful information about the organization of the *unc-54* gene and the function of the protein in the assembly and contraction of muscle. But to do so requires further information about the nature of the molecular defects. Because of the large size of the myosin gene and its product, a direct chemical analysis of these mutations would be difficult. An indirect approach, however, is provided by the technique of genetic fine-structure mapping.

In fine-structure mapping, the relative positions of mutations within a gene may be determined by analyzing the nature of the recombinant events that occur to produce a wild-type gene. As illustrated in Figure 1, these methods allow one to place the mutation *a* to the left of *b* and the mutation *c* to the right of *b*. With this technique, using the genes *lev-11* and *let-50* as outside markers, 30 *unc-54* mutations are now ordered as shown in Figure 2. More than 700 exceptional chromosomes from almost 100 pair-wise combinations were analyzed to construct this map (Waterston et al. 1982a).

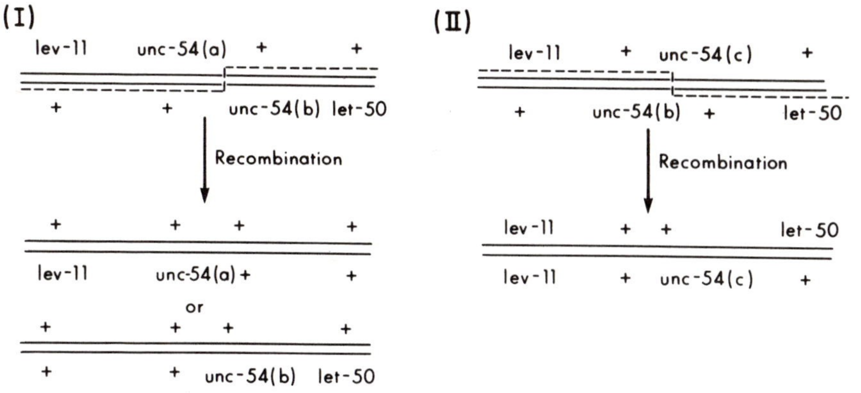

FIGURE 1

Diagram of the recombination events used to order the *unc-54* alleles on the fine-structure map. Heteroallelic strains, as shown in *I* or *II*, are constructed using standard genetic methods (Brenner 1974). Individual heteroallelic animals are placed on plates and allowed to self for several generations to produce more than 10^4 progeny/plate. Because of the nature of the marker mutations, only heterozygous animals will grow and reproduce well, so that most of the progeny, even after several generations, are heteroallelic. The plates are screened for non-Unc-54 animals, and the genotype of the recombinant chromosome is determined. (*I*) Illustration of the expected recombinant chromosome, if the *unc-54(a)* mutation is coupled with *lev-11* and is to the left of *unc-54(b)*. The + + + + chromosome may be recovered over either parental chromosome. (*II*) Illustration of the expected recombination event, if the *unc-54(c)* mutation is coupled with *lev-11* and is to the right of *unc-54(b)*. The *lev-11* + + *let-50* chromosome will be recovered only over the *lev-11* + *unc-54(c)* + chromosome, since *let-50* homozygotes are inviable.

Most of the mutations mapped are null alleles, as they have quite uniform phenotypes and are easy to work with. In addition, representative alleles of each of the other three classes were mapped. The types of exceptional events obtained and their frequency are similar to those reported for fine-structure mapping in other organisms (Waterston et al. 1982a).

Interpretation of the Genetic Map in Physical Terms

From the map itself, it is possible to draw certain inferences about the various mutation types. For example, the mutations altering contraction with little or no effect on muscle structure, *s77, s74,* and *s95,* are clustered near the right end of the genetic map. To learn more about the mutations, we have begun to interpret the genetic map in physical terms.

Nonsense Mutations

A subset of the null mutations are suppressible by the informational suppressors *sup-7 X* and *sup-5 III* (R.H. Waterston et al., in prep.) which, from

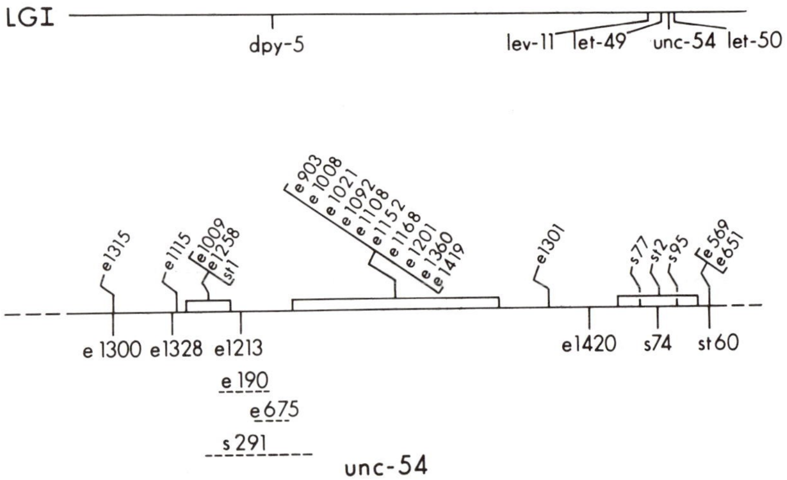

FIGURE 2
Fine-structure map of the *unc-54* gene. (*Top*) The position of the *unc-54* gene and flanking genes on linkage group I (LGI). (*Bottom*) The relative positions of the *unc-54* mutations. The mutations shown below the bottom line have been ordered by reciprocal crosses, and their positions are firmly established. The mutations shown above the line have been ordered only relative to the mutations below the line. For example, *e1009, e1258,* and *st1* lie between *e1328* and *e1213* but have not been ordered relative to one another. (----) The positions and relative extents of the deficiencies *e190, e675,* and *s291.* (Data from Waterston et al. 1982a.)

preliminary results, are probably altered tRNA molecules that act to suppress premature termination at UAG nonsense mutations (R.H. Waterston et al., unpubl.). Thus, these *unc-54* mutations can be assigned as nonsense mutations. Because the size of the truncated polypeptide produced by premature termination at a nonsense mutation should reflect accurately the position of the nonsense mutation in the coding sequence, these mutations may serve as an excellent means of making genetic-physical correlations, if the truncated polypeptides can be isolated or analyzed (see below). In general, such truncated products have not been found in vivo in *C. elegans* (Macleod et al. 1977b).

However, an exception to this general rule has been recognized. The mutation *e1300* is suppressible by both *sup-5* and *sup-7* (Waterston and Brenner 1978; Waterston 1981), and a consideration of its map position indicates that it is probably near the left end of the coding sequence. A careful analysis of SDS-polyacrylamide gels of actomyosin fractions from this mutant reveals a unique minor peptide about 7000 daltons smaller than the *unc-54* product itself (Fig. 3a). The polypeptide copurifies with myosin, and analysis of cyanogen bromide cleavage products on two-dimensional gel

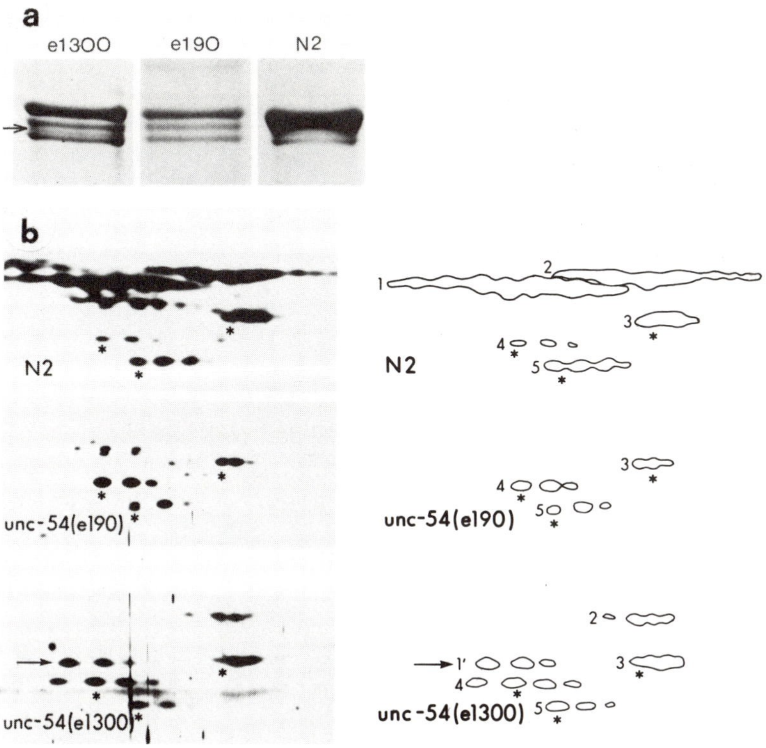

FIGURE 3

Analysis of myosin present in *unc-54(e1300)*. (*a*) Actomyosins prepared from the wild-type N2, from the *unc-54(e190)* null allele, and from the *unc-54(e1300)* suppressible allele were analyzed by SDS-gel electrophoresis using a 4% acrylamide gel, with the buffer system of Chua and Bennoun (1975). This system resolves the myosin heavy chains into several distinct bands, which differ by less than 5000 daltons out of a total of more than 200,000 daltons (Waterston et al. 1982b). In N2, the *unc-54* polypeptide predominates, but two other much weaker bands are also present. In *unc-54(e190)*, three bands are still present, but no *unc-54*-specific myosin is detectable in this allele (MacLeod et al. 1977b). The *unc-54(e1300)* has the three bands present in *e190* but, in addition, has a faint band running between the lower two bands (→). From its increased mobility relative to wild-type *unc-54* myosin, this band is estimated to be about 7000 daltons smaller than the *unc-54* protein. (*b*) Myosins were prepared from N2, *e190*, and *e1300* animals and cleaved with cyanogen bromide. The resultant peptides were separated by two-dimensional gel electrophoresis (O'Farrell 1975), and the gels were stained with silver nitrate (Switzer et al. 1979; Oakley et al. 1980). The same area from gels of each of the three strains is shown. The cyanogen bromide cleavage results in a series of isoelectric forms for each peptide. Peptide 1 of N2 is about 37,000 daltons and is the COOH-terminal or penultimate COOH-terminal peptide (MacLeod et al. 1977; Karn et al., this volume). It is absent in *e190*, and a new peptide of 34,000 daltons is present in *e1300* (1′ [→]). This peptide is presumed to be related to the 37,000-dalton peptide of N2, since only this *unc-54* peptide is absent in *e1300*. For example, peptide 2 is present in N2, absent in *e190*, and present in *e1300*. Both 1′ and 2 are present in much lower amounts in *e1300* than in N2, relative to peptides 3, 4, and 5, which are derived from non-*unc-54* polypeptides. (*) Similar positions in each of the three gels.

electrophoresis shows that the minor polypeptide is an *unc-54* product, which differs from the *unc-54* wild-type product in that its penultimate carboxyterminal peptide is about 3300 daltons smaller than the wild type (Fig. 3b). Thus, the minor polypetide is the truncated fragment produced by the *e1300* mutation and must lie about 150 bp from the end of the coding sequence. Therefore, the amino terminus (the 5′ end of the gene) must be at the right end of the map shown in Figure 2.

Internal Deficiencies

The positions of the internal deficiences of *e675* and *s291* have been determined at both the protein and nucleic acid level (MacLeod et al. 1981; D.G. Moerman et al., in prep.). *e675* is a deletion of about 300 bp in the DNA, encoding 10,000 daltons in a region about 40,000–50,000 daltons from the carboxyl terminus of the protein. The *s291* deficiency is larger, representing about 1300 bp of DNA sequence or 43,000 daltons in the protein, and overlaps the *e675* deficiency. The position of these mutations to the right of *e1300* places the 3′ end of the gene or the COOH end of the protein sequence at the left end of the map, in agreement with the assignment made previously, based on the analysis of the *e1300* UAG nonsense mutation.

Missense Mutations

In addition to these mutations, one other mutation, *s74*, has been localized to a specific region of the *unc-54* region. The *s74* allele contains a missense mutation affecting contractility but not structure. Myosin purified from animals homozygous for *s74* has been cleaved with cyanogen bromide and analyzed by two-dimensional gels (Fig. 4). When compared with wild type, one and only one peptide of 21,000 daltons has an altered isoelectric point. This peptide is not present in the 140,000-dalton carboxyterminal cyanylation peptide and thus must lie in the amino half of the protein. The nucleotide sequence of the gene, as determined by Karn et al. (this volume), shows that only one such peptide is present in this region, and this peptide lies within about 10,000 daltons of the amino terminus. Thus, the *s74* mutation is probably a missense mutation that lies within 20% of the amino terminus.

CONCLUSIONS

In sum, these correlations unambiguously assign the orientation of the gene on the chromosome and indicate that the genetic map includes at least 80% of the coding sequences of the gene. The resolution in physical terms is limited at present, but an analysis of the polypeptides produced by the suppressible alleles and by other null mutations, either in vivo or by in vitro

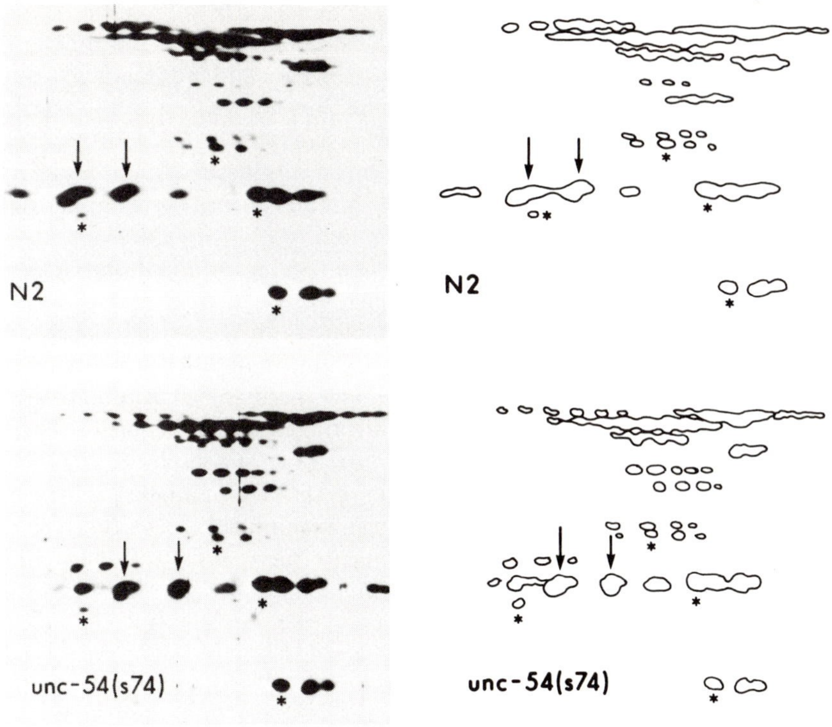

FIGURE 4
Analysis of *unc-54(s74)* myosin heavy chain. Myosins purified from N2 and from *unc-54(s74)* were cleaved with cyanogen bromide, the resultant peptides separated by two-dimensional gel electrophoresis, and the gel stained with silver as described in the legend to Fig. 3. Of the *unc-54* peptides, only one set is altered in isoelectric point in *s74*, with the two major spots of the set marked by arrows in both N2 and *s74*. This peptide is about 21,000 daltons from its relative mobility. The non-*unc-54* peptides are more prominent in the purified *s74* myosin. Whether this reflects the in vivo situation or a selective loss in purification of the *unc-54* heavy chain has not been determined. (*) Identical positions in each gel. Isoelectric focusing separates peptides in the horizontal direction, with the basic end at the left, and SDS-gel electrophoresis separates peptides in the vertical dimension. Peptides shown range from 39,000 daltons to 16,000 daltons.

translation of hybrid-selected *unc-54* RNAs, should provide a ready means of increasing the resolution.

Even in its present state, the map gives some insight into the nature of the mutations available. The null mutations analyzed appear to be scattered throughout the map and are probably limited to the coding region of the gene (or its introns). The missense mutations *s74*, *s77*, and *s95*, affecting contractility but not muscle structure, are limited to a discrete region near the aminoterminal end in the myosin head. The position of the two missense

mutations that alter muscle structure, *e1152* and *e1301*, to the right of the deficiencies, is intriguing, but better resolution of the map in this region is necessary before these mutations can be assigned to specific functional domains of the myosin molecule. With such improvements, it should be possible to determine exactly the nature of mutations like *s74* or *e1152*. In turn, by defining more precisely the functional defect of these mutant myosins through in vivo or in vitro studies, we may be able to learn more about how myosin functions in assembly and contraction.

On the other hand, we have not yet found regulatory mutations closely linked to but outside the structural sequence, which influence the levels of expression of *unc-54*. Perhaps such control regions are too small to be easily detected, or perhaps they are insensitive to point mutations, which would be expected to predominate with the mutagens used so far. Alternatively, control sequences could possibly lie within the structural sequence itself. Since other myosin genes are present in *C. elegans*, a comparison of their structure with that of the *unc-54* structure might be illuminating. Furthermore, it may prove feasible in the future to manipulate these other genes and thereby gain some insight into how their structural organization contributes to their regulation in space and time.

ACKNOWLEDGMENTS

We thank Michelene Harris for technical assistance, Carla Strubhart for typing the manuscript, and Marc Davis for photography. R.H.W. is an Established Investigator of the American Heart Association, and D.G.M. is a postdoctoral fellow of the Medical Research Council (Canada). This work was supported by a U.S. Public Health Service grant and the Muscular Dystrophy Association of America.

REFERENCES

Brenner, S. 1974. The genetics of *Caenorhabditis elegans. Genetics* **77:** 71.

Chua, N.H. and P. Bennoun. 1975. Thykaloid membrane polypeptide of *Chlamydomonas rheinhardi*: Wild type and mutant strains deficient in photo system II reaction center. *Proc. Natl. Acad. Sci.* **72:** 2175.

Epstein, H.F., R.H. Waterston, and S. Brenner. 1974. A mutant affecting the heavy chain of myosin in *Caenorhabditis elegans. J. Mol. Biol.* **90:** 291.

Karn, J., S. Brenner, L. Barnett, and G. Cesareni 1980. Novel bacteriophage λ cloning vector. *Proc. Natl. Acad. Sci.* **77:** 5172.

Mackenzie, J.M., Jr. and H.F. Epstein. 1980. Paramyosin is necessary for determination of nematode thick filament length *in vivo. Cell* **22:** 747.

MacLeod, A.R., J. Karn, and S. Brenner. 1981. Molecular analysis of the *unc-54* myosin heavy chain gene of *Caenorhabditis elegans. Nature* **291:** 386.

MacLeod, A.R., R.H. Waterston, and S. Brenner. 1977a. An internal deletion mutant of a myosin heavy chain in *C. elegans. Proc. Natl. Acad. Sci.* **74:** 5336.

MacLeod, A.R., R.H. Waterston, R.M. Fishpool, and S. Brenner. 1977b. Identification of the structural gene for a myosin heavy chain in *C. elegans. J. Mol. Biol.* **114:** 133.

Moerman, D.G., S. Plurad, R.H. Waterston, and D.L. Baillie. 1982. Mutations in the *unc-54* myosin heavy chain gene of *Caenorhabditis elegans* that alter contractility but not muscle structure. *Cell* **29:** 773.

Oakley, B.R., D.R. Kirsch, and N.R. Morris. 1980. A simplified ultrasensitive silver stain for detecting proteins in polyacrylamide gels. *Anal. Biochem.* **105:** 361.

O'Farrell, P.H. 1975. High resolution two-dimensional electrophoresis of proteins. *Proc. Natl. Acad. Sci.* **72:** 4007.

Schachat, F., R.L. Garcea, and H.F. Epstein. 1978. Myosins exist as homodimers of heavy chains: Demonstration with specific antibody purified by nematode mutant myosin affinity chromatography. *Cell* **15:** 405.

Switzer, R.C.,III, C.R. Merril, and S. Shifrin. 1979. A highly sensitive silver stain for detecting proteins and peptides in polyacrylamide gels. *Anal. Biochem.* **98:** 231.

Waterston, R.H. 1981. A second informational suppressor *sup-7 X* in *Caenorhabditis elegans. Genetics* **97:** 307.

Waterston, R.H. and S. Brenner. 1978. A suppressor mutation in the nematode acting on specific alleles of many genes. *Nature* **275:** 715.

Waterston, R.H., R.M. Fishpool, and S. Brenner. 1977. Mutants affecting paramyosin in *Caenorhabditis elegans. J. Mol. Biol.* **117:** 679.

Waterston, R.H., J.N. Thomson, and S. Brenner. 1980. Mutants with altered muscle structure in *Caenorhabditis elegans. Dev. Biol.* **77:** 271.

Waterston, R.H., K.C. Smith, and D.G. Moerman. 1982a. Genetic fine structure analysis of the myosin heavy chain gene *unc-54* of *Caenorhabditis elegans. J. Mol. Biol.* **158:** 1.

Waterston, R.H., D.G. Moerman, D.L. Baillie, and T.R. Lane. 1982b. Mutations affecting myosin heavy chain accumulation and function in the nematode *Caenorhabditis elegans.* In *Diseases of the motor unit* (ed. D. Schotland), p. 747. Houghton Mifflin, New York.

Zengel, J.M. and H.F. Epstein. 1980. Identification of genetic elements associated with muscle structure in the nematode *Caenorhabditis elegans. Cell Motil.* **1:**73.

unc-54 *Myosin Heavy-chain Gene of* Caenorhabditis elegans: *Genetics, Sequence, and Structure*

JONATHAN KARN
ANDREW D. McLACHLAN
LESLIE BARNETT
Medical Research Council
Laboratory of Molecular Biology
University Postgraduate Medical School
Cambridge, CB2 2QH England

The small soil nematode *Caenorhabditis elegans* is an attractive organism for the molecular study of muscle function and development because of its anatomical simplicity and suitability for genetic and biochemical analysis (Brenner 1974; Sulston and Horvitz 1977). The body-wall musculature of *C. elegans* is composed of 95 cells disposed in four quadrants, which run the length of the animal beneath the cuticle. The musculature is obliquely striated, and the sarcomeres are oriented parallel to the long axis of the animal. Since these cells represent a large fraction of the animal mass, isolation of contractile proteins is comparatively simple (Epstein et al. 1974; Waterston et al. 1974, 1977a; Harris and Epstein 1977; Mackenzie and Epstein 1980). Mutants affecting the characteristic pattern of motility of *C. elegans* can be easily identified, and microscopic examination of these "uncoordinated," or *unc* strains, in the living animal by polarized light microscopy or, more carefully, by electron microscopy has led to the identification of 22 genes that produce altered muscle phenotypes (Waterston et al. 1980; Zengel and Epstein 1980). Of these, two are known to code for major structural proteins of muscle: The *unc-54* gene codes for the major heavy chain of myosin (Epstein et al. 1974; MacLeod et al. 1977b), whereas the *unc-15* gene

codes for paramyosin, the core protein of the thick filaments (Waterston et al. 1974; MacLeod et al. 1977a; Harris and Epstein 1977).

In this paper we describe the molecular cloning and nucleotide sequence of the *unc-54* myosin heavy-chain gene (Karn et al. 1980; MacLeod et al. 1981). Knowledge of the sequence of this gene has allowed prediction of the amino acid sequence of a myosin heavy chain and construction of a model for the structure of the rod portion of the molecule. This detailed molecular information provides a basis for the analysis of the numerous *unc-54* mutants that have been isolated, and the phenotypes of two small deletion mutants are described.

unc-54 *GENE*

All *unc-54* mutants are paralyzed and most have severely disorganized body-wall muscle. However, these animals are able to survive, since the pharyngeal muscle cells are unaffected, allowing normal feeding. Examination of cross sections in the electron microscope has shown that the disruption of the body-wall muscle in *unc-54* mutants results from a loss of 75–80% of the thick filaments. These observations suggested to Epstein et al. (1974) that the *unc-54* gene might specify one of several myosin heavy chains. At least two other heavy chains should also be present in nematode muscle, corresponding to the heavy chains for the pharyngeal muscle and the residual thick filaments of the body-wall muscle. Analysis of the myosin heavy chains from *unc-54* mutants has confirmed this conjecture.

The heavy chains present in the total myosin fraction from wild-type (N2) animals and two mutants, E675 and E190[1], are characterized by SDS-polyacrylamide gel electrophoresis in Figure 1A. Also shown are the patterns of in vitro heavy-chain synthesis in a wheat germ cell-free system programmed with RNA from these strains.

The major heavy chain has a molecular weight (m.w.) of approximately 226,000 and can be identified as the *unc-54* heavy chain since this myosin specifically shows altered patterns of synthesis in the *unc-54* mutants. In E675, the wild-type heavy chain is replaced by an equally abundant but novel polypeptide of 216,000 m.w. Chemical cleavage studies by MacLeod et al. (1977b) have demonstrated that the *e675* myosin heavy chain arises from an internal deletion of 100 amino acids near the carboxyl terminus of the heavy chain. In E190, no *unc-54* myosin heavy chain or fragment is detectable in the isolated protein fraction. However, the cell-free translation assay reveals the presence of a novel polypeptide of 170,000 m.w., detectable in low yield. This suggests that the *e190* mutation produces low amounts of active mRNA for this heavy chain. As is shown below, *e190* is also a small deletion.

[1]In the accepted nomenclature for *C. elegans*, uppercase letters and numbers refer to the strains (i.e., E675, E190), whereas lowercase, italicized letters and numbers refer to alleles (i.e., *e675, e190*).

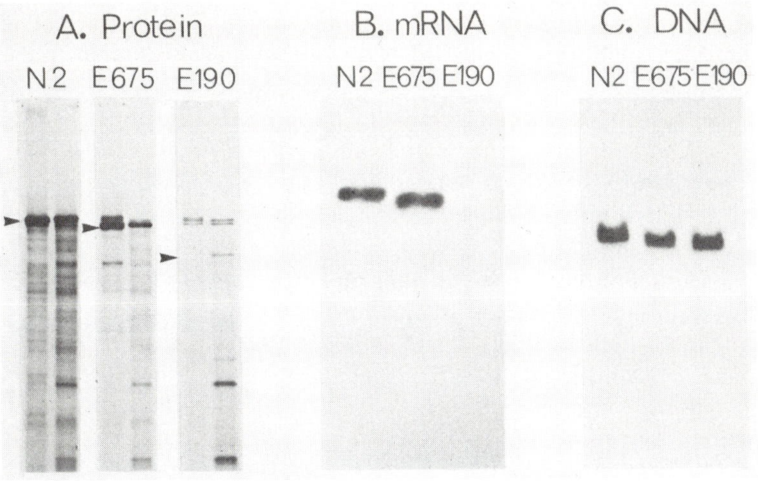

FIGURE 1
Identification of the *unc-54* myosin heavy-chain, mRNA, and DNA sequences. (*A*) ^{35}S-labeled myosin heavy chains, synthesized in vivo and in vitro using a wheat germ cell-free system, were fractionated by SDS-polyacrylamide gel electrophoresis after purification by indirect immunoprecipitation. For each strain, the left gel shows in vivo products and the right gel shows in vitro products. (►) The *unc-54* heavy chain or *unc-54*-specific heavy-chain fragments. Note the absence of any immunoprecipitable fragments in the in vivo myosin fraction isolated from E190. (*B*) RNAs from N2, E675, and E190 were fractionated on a 1.5% agarose gel, transferred to diazotized paper, and hybridized to nick-translated pMbo7. (*C*) *Bam*HI-digested DNAs from N2, E675, and E190 were fractionated on a 0.8% agarose gel, transferred to nitrocellulose, and hybridized to the pMbo7 probe. (Data from MacLeod et al. 1981.)

In addition to the *unc-54* product, at least two other myosin heavy chains are synthesized in the nematode. These are the pharyngeal (221,000 m.w.) and body-wall (217,000 m.w.) heavy chains. Although the *unc-54* heavy chain sometimes obscures these proteins in SDS gels, examination of myosins isolated from numerous mutants has shown that these chains are unaltered in *unc-54* mutants.

MOLECULAR CLONING OF THE unc-54 *GENE*

To analyze these and other myosin mutants in detail, we undertook the molecular cloning and nucleotide sequencing of the *unc-54* gene. Since the presence of multiple myosin heavy-chain genes might confuse the identification of *unc-54* -specific clones, we devised assays based on the properties of *e190* and *e675* to assist in the identification of clones (MacLeod et al. 1981).

The cell-free translation experiment described above demonstrated that the *unc-54* mRNA could be identified as an mRNA that programmed the synthesis of polypeptides of altered molecular weight in E675 and E190. We reasoned that since *e675* was an internal deletion mutant, both the *unc-54* mRNA itself and restriction fragments from the *unc-54* gene should be of altered length in E675. In E190, a reduced level of mRNA should be observed since the translation polypeptide is seen in reduced yield. Thus, clones that specifically hybridize to *unc-54* sequences should be identifiable as clones that detected strain differences between *unc-54* mutants in hybridization experiments. Figure 1B shows the hybridization pattern obtained when a plasmid (pMbo7), which contains *unc-54* sequences, was used as a probe against N2, E675, and E190 RNA fractionated by agarose gel electrophoresis. A single RNA species of approximately 6 kb is detected in the wild-type fraction, and this species shows the expected reduction of 300 nucleotides in E675. A corresponding mRNA is absent from E190. Because the genetic evidence unambiguously assigns *e675* and *e190* to a single locus, this hybridization experiment demonstrates that an *unc-54*-specific probe is present in pMbo7.

This cDNA plasmid was also used to identify restriction fragments of nematode genomic DNA that contained *unc-54*-specific sequences. Figure 1C shows the pattern obtained when *Bam*HI-digested DNA from N2, E675, and E190 was hybridized to pMbo7. A single restriction fragment of 5.5 kb was detected in the wild-type DNA. This was reduced in size in E675 by approximately 300 bp, consistent with the size of the internal deletion in *e675*. Surprisingly, this fragment also shows a reduction in size of approximately 600 bp in E190, demonstrating that this mutation is also a small deletion. Since the sequence in the pMbo7 probe is present in the genome but undetectable in the *e190* mRNA fraction, this result suggests that the *e190* mutation produces a more complicated alteration of the *unc-54* transcription unit than might be expected to result from a single deletion of part of the coding sequence as in *e675*.

The cDNA probe was then used to identify *unc-54*-specific bacteriophage λ clones, which harbored 15–20-kb inserts of randomly fragmented nematode DNA (Karn et al. 1980; MacLeod et al. 1981). Twenty-two recombinants were picked, and these phage were found to contain a series of overlapping restriction fragments that could be ordered to generate a 30-kb restriction map of the *unc-54* gene and flanking sequences.

STRUCTURE OF THE MYOSIN HEAVY-CHAIN unc-54 *GENE*

The structure of the *unc-54* gene has been determined by DNA sequencing (Sanger et al. 1977, 1980). The gene (Fig. 2) is split by at least seven intervening sequences. The coding sequence so far determined is a sequence of

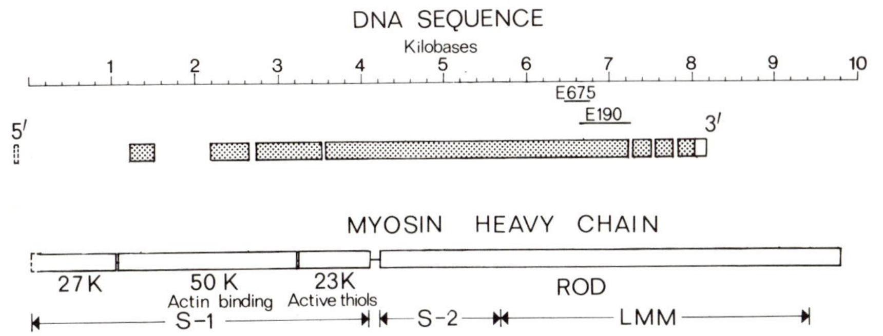

FIGURE 2
Structure of the *unc-54* gene and myosin heavy chain determined by DNA sequencing. Coding sequences are shown by bars, and the intervening sequences are indicated by the spaces between the bars. The 3′-untranslated region is shown as an open bar. The exon containing the amino terminus of the *unc-54* heavy chain has not been sequenced but is indicated by the broken bar. This exon is separated by an intervening sequence of at least 1.4 kb. A diagram of the myosin heavy-chain protein sequence is aligned below the DNA sequence diagram (see also Fig. 3). The major targets of limited proteolysis are indicated. Note the lack of obvious correspondence between the protein-structural domains and the positions of the introns in the DNA sequence.

1958 amino acids. This corresponds to a protein chain of 223,709 m.w., but the exon containing the amino terminus is missing and an estimated 20 amino acids, or 2240 m.w., remain to be sequenced. The deduced protein sequence is shown in Figure 3.

The nematode myosin heavy-chain sequence shows approximately 50% homology to sequences from other myosin heavy chains, including the fragments of rabbit skeletal myosin analyzed by Elzinga and his colleagues by protein chemical methods (Elzinga and Collins 1977; Elzinga and Trus 1980; Capony and Elzinga 1981; M. Elzinga, pers. comm.). Representative peptides are aligned in Figure 3.

For many years, biochemical studies of myosin have concentrated on subfragments generated by enzymatic digestion (Szent-Gyorgyi 1953; Mueller and Perry 1962; Huxley 1969; Weeds and Pope 1977; Cardinaud 1979; Mornet et al. 1981; Walser et al. 1981). Limited proteolysis of myosin splits the globular head of the molecule that performs the enzymatic functions of myosin (S-1) from the α-helical coiled-coil rod (S-2 and light meromyosin [LMM]). Tryptic digestion of S-1 produces three discrete fragments of 27,000 m.w., 51,000 m.w., and 23,000 m.w., which are aligned in this order within the heavy-chain polypeptide. The 23,000-m.w. peptide is known to contain two cysteinyl residues, SH-1 and SH-2, which may be

```
        27 K PEPTIDE (INCOMPLETE)
                10         20         30         40         50         60
NEM     DQSKPYDSKK NVWIPDPEEG YLAGEITATK GDQVTIVTAR EMRFVFKLMI TVFNVGVIQV
                70         80         90        100      ▼ 110        120
NEM     TLKKELVQEM NPPKFEKTED MSNLSFLNDA SVLHNLRSRY AAMLIYTYSG LFCVVINPYK
RBT     SNKKPDLIEM NPPKYDKIQD M                       MIYTYSG LFCVTVNPYK
               130        140        150        160        170        180
NEM     RLPIYTDSCA RMFMGKRKTE MPPHLFAVSD EAYRNMLQDH ENQSMLITGE SGAGKTENTK
RBT     WLPVYNAQVV TAYRGKKRQE APPHIFSISD NAYQFMLTDR ENQSILITGE SGAGKQVNTI
                                   50 K PEPTIDE
               190        200        210        220        230        240
NEM     KVICYFAAVG ASQQEGGAEV DPNKKKVTLE DQIVQTNPVL EAFGNAKTVR NNNSSRFGKF
RBT     T
               250      ▼            270        280        290        300
NEM     IRIHFNKHGR LASCDIEHDL LEKSRVIRQA PGERCYHIFY QIYSDFRPEL KKELLLDLPI
RBT                                                                MLLITNP
               310        320        330        340        350        360
NEM     KDYWFVAQAE LIIDGIDDVE EFQLTDEAFD ILNFSAVEKQ DCYRLMSAHM HMGNMKFKQR
RBT     YDYAFVSQGE ITVPSIDDQE ELM
               370        380        390        400        410        420
NEM     PREEQAEPDG TVEAEKASNM YGIGCEEFLK ALTKPRVKVG TEWVSKGQNC EQVNWAVGAM
               430        440        450        460        470        480
NEM     AKGLYSRVFN WLVKKCNLTL DQKGIDRDYF IGVLDIAGFE IFDFNSFEQL WINFVNEKLQ
RBT                MVTRINQQL DTK-QPRQYF IGVLDIAGFE SFDFQKLEQ
               490        500        510        520▼       530        540
NEM     QFFNHHMFVL EQEEYAREGI QWVFIDFGLD LQACIELIEK PLGIISMLDE ECIVPKATDL
RBT                                                MGIFSILEE ECMFPKATDT
               550        560        570        580        590        600
NEM     TLASKLVDQH LGKHPNFEKP KPPKGKQGEA HFAMRHYAGT VRYNCLNWLE KNKDPLNDTV
RBT     SFKNKLYEQ- LGKNNNFQKP KPAKGK
                                                 23 K PEPTIDE
               610        620        630        640        650        660
NEM     VSAMKQSKGN DLLVEIWQDY TTQEEAAAKA KEGGGGGKKK GKSGSFMTVS MLYRESLNNL
RBT                       KTL AFLFSGAQAG EEGGGGKKGG KKKGAYF       RENLNKL
                                                 ACTIVE THIOLS
               670        680        690        700        710        720
NEM     MTMLNKTHPH FIRCIIPNEK KQSGMIDAAL VLNQLTCNGV LEGIRICRKG FPNRTLHPDF
RBT     MTNLRSTHPH FVRCIIPNET KTPGAMEHEL VLHQLRCNGV LEGIRICRKG FPSRILYADF
               730        740        750        760        770        780
NEM     VQRYAILAAK EAKSDDDKKK CAEAIMSKLV NDGSLSEEMF RIGLTKVFFK AGVLAHLEDI
RBT     KQRYKVLNA- -SAIPEGQFI DSKKASEKLL GSIDVDHQTY KFGHTKVFFK AGLLGLLEEM
                              END 23K
               790        800        810        820        830        840
NEM     RDEKLATILT GFQSQIRWHL GLKDRKRRME QRAGLLIVQR NVRSWCTLRT WEWFKLYGKV
RBT     RDDKLAQILT RTQAICRGFL ARVEYKKMME -R                         MKLYFKI
          ROD (S-2)
           *   850      * 860        870        880      * 890        900
NEM     KPMLKAGKEA EELEKINDKV KALEDSLAKE EKLRKELEES SAKLVEEKTS LFTNLESTKT
RBT     KPLLKSAETE KEMANMKEEF EKTKESLAKA EAKRKELEEK MVALMQEKND LQLQVQAEAD
               910      * 920        930          *        950        960
NEM     QLSDAEERLA KLEAQQKDAS KQLSELNDQL ADNEDRTADV QRAKKKIEAE VEALKKQIQD
RBT     SLADAEERQD LIKTKIQLEA K-IKEVTERA EDEEEINAEL TAKKRKLEDE CSELKKDIDD
                 *        980        990      *1000       1010       1020
NEM     LEMSLRKAES EKQSKDHQIR SLQDEMQQQD EAIAKLNKEK KHQEEINRKL MEDLQSEEDK
RBT     LELTLAKVEK EKHATENKVK NLTEEMAGLD ENIAKLTKEK KALQEAHQQT LDDLQAEEDK
```

FIGURE 3

Amino acid sequence of the *unc-54* myosin heavy chain. Homologous sequences from rabbit skeletal myosin (Elzinga and Collins 1977; Elzinga and Trus 1980; Capony and Elzinga 1981; M. Elzinga, pers. comm.) and rat skeletal myosin (Nadal-Ginard et al., this volume and pers. comm.) are aligned beneath the nematode sequence. In the rod portion of the molecule,

```
           * 1030       1040       1050  *    1060       1070          *
NEM  GNHQNKVKAK LEQTLDDLED SLEREKRARA DLDKQKRKVE GELKIAQENI DESGRQRHDL
RBT  VNTLTKAKTK LAQQVDDLEG SLEQEKKIPM DLERAKRKLE GDLKILQEST MDIENDKQQL
                                                END S-2      LMM
           1090       1100          *       1120       1130      *1140
NEM  ENNLKKKESE LHSVSSRLED EQALVSKLQR QIKDGQSRIS ELEEELENER QSRSKADRAK
RBT  DEKLKKKEF- MTNLQSKIED EQALM              IE ELEEEIEAER
           1150       1160    *  1170       1180       1190  +*   1200
NEM  SDLQRELEEL GEKLDEQGGA TAAQVEVNKK REAELAKLRR DLEEANMNHE NQLGGLRKKH
RBT                             MNKK REAEFQKMRR DLEEATLQHE ATAAALRKKH
           1210       1220 *     1230       1240          *       1260
NEM  TDAVAELTDQ LDQLNKAKAK VEKDKAQAVR DAEDLAAQLD QETSGKLNNE KLAKQFELQL
RBT  ADSVAELGEQ IDNLQRVKQK LEKEKSELKM EIDDLAGNME TVSKAKGNLE   KRTLEDQL
           1270          *       1290       1300     * 1310       1320
NEM  TELQSKADEQ SRQLQDFTSL KGRLHSENGD LVRQLEDAES QVNQLTRLKS QLTSQLEEAR
RBT  SEVRTKEEEH QRLINELSAQ KARLHTESGE FSRQLLNKDA M
           1330   *   1340       1350       1360 *     1370       1380
NEM  RTADEEARER QTVAAQAKNY QHEAEQLQES LEEEIEGKNE ILRQLSKANA DIQQWKARFE
             +*       1400       1410          *       1430       1440
NEM  GEGLLKADEL EDAKRRQAQK INELQEALDA ANSKNASLEK TKSRLVGDLD DAQVDVERAN
          *1450       1460       1470    *  1480       1490       1500
NEM  GVASALEKKQ KGFDKIIDEW RKKTDDLAAE LDGAQRDLRN TSTDLFKAKN AQEELAEVVE
       *   1510       1520          *       1540       1550          *
NEM  GLRRENKSLS QEIKDLTDQL GEGGRSVHEM QKIIRRLEIE KEELQHALDE AEAALEAEES
RAT                 DLTEQL GEGGKNVHEL EKIRKQLEVE KLELQSALEE AEASLEHEEG
           1570       1580         +*       1600       1610     * 1620
NEM  KVLRAQVEVS QIRSEIEKRI QEKEEEFENT RKNHARALES MQASLETEAK GKAELLRIKK
RAT  KILRAQLEFN QIKAEIESKL AEKDEEMEQA KRNHLRVVDS LQTSLDAETR SRNEALRVKK
           1630       1640    *  1650       1660       1670 *     1680
NEM  KLEGDINELE IALDHANKAN ADAQKNLKRY QEQVRELQLQ VEEEQRNGAD TREQFFNAEK
RBT          ME IQLNHANRQA AEAIKNLRNT QGILKDTQLH LDDAVRGQDD LKEQLAMVER
RAT  KMEGDLNEME IQLSQANRIA SEAQLHLKNA QAHLKDTQLQ LDDAVRANDD LKQNIAIVER
           1690          *       1710     △1720          *       1740
NEM  RATLLQSEKE ELLVANEAAE RARKQAEYEA ADARDQANEA NAQVSSLTSA KRKLEGEIQA
RBT  RANLLQAEIE ELRATLEQTE RSRKVAEQEL LDASERVQLL HTQNTSLINT KKKLETDISQ
RAT  RNTLLQAELE ELRAVVEQTE RSRKLAEQEL IETSERVQLL HSQNNSLINQ KKKMDADLSQ
      ▼    1750     * 1760       1770       1780    *△ 1790       1800
NEM  IHADLDETLN EYKAAEERSK KAIADATRLA EELRQEQEHS QHVDRLRKGL EQQLKEIQVR
RBT  IQGEMEDIVQ EARNAEEKAK KSITDAAMMA EELKKEQDTS AHLERMKKNM EQTVKDLQHR
RAT  LQTEVEEAVQ ECRNAEEKAK KAITDAAMMA EELKKEQDTS AHLERMKKNM EQTIKDLQHR
           1810 +* ▼ 1820     △1830          *       1850       1860
NEM  LDEAEAAALK GGKKVIAKLE QRVRELESEL DGEQRRFQDA NKNLGRADRR VRELQFQVDE
RBT  LDEAEQLALK GGKKQIQKLE ARVRELENEV ESEQKRNVEA VKGLRKHERR VKELTYQTEE
RAT  LDEAEQIALK GGKKQLQKLE ARVRELENEL EAEQKRNAES VKGMRKSERR IKELNYQTEE
                 END LMM △
             *       1880          ▼    *       1910       1920
NEM  DKKNFERLQD LLDKLQQKLK TQKKQVEEAE ELANLNLQKY KQLTHQLEDA EERADQAENS
RBT  DRKNVRLRQD LVDKLQSKVK AYKRQAEEAE EQSNVNLSKF RKLQHELEEA EERADIAESQ
RAT  DKKNLVRLQD LVNKLQLKVK AYKRKAEEAE EQANTNLSKF RKVQHELDEA EERADIAESQ
        △*  1930       1940       1950  *
NEM  LSKMRSKSRA SASVAPGLQS SASAAVIRSP SRARASDF*
RBT  VNKLRVKSRE VHTKVISEE*
RAT  VNKLRVKSRD IGAKQKIHDE E*
```

FIGURE 3 (Continued)

* above a residue indicates the start of one of 28 amino acid repeat sequences, and + indicates four additional residues that interrupt the repeating sequence structure. (▼) The positons of intervening sequences in the nematode sequences. (△) The positions of intervening sequences in the rat skeletal sequence.

selectively alkylated and are required for myosin ATPase activity (Yount et al. 1971; Schaub and Watterson 1972; Elzinga and Collins 1977; Walser et al. 1981).

The homology between the nematode and rabbit sequences is sufficiently strong to allow tentative assignments of the major proteolytic cleavage sites. Potential tryptic cleavage sites in S-1 are present at residues 204–206 and 637–670. These sequences are characterized by three tandem lysines in a region not likely to form either α-helical or β-sheet structures. Cleavage at these residues would generate peptides of 23,240 m.w. (plus the estimated 2200 m.w. missing), 49,850 m.w., and 23,229 m.w. The sequences from the 23,000-m.w. peptide surrounding the active thiols are among the best conserved in the myosin head, which suggests that this region does in fact play an important role in myosin activity.

It is interesting to note that the various domains of the myosin molecule defined by proteolysis are not separated in the gene by intervening sequences. The four known intervening sequences in the S-1 region do not separate the tryptic fragments, nor do they separate regions of sequence that might be expected to form different types of secondary structure. No intervening sequences are found between the rod and the beginning of the globular head. Three intervening sequences are present toward the carboxyl end of the rod portion of the molecule. As is discussed below, this sequence shows an underlying repetitive structure of 28 amino acids, but the disposition of intervening sequences in this region appears random with respect to this feature. We conclude that in myosin, intervening sequences are not used to separate domains in the protein. This view is reinforced by a comparison of the distribution of intervening sequences in the rod of the *unc-54* heavy chain and in the rat skeletal myosin sequences studied by B. Nadal-Ginard and his colleagues (this volume and pers. comm.). In rat, there are at least 18 intervening sequences in the gene, and 11 of these are in the rod. Five of the sequences have been located by DNA sequencing and, as is shown in Figure 3, although the protein sequence homology is high, these are all in different positions in the sequence compared to the positions of the introns in the nematode sequence.

One other feature about the intervening sequences in nematode deserves mention. The signal sequences at the intron-exon boundaries are similar to those found in mammalian sequences but are uniquely well conserved. The 5′ end of each intron is defined by the consensus sequence G(6)-T(6)A(5)A(6)G(5)T(4)T(5), whereas the 3′ end of the intron is defined by the consensus sequence A(7)NT(5)NT(6)NNNNNT(7)T(7)T(6)C(7)A(7)G(7), where the numbers in parentheses refer to the frequency with which a given nucleotide appears in the six known 5′ sequences or seven known 3′-end sequences. It is possible that the splicing mechanism in nematode is somewhat simpler than that used in mammals and that only one type of intervening sequence junction is used.

STRUCTURAL FEATURES OF THE MYOSIN ROD

The rod portion of the myosin sequence begins at residue 843 of the existing nematode sequence. The rod shows a number of periodic structural features, which are shown diagrammatically in Figure 4. As with other α-fibrous sequences that form α-helical coiled-coil structures, there is a regular disposition of hydrophobic residues in an alternating 3–4-heptapeptide repeat (Crick 1952; McLachlan and Stewart 1975; McLachlan et al. 1975; Parry et al. 1977). In the histogram in Figure 4A, these positions are marked by stars, and in the helical-net diagrams (Fig. 4B), hydrophobic residues can be seen occupying the internal positions a and d. Not all of the hydrophobic positions in the rod sequences are equally well occupied, and so the 3–4-heptapeptide repeat does not appear as strictly maintained as the corresponding repeat in tropomyosin. The charged amino acids show a period of 28 residues and like charges tend to be clustered. As shown in Figure 4A, the major peak of positive charge is separated from the major peak of negative charge by 14 residues. In three dimensions, the positive- and negative-charge clusters appear as well-separated bands along the outer surface of the rod. This is shown diagrammatically in Figure 4B, which plots the frequency of charged residues within the 28-amino-acid repeat (proportional to the diameter of the circles) on a helical net representing the two strands of the coiled coil.

Unlike the tropomyosin and α-keratin sequences, four discontinuities interrupt the repeat structure. At residues 1192, 1389, 1586, and 1811, a single residue is inserted, which introduces a shift in the phasing of the 28-residue repeat. The carboxyterminal fragment shown in Figure 3 illustrates two of these discontinuities, at residues 1586 and 1811.

MODEL BUILDING

The structural features of the rod described above have been incorporated into a model for the myosin tail and the thick filament. We wondered whether ionic interactions between two myosin molecules might account for some of the features of myosin packing in the thick filament, which must involve both parallel and antiparallel interaction. Pairs of myosin molecules should interact optimally when they have a relative axial stagger of multiples of 14 residues, or $n \times 1.5$ Å/residue $\times$ 14 residues, given the periodicity observed in Figure 4A for charged residues. We calculated the value of n for both parallel and antiparallel sequences using a simple one-dimensional model for the rod. This is shown in Figure 5 for the parallel interactions. Two myosin rod sequences were aligned with a shift of s residues, and the number of charged residues in a zone of 99 residues about each amino acid was calculated. A score of $+1$ was assigned for each pair of opposite charges within ± 2 residues on the parallel strand and within the zone. A score of -1 was assigned for pairs of like charges. Graphs of these

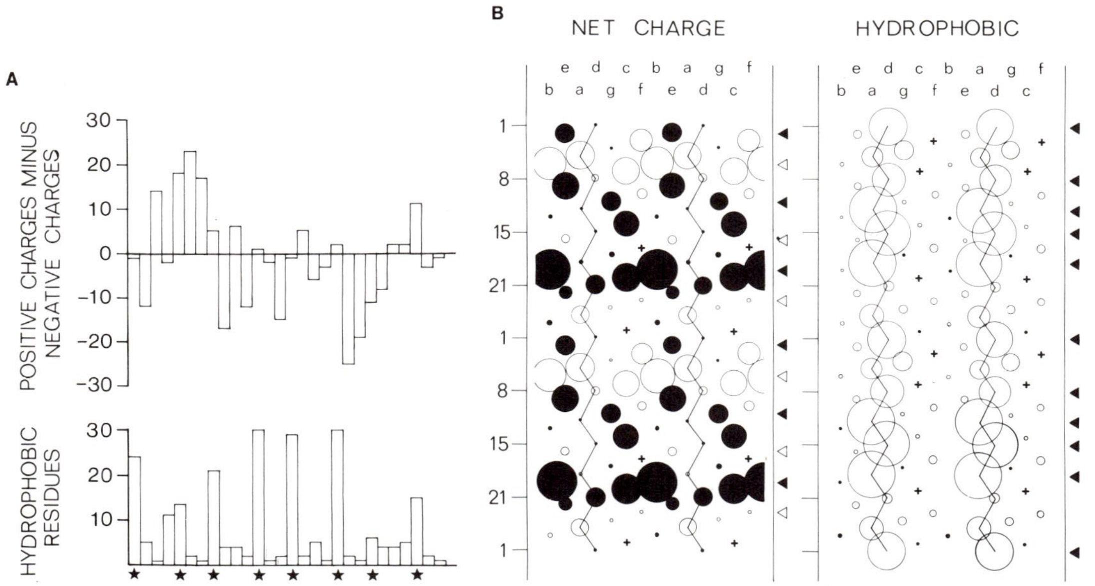

FIGURE 4

Distribution of amino acid residues in the 28 amino acid repeats of the rod. (*A*) Histogram plotting the number of hydrophobic and charged residues within 30 of the 28 amino acid repeats of the rod sequence. Note the high occupancy of hydrophobic positions at every alternating third and fourth residue (★). The major peak of positive charge is separated from the major peak of negative charge by 14 residues. (*B*) Helical-net diagram showing distributions of charged residues and hydrophobic residues. The two chains of the coiled coil in a myosin dimer have been included. The net is drawn with 3.5 residues per turn. (*a, d*) Inner positions; (*b,c,f*) outer surface of the superhelix. The diameter of the circles is proportional to the number of times a given type of amino acid appears in the rod sequence. (●) Negative charges; (○) either positively charged amino acids or hydrophobic residues. Two successive normalized 28-amino-acid repeats are shown.

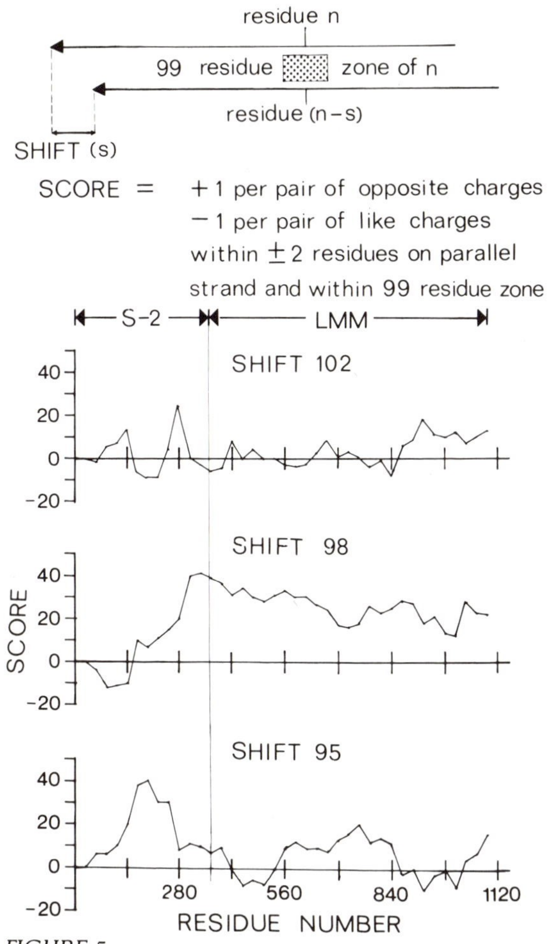

FIGURE 5
Calculation of the optimal stagger between two parallel myosin heavy-chain rods. The graphs plot scores for potential charge interactions calculated by the model along the length of the rod sequence. The S-2/LMM junction defined by proteolysis of rabbit myosin is indicated. Plots are drawn for three staggers corresponding to shifts of 102, 98, and 95 amino acids. See text for details.

potential interaction scores along the length of the rod sequence are shown in Figure 5. Optimal potential parallel interactions were obtained with a stagger of 98 residues (or n = 7). This corresponds to a stagger of 146 Å or near to the well-known 143-Å stagger of myosin heads in thick filaments.

The calculation above also provides an explanation for the suggestion

that the S-2/LMM junction might act as a second hinge during muscle contraction (Huxley 1969), although we have found no obvious structural discontinuity in the rod at this region. However, we have observed that when two myosin sequences are aligned in a parallel configuration, the potential ionic interaction between the two chains begins at about residue 290 of the rod and continues until the end of the chain. In contrast, in the S-2 region, these potential interactions are low. Therefore, we speculate that this lack of ionic interactions may allow the S-2 region to be pulled away from the thick filament during contraction and therefore act as a "hinge."

Similar calculations were performed on molecules aligned in an antiparallel configuration. Optimal electrostatic interactions occur with nearly complete overlap of the rods. Harris and Epstein (1977) showed that purified nematode myosin could form bipolar aggregates upon rapid lowering of ionic strength, in which antiparallel rods are overlapped throughout most of their lengths.

FUTURE PROSPECTS

The detailed molecular information we have obtained about the *unc-54* gene should provide a basis for the understanding of the numerous mutants affecting myosin in *C. elegans*. For example, the phenotypes of two deletions, *e190* and *e675*, may be explained as follows. The deletion *e675* is in phase, and this gives rise to a truncated mRNA and a shortened protein product. It seems likely that the deletion introduces a phase shift in the 28-amino-acid repeat and therefore disrupts the normal charge interactions of myosin. The deletion *e190* maps near *e675* but does not give rise to a stable mRNA or a protein product. The *e190* mRNA lacks the 3′-terminal sequence, even though this sequence is present in the gene. The most probable explanation is that the deletion extends into one of the 3′-intervening sequences, and this creates a novel junction that gives rise to a prematurely terminated mRNA. By cloning and sequencing these and other *unc-54* alleles, it will be possible to locate quickly the altered sequences and to deduce from this mutational evidence which amino acid residues are required for the multiple enzymatic and structural function of myosin and which signal sequences may be used for mRNA maturation.

ACKNOWLEDGMENTS

This work continues the genetic analysis of nerve and muscle started by Sydney Brenner nearly 15 years ago. We are indebted to him for his enthusiastic support and his many gifts of strains. We also thank our colleagues who generously supplied us with unpublished sequence information.

REFERENCES

Brenner, S. 1974. The genetics of *Caenorhabditis elegans*. *Genetics* **77:** 71.

Capony, J.P. and M. Elzinga. 1981. The amino acid squence of a 37,000 dalton fragment from S-2 of myosin. *Biophys. J.* **33:** 148a.

Cardinaud, L. 1979. Proteolytic fragmentation of myosin: Location of SH-1 and SH-2 thiols. *Biochimie* **61:** 807.

Crick, F.H.C. 1952. Is α-keratin a coiled-coil? *Nature* **170:** 882.

Elzinga, M. and J.H. Collins. 1977. Amino acid sequence of a myosin fragment that contains SM-1, SM-2 and 3-methylhistidine. *Proc. Natl. Acad. Sci.* **74:** 4281.

Elzinga, M. and B. Trus. 1980. Sequence and proposed structure of a 17,000 dalton fragment of myosin. In *Methods in peptide and protein sequence analysis* (ed. C. Burr), p. 213. Elsevier/North-Holland Biomedical Press, Amsterdam.

Epstein, H.F., R.H. Waterston, and S. Brenner. 1974. A mutant affecting the heavy chain of myosin in *-aenorhabditis elegans. J. Mol. Biol.* **90:** 291.

Harris, H.E. and H.F. Epstein. 1977. Myosin and paramyosin of *Caenorhabditis elegans:* Biochemical and structural properties of wild-type and mutant proteins. *Cell* **10:** 709.

Huxley, H.E. 1969. The mechanism of muscular contraction. *Science* **164:** 1356.

Karn, J., S. Brenner, L. Barnett, and G. Cesareni. 1980. Novel bacteriophage lambda cloning vector. *Proc. Natl. Acad. Sci.* **77:** 5172.

Mackenzie, J.M. and H.F. Epstein. 1980. Paramyosin is necessary for determination of nematode thick filament length *in vivo*. *Cell* **22:** 747.

MacLeod, A.R., J. Karn, and S. Brenner. 1981. Molecular analysis of the *unc-54* myosin heavy chain gene of *Caenorhabditis elegans. Nature* **291:** 386.

MacLeod, A.R., R.H. Waterston, and S. Brenner. 1977a. An internal deletion mutant of a myosin heavy chain in *Caenorhabditis elegans. Proc. Natl. Acad. Sci.* **74:** 5336.

MacLeod, A.R., R.H. Waterston, R.M. Fishpool, and S. Brenner. 1977b. Identification of the structural gene for a myosin heavy chain in *Caenorhabditis elegans. J. Mol. Biol.* **114:** 133.

McLachlan, A.D. and M. Stewart. 1975. Tropomyosin coiled-coil interactions: Evidence for an unstaggered structure. *J. Mol. Biol.* **98:** 293.

McLachlan, A.D., M. Stewart, and L.B. Smillie. 1975. Sequence repeats in α-tropomyosin. *J. Mol. Biol.* **98:** 281.

Mornet, D., R. Bertrand, P. Pantel, E. Audemard, and R. Kassab. 1981. Proteolytic approach to structure and function of actin recognition site in myosin heads. *Biochemistry* **20:** 2110.

Mueller, M. and S.V. Perry. 1962. Degradation of heavy meromyosin by trypsin. *Biochem. J.* **85:** 431.

Parry, D.A.D., W.G. Crewther, R.D.B. Fraser, and T.P. MacRae. 1977. Structure of α-keratin: Structural implications of the amino acid sequences of the type I and type II chain segments. *J. Mol. Biol.* **113:** 449.

Sanger, F., S. Nichlen, and A.R. Coulson. 1977. DNA sequencing with chain-terminating inhibitors. *Proc. Natl. Acad. Sci.* **74:** 5463.

Sanger, F., A.R. Coulson, B.G. Barrell, A.J.H. Smith, and B.A. Roe. 1980. Cloning on single strand bacteriophage as an aid to rapid DNA sequencing. *J. Mol. Biol.* **143:** 161.

Schaub, M.C. and J.G. Watterson. 1973. Possible differentiation between rigor type interactions of the two myosin heads with actin. *Cold Spring Harbor Symp. Quant. Biol.* **37:** 153.

Sulston, J.E. and R.H. Horvitz. 1977. Post-embryonic cell lineages of the nematode *Caenorhabditis elegans. Dev. Biol.* **56:** 100.

Szent-Györgyi, A.G. 1953. Meromyosins, the subunits of myosin. *Arch. Biochem. Biophys.* **42:** 305.

Walser, J.T., J.G. Watterson, and M.C. Schaub. 1981. Cyanylation and cleavage of myosin heavy chains at reactive thiol groups: Direct localization of thiol-1 and thiol-2 groups. **20:** 1169.

Waterston, R.H., H.F. Epstein, and S. Brenner. 1974. Paramyosin of *Caenorhabditis elegans. J. Mol. Biol.* **90:** 285.

Waterston, R.H., R.M. Fishpool, and S. Brenner. 1977. Mutants affecting paramyosin in *Caenorhabditis elegans. J. Mol. Biol.* **117:** 825.

Waterston, R.H., J.N. Thomson, and S. Brenner. 1980. Mutants with altered muscle structure in *Caenorhabditis elegans. Dev. Biol.* **77:** 271.

Weeds, A.G. and B. Pope. 1977. Studies on the chymotryptic digestion of myosin. *J. Mol. Biol.* **111:** 129.

Yount, R.G., D. Ojala, and D. Babcock. 1971. Interaction of P-N-P and P-C-P analog of adenosine triphosphate with heavy meromyosin, myosin and actomyosin. *Biochemistry* **10:** 2490.

Zengel, J.M. and H.F. Epstein. 1980. Identification of genetic elements associated with muscle structure in the nematode *Caenorhabditis elegans. Cell Motil.* **1:** 73.

Structure and Regulation of a Mammalian Sarcomeric Myosin Heavy-chain Gene

BERNARDO NADAL-GINARD
RUSSELL M. MEDFORD
HANH T. NGUYEN
MUTHU PERIASAMY
ROBERT M. WYDRO
DANUTA HORNIG
RUTH GUBITS
LEONARD I. GARFINKEL
DAVID WEICZOREK
EVA BEKESI
VIJAK MAHDAVI
Department of Pediatrics
Harvard Medical School
Department of Cardiology
Children's Hospital Medical Center
Boston, Massachusetts 02115

Myogenesis in vitro and in vivo is characterized by a well-defined sequence of cellular events leading to the conversion of cycling undifferentiated myoblasts to a terminally differentiated myotube (Okazaki and Holtzer 1966; Coleman and Coleman 1968; Yaffe 1968, 1974; Paterson and Strohman 1972; Yaffe and Dym 1973; Chi et al. 1975; Prives and Paterson 1976; Nadal-Ginard 1978). These cellular events, which are amenable to experimental manipulation, include cessation of DNA synthesis with irreversible withdrawal from the cell cycle and fusion of the cell membranes to form the multinucleated myotubes. Concurrent with these cellular changes is both an induction of a large battery of genes as well as repression of others (for review, see Merlie et al. 1977; Shani et al. 1981a; Garfinkel et al. 1982; Hastings and Emerson 1982; R.M. Medford; H.T. Nguyen et al.; both

in prep.). The main components of the myofibril include myosin heavy chain (MHC); myosin light chain (MLC) 1, 2, and 3; α-actin, α-tropomyosin; and troponin I, C, and T. These proteins are induced at, or just before, the onset of myotube formation. The genes coding for the myofibrillar proteins are coordinately regulated (Garfinkel et al. 1982; Hastings and Emerson 1982; L. Garfinkel and B. Nadal-Ginard; R.M. Medford et al.; H.T. Nguyen et al.; all in prep.) to accumulate their protein products in precise stoichiometric amounts (Potter 1974).

Most of the myofibrillar proteins are encoded by multigene families (for review, see Engel et al. 1981; Firtel 1981; Nguyen et al. 1982). Individual members of these gene families exhibit tissue-specific and developmental regulation (Huszar 1972; Masaki and Yoshizaki 1974; Hoh and Yeoh 1979; Lowey et al. 1979; Whalen and Sell 1980; Shani et al. 1981b; Umeda et al. 1981; Whalen et al. 1981; Mahdavi et al. 1982; Wydro et al. 1982).

Analysis of the interrelationship between the cellular events of myogenesis and the expression of the myofibrillar genes requires both a molecular approach and a genetic dissection of the myogenic pathway. This dissection can be accomplished by the isolation and molecular analysis of mutant myogenic cells defective at different stages of differentiation. We have used this combined approach—cellular, genetic, and molecular—to analyze some of the events involved in MHC gene expression during differentiation of the myogenic L6E9 cell line (Nadal-Ginard 1978) and its temperature-sensitive variants (H.T. Nguyen et al., in prep.).

RESULTS AND DISCUSSION

MHC Multigene Family: Structure and Organization

SARCOMERIC MHC IS ENCODED BY A MULTIGENE FAMILY. The MHC cDNA clone (pMHC-25) (Medford et al. 1980), isolated from the myogenic cell line L6E9, "codes" for 207 amino acids of the light meromyosin (LMM) portion of an embryonic skeletal MHC (M. Periasamy et al., in prep.). As shown in Figure 1, this embryonic MHC sequence hybridizes at high stringency to MHC mRNAs from all rat striated-muscle tissues so far tested (i.e., cardiac and skeletal muscles at different stages of development) but not to RNA from smooth muscle or nonmuscle tissues. This result suggests either (1) that the embryonic skeletal MHC sequences represented by pMHC-25 are expressed, at different levels, in all striated-muscle tissues but not in nonstriated-muscle tissues or (2) that distinct tissue-specific MHC mRNAs share a different degree of homology with the pMHC-25 insert.

Similar ambiguity becomes apparent when pMHC-25 DNA is hybridized to genomic DNA. pMHC-25 that does not contain *Eco*RI, *Bam*HI, and *Hind*-III restriction sites detects a minimum of eight distinct size fragments in the genomic DNA restricted with any of these enzymes (Fig. 2). This result sug-

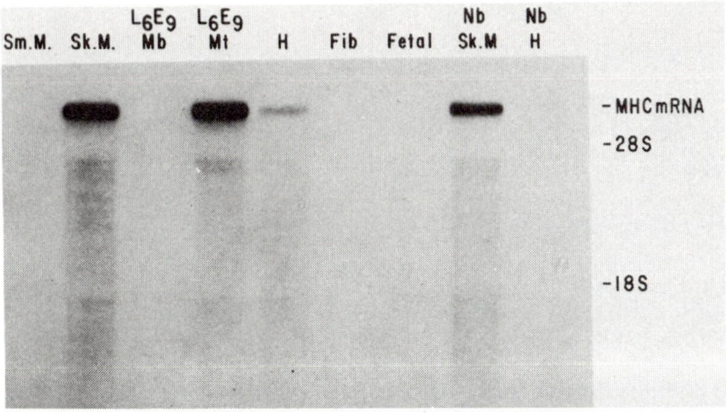

FIGURE 1
Detection of sarcomeric MHC mRNA sequences by plasmid pMHC-25. 10 μg of total RNA from myogenic and nonmyogenic sources was size-fractionated on 0.8% denaturing agarose gels, transferred to nitrocellulose filters, and hybridized to ^{32}P-labeled pMHC-25 DNA. (Sm.M.) Smooth muscle from uterine tissue; (Sk.M.) adult rat skeletal muscle; (L_6E_9 Mb) logarithmically growing L6E9 myoblasts; (L_6E_9 Mt) L6E9 myotubes; (H) adult rat cardiac muscle (ventricles); (Fib) rat primary fibroblasts; (Fetal) pooled rat skeletal muscles from day 16 of gestation period; (Nb Sk.M) newborn rat skeletal muscle (1 day old); (Nb H) newborn rat cardiac-muscle tissues (1 day old).

gests either that the embryonic skeletal MHC gene is interrupted by multiple intervening sequences containing these restriction sites or that distinct MHC genes have sequence homology with the pMHC-25 probe. To distinguish between these alternatives, pMHC-25 DNA was used to screen cDNA libraries from skeletal- and cardiac-muscle tissue and a rat genomic library. From these libraries, we have isolated several MHC cDNA clones different from pMHC-25 (Mahdavi et al. 1982; D. Hornig and B. Nadal-Ginard; M. Periasamy et al.; both in prep.), which comprise most of the MHC genomic fragments detected by pMHC-25. Further characterization of the genomic and cDNA MHC clones identified a minimum of seven sarcomeric MHC genes that are closely related but are distinguishable by their tissue or developmental specificity. These observations demonstrate that sarcomeric MHC mRNAs have at least partial sequence homology, which is not shared by the nonsarcomeric MHC mRNAs, within the sequences contained in the pMHC-25 probe.

The MHC gene sequences analyzed to date include those encoding four skeletal-muscle MHCs (one fetal, one embryonic, and two adult fast MHCs) and two adult ventricular cardiac MHCs (Mahdavi et al. 1982; D. Horning and B. Nadal-Ginard; M. Periasamy et al.; both in prep.). These clones, however, do not represent the complete sarcomeric MHC gene family,

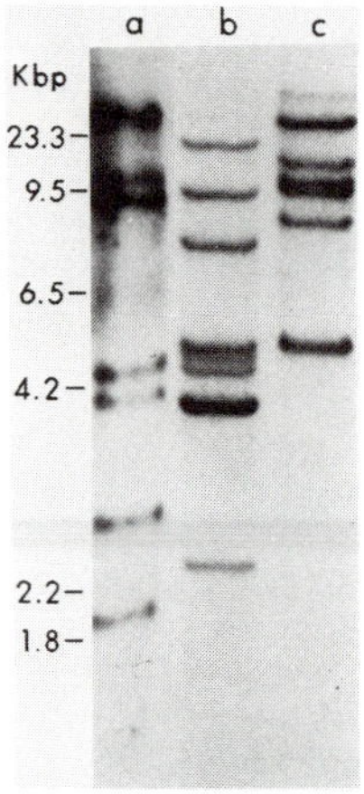

FIGURE 2
Hybridization of plasmid pMHC-25 to rat genomic DNA. 10 μg of genomic DNA, isolated from L6E9 myotubes, was digested with endonuclease *Eco*RI (*a*), *Hin*dIII (*b*), or *Bam*HI (*c*). The digested DNA was size-fractionated on agarose gels, transferred to nitrocellulose filters, and hybridized to ^{32}P-labeled pMHC-25 DNA.

because we have not identified the MHC sequences expressed in slow skeletal muscle and fetal ventricular and atrial tissues. This analysis, combined with the protein data (Lompre et al. 1981; Whalen et al., this volume), suggest that a minimum of 13 sarcomeric MHC genes are expressed in the rat.

SARCOMERIC mRNA SEQUENCES ARE HIGHLY CONSERVED. To elucidate the relationship between different MHCs expressed in the same species, we determined the complete nucleotide sequence of the adult cardiac-muscle and the embryonic, fetal, and adult skeletal-muscle MHC cDNA clones. This analysis identified gene sequences specific for these tissues and developmental stages. The sequences (800–2100 nucleotides long) code for a portion of the LMM and the 3′-untranslated region of six different MHC mRNAs (Mahdavi et al. 1982; D. Hornig and B. Nadal-Ginard; M. Periasamy et al.; both in prep.).

The sarcomeric MHC cDNA sequences analyzed are highly conserved at the nucleotide and amino acid level. The nucleotide sequence homology is close to 90% between two adult skeletal and cardiac MHCs. This high degree of sequence homology strongly suggests that the sarcomeric MHC genes have evolved by duplication of a common ancestral gene. However, this sequence conservation is not uniform throughout the length of the region coding for the LMMs. This pattern is evident in Figure 3, where nucleotide sequence comparison between adult skeletal and cardiac MHCs is represented graphically by computer (Mahdavi et al. 1982). Highly conserved regions up to a few hundred nucleotides long, illustrated in Figure 3 by continuous segments on the homology axis, are interspersed with divergent sequences of similar length, represented by interruptions in the diagonal line. This high degree of sequence homology among the members of the sarcomeric MHC gene family persumably accounts for the level of cross hybridization among them, as shown in Figures 1 and 2.

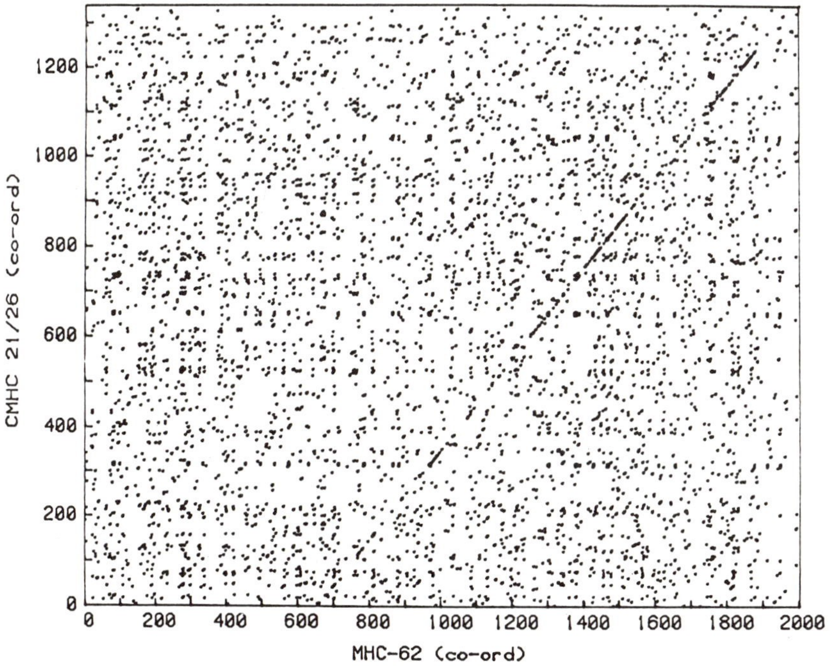

FIGURE 3
Computer graphic representation of nucleotide sequence homology between adult rat skeletal MHC (MHC-62) and cardiac MHC (CMHC 21/26) cDNA clones. The homology computer program compares sequentially 6 nucleotides in a row from the sequence displayed on the x axis to each position in the sequence on the y axis. Homology generates a dot at the corresponding coordinate. Continuous homology is represented by a diagonal segment covering the length of the entire corresponding sequence. CMHC 21/26 sequence covers nucleotides 630–2000 of the MHC-62 sequences. Homology of 5 nucleotides in a row is required for a dot placement (Mahdari et al. 1982).

In contrast to the high degree of sequence homology in large segments of the coding region, there is a continuous portion where complete divergence among the MHC sequences has been observed. This gene-specific region comprises the last few amino acids at the carboxyterminal end of the coding segment and the entire 3′-untranslated ends of the mRNAs (Fig. 4). The gene-specific sequences are preceded by a region of high homology that extends over 210 nucleotides. All of the 3′-untranslated sequences of the MHC mRNA analyzed are short, ranging from 55 to 141 nucleotides (Fig. 4), whereas MHC mRNA is over 7000 nucleotides long. The drastic differences in the nucleotide sequence and length of the 3′ ends of the MHC mRNAs confirm unambiguously their gene specificity.

From this analysis, it is evident that the degree of sequence conservation in sarcomeric MHCs makes it very difficult to characterize the correspond-

```
                    1950                          1955                    1960
a                    |                             |                       |
GTC AAG AGC CGT GAC ATT GGC GCC AAG GGC CTG AAT GAA GAG TAGATCTTGCTCTACCCAACCCTAAGGATGCCTGTGAAGCCCTGAGACCTGGAGCCTTTGAAACAGCACCTTCAGGCAGAAACACAATAAAGCAATTTTCCTTCAAGCC [A]37
Val Lys Ser Arg Asp Ile Gly Ala Lys Gly Leu Asn Glu Glu END

b
GTC AAG AGC CGT GAC ATT GGC GCC AAG CAG AAA ATC CAC GAT GAG GAA TAACCTGTCCAGCAGAAAGAGCCTCGCCGTTGCCATCCCACAATAAATACGAATGTTCGATTTGCCTGC [A]93
Val Lys Ser Arg Asp Ile Gly Ala Lys Gln Lys Ile His Asp Glu Glu END

c
GTG AAG AGC CGA GAG GTT CAC ACC AAA GTC ATA AGC GAA GAA TAGCTCAATTCCTTCTGTTGAAAGGTGACAGAAGAAATCACACAATGTGACGTTCTTTGTCACTGTCCTGTATATCAAGGAAATAAAGCTGCAGATAATTTTGC [A]90
Val Lys Ser Arg Glu Val His Thr Lys Val Ile Ser Glu Glu END

d
GTA AAG AGC CGC GAG GTT CAC ACT AAA GTC ATA AGT GAA GAG TAAGGCAGCTCTGATGCTGTAGAATGACCGAAGAAAGGCACAAAATGTGAAGCCTTTGGTCATGCCCCCATGTGATTCTATTTAATCCTATTGTAAGGAAATAAAGAGCCCAAGTTCTTGCAAGC [A]70
Val Lys Ser Arg Glu Val His Thr Lys Val Ile Ser Glu Glu END

e
GTG AAG AGC CGC GAG GTT CAC ACC AAA ATC AGT GCA GAG TAAACGCATCTTGAGGAGGCCGCCAAGTGGCTGAAGGAAAGGCACAGAATGTGCTGCCTTGGGTCGCTTGCTGGGTCGCTTGCCTCTCGTGTTTACTTTTCTCCCACTGCTGACTGAATAAAACCACAACTCATTGTAATT [A]
Val Lys Ser Arg Glu Val His Thr Lys Ile Ser Ala Glu END
f
GCT AAA ACC CGG GAC TTC ACC TCT AGC CGG ATG GTG GTC CAT GAA AGT GAA GAA TGAGCATGTCCTCTTGGTGAGGGGCAGAAGATATGCAGAATGTATGTTTTCCGTGGCCTCCTGACCACCTGCTTAATTTCCACGTAACCCCTTTCCACATGCAATAAATTTGCCTTGTTCAAG.
Ala Lys Thr Arg Asp Phe Thr Ser Ser Arg Met Val Val His Glu Ser Glu Glu END
```

FIGURE 4

Protein carboxyl termini and the 3′-untranslated regions of rat MHC cDNA clones. Numbers above the sequences represent amino acid positions in the protein. (*a*) Adult cardiac (ventricular); (*b*) adult cardiac (ventricular); (*c*) adult skeletal (fast); (*d*) adult skeletal (fast); (*e*) fetal skeletal; (*f*) embryonic skeletal.

ing genomic sequences. However, the availability of gene-specific fragments at the 3′ end of the MHC mRNAs does make it possible to identify the individual MHC genes, assign their tissue specificity, and study their developmental expression.

Identification and Structural Analysis of an MHC Expressed in L6E9 Myotubes

MHC GENE SEQUENCES IN GENOMIC CLONE 287 A-3 ARE EXPRESSED IN L6E9 MYOTUBES. The identification of a particular MHC gene, which is expressed in L6E9 cells, is a necessary preliminary step in the study of the molecular mechanisms responsible for its expression during myogenesis in this cell line. It has been shown that one of the genomic clones isolated with pMHC-25, clone 287 A-3, hybridizes preferentially to MHC mRNA from L6E9 myotubes (Wydro et al. 1982). To demonstrate that sequences in clone 287 A-3 are indeed transcribed in L6E9 myotubes, a gene-specific fragment from this clone was identified. This fragment, designated pA3I/6, corresponds to an intron as determined from DNA sequencing analysis and hybridizes to a single band in a Southern blot of *Eco*RI-digested DNA from L6E9 myotubes (M. Periasamy et al., in prep.). In a Northern blot of size-fractionated L6E9 myotube RNA, pA3I/6 hybridizes to a high-molecular-weight nuclear RNA species in the nuclear but not in the cytoplasmic RNA fraction (Fig. 5A). In contrast, pMHC-25 hybridizes to the mature MHC mRNA in the cytoplasmic and nuclear RNA fractions as well as the higher-molecular-weight band in the nuclear fraction, detected previously by pA3I/6 (Fig. 5B). This higher-molecular-weight nuclear RNA species most probably represents a nuclear precursor of the mature cytoplasmic MHC

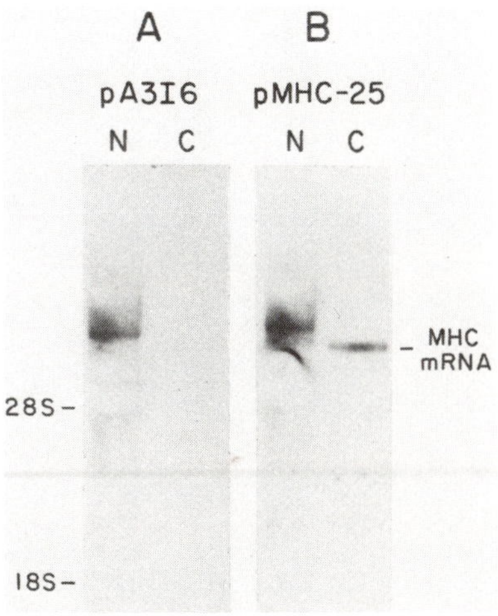

FIGURE 5
Identification of a putative MHC nuclear RNA precursor. 15 μg of total nuclear (N) RNA and 5 μg of total cytoplasmic (C) RNA from differentiated L6E9 myotubes were size-fractionated in denaturing gels and transferred to nitrocellulose filters. (*A*) RNAs were hybridized to ^{32}P-labeled pA3I/6 DNA. (*B*) RNAs were hybridized to ^{32}P-labeled pMHC-25 cDNA.

mRNA, although a precursor-product relationship between the two molecules has not been demonstrated directly. Recently, DNA sequence analysis has shown that clone 287 A-3 contains the entire pMHC-25 insert as well as the 3′-untranslated sequence of the corresponding mRNA (M. Periasamy et al., in prep.). These results strongly suggest that clone 287 A-3 contains sequences of the embryonic MHC gene expressed in L6E9 myotubes.

THE EMBRYONIC SARCOMERIC GENE EXPRESSED IN L6E9 MYOTUBES IS INTERRUPTED BY MANY INTERVENING SEQUENCES. The structural organization of the embryonic MHC gene expressed in the L6E9 cells was investigated by R-loop formation of genomic clone 287 A-3 to poly(A)$^+$ RNA from L6E9 myotubes. Analysis of a number of hybrid molecules, one of which is represented in Figure 6A, yields the consensus molecule shown in Figure 6B. The 18 introns in the MHC gene sequence in clone 287 A-3 are variable in length, ranging from 100 bp to 1000 bp. In contrast, the exons are uniformly short, with an average length of approximately 200 bp. The total exon

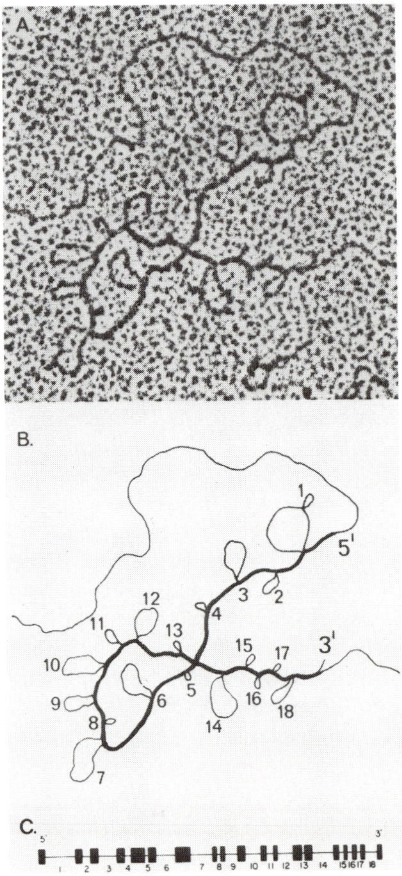

FIGURE 6
R-loop analysis of genomic clone 287 A-3 with RNA from L6E9 myotubes. R loops were formed between DNA from genomic clone 287 A-3 and poly(A)$^+$ from L6E9 myotubes. (*A*) Electron micrograph of a representative hybrid molecule; (*B*) line drawing of the hybrid molecule depicted in *A*; (*C*) summary diagram of exon (■) and intron (——) locations in clone 287 A-3.

length contained in clone 287 A-3 is approximately 4600 bp, thus accounting for 65% of the MHC mRNA. On the basis of protein data (Elzinga and Trus 1980), this region should code for most of the S-2 fragment, the entire LMM region, and the carboxyl terminus of the MHC molecule.

Regulation of Embryonic MHC Gene Expression during L6E9 Muscle Cell Differentiation

With the isolation of tissue and developmentally specific MHC cDNA and genomic clones, several important questions may be addressed regarding the mechanism of MHC gene expression and its regulation during muscle cell differentiation. Recombinant plasmid pMHC-25 (Medford et al. 1980), which is the cDNA analog to the 287 A-3 genomic clone, was used to assay specifically for embryonic MHC nuclear RNA and cytoplasmic mRNA sequences expressed during the course of myogenesis in the L6E9 rat skeletal-muscle cell line. Pulse-chase labeling of RNA in L6E9 cells and in vitro labeling of nuclear RNA nascent transcripts were employed to analyze critically the kinetic parameters that regulate MHC mRNA accumulation during myogenesis and to determine the relative contribution of transcriptional and posttranscriptional processes to the induction of MHC gene expression.

MHC PROTEIN SYNTHESIS IS DIRECTLY CORRELATED WITH THE CYTOPLASMIC ACCUMULATION OF MHC mRNA. Expression of the MHC gene during L6E9 myogenesis was studied at the level of MHC mRNA accumulation to define the relationship between MHC protein synthesis and the cytoplasmic content of MHC mRNA. MHC mRNA content was assayed by RNA (Northern) blot hybridization using the pMHC-25 clone (Medford et al. 1980) and by solution hybridization using a purified MHC cDNA (Benoff and Nadal-Ginard 1980). As shown in Figures 7 and 8, MHC mRNA is not present in growing myoblasts ($<$10 MHC mRNA molecules per cell) and remains undetectable until the third day of differentiation. At this time, MHC mRNA accumulates rapidly and attains a maximum level of 50,000 molecules per cell-equivalent by the seventh to eighth day of differentiation. Over 90% of the MHC mRNA molecules are found on EDTA-dissociable heavy polysomes throughout differentiation (R.M. Medford and B. Nadal-Ginard, unpubl.).

The accumulation of polysome-associated MHC mRNA closely parallels the induction of MHC protein synthesis, as well as the appearance of the classical biochemical and morphological features of L6E9 myogenesis (Nadal-Ginard 1978; Benoff and Nadal-Ginard 1980). This result strongly suggests that the rate of MHC protein synthesis is regulated by the cytoplasmic concentration of MHC mRNA. The same relationship between mRNA concentration and protein synthesis has been observed for all of the other protein components of the myofibrillar contractile apparatus in a variety of muscle cell systems using in vitro translational assays, cDNA

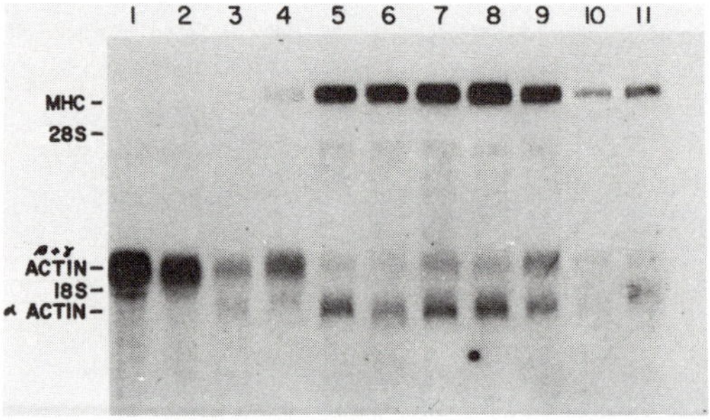

FIGURE 7
Cytoplasmic accumulation of MHC and actin mRNA during L6E9 myogenesis. Undifferentiated L6E9 myoblasts were allowed to undergo myogenic differentiation by plating in differentiation medium (Medford et al. 1980). Each day during the course of myogenesis, cells were harvested and total RNA was extracted. 10 μg of RNA was size-fractionated by electrophoresis on 1% agarose–3% formaldehyde gels, transferred to nitrocellulose filter paper, and hybridized to ^{32}P-labeled, nick-translated pMHC-25 and pAC269 (an actin probe) DNA. Lane numbers correspond to days after plating in differentiation medium.

probes, and recombinant cDNA plasmids (John et al. 1977; Strohman et al. 1977; Devlin and Emerson 1979; Affara et al. 1980; Moss and Schwartz 1981; Garfinkel et al. 1982, Hastings and Emerson 1982; L. Garfinkel and B. Nadal-Ginard, in prep.). The absence of detectable MHC mRNA until the third day of differentiation argues strongly against significant amounts of stored MHC mRNA residing in postpolysomal particles in myoblasts (Heywood et al. 1973; Bag and Sarkar 1976). These results obviate the need for translational control RNA (Heywood and Kennedy 1976) to regulate MHC gene expression.

MHC mRNA IS A HIGHLY STABLE MOLECULE IN BOTH MYOBLASTS AND MYOTUBES. The regulation of MHC gene expression, at the level of MHC mRNA accumulation, could be due to an increase in the rate of MHC mRNA synthesis, an increase in the stability of the MHC mRNA molecule, or a combination of both. As a first step in distinguishing between these possibilities, MHC mRNA stability was measured directly in both myoblasts and myotubes. As summarized in Table 1, MHC mRNA is a highly stable molecule in fully differentiated myotubes with an intrinsic half-life of 55–60 hours. MHC mRNA has the same stability when the molecule is first detected in myoblasts. Thus, the intrinsic stability of MHC mRNA does not change during myogenesis and does not contribute directly to the induction of MHC gene expression. The same experimental approach was used to show that the total cytoplasmic poly(A)$^+$ mRNA population in both myoblasts and myotubes consists of two classes of mRNA: (1) a stable mRNA class with a

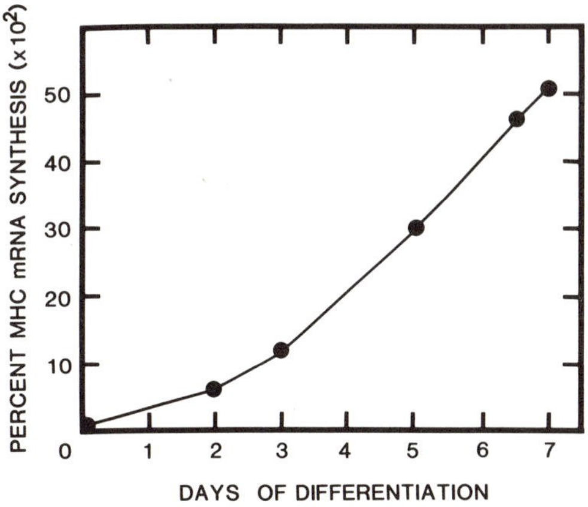

FIGURE 8
Induction of MHC mRNA synthesis during myogenesis. Undifferentiated L6E9 myoblasts were allowed to undergo myogenic differentiation by plating in differentiation medium. At various times after plating, a portion of the cell population was labeled in vivo with [^{3}H]uridine for 5 hr. Total cytoplasmic RNA was extracted, and poly(A)$^+$ mRNA was isolated by oligo(dT) affinity chromatography. Both total and poly(A)$^+$ mRNA were prepared for filter hybridization and hybridized to 8–10 μg of filter-bound pMHC-25 DNA. The percentage of MHC mRNA bound to the filter was calculated as described previously (Medford et al. 1980). MHC mRNA synthesis was expressed as the percentage of labeled poly(A)$^+$ mRNA during each stage of myogenesis. The percentage of measured MHC mRNA on day 0 (growing myoblasts) was 0.005%.

half-life comparable to that of MHC mRNA (50 hr) and (2) an unstable mRNA class with a half-life of 5 hr. Like MHC mRNA, the stability of both mRNA classes remains constant during myogenesis (Table 1) and therefore does not contribute directly to the fivefold increase in mRNA content per cell that has been observed consistently during muscle cell differentiation (Benoff and Nadal-Ginard 1980; Moss and Schwartz 1981).

MHC mRNA SYNTHESIS AND GENE TRANSCRIPTION ARE INDUCED DURING MYOGENESIS. In contrast to the constancy of MHC mRNA stability, MHC mRNA synthesis increases dramatically during myogenesis. The relative rate of MHC mRNA synthesis was assayed by pMHC-25 DNA-excess filter hybridization and was shown to increase from less than 0.007% of total poly(A)$^+$ mRNA synthesis in myoblasts to over 0.5% of synthesis in myotubes (Fig. 8). The kinetics of the increase in rate of synthesis correlate well with the accumulation of MHC mRNA during the same time period (Fig. 8). The absolute rate of MHC mRNA synthesis was calculated by directly measuring the specific activity of the UTP pool at short intervals during the labeling (R.M. Medford et al., in prep.). As summarized in Table

TABLE 1
MHC and Cytoplasmic Poly(A)$^+$ mRNA Population Dynamics during Myogenesis

	Myoblast	Myotube	Myotube/ myoblast
Whole Cell			
Cell generation time (hr)	16	∞	—
MHC			
Intrinsic half-life (hr)[a]	60	55	0.92
Effective half-life (hr)[b]	12.3	55	4.5
Rate of synthesis (m/n/m)[c]	<0.047	5.35	>140
Cytoplasmic concentration (m/n)[d]	<50	25,477	>505
MHC Transcription			
Total nuclear incorporation (cpm/nucleus)	10.0	5.0	0.5
Net pMHC-25 filter signal (ppm)[e]	<0.05	4.75	>95.0

[a]Cells were labeled for 5 hr with [^{3}H]uridine and effectively chased with excess nonradioactive uridine and cytidine for up to 4 days. At various times during the chase, either ^{3}H-labeled total cytoplasmic RNA or ^{3}H-labeled poly(A)$^+$ mRNA was hybridized to excess filter-bound pMHC-25 DNA. Because rRNA decay in both L6E9 myoblasts and myotubes is not detectable (Krauter et al. 1979), the intrinsic stability of MHC mRNA and poly(A)$^+$ mRNA was measured as the changing proportion of ^{3}H-labeled MHC and poly(A)$^+$ mRNA relative to ^{3}H-labeled rRNA.

[b]The effective half-life (t_e) is calculated by

$$t_e = \ln 2/(K_i + K_g)$$

where K_i and K_g are related to the intrinsic half-life (t_i) and cell generation time (t_g), respectively, by

$$K = \ln 2/t$$

1, the absolute rate of MHC mRNA synthesis increases at least 140-fold, from less than 0.05 mRNA molecules per myoblast nucleus per minute to 5.35 molecules per myotube nucleus per minute. The relative rate of MHC gene transcription was also measured by incubating isolated nuclei with [^{32}P]uridine triphosphate (UTP) for 20 minutes (R.M. Medford et al., in prep.). The rate of nascent MHC nuclear RNA synthesis was assayed by pMHC-25 DNA filter hybridization and shown to be 50–100-fold greater in myotube nuclei than in myoblast nuclei (Table 1). From this data, it is clear that an induction of MHC gene transcription plays a fundamental role in the accumulation of MHC mRNA during L6E9 myogenesis.

To put these changes of MHC gene expression into a metabolic context, the kinetic parameters of the poly(A)$^+$ mRNA population were measured directly. Overall, poly(A)$^+$ mRNA synthesis increases approximately twofold during myogenesis (Table 1). Although overall mRNA synthesis is induced during myogenesis, the ratio of unstable to stable mRNA synthesis

TABLE 1
(Continued)

	Myoblast	*Myotube*	*Myotube/ myoblast*
	Cytoplasmic Poly(A)$^+$ mRNA		
Unstable mRNA			
intrinsic half-life (hr)[a]	5	5	1
effective half-life (hr)[b]	3.8	5	1.3
rate of synthesis (m/n/m)[c]	440	1,050	2.4
cytoplasmic concentration (m/n)[d]	150,000	460,000	3.1
Stable mRNA			
intrinsic half-life (hr)[a]	50	50	1
effective half-life (hr)[b]	12.1	50	4.2
rate of synthesis (m/n/m)[c]	180	340	1.9
cytoplasmic concentration (m/n)[d]	190,000	1,460,000	7.7

[c]The rate of synthesis is expressed as mRNA molecules per nucleus-equivalent per minute (m/n/m). MHC mRNA synthesis is corrected for a 7100-nucleotide molecule. Otherwise, the rate is expressed for an average-size 2000-nucleotide mRNA. The rate of synthesis (K_s) was computed as the rate of accumulation, M_t, of MHC mRNA molecules during a 5-hr labeling (t) corrected for MHC mRNA turnover ($K_i + K_g$), such that

$$M_t = \frac{K_s}{K_i + K_g}\left[1 - e^{-(K_i + K_g)^t}\right]$$

[d]The cytoplasmic concentration (M) is expressed in mRNA molecules per nucleus-equivalent (m/n), corrected for the appropriate size of the mRNA (see c), and calculated by

$$M = \frac{K_s}{K_i + K_g}$$

[e]Filter signals were fivefold to tenfold greater than background. Signals <0.075 ppm were not considered significantly greater than background.

remains unchanged. The accumulation of MHC mRNA far exceeds the overall accumulation of poly(A)$^+$ mRNA during myogenesis, due to the significantly greater increase in the rate of MHC mRNA synthesis (~140-fold) compared with poly(A)$^+$ mRNA synthesis (~2-fold).

WITHDRAWAL FROM THE CELL CYCLE MODULATES MHC GENE EXPRESSION DURING MYOGENESIS. Withdrawal from the cell cycle is a fundamental feature of terminal myogenic differentiation and plays an essential role in MHC gene expression. In rapidly dividing myoblast cells, each mitosis depletes cytoplasmic mRNA content by dilution. Therefore, the *effective* half-life of an mRNA within a cell is determined by two factors: (1) the intrinsic stability of the mRNA and (2) the cell generation time. The effective half-life of an mRNA within a cell cannot exceed the time interval between mitoses, i.e.,

the cell generation time. The effective half-life of the mRNA and the cell generation time may be used to calculate the mRNA cytoplasmic content by

$$M = \frac{K_s}{K_i + K_g}$$

where M is the mRNA content (molecules per cell), K_s is the rate of mRNA synthesis (molecules per cell per minute), K_i is the intrinsic rate of mRNA decay (minutes^{-1}), and K_g is the cell growth rate (minutes^{-1}) (see footnote to Table 1 for derivation of K_i and K_g) (Tobin 1979).

From the equation above, it is clear that the cytoplasmic content of highly stable mRNA ($K_i << K_g$) will be affected preferentially by the cell generation time; in comparison, unstable mRNA ($K_i >> K_g$) will be unaffected. Differential regulation of mRNA accumulation by the cell cycle undergoes a dramatic change when the myoblast cell withdraws from the cell cycle ($K_g = 0$). The net result of this change in growth rate is to increase significantly, by fourfold to fivefold, the effective stability of highly stable mRNA, such as MHC mRNA and, hence, its cytoplasmic concentration.

During myogenesis, cytoplasmic MHC mRNA content is induced at least 500-fold, but MHC mRNA synthesis is induced only 140-fold. Transcriptional induction of the MHC gene appears to account for only 25% of MHC mRNA accumulation. Although the intrinsic stability of MHC mRNA does not change during myogenesis, the effective half-life of MHC mRNA does increase over fourfold from 12 hours in the myoblast to over 50 hours in the myotube (Table 1). This mechanism of preferential accumulation of stable mRNA due to withdrawal from the cell cycle (4–5-fold), combined with the transcriptional induction of the MHC gene (~100-fold), accounts completely for the 500-fold accumulation of cytoplasmic MHC mRNA during L6E9 myogenesis. These results demonstrate clearly that both transcriptional and cell-cycle-mediated regulation of MHC gene expression are necessary—but either one alone is not sufficient—to produce the differentiated muscle cell phenotype.

Quantitative changes in the total poly(A)$^+$ mRNA population that are detected during myogenesis are also explained by cell-cycle-mediated regulation of gene expression. Withdrawal from the cell cycle is responsible for an increase of threefold to fourfold in mRNA content in myotubes. This mechanism accounts for a change in mRNA composition such that stable mRNA increases from 56% of total mRNA in myoblasts to over 76% in myotubes (Table 1). Changes in the relative abundance of specific mRNA classes observed by RNA-cDNA hybridization analysis (Paterson and Bishop 1977; Leibovitch et al. 1979) may also be explained by this concept of changing effective half-life during terminal differentiation.

TRANSCRIPTIONAL INDUCTION OF THE EMBRYONIC MHC GENE EXPRESSED IN L6E9 CELLS IS CORRELATED WITH CHANGES IN CHROMATIN CONFORMATION. The molecular mechanisms responsible for the transcriptional regulation of

eukaryotic genes are poorly understood. However, there is undoubtedly a relationship between transcriptional regulation and chromatin structure (Weisbrod 1982). Actively transcribed genes have been shown to be preferentially sensitive to digestion by pancreatic DNase I (Weintraub and Groudine 1976; Bellard et al. 1980; Weintraub et al. 1982), which suggests a structural difference at the chromatin level between active and inactive genes. DNase-I sensitivity has also been shown to precede the active transcription of certain genes (Groudine and Weintraub 1981). Thus, this technique provides a potential assay for factors determining the inheritable, but repressed, transcriptional potential of specific genes in precursor cells committed to a particular cell lineage.

Changes in chromatin packaging necessary for the expression of the embryonic MHC gene expressed in L6E9 cells may have already occurred in the cycling myoblasts. Alternatively, these changes may be induced in vitro when the cells are stimulated to differentiate. To distinguish between these two possibilities, we have used recombinant cDNA and genomic clones to determine the DNase-I sensitivity of the embryonic MHC gene (represented by genomic clone 287 A-3 in myoblasts and myotubes). As shown in Figure 9, we have detected a change in the DNase-I sensitivity of the embryonic MHC gene from an insensitive chromatin conformation in myoblasts to a sensitive structure in fully induced myotubes. This result corroborates well with the biochemical data presented above, indicating that in myoblasts the MHC gene is transcriptionally inactive and becomes induced during differentiation. This transition from a "closed" to an "open" chromatin structure accompanying transcriptional induction has been observed in several other systems (Wu et al. 1979; Bellard et al. 1980; Groudine and Weintraub 1981) and for α-actin and MLC2 genes in L6 myoblasts (Carmon et al. 1982). However, during myogenesis, this transition can be induced in vitro. Thus, it should be possible to dissect the earliest events leading to muscle-specific gene induction during differentiation.

Commitment and Induction of Muscle-specific Contractile Protein Genes in Myoblasts Mutants

Many of the biochemical parameters involved in the induction and accumulation of MHC mRNA have been elucidated and described above. However, little is known about the cellular events leading to the induction of this gene. In myogenesis, cell growth and differentiation are mutually exclusive phenomena. It has been postulated that the transition from proliferating undifferentiated myoblasts to postmitotic differentiated myotubes requires a unique mitotic event, designated "quantal mitosis," immediately prior to which changes in the genetic program, presumably needed for differentiation, are produced (Dienstman and Holtzer 1977). However, in L6E9 cells (Nadal-Ginard 1978), as well as in other myogenic systems (Buckley and Konigsberg 1974; Doehring and Fischman 1974), DNA syn-

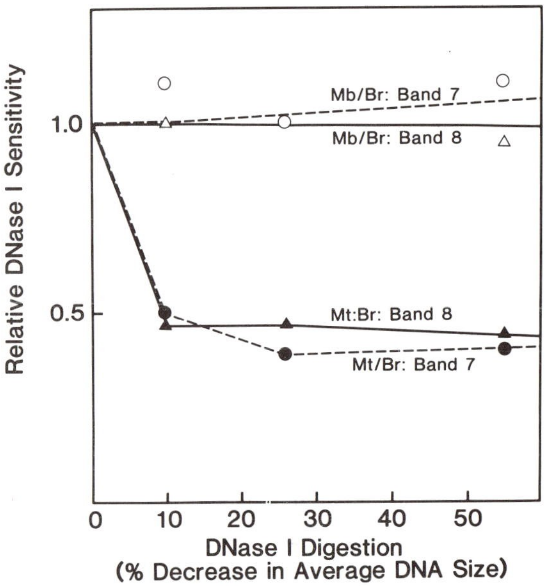

FIGURE 9

Relative DNase-I sensitivity of the embryonic MHC gene expressed in L6E9 cells. The percentage of decrease in average size of bulk DNA after DNase-I digestion was plotted against the ratio of the intensity of two *Eco*RI fragments (bands 7 [Δ] and 8 [o]) of the embryonic MHC gene represented by clone 287 A-3. The intensity of these bands in myoblast (Mb) and myotube (Mt) DNAs, as determined by Southern blot hybridization, was compared to their intensity in rat brain (Br) DNA digested to the same extent.

thesis is not required to switch from the growing to the differentiated state. Furthermore, it has been shown that L6E9 cells "commit" to the differentiation pathway prior to and not as a consequence of fusion (Nadal-Ginard 1978). In this system, commitment has been defined operationally as permanent withdrawal from the cell cycle with the concomitant loss of proliferative capacity. The kinetics of the loss of proliferative capacity is suggestive of a stochastic process, the probability of which is inversely correlated with the concentration of growth factors in the medium. Following commitment, the L6E9 cells can differentiate in the absence of inducing stimuli from the environment.

The close temporal correlation between cell fusion and the initiation of muscle-specific protein synthesis has led to the suggestion that fusion "triggers" the expression of these genes (Yaffe and Dym 1973; Turner et al. 1976). However, mononucleated postmitotic somitic myoblasts accumulate

myosin prior to fusion (Holtzer et al. 1972). Studies in which fusion has been inhibited by EGTA or phospholipase C have shown that myosin synthesis and increased levels of creatine phosphokinase activity are induced in these mononucleated cells to levels that are similar to those detected in fused myotubes (Emerson and Beckner 1975; Morris et al. 1976; Moss and Strohman 1976). Thick and thin filaments have been observed in the fusion-inhibited cultures (Emerson and Beckner 1975; Trotter and Nameroff 1976). Moreover, EGTA-treated cells accumulate muscle-specific mRNAs to levels observed in untreated cells (B. Nadal-Ginard et al., in prep.).

3b-2 CELLS ARE TEMPERATURE SENSITIVE FOR MORPHOLOGICAL DIFFERENTIATION. The question of whether muscle-specific gene induction occurs concomitant with, or as a consequence of, commitment (permanent withdrawal from the cell cycle) can be addressed by using cell lines that are temperature sensitive for commitment. Using such cell lines, it should be possible to establish cause-effect correlations between growth and muscle-specific gene expression.

Five different mutant cell lines were isolated, each of which was temperature sensitive for fusion and commitment while expressing a wild-type phenotype at 33°C. The isolation procedure included mutagenesis of the parental L6E9 cells (Nadal-Ginard 1978) with ethylmethanesulfonate (EMS) and selection protocols against commitment at 40°C. One of these mutants, designated 3b-2, has been studied at cellular and biochemical levels. Under standard differentiation conditions at 33°C, 3b-2 cells fuse to form myotubes to the same extent as L6E9 cells. In contrast, although L6E9 cells differentiate normally, 3b-2 cells remain mononucleated and do not form myotubes at 40°C. The inability of 3b-2 cells to fuse at 40°C is not affected by changes in serum conditions.

3b-2 CELLS DIFFERENTIATE BIOCHEMICALLY IN THE ABSENCE OF MORPHOLOGICAL DIFFERENTIATION. The ability of 3b-2 cells to express the differentiated biochemical phenotype at the nonpermissive temperature (40°C) for morphological differentiation was analyzed. The cytoplasmic mRNA content was determined for several muscle-specific genes coding for components of thick and thin filaments (MHC, MLC 2, actin), as well as a regulatory protein (troponin T). Despite the dramatic difference in morphological phenotype at the two temperatures, 3b-2 cells accumulate similar amounts of muscle-specific mRNAs (Fig. 10). These levels are similar to those observed in L6E9 cells at both temperatures. Moreover, their rates of synthesis are comparable (Fig. 11). These results demonstrate clearly that 3b-2 cells are able to differentiate biochemically, as defined by the synthesis and accumulation of muscle-specific mRNAs, in the absence of morphological differentiation.

The muscle-specific mRNAs accumulated in 3b-2 cells at 40°C are functional, as determined by their location on polysomes. Moreover, these cells accumulate the corresponding proteins to levels similar to those observed in

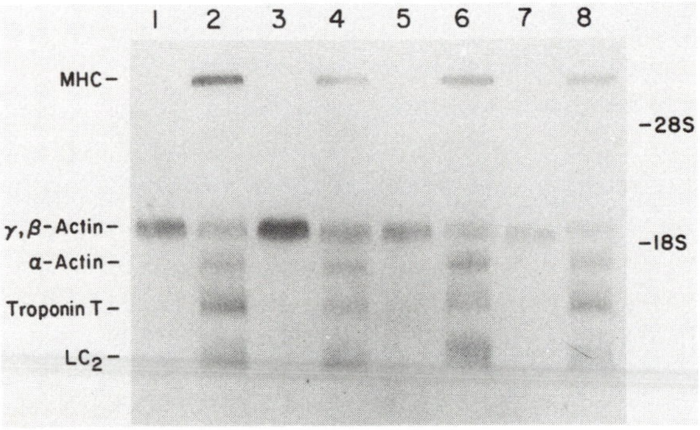

FIGURE 10
3b-2 cells accumulate normal levels of muscle-specific mRNAs in the absence of morphological differentiation. 10 μg of total cytoplasmic RNAs were extracted from 3b-2 and L6E9 cells, size-fractionated in denaturing gels, transferred to nitrocellulose filters, and hybridized to ^{32}P-labeled muscle-specific cDNA clones. (*1*) Undifferentiated 3b-2 cells at 33°C; (*2*) differentiated 3b-2 cells at 33°C; (*3*) undifferentiated 3b-2 cells at 40°C; (*4*) differentiated 3b-2 cells at 40°C; (*5*) undifferentiated L6E9 cells at 33°C; (*6*) differentiated L6E9 cells at 33°C; (*7*) undifferentiated L6E9 cells at 40°C; (*8*) differentiated L6E9 cells at 40°C.

L6E9 cells at both temperatures. However, 3b-2 cells do not assemble myofibrils at 40°C, whereas they do at 33°C. The mechanisms responsible for the lack of myofibril formation at 40°C in 3b-2 cells is not known.

3b-2 CELLS DIFFERENTIATE BIOCHEMICALLY IN THE ABSENCE OF COMMITMENT. The selection scheme used to isolate the 3b-2 cells was designed to select for mutants temperature sensitive for commitment. Cloning efficiency studies have shown that 3b-2 cells do not lose their proliferative capacity at 40°C when plated under differentiation conditions, whereas they do at 33°C. In contrast, under similar conditions, L6E9 cells lose their proliferative capacity (commit) at both temperatures. [^{3}H]Thymidine pulse-labeling and flow microfluorometry analyses show that biochemically differentiated 3b-2 cells do not commit at 40°C and can be induced to reenter the cell cycle upon growth stimulation. At 40°C, the noncommitted but biochemically differentiated 3b-2 cells are reversibly arrested in the G_1 phase of the cell cycle.

The phenotype of the 3b-2 cells demonstrates clearly that muscle-specific gene expression can be induced in the absence of permanent withdrawal from the cell cycle. This observation strongly suggests that biochemical differentiation and commitment, although correlated temporally in the normal

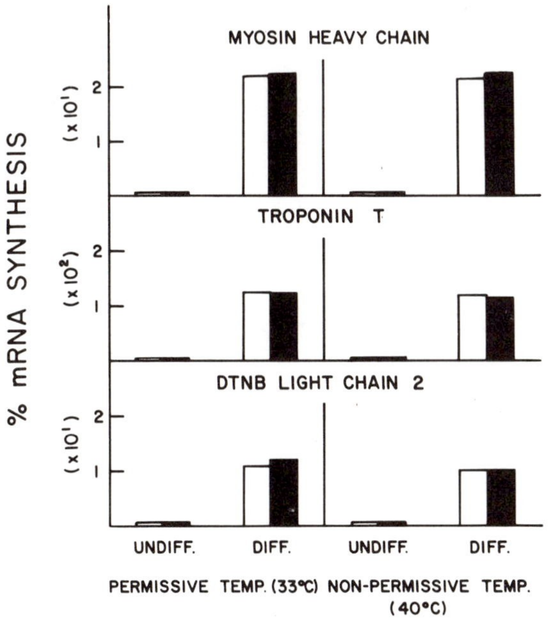

FIGURE 11
3b-2 cells synthesize muscle-specific mRNAs at normal rates, in the absence of morphological differentiation. Undifferentiated and differentiated cultures of 3b-2 and L6E9 cells were labeled with [^{3}H]uridine for 5 hr. Total cytoplasmic RNAs were extracted and processed for filter hybridization with filter-bound plasmid DNA. The plasmid DNAs used were muscle-specific cDNA clones. The percentage of muscle-specific cytoplasmic RNA bound to each type of DNA filter was calculated. The synthesis of each type of muscle-specific mRNA was expressed as a percentage of labeled poly(A)$^+$ mRNA, which is determined by oligo(dT) affinity chromatography. The sources of the labeled RNAs are L6E9 cells at 33°C and 40°C (■) and 3b-2 cells at 33°C and 40°C (□).

differentiation pathway, are two independent events with no cause-effect relationship between them.

IN THE ABSENCE OF COMMITMENT, THE BIOCHEMICAL DIFFERENTIATED PHENOTYPE IS DEINDUCED UPON GROWTH STIMULATION. Myogenesis is a good example of terminal differentiation. Once muscle-specific genes have been activated, they are constitutively expressed throughout the life span of the muscle cell. Thus, it was of interest to determine the fate of the biochemical-

ly differentiated phenotype of 3b-2 cells at 40°C after they have been induced to reenter the cell cycle. It is possible that upon reentering the cell cycle, the biochemically differentiated phenotype could remain expressed, thereby producing a cell line of constitutively differentiated myoblasts. Alternatively, the differentiated phenotype could be deinduced, thus providing an example of dedifferentiation. For this reason, the cytoplasmic levels and the rates of synthesis of muscle-specific mRNAs were measured in the biochemically differentiated 3b-2 cells that had been induced to reenter the cell cycle.

Nine hours after biochemically differentiated 3b-2 cells were transferred to growth-stimulating medium, the cytoplasmic levels of MHC, α-actin, troponin T, and MLC 2 mRNAs were decreased (Fig. 12). These mRNAs were no longer detectable 24 hours and 48 hours after growth stimulation. Under similar conditions, 3b-2 cells differentiated at 33°C and L6E9 cells differentiated at either 33°C or 40°C did not exhibit any detectable changes in cytoplasmic level of these muscle-specific mRNAs. The biochemically differentiated 3b-2 cells at 40°C did not initiate DNA synthesis until after a minimum of 15 hours in growth-stimulating medium. It is therefore unlikely that dilution resulting from cell division plays a role in the decrease of the cytoplasmic levels of the muscle-specific mRNAs. Neither is this decrease due to nonspecific mRNA degradation, because the level of constitutively expressed "housekeeping" mRNAs remains constant in these cells.

Concomitant with growth stimulation of 3b-2 cells (by changing to "growth medium" while maintaining the temperature at 40°C) is a rapid cessation of muscle-specific mRNA synthesis (Fig. 13). Under similar condi-

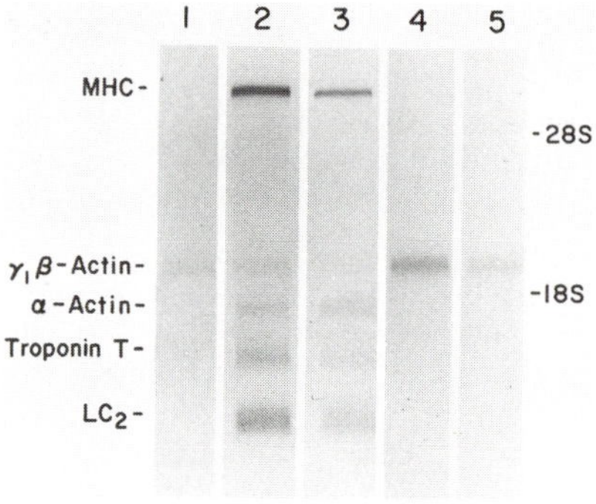

FIGURE 12
The muscle-specific mRNAs in the biochemically differentiated 3b-2 cells at 40°C are repressed upon growth stimulation. Total cytoplasmic RNAs were extracted from 3b-2 cells at 40°C, which had been switched to growth medium for different lengths of time. The RNAs were size-fractionated in denaturing gels, transferred to nitrocellulose filters, and hybridized to ^{32}P-labeled muscle-specific cDNA. The sources of RNAs are undifferentiated 3b-2 cells (*1*); differentiated 3b-2 cells (*2*); differentiated 3b-2 cells that have been switched to growth medium for 9 hr (*3*); differentiated 3b-2 cells that have been switched to growth medium for 24 hr (*4*); differentiated 3b-2 cells that have been switched to growth medium for 48 hr (*5*).

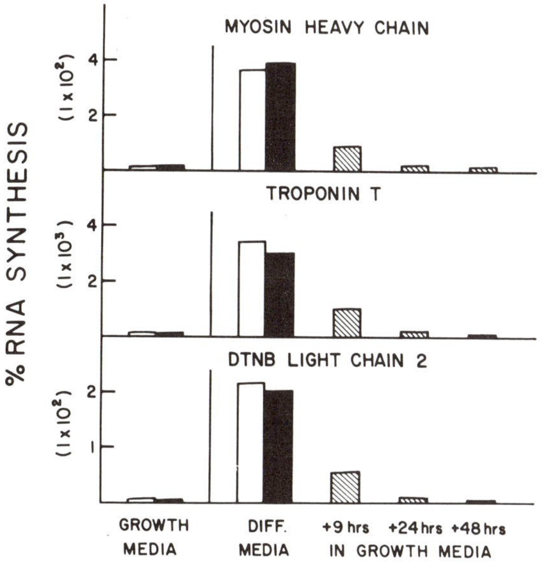

FIGURE 13
Muscle-specific mRNA synthesis is repressed in biochemically differentiated 3b-2 cells at 40°C upon growth stimulation. Undifferentiated and differentiated cultures of 3b-2 cells at 33°C and 40°C were labeled with [^{3}H]uridine for 5 hr. Differentiated cultures of 3b-2 cells that had been switched to growth medium for different lengths of time were also labeled. Total cytoplasmic RNAs were extracted and processed for filter hybridization with filter-bound plasmid DNA. The plasmid DNAs used were muscle-specific cDNA clones. The synthesis of each type of muscle-specific mRNA was expressed as a percentage of labeled total RNA. The sources of the labeled RNAs are 3b-2 cells at 33°C (□); 3b-2 cells at 40°C (■); biochemically differentiated 3b-2 cells at 40°C that were switched to growth medium for indicated length of time (▨).

tions, there are no detectable changes in the rates of synthesis of the muscle-specific mRNAs in the 3b-2 cells differentiated at 33°C or in the L6E9 cells differentiated either at 33°C or 40°C. However, the cessation of synthesis of these muscle-specific mRNAs alone cannot account for the dramatic decrease of their cytoplasmic levels upon growth stimulation of biochemically differentiated 3b-2 cells at 40°C. mRNA metabolic studies have shown that the half-lives of MHC, MLC 2, and troponin T mRNAs are greater than 50 hours (L. Garfinkel et al.; R.M. Medford et al.; both in

prep.). The highly stable nature of these mRNAs would dictate a significantly longer time course of decay. The observed rapid rate of decay suggests a preferential destabilization of these mRNAs. This preferential destabilization, combined with the cessation of the synthesis of the muscle-specific mRNAs, appears to be responsible for the rapid disappearance of these mRNAs from the cytoplasm.

DEDIFFERENTIATED 3b-2 CELLS HAVE NOT LOST THEIR CAPACITY TO REDIFFERENTIATE. Dedifferentiated 3b-2 cells have the same biochemical and morphological characteristics as undifferentiated myoblasts. When plated in "differentiation medium," these cells undergo morphological and biochemical differentiation at 33°C. At 40°C, the cells redifferentiate biochemically in the absence of commitment and morphological differentiation. The kinetics and extent of biochemical differentiation observed in the reinduced 3b-2 cells are similar to those that have not been induced previously. The same cell populations can undergo repeated biochemical induction and repression. These results demonstrate that in the absence of irreversible withdrawal from the cell cycle (commitment), the differentiated phenotype is reversible and can be reinduced upon withdrawal of mitogenic factors from the media. Thus, in the absence of commitment to terminal differentiation, muscle-specific genes can be reversibly induced, similar to other inducible genes in a variety of systems (e.g., hormally inducible genes).

The phenotype of this conditional mutant cell line demonstrates clearly that the cellular and biochemical events that characterize the differentiated muscle phenotype can be dissociated. The muscle-specific genes can be expressed in the absence of permanent withdrawal from the cell cycle (commitment). Consequently, biochemical differentiation occurs concomitantly and not as a consequence of commitment. However, in the absence of commitment, the expression of the muscle-specific genes is deinducible upon growth stimulation. The production of the terminally differentiated muscle phenotype therefore requires the coupling of the biochemical differentiation program and permanent withdrawal from the cell cycle to lock the cell in the terminally differentiated state. This mutant cell line, combined with the availability of recombinant DNA clones containing structural gene sequences coding for muscle-specific and housekeeping proteins, provides a model system for the study of the molecular and cellular mechanisms involved in muscle-specific gene induction and repression during myogenesis.

SUMMARY

Sarcomeric MHC is encoded by a highly conserved multigene family. The structure and expression of one member of this family, expressed in the L6E9 rat myogenic cell line, has been analyzed. This gene is interrupted by

multiple intervening sequences. The exons contain highly conserved sequences interspersed with gene-specific sequences. Transcriptional induction of this gene during L6E9 cell myogenesis combined with changes in cell-cycle parameters can account for the accumulation of MHC mRNA in these cells. Mutant cell lines temperature sensitive for commitment and fusion demonstrate that muscle-specific genes, including MHC, can be induced in the absence of permanent withdrawal from the cell cycle. However, under these conditions, the myofibrillar genes are deinducible upon growth stimulation. These results demonstrate that transcriptional induction of myofibrillar genes and permanent withdrawal from the cell cycle are independent events, but both are required—and neither alone is sufficient—to produce a stable muscle phenotype.

ACKNOWLEDGMENTS

The research reported in this paper was supported by grants from the National Institutes of Health, the American Cancer Society, the Muscular Dystrophy Association of America, and the New York Heart Association.

REFERENCES

Affara, N.A., B. Rober, M. Jacquet, M.E. Buckingham, and F. Gros. 1980. Changes in gene expression during myogenic differentiation. l. Regulation of mRNA sequences expressed during myotube formation. *J. Mol. Biol.* **140:** 441.

Bag, J. and S. Sarkar. 1976. Studies on a non-polysomal ribonucleoprotein coding for myosin heavy chain from chick embryonic muscle. *J. Biol. Chem.* **251:** 7600.

Bellard, M., M.T. Kus, N. Dretzedn, and P. Chambon. 1980. Differential nuclease sensitivity of the ovalbumin and beta globin chromatin regions in erythrocytes and oviduct cells of laying hens. *Nucleic Acids Res.* **8:** 2737.

Benoff, S. and B. Nadal-Ginard. 1980. Transient induction of poly(A)-short myosin heavy chain messenger RNA during terminal differentiation of L_6E_9 myoblasts. *J. Mol. Biol.* **140:** 283.

Buckley, P.A. and I.R. Konigsberg. 1974. Myogenic fusion and the duration of the mitotic gap. *Dev. Biol.* **37:** 193.

Carmon, Y., H. Czosnek, U. Nudel, M. Shani, and D. Yaffe. 1982. DNase I sensitivity of genes expressed during myogenesis. *Nucleic Acids Res.* **10:** 3085.

Chi, J.C.H., H. Rubinstein, K. Strahs, and H. Holtzer. 1975. Synthesis of myosin heavy chain and light chain in muscle cultures. *J. Cell Biol.* **67:** 523.

Coleman, J.R. and A.W. Coleman. 1968. Muscle differentiation and macromolecular synthesis. *J. Cell Physiol. (Suppl.)* **72:** 19.

Devlin, R.B. and C.P. Emerson, Jr. 1979. Co-ordinate accumulation of contractile protein mRNA during myoblast differentiation. *Dev. Biol.* **69:** 202.

Dienstman, S.R. and H. Holtzer. 1977. Skeletal myogenesis: Control of proliferation in a normal cell lineage. *Exp. Cell Res.* **107:** 355.

Doehring, J.L. and D.A. Fischman. 1974. The *in vitro* cell fusion of embryonic chick muscle without DNA synthesis. *Dev. Biol.* **36:** 225.

Elzinga, M. and B. Trus. 1980. Sequence and proposed structure of a 17,000 dalton fragment of myosin. In *Methods in peptide and protein sequence analysis* (ed. C. Birr), p. 213. Elsevier, Amsterdam.

Emerson, C.P. and S.K. Beckner. 1975. Activation of myosin synthesis in fusing and mononucleated myoblasts. *J. Mol. Biol.* **93:** 431.

Engel, J.N., P.W. Gunning, and L. Kedes. 1981. Isolation and characterization of human actin genes. *Proc. Natl. Acad. Sci.* **78:** 4674.

Firtel, R.A. 1981. Multigene families encoding actin and tubulin. *Cell* **24:** 6.

Garfinkel, L., M. Periasamy, and B. Nadal-Ginard. 1982. Cloning, identification and characterization of α-actin, myosin light chain 1, 2 and 3, α-tropomyosin, and troponin C and T. *J. Biol. Chem.* **257:** 11078.

Groudine, M. and H. Weintraub. 1981. Activation of globin genes during chicken development. *Cell* **24:** 393.

Hastings, K.E. and C.P. Emerson. 1982. cDNA clone analysis of six co-regulated mRNAs encoding skeletal muscle contractile proteins. *Proc. Natl. Acad. Sci.* **79:** 1553.

Heywood, S.M. and D.S. Kennedy. 1976. Purification of myosin translational control RNA and its interaction with myosin messenger RNA. *Biochemistry* **15:** 3314.

Heywood, S.M., D.S. Kennedy, and A.J. Bester. 1973. Stored myosin messenger in embryonic chick muscle. *FEBS Lett.* **53:** 69.

Hoh, J.F.Y. and G.P.S. Yeoh. 1979. Rabbit skeletal myosin isoenzymes from fetal, fast-twitch and slow-twitch muscles. *Nature.* **280:** 321.

Holtzer, H., H. Weintraub, R. Mayne, and B. Mochan. 1972. The cell cycle, cell lineages, and cell differentiation. *Curr. Top. Dev. Biol.* **7:** 229.

Huszar, G. 1972. Developmental changes of the primary structure and histidine methylation in rabbit skeletal muscle myosin. *Nat. New Biol.* **240:** 260.

John, H.A., M. Patrinou-Georgoulas, and K.W. Jones. 1977. Detection of myosin heavy chain mRNA during myogenesis in tissue culture by in vitro and in situ hybridization. *Cell* **12:** 501.

Krauter, K.S., R. Soeiro, and B. Nadal-Ginard. 1979. Transcriptional regulation of ribosomal RNA accumulation during L_6E_9 myoblast differentiation. *J. Mol. Biol.* **134:** 727.

Leibovitch, M.-P., S.-A. Leibovitch, J. Harel, and J. Kruh. 1979. Changes in the frequency and diversity of messenger RNA population in the course of myogenic differentiation. *Eur. J. Biochem.* **97:** 321.

Lompre, A.M., J.G. Mercadier, C. Wisnewsky, P. Bouveret, D. Pantaloni, A. D'Albis, and K. Schwartz. 1981. Species and age-dependent changes in the relative amounts of cardiac myosin isoenzymes in mammals. *Dev. Biol.* **84:** 286.

Lowey, S., P.A. Benfield, L. Silberstein, and L.M. Lang. 1979. Distribution of light chains in fast skeletal myosin. *Nature* **282:** 522.

Mahdavi, V., M. Periasamy, and B. Nadal-Ginard. 1982. Molecular characterization of two myosin heavy chain genes expressed in the adult heart. *Nature* **297:** 659.

Masaki, T. and C. Yoshizaki, 1974. Differentiation of myosin in chick embryo. *J. Biochem.* **76:** 123.

Medford, R.M., R.M. Wydro, H.T. Nguyen, and B. Nadal-Ginard. 1980. Cytoplasmic processing of myosin heavy chain messenger RNA: Evidence provided by using a recombinant DNA plasmid. *Proc. Natl. Acad. Sci.* **77:** 5749.

Merlie, J.P., M.E. Buckingham, and R.G. Whalen. 1977. Molecular aspects of myogenesis. *Cur. Top. Dev. Biol.* **77:** 61.

Morris, G.E., M. Piper, and P. Cole. 1976. Differential effects of calcium ion concentration on cell fusion, cell division and creatine kinase activity in muscle cell cultures. *Exp. Cell Res.* **99:** 106.

Moss, M. and R. Schwartz. 1981. Regulation of tropomyosin gene expression during myogenesis. *Mol. Cell. Biol.* **1:** 289.

Moss, P.S. and R.C. Strohman. 1976. Myosin synthesis by fusion-arrested chick embryo myoblasts in cell culture. *Dev. Biol.* **48:** 431.

Nadal-Ginard, B. 1978. Commitment, fusion and biochemical differentiation of a myogenic cell line in the absence of DNA synthesis. *Cell* **15:** 855.

Nguyen, H.T., R.M. Gubits, R.M. Wydro, and B. Nadal-Ginard. 1982. Sarcomeric myosin heavy chain is coded by a highly conserved multigene family. *Proc. Natl. Acad. Sci.* **79:** 5230.

Okazaki, K. and H. Holtzer. 1966. Myogenesis: Fusion, myosin synthesis and the mitotic cycle. *Proc. Natl. Acad. Sci.* **56:** 1484.

Paterson, B.M. and J.O. Bishop. 1977. Changes in the mRNA population of chick myoblasts during myogenesis in vitro. *Cell* **12:** 751.

Paterson, B.M. and R.C. Strohman. 1972. Myosin synthesis in cultures of differentiating chicken embryo skeletal muscle. *Dev. Biol.* **29:** 113.

Potter, J.D. 1974. The content of troponin, tropomyosin, actin, and myosin in rabbit skeletal muscle myofibrils. *Arch. Biochem. Biophys.* **162:** 436.

Prives, J. and B. Paterson. 1976. Differentiation of cell membranes in cultures of embryonic chick breast muscle. *Proc. Natl. Acad. Sci.* **71:** 3209.

Shani, M., D. Zevin-Sonkin, O. Saxel, Y. Carmon, D. Katcoff, U. Nudel, and D. Yaffe. 1981. The correlation between the synthesis of skeletal muscle actin, myosin heavy chain and myosin corresponding mRNA sequences during myogenesis. *Dev. Biol.* **86:** 483.

Shani, M., U. Nudel, D. Zevin-Sonkin, R. Zakut, D. Givol, D. Katcoff, Y. Carmon, J. Reiter, A.M. Frischauf, and D. Yaffe. 1981. Skeletal muscle actin mRNA. Characterization of the 3′ untranslated region. *Nucleic Acids Res.* **9:** 579.

Strohman, R.C., P.S. Moss, J. Micow-Eastwood, P. Spectro, A. Przbyla, and B. Paterson. 1977. Messenger RNA for myosin polypeptides; isolation from single myogenic cell cultures. *Cell* **10:** 265.

Tobin, A.J. 1979. Evaluating the contribution of posttranscriptional processing to differential gene expression. *Dev. Biol.* **68:** 47.

Trotter, J.A. and M. Nameroff. 1976. Myoblast differentiation *in vitro*. Morphological differentiation of mononucleated myoblasts. *Dev. Biol.* **49:** 548.

Turner, D.C., V. Maier, and H.M. Eppenberger. 1976. Creatine kinase and aldolase isoenzyme transitions in cultures of chick skeletal muscle cells. *Dev. Biol.* **37:** 63.

Umeda, P.K., A.M. Sinha, S. Jakoveis, S. Merten, H.-J. Hsu, K.N. Subramanian, R. Zak, and M. Rabinowitz. 1981. Molecular cloning of two fast myosin heavy chain cDNAs from chicken embryo skeletal muscle. *Proc. Natl. Acad. Sci.* **78:** 2843.

Weintraub, H. and M. Groudine. 1976. Chromosomal subunits in active genes have an altered conformation. *Science* **193:** 848.

Weintraub, H., H. Beng, H. Groudine, and T. Graf. 1982. Temperature-sensitive changes in the structure of globin chromatin in lines of red cell precursors transformed by ts-AEV. *Cell* **28:** 931.

Weisbrod, S. 1982. Active chromatin. *Nature* **297:** 289.

Whalen, R.G. and S.M. Sell. 1980. Myosin from fetal hearts contains the skeletal muscle embryonic light chain. *Nature* **286:** 731.

Whalen, R.G., S.M. Sell, G. Butler-Browne, K. Schwartz, P. Bouveret, and I. Pinset-Härström. 1981. Three myosin heavy-chain isozymes appear sequentially in rat muscle development. *Nature* **292:** 805.

Wu, C., Y.C. Wong, and S.C.R. Elgin. 1979. The chromatin structure of specific genes. II. Disruption of chromatin structure during gene activity. *Cell* **16:** 807.

Wydro, R.M., H.T. Nguyen, R.M. Gubits, and B. Nadal-Ginard. 1982. Characterization of sarcomeric myosin heavy chain genes. *J. Biol. Chem.* (in press).

Yaffe, D. 1968. Retention of differentiation potential during prolonged cultivation of myogenic cells. *Proc. Natl. Acad. Sci.* **61:** 477.

——. 1974. Developmental changes preceding cell fusion during muscle differentiation *in vitro. Exp. Cell Res.* **66:** 33.

Yaffe, D. and M. Dym. 1973. Gene expression during differentiation of contractile muscle fibers. *Cold Spring Harbor Symp. Quant. Biol.* **37:** 543.

Differential Expression of Two Fast Myosin Heavy-chain mRNAs during Chicken Skeletal-muscle Development

PATRICK K. UMEDA
CLIFFORD J. KAVINSKY
ACHYUT M. SINHA
HUEY-JUANG HSU
SMILJA JAKOVCIC
MURRAY RABINOWITZ
Departments of Medicine and Biochemistry
The University of Chicago
Chicago, Illinois 60637

Multiple molecular forms of myosin heavy chains (HCs) are present in fast and slow skeletal muscle and in cardiac tissue. These variants apparently are products of different myosin HC genes, whose expression is developmentally determined, since embryonic (Huszar 1972; Whalen et al. 1979; Benfield et al. 1980), neonatal (Whalen et al. 1981), and adult forms have been detected in skeletal muscle. Similarly, in cardiac tissue, two ventricular forms of myosin HCs have been identified. The expression of these forms varies during development and can be altered by environmental factors, such as the administration of thyroid hormone (Flink et al. 1979; Hoh et al. 1979; Chizzonite et al. 1982).

To examine the structural organization of the myosin HC gene family and to evaluate the control of transcription during development and altered physiologic states, we have cloned myosin HC cDNA sequences from chick skeletal and rabbit cardiac muscle. This paper describes the characterization of some of our cloned cDNA sequences and preliminary studies on the use of these clones to define unequivocally the expression of variant myosin HC genes during development.

RESULTS AND DISCUSSION

To obtain recombinant clones for myosin HC cDNA sequences, we first purified myosin HC mRNA from 14-day-old embryonic chick leg muscle by successive sucrose gradient centrifugations. Double-stranded cDNA was inserted into the *Pst*I site of plasmid pBR322, and recombinant clones were screened with labeled cDNA synthesized from highly purified myosin HC mRNA that was isolated by immunoprecipitation of polysomes with specific antibodies (Umeda et al. 1980). Most of the positive clones were verified to contain myosin HC cDNA sequences (Umeda et al. 1981).

Restriction enzyme mapping of the cDNA inserts indicated that the clones contained sequences for five distinct myosin HC mRNAs. Only clones for two classes, illustrated by clones 251 and 110, respectively, are shown in Figure 1. These two clones showed distinct homology but were clearly different, in that 6 out of 17 restriction sites were nonidentical. The class represented by clone 251 was more abundant than that represented by clone 110 (data not shown). Since Southern blot hybridization indicated that the left *Pst*I fragment of clone 184 hybridized only to the right *Pst*I fragment of clone 60 (Fig. 1), we now have cloned cDNA sequences for about 3.5 kb or 50% of the 7.0-kb myosin HC mRNA.

Sequencing of selective regions of clone 251 conclusively established the myosin HC origin of our clones. Comparison of the derived amino acid sequence with that of fast myosin HC from rabbit skeletal muscle (Elzinga et al. 1980; S.W. Tong and M. Elzinga, pers. comm.) revealed 89% homology

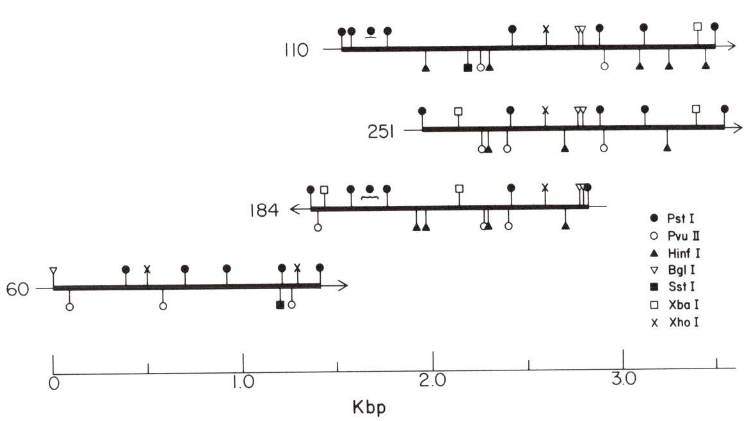

FIGURE 1
Restriction enzyme maps of the cDNA inserts of the myosin HC clones. The cDNA inserts are represented by the solid bars, with arrows indicating the direction of transcription of the β-lactamase gene of pBR322. The *Hinf*I sites of clone 60 have not been mapped, and there are no restriction sites in the inserts for *Eco*RI, *Bam*HI, *Hin*dIII, and *Sal*I.

between the amino acid sequences encoded in the embryonic chicken clones and those determined directly for adult rabbit myosin HC. Of the 285 amino acids compared in this analysis, only 31 were divergent. The data localized the carboxyl terminus of myosin HC to about 220 nucleotides in from the right end of the restriction map (Fig. 1). Therefore, the region to the right of the carboxyl terminus corresponds to 3′-nontranslated sequences (Fig. 2).

Since leg muscle is a mixed muscle containing different fiber types, fast, slow, and embryonic HC forms may be expressed at 14 days in ovo. To determine whether the cloned cDNAs correspond to sequences for fast or slow myosin HC mRNAs, hybrid melting experiments were performed (Umeda et al. 1981). We hybridized filter-bound DNAs for plasmids 110 and 251 with labeled cDNA probes synthesized from fast or slow myosin HC mRNAs. These mRNAs, fractionated by indirect immunoprecipitation of polysomes, were more than 90% enriched with respect to myosin type. After hybridization with either cDNA probe, the melting curves of plasmids 110 and 251 were similar, with identical melting temperatures (T_ms). The T_m of hybrids formed with fast myosin HC cDNA was 14°C higher than that for hybrids obtained with slow myosin HC cDNA. The T_m of hybrids with cardiac cDNA was even lower. Therefore, the cDNA sequences in both plasmids 110 and 251 correspond to fast-type myosin HC mRNAs.

```
251
                                                                   *             *      *
Asp Ile Ala Glu Ser Gln Val Asn Lys Leu Arg Ala Lys Ser Arg Glu         Ile Gly Lys Lys Ala Glu Ser Glu Glu
GAC ATT GCA GAG TCG CAA GTC AAC AAG CTC CGA GCA AAG AGC CGT GAA         ATA GGC AAG AAG GCA GAA AGT GAA GAG
                HinfI                                                                 C-TERMINUS
GAC ATT GCA GAG TCA CAG GTC AAC AAG CTC CGA GCA AAG AGC CGG GAG TTT CAT GGC AAG AAG ATA GAA GAG GAA GAG
Asp Ile Ala Glu Ser Gln Val Asn Lys Leu Arg Ala Lys Ser Arg Glu Phe His Gly Lys Lys Ile Glu Glu Glu Glu
110                                                                *             *      *

3' NON-TRANSLATED SEQUENCES

TAGATGCCTCAAGCGGTGCAAAGTGAAATAGAATTGCACAAAATGTGAAATTCTATCACTTTGATTTGTAATTACTGCTTAGTTCT
TGAAGATATCATCTGACAGCAAAGTGGCCTGAGGAGTGCACAAAATGTGAACCCTCTGTTGCTATCACTTATAATTTATCTTTATA

TCAACTATCTAGATAATTAATATTTAGATAATAAAAATTGTAGAGATTTTCCCATGG(A)58
ACCACCTCAATGTCTAGAGAATAAAGACCATAGATTCCTCTGCATAT(A)30
```

FIGURE 2

Comparison of DNA and derived amino acid sequences of clones 251 and 110. Sequences shown correspond to the carboxyterminal portion of the translated region and include the 3′-nontranslated sequences. Only sequences analogous to the mRNA strand are shown. Nucleotide mismatches are enclosed within boxes, whereas amino acid differences are indicated by asterisks (*). The break in the 251 sequence, 10 amino acids from the carboxyl terminus, indicates the site of a proposed triplet insertion in clone 110. Regions of homology within the 3′-nontranslated sequences are denoted by blocks.

The two classes of fast myosin HC clones could represent an embryonic and an adult form, multiple embryonic or adult species, or even alleles of a single gene. To exclude the possibility of allelism, genomic blotting experiments were carried out using DNA from six different animals. After cleavage with several restriction enzymes, DNA blots were probed with labeled clone 251 DNA. Multiple restriction bands were observed and were identical in all six animals. Similar results were obtained with clone 110 and also when smaller restriction probes representing primarily nontranslated regions were used (data not shown). These results argue against the presence of allelic variants.

To examine the sequence divergence of the two clones, DNA sequence analysis was performed. A comparison of the nucleotide and derived amino acid sequences of the carboxyterminal portions of clones 110 and 251 is shown in Figure 2. Sequences of the 3′-nontranslated regions are also shown. There is 83% homology in the nucleotide sequences within the translated portions of these two clones, with 88% homology in the corresponding amino acids. Furthermore, clone 110 contains an insertion triplet coding for phenylalanine ten amino acids from the carboxyl terminus. In contrast, the 3′-nontranslated regions reveal much greater sequence divergence in excess of 50%. However, sequence homologies were still observed. The presence of a poly(A) sequence in both clones indicated that we have cloned all of the 3′ ends of these mRNAs.

To establish further that the two classes of clones represent different genes, we examined the expression of the corresponding mRNAs in different developmental stages using a nuclease S1 mapping procedure (Weaver and Weissman 1979). A DNA strand complementary to the mRNA was isolated from our clones and labeled at the *Hin*fI site designated in Figure 2. This strand, which extends to the 3′ end of the mRNA, was then hybridized to various muscle mRNAs, and the hybrids were treated with nuclease S1. Since S1 cleaves single-stranded regions in DNA or regions where sequence divergence exists, the single-stranded DNA probe should be protected by hybridization to the homologous mRNA. For clone 251, in which the probe extends into the vector sequences (Fig. 3A), hybridization to the homologous mRNA should protect a 270–280-nucleotide fragment (corresponding to the distance from the *Hin*fI site to the end of the poly[A] sequence). On the other hand, hybridization to heterologous (clone 110) mRNA would result in a labeled fragment extending from the *Hin*fI site to the site of sequence divergence. This fragment could be 4 or 7 nucleotides in size, corresponding to the single nucleotide substitutions at these sites, or 34 nucleotides, corresponding to the beginning of a region of marked sequence divergence, including an insertion triplet in one mRNA.

The results with the probe from clone 251 are shown in Figure 3B. Hybridization with *embryonic* breast and leg muscle RNAs protects a 210-nucleotide fragment. Although this fragment is smaller than expected, it corresponds to the distance from the *Hin*fI site to the beginning of the

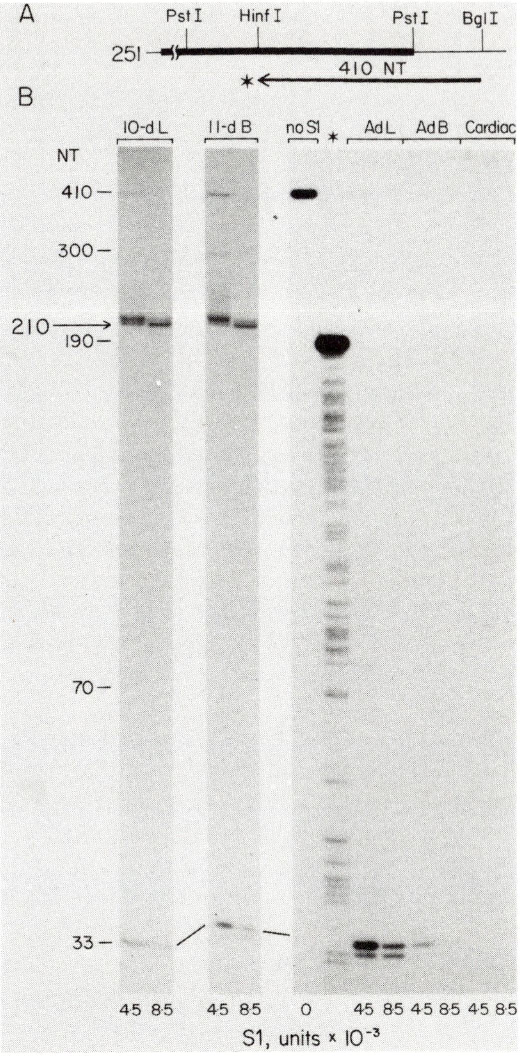

FIGURE 3
Expression of clone 251 mRNA sequences during development. (*A*) Nuclease S1 mapping was performed as described by Orkin and Goff (1981) with a ^{32}P-labeled 3′-end-labeled, single-stranded probe from clone 251. (*) The labeled end; (→) the 5′ → 3′ direction of the labeled strand. (*B*) After nuclease S1 treatment, the products were denatured and analyzed on a 6% polyacrylamide sequencing gel. Hybridizations were carried out with total RNA from 10-day-old *embryonic* leg muscle (10-d L), 11-day-old *embryonic* breast muscle (11-d B), *adult* leg muscle (Ad L), *adult* breast muscle (Ad B), and 14-day-old *embryonic* cardiac muscle (Cardiac). (*) A G+A sequencing ladder of the 190-nucleotide probe from clone 110 (Fig. 4A).

poly(A) sequences. The poly(A) region is evidently cleaved by nuclease S1, since a smear of fragments between 210 and 300 nucleotides is observed at lower enzyme concentrations (Fig. 3B). A much smaller amount of low-molecular-weight fragments of 32 and 33 nucleotides are also present. In contrast, hybridization with *adult* leg (anterior tibialis) and breast muscle RNAs shows only low-molecular-weight forms (Fig. 3B). There were no protected fragments when the hybridization was carried out with cardiac RNA. We conclude that clone 251 is transcribed only in embryonic fast muscle and that its mRNA represents the predominant form transcribed at this time.

A single-stranded probe from clone 110 was prepared in a similar manner. In this case, however, the probe is 190 nucleotides long and ends in the 3′-nontranslated region (Fig. 4A). Therefore, hybridization to mRNA corresponding to clone 110 should result in a 190-nucleotide protected fragment, whereas hybridization to mRNA of the clone 251 class should result in a 34-nucleotide fragment.

Figure 4B shows that hybridization with RNAs from *embryonic* tissue (10- and 14-day-old leg muscle and 11-day-old breast muscle) results in the protection of a 190-nucleotide fragment and a greater amount of fragments of 32 and 33 nucleotides. The slight difference in size from 190 nucleotides with the RNA from 14-day-old embryonic leg muscle could be due to trimming the ends of the fragment by nuclease S1 or to the appearance of a different mRNA, reflecting either a different gene or an allelic variant. Hydridization with adult breast and leg muscle RNA also results in the protection of a 190-nucleotide fragment but virtually no bands at 33 nucleotides. On the other hand, with *adult* leg muscle RNA, another group of fragments, larger than 34 nucleotides, is present. These new bands are not the result of partial homology to the slow myosin HC mRNA, since no protected fragments are seen on hybridization with RNA from adult slow muscle (anterior lattissimus dorsi [ALD]). This result indicates that a different fast myosin HC mRNA is expressed in adult leg muscle that is not present in adult breast muscle.

From these hybridization experiments, we conclude that clones 251 and 110 represent two different genes that are expressed differentially in different skeletal-muscle tissues at different stages of development. Clone 251 is expressed only in embryonic fast muscle where its transcripts appear to be the predominant species. Clone 110, on the other hand, is expressed both in adult and embryonic fast muscle. Since the amount of myosin HC mRNAs in all hybridizations was approximately the same, the low amount of labeled fragments with RNA from adult breast muscle strongly suggests that the mRNA represented by clone 110 is not the major adult species in breast muscle. Clone 110 could correspond to a yet undetected myosin HC species or to a second embryonic species (Lowey et al., this volume) or neonatal form (Whalen et al. 1981) that is also partially expressed in the adult.

We are also currently analyzing the expression of myosin HC genes in rabbit cardiac muscle. It has been established that there are at least two molecular forms of myosin HCs in rabbit ventricle (Flink et al. 1979). Expression of these forms varies during development, and the administration of thyroid hormone greatly enhances the synthesis of one of the variants. We have isolated a clone from adult rabbit heart that hybridizes to an RNA band corresponding in migration to cardiac myosin HC mRNA. Restriction enzyme mapping of this clone shows that the insert is approximately 2 kb in size and contains multiple *Pst*I sites similar to our chicken skeletal clones. Clones representing myosin HC mRNAs transcribed after thyroid hormone administration are also being characterized.

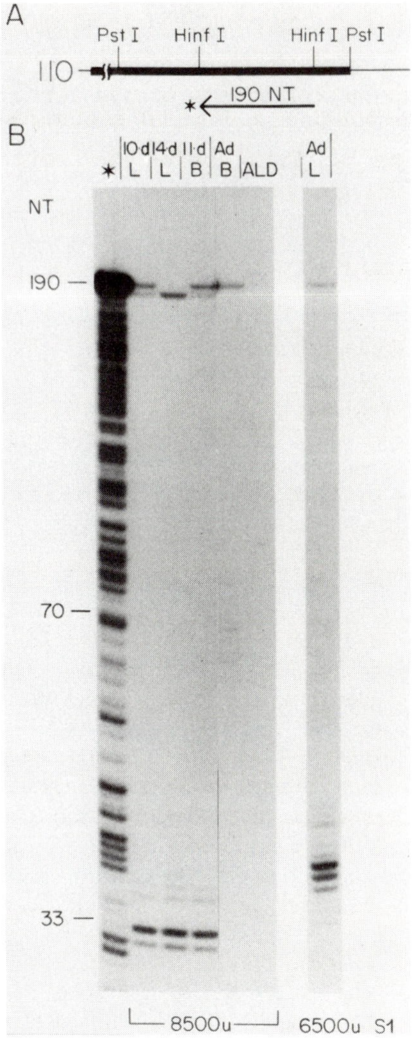

FIGURE 4
Expression of clone 110 mRNA sequences during development. (*A*) The ^{32}P-labeled 3′-end-labeled probe from clone 110. (*B*) The nuclease-S1-resistant products were analyzed on an 8% polyacrylamide sequencing gel. Hybridizations were carried out with total RNA from muscle tissues indicated in the legend to Fig. 3 and also from 14-day-old *embryonic* leg muscle (14-d L) and *adult* ALD muscle. (*) A G+A sequencing ladder of the 190-nucleotide probe from clone 110.

SUMMARY

We have characterized clones representing sequences of two types of fast myosin HCs from chicken skeletal muscle as well as clones for myosin HC from rabbit cardiac muscle. These clones and others that can be similarly isolated offer the opportunity for detailed analysis of transcription of myosin HC both in skeletal and cardiac muscle. Fully characterized cDNA clones will be crucial in the identification and characterization of individual genes in the myosin HC gene family currently in progress in our laboratory.

ACKNOWLEDGMENTS

We thank Dr. M. Elzinga for providing us with his unpublished data on the amino acid sequence of myosin HC from rabbit skeletal muscle. This work was supported by grants from the National Institutes of Health (HL-20592 and HL-09172) and by the Louis Block Foundation, The University of Chicago.

REFERENCES

Benfield, P.A., D.D. LeBlanc, and S. Lowey. 1980. Characterization of embryonic myosin from chicken pectoralis muscle. *Fed. Proc. Fed. Am. Soc. Ex. Biol.* **39:** 2169. (Abstr.).

Chizzonite, R.A., A.W. Everett, W.A. Clark, S. Jakovcic, M. Rabinowitz, and R. Zak. 1982. Isolation and characterization of two molecular variants of myosin heavy chain from rabbit ventricle: Change in their content during normal growth and after treatment with thyroid hormone. *J. Biol. Chem.* **257:** 2056.

Elzinga, M., K. Behar, and G. Walton. 1980. Sequence and proposed structure of a 17,000 dalton CNBr fragment from the C-terminus of myosin. *Fed. Proc. Fed. Am. Soc. Exp. Biol.* **39:** 2168. (Abstr.).

Flink, I.L., J.H. Rader, and E. Morkin. 1979. Thyroid hormone stimulates synthesis of a cardiac myosin isozyme. *J. Biol. Chem.* **254:** 3105.

Hoh, J.F.Y., G.P.S. Yeoh, M.A.W. Thomas, and L. Higgenbottom. 1979. Structure differences in the heavy chains of rat ventricular myosin isozymes. *FEBS Lett.* **97:** 330.

Huszar, G. 1972. Developmental changes of the primary structure and histidine methylation of rabbit sketal muscle myosin. *Nat. New Biol.* **240:** 260.

Orkin, S.H. and S.C. Goff. 1981. The duplicated human α-globin genes: Their relative expression as measured by RNA analysis. *Cell* **24:** 345.

Umeda, P.K., R. Zak, and M. Rabinowitz. 1980. Purification of messenger ribonucleic acids for fast and slow myosin heavy chains by indirect immunoprecipitation of polysomes from embryonic chick skeletal muscle. *Biochemistry* **19:** 1955.

Umeda, P.K., A.M. Sinha, S. Jakovcic, S. Merten, H.-J. Hsu, K.N. Subramanian, R. Zak, and M. Rabinowitz. 1981. Molecular cloning of two fast myosin heavy chain cDNAs from chicken embryo skeletal muscle. *Proc. Natl. Acad. Sci.* **78:** 2843.

Weaver, R.F. and C. Weissman. 1979. Mapping of RNA by a modification of the Berk-Sharp procedure: The 5′ termini of 15S β-globin mRNA precursor and mature 10S β-globin mRNA have identical map coordinates. *Nucleic Acids Res.* **7:** 1175.

Whalen, R.G., K. Schwartz, P. Bouveret, S. Sell, and F. Gros. 1979. Contractile protein isozymes in muscle development: Identification of an embryonic form of myosin heavy chain. *Proc. Natl. Acad. Sci.* **76:** 5197.

Whalen, R.G., S.M. Sell, G.S. Butler-Browne, K. Schwartz, P. Bouveret, and I. Pinset-Härström. 1981. Three myosin heavy-chain isozymes appear sequentially in rat muscle development. *Nature* **292:** 805.

Isolation and Structural Analysis of the Genes Coding for Rat Skeletal-muscle Actin and for Cytoplasmic Actin

URI NUDEL
RINA ZAKUT
DON KATCOFF
YORAM CARMON
HENRYK CZOSNEK
MOSHE SHANI
DAVID YAFFE
Department of Cell Biology
The Weizman Institute of Science
Rehovot 76100, Israel

It has become evident in recent years that many structural proteins and enzymes are members of small families of isoproteins. Some of the isoproteins are tissue-specific and their synthesis is developmentally regulated (Hunter and Garrels 1977; Carmon et al. 1979; Waring and Mahowald 1979; Long and Dawid 1980; Maniatis et al. 1980; Shani et al. 1981a; Whalen et al. 1981). The genes coding for the isoproteins arose, most probably, from duplications of ancestral genes, which later diverged during evolution. From the functional point of view, the significance of the existence of isoproteins is obvious for some of these famlies (e.g., the β and γ globins in man. Each of them is best fitted to function in a different environment). However, in some other cases, no functional advantages could be assigned to one of the isoproteins over another (e.g., β-minor and β-major globins in mouse; β and δ globins in man). It is possible, therefore, that the existence of isogenes is related to the mode of control of their expression. Studies on the structure function relationship of isogenes may prove to be very helpful in understanding eukaryotic gene evolution and regulation of expression.

The actin gene family is most suitable for this type of study. Actins are among the most prevalent proteins in eukaryotic cells. They are found in all eukaryotes, from unicellular organisms to mammals. The amino acid sequences of the different actins have been highly conserved during evolution, and even actins from remote evolutionary sources, such as rabbit skeletal muscle and yeasts, differ by only 10–12% (Gallwitz and Sures 1980). Six different actins, encoded by at least six different genes, have been identified by amino acid sequence analysis in mammals (Vandekerckhove and Weber 1978, 1979). The expression of some of these genes is tissue specific and developmentally regulated (Hunter and Garrels 1977; Vandekerckhove and Weber 1978, 1979; Shani et al. 1981a).

This paper describes the isolation, characterization, and structural analysis of two of the rat actin genes: the gene coding for the skeletal-muscle actin and the gene coding for a cytoplasmic actin. The structure of the rat skeletal-muscle and cytoplasmic actin genes is compared with that of other actin genes, and the evolution of the actin genes is discussed.

RESULTS

Actin Genes in the Rat Genome

We have previously described the construction and characterization of two recombinant plasmids containing rat skeletal-muscle actin cDNA sequences (Katcoff et al. 1980; Shani et al. 1981b). Plasmid p749 specifies codons 171–360 of the mRNA; plasmid p106 specifies the codon for the last amino acid, the termination codon, and the entire 3′-untranslated region of rat striated-muscle α-actin mRNA. We compared the sequence homology among plasmids p749, p106, and actin mRNA from various sources and found that whereas p749 hybridized to muscle and nonmuscle actin mRNA of several mammals and of chicken, p106 hybridized specifically to mRNA of striated-muscle actins of mammals. It did not hybridize to rat smooth-muscle and nonmuscle actin mRNAs. An additional plasmid used in this study is p72, which was constructed by Ginsburg et al. (1980), using rat brain mRNA as the template for the cDNA synthesis. Under stringent conditions, this plasmid binds β-actin mRNA only. Nucleotide sequence analysis indicated that p72 contains an insert, approximately 250 nucleotides long, homologous to a part of the 3′-untranslated region of a rat β-like actin gene (Nudel et al. 1982).

The inserts of plasmids p749, p106, and p72 were used to probe Southern blots (Southern 1975) of rat spleen DNA, restricted by *Eco*RI (Fig. 1). Plasmid p749 hybridized to at least 12 DNA fragments. Two of the hybrids (5.7-kb and 6.9-kb fragments) were stable at 72°C in 0.1× SSC (for 30 min). Only the 5.7-kb hybrid was stable at 76°C (15 min in 0.1× SSC). p106 hybridized at low stringency only to the 5.7-kb and the 6.9-kb

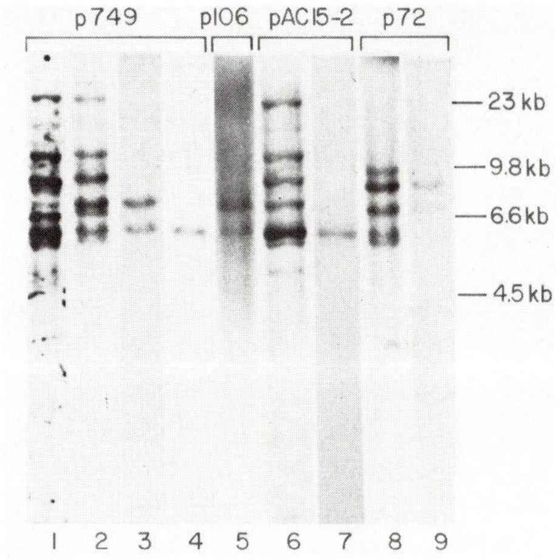

FIGURE 1
Southern blots of rat spleen DNA hybridized to various actin DNA probes. 15 μg of rat spleen DNA samples were restricted with *Eco*RI, fractionated on 0.7% horizontal agarose gels, transferred to nitrocellulose filters, and hybridized with nick-translated DNA inserts, as indicated. Hybridization was done in 50% formamide, 5× SSC, 5× Denhardt's solution, and 200 μg/ml of *Escherichia coli* DNA at 42°C for 40 hr. After the hybridization, the blots were washed for 90 min in 2× SSC at room temperature and were then washed as follows: (*1,5,6*) 90 min in 2× SSC, 50°C; (*2,8*) 60 min in 0.1× SSC, 50°C; (*3,7,9*) 30 min in 0.1× SSC, 72°C; (*4*) 15 min in 0.1× SSC, 76°C. Two DNA preparations (each from several spleens) were used for the Southern blots. One was used in *1*, and the other, in *2–9*. The variation in some bands may be due to genetic polymorphism.

fragments. p72 hybridized to several DNA fragments. Only two of the hybrids (8.0-kb and 6.9-kb DNA fragments) were stable at 72°C in 0.1× SSC (for 30 min).

Two rat genomic DNA libraries (both gifts from T. Sargent and J. Bonner, California Institute of Technology, Pasadena) were screened for clones containing actin gene sequences. Twenty recombinant phages were isolated and analyzed. Their restriction pattern indicated that they originated from eight different actin genes. Thus, our results indicate the existence of at least eight, and probably more, actin genes in rats. Some of these may represent functionally inactive pseudogenes.

On the basis of the hybridization experiments, we concluded that the 5.7-kb *Eco*RI-restricted DNA fragment contains a skeletal-muscle actin gene, and the 8-kb *Eco*RI-restricted DNA fragment contains a β-like actin gene. Experiments in which we measured the DNase-I sensitivity of chromatin had shown that the 5.7-kb DNA fragment contained an active skeletal-muscle actin gene and the 8-kb DNA fragment contained an active cytoplasmic actin gene (Carmon et al. 1982).

Structure of Rat Skeletal-muscle Actin Gene

One of the rat genomic DNA clones, Act 15, was identified by nucleotide sequence analysis as containing the rat skeletal-muscle α-actin gene. It has all the codons that are specific to striated-muscle actin (Nudel et al. 1982; Zakut et al. 1982). The rat DNA insert of Act 15 is 14.7 kb long (Fig. 2A), and the actin structural-gene sequences are within the terminal 3.7-kb *Eco*RI DNA fragment. When this 3.7-kb DNA fragment was used to probe a Southern blot of rat genomic DNA, it hybridized most strongly to the 5.7-kb *Eco*RI DNA fragment, and the hybrid was stable at 76°C (15 min in 0.1× SSC; Fig. 1). The 3.7-kb *Eco*RI DNA fragment of Act 15 is smaller than the 5.7-kb *Eco*RI DNA fragment observed in the Southern blot, because Act 15 was isolated from a partial *Hae*III-restricted DNA library, in which new terminal *Eco*RI sites were introduced using *Eco*RI linkers. The new terminal *Eco*RI site in Act 15 replaced the *Hae*III site at nucleotide 75 of the 3′-untranslated end of the gene. The 3′-terminal 165 nucleotides of the α-actin mRNA (Shani et al. 1981b) are missing from the clone Act 15.

The 3.7-kb *Eco*RI-restricted DNA fragment of Act 15 was subcloned into the *Eco*RI site of pBR322, and the recombinant plasmid pAC15.2 was used for further structural analysis. Examination by electron microscope of the R loops formed between pAC15.2 DNA and rat muscle mRNA (Fig. 3) indicated the existence of at least five introns, some of which are very small. The largest intron is located very close to the 5′ end of the gene. The 5′-end terminal exon is too small to form a visible R loop, but its existence was

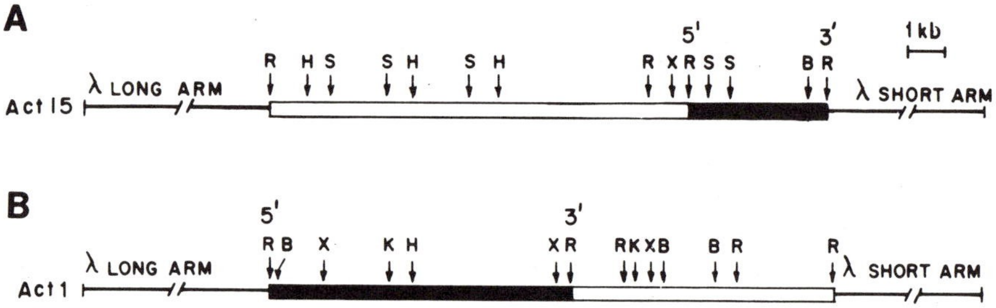

FIGURE 2
Restriction maps of the DNA inserts in the recombinant bacteriophages Act 15 (*A*) and Act 1 (*B*). The maps were constructed by single and double digests of the recombinant bacteriophages Act 15 and Act 1 with the indicated restriction enzymes. The sizes of the restricted DNA fragments were analyzed by agarose gel electrophoresis. Bars represent rat DNA. Lines represent λ Charon 4A arms. Black bars represent the fragments of DNA that were subcloned in pBR322. 5′–3′ indicates the orientation of the structural genes. The restriction enzymes used were *Eco*RI (R), *Hind*III (H), *Bam*HI (B), *Sac*I (S), *Kpn*I (K), and *Xba*I (X).

inferred from the large double-strand loop corresponding to the 5′-last intron.

Nucleotide sequence analysis of the entire α-actin gene allowed us to determine the size and location of all of the exons and introns of the gene. Introns interrupt the α-actin gene at codons 41, 150, 204, 267, and 327. The large intron at the 5′ end is located in the noncoding part of the gene, 12 nucleotides upstream from the initiator ATG. The sizes of the other introns (5′–3′ direction) are 94 bp, 189 bp, 96 bp, 136 bp, and 76 bp.

On the basis of these data, we propose the structure of rat skeletal-muscle actin gene as presented in Figure 4.

Gross Structure of a Nonmuscle Actin Gene

One of the recombinant phages (Act 1), which was isolated from the partial *Eco*RI digest library, hybridized to plasmids p749 and p72 but not to plasmid p106. Restriction mapping and partial DNA sequencing indicated that the inserted DNA contained an entire actin structural gene and additional DNA flanking its 5′ and 3′ ends (Fig. 2B). The 8-kb *Eco*RI fragment containing the structural gene was subcloned in the *Eco*RI site of pBR322 for further structural analysis. The plasmid containing this fragment was designated pACR1.

Partial DNA sequencing has shown that Act 1 is a gene coding for a cytoplasmic actin (Nudel et al. 1982). However, in the region sequenced so far, the amino acid sequences of β- and γ-actin are identical. A fragment of 105 nucleotides from plasmid p72 was sequenced and was found to be almost identical to a segment in the middle of the 3′-noncoding region of Act 1. Since plasmid p72, under stringent conditions, specifically binds β-actin mRNA, the cloned gene was tentatively identified as a rat β-like actin gene (see Note Added in Proof).

Examination by electron microscope of the R loops formed between the plasmid pACR1 and the actin-enriched mRNA from proliferating myoblasts indicates the existence of at least five introns (Fig. 3). Similarly to the α-actin gene, the largest intron (820 $\pm$ 45 bp) is located very close to the 5′ end of the transcribed region, and the 5′-terminal exon was too small to be measured. The lengths of the other exons were calculated from the measurement of 19 hybrid molecules (5′–3′): 100 ($\pm$10), 200 ($\pm$25), 490 ($\pm$70), and 710 ($\pm$80).

Partial DNA sequence analysis of the β-like actin gene led to the assignment of three of the introns to codons 120, 267, and 327. Interestingly, no introns were found at codons 150 and 204. On the basis of endonuclease S1 mapping (Berk and Sharp 1977), one additional intron was assigned to the 5′-untranslated region, and another intron was assigned to the region of codons 40–42 (see Note Added in Proof).

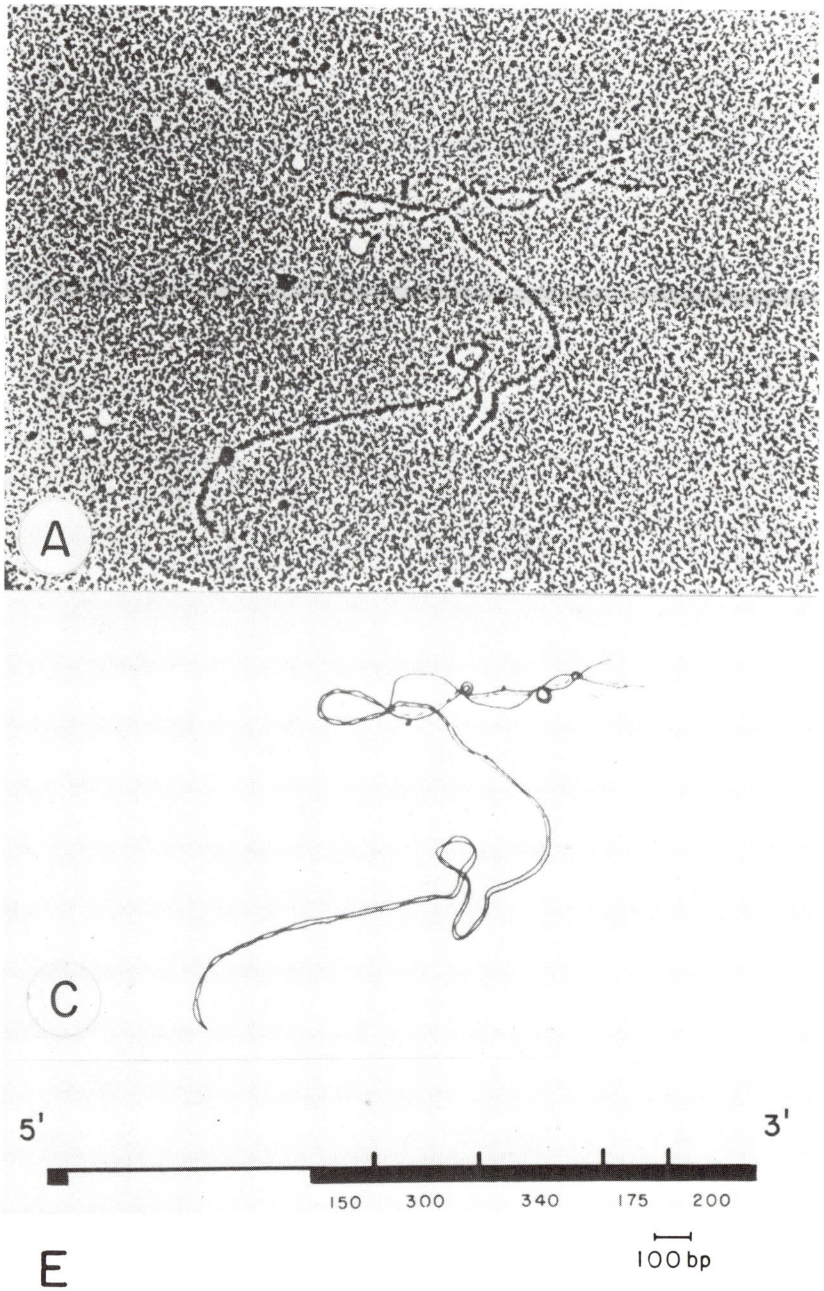

FIGURE 3
Electron micrographs of the R loops formed between plasmid pAC15.2 and rat muscle mRNA (*A*) and between plasmid pACR1 and mRNA from dividing myoblasts (*B*). (*C,D*) The interpretative drawings of the molecules shown in *A* and *B*, respectively. (Single lines) Single-stranded DNA; (double lines) double-stranded DNA;

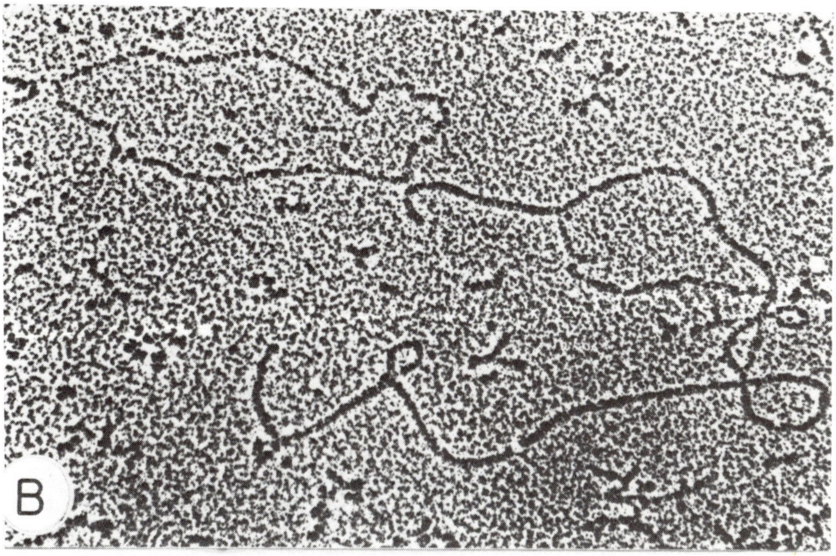

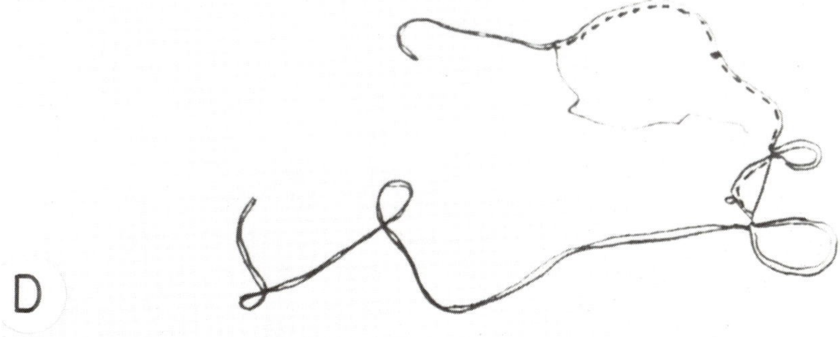

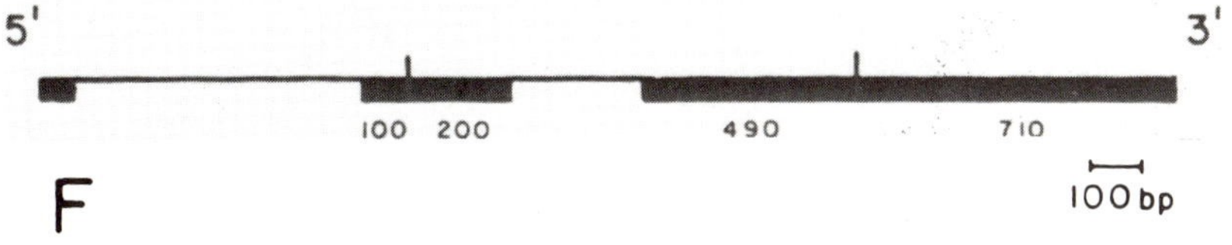

FIGURE 3 (Continued)
(dashed lines) RNA. (*E,F*) Schematic representations, obtained from the R-loop patterns of the structure of the α-actin gene and the β-like actin gene, respectively. (▬) Exons; (——) introns. (⊥) Introns too small to be measured. (*E*) Measurement of 31 molecules; (F) measurement of 19 molecules.

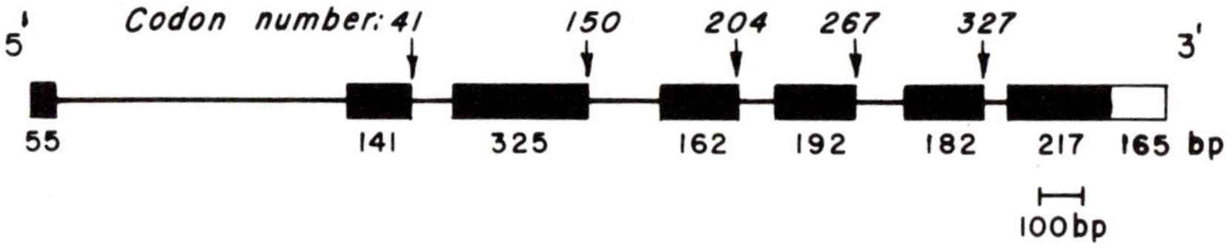

FIGURE 4
The structure of rat skeletal-muscle α-actin gene. (■) Exons; (□) the part of the 3′-untranslated region that is missing from Act 15; (——) introns. Numbers above the scheme indicate positions of introns (codon number). Numbers below the scheme indicate sizes of exons determined from the nucleotide sequence. The size of the exon at the 5′ end was determined by the endonuclease S1 mapping technique (Berk and Sharp 1977).

DISCUSSION

In this paper we describe the isolation and characterization of the genes coding for rat skeletal-muscle actin and for rat cytoplasmic actin. We showed that the skeletal-muscle actin gene is split by six introns, five of which are in the coding region, at codons 41, 150, 204, 267, and 327. The sixth intron, which is the largest, is at the 5′-noncoding part of the gene. The β-like actin gene was partially characterized. The gene consists of at least six exons. The positions of three of the introns were assigned to codons 121, 267, and 327. Like the α-actin gene, the cytoplasmic actin gene is split by a large intron located very close to its 5′ end.

The evolution of the sequences and sites of introns has been analyzed, so far, in several gene families, e.g., globin genes, ovalbumin genes, and the immunoglobulin genes (for review, see Breathnach and Chambon 1981). The main conclusion from these studies was that although the nucleotide sequence of introns changes rapidly during evolution, the location of introns in relation to the coding sequence is highly conserved. In previous studies, however, the comparisons were done between genes from vertebrates only. The data on the actin genes offer a wider comparison of gene structure across a much greater evolutionary range, i.e., from unicellular organisms to fungi, insects, and vertebrates. From the data available so far, it is apparent that in spite of the great conservation of the amino acid sequence of actins, the number of introns and their positions can vary greatly (Table 1).

It is evident that there is no similarity in the number and location of introns among fungi, protostomes, and the known actin genes of deuterostomes (except for the possible homology regarding the intron in the 5′-noncoding region of rat α- and β-like actin genes and the *Drosophila* actin gene DmA2). However, within the deuterostomes, all known intron positions found in sea urchin actin gene J (41, 121, 203, 267) are found in the rat skeletal-muscle gene, or in the cytoplasmic actin gene, or in both. But these genes do contain additional introns that are not found in the sea urchin actin gene J. The posi-

TABLE 1

Positions of Introns in Actin Genes

Source and type of actin gene	Intron location									
Dictyostelium (several genes)[1]					none					
Yeast[2]		4								
Drosophila[3]										
DmA6			—						307	
DmA4			13						—	
DmA2	5′UT		—						—	
Sea urchin gene J[4]				41	121	—	203	267		—
Chicken skeletal muscle[5]				41	—	150	203	267		327
Rat skeletal muscle[6]	−12			41	—	150	204	267		327
Rat β-like[6]	(5′UT)			(41)	121	—	—	267		327

The numbers indicate the codons at which introns are located. No introns were found in the sites indicated by dashes. Numbers in parentheses indicate approximate location of introns, as interpreted from endonuclease S1 mapping and electron microscopy.

Data were taken from [1]Firtel et al. (1979); [2]Gallwitz and Sures (1980) and Ng and Abelson (1980); [3]Fyrberg et al. (1981); [4]Scheller et al. (1981); [5]C. Ordahl (pers. comm.); [6]Nudel et al. (1982). (See Note Added in Proof.)

tions of introns in the coding region of the chicken muscle actin gene are identical to those in the rat muscle actin gene.

It was suggested by Gilbert (1978) that split genes were built by rearrangement of DNA fragments containing coding sequences. Noncoding flanking regions of these fragments became introns. This suggests that the primordial actin gene was split in at least all those positions in which introns are found in the various actin genes and that the variability in intron sites was formed by deletion of introns during evolution. Accordingly, the absence of introns in *Dictyostelium* actin and the presence of only one intron in yeast may perhaps indicate that in unicellular organisms, which have high reproduction rates, there is a strong selective pressure to minimize their genome size by deleting DNA sequences that have become nonfunctional. However, an alternative possibility may be considered as well, namely that the variability in intron positions and numbers resulted from insertion and deletion of introns, possibly as transposable elements. It should be pointed out that the two possible explanations for the diversity of intron number and locations are not mutually exclusive; both could have contributed to the structure of the present-day actin genes.

In spite of the fragmentary nature of the data available, the homology of intron sites may be used as an indication of the evolutionary relatedness of actin genes. It seems that differences between intron sites in an actin isogene family in the same organism can be taken as an indication of an ancient separation of the genes during evolution. Thus, the differences in intron sites among the *Drosophila* actin genes suggest that these genes have separated from one another very early in the evolution of the arthropods. On the other hand, the available data (Table 1) on the intron sites in deuterostomes suggest that the vertebrate skeletal-muscle actin gene has separated from the nonmuscle actin genes along the branch of evolution leading to the vertebrates. It was reported that none of the six *Drosophila* actin genes that have been isolated and partially sequenced codes for the amino acid sequences typical of vertebrate muscle actin (Fyrberg et al. 1981). This would imply that the vertebrate-muscle actin genes and the insect-muscle actin genes have evolved independently from ancient nonmuscle actin genes.

SUMMARY

The genes coding for skeletal-muscle α-actin and for a cytoplasmic actin were isolated from a rat genomic DNA library. The skeletal-muscle α-actin gene was sequenced completely. The gene is split by six introns, five of them at codons 41, 150, 204, 267, and 327, and a large intron is found in the 5′-noncoding region of the gene, 12 nucleotides upstream from the initiator ATG. The cytoplasmic β-like actin gene was partially sequenced. In the sequenced region (codons 121–327), it is split by introns at codons 121, 267,

and 327. No introns were found at codons 150 and 203. Like the α-actin gene, the cytoplasmic actin gene contains a large intron near its 5′ end. The structures of the rat skeletal-muscle and cytoplasmic actin genes were compared with that of actin genes from other organisms. The evolutionary pattern of actin genes was discussed.

ACKNOWLEDGMENTS

We wish to thank Ms. Zehava Levi and Sara Neuman for their excellent technical assistance. The work was supported by a grant from the Muscular Dystrophy Association of America and by grant RO1-GM-22767 from the National Institutes of Health, Department of Health and Human Services. U.N. is the incumbent of the A. and E. Blum Career Development Chair in Cancer Research.

NOTE ADDED IN PROOF

Since submitting this manuscript, the cytoplasmic actin gene has been sequenced completely. It was identified as the gene coding for rat β-actin. The gene has five introns at the following locations: The largest intron is at the 5′-noncoding part of the gene, 6 bp upstream from the initiator ATG, and the other four introns are at codons 41, 121, 267, and 327.

The sequence of a soybean actin gene was recently published (Shah et al. 1982). This gene has introns at codons 20, 150, and 355. These data suggest that the intron at codon 150 of actin gene is not specific to vertebrate skeletal-muscle actin gene, but rather, a very old intron that might be found in additonal actin genes (and probably lost, by a single or several deletion events, from other actin genes).

REFERENCES

Berk, A.J. and P.A. Sharp. 1977. Sizing and mapping of early adenovirus mRNAs by gel electrophoresis of S1 endonuclease-digested hybrids. *Cell* **12:** 721.

Breathnach, R. and P. Chambon. 1981. Organization and expression of eukaryotic split genes coding for proteins. *Annu. Rev. Biochem.* **50:** 349.

Carmon, Y., S. Neuman, and D. Yaffe. 1979. Synthesis of tropomyosin in myogenic cell cultures and in RNA-directed heterologous cell-free system. Qualitative changes in the polypeptides associated with differentiation. *Cell* **14:** 383.

Carmon, Y., H. Czosnek, U. Nudel, M. Shani, and D. Yaffe. 1982., DNase sensitivity of genes expressed during myogenesis. *Nucleic Acids Res.* (in press).

Firtel, R.A., R. Timm, A.R. Kimmel, and M. McKeown. 1979. Unusual nucleotide sequences at the 5′ end of actin genes in *Dictyostelium discoideum. Proc. Natl. Acad. Sci.* **76:** 6206.

Fyrberg, E.A., B.J. Bond, N.D. Hershel, K.S. Mixter, and N. Davidson. 1981. The actin genes of *Drosophila:* Protein coding regions are highly conserved but intron positions are not. *Cell* **24:** 107.

Gallwitz, D. and I. Sures. 1980. Structure of a split yeast gene: Complete nucleotide sequence of the actin gene in *Saccharomyces cerevisiae. Proc. Natl. Acad. Sci.* **77:** 2546.

Gilbert, W. 1978. Why genes in pieces. *Nature* **271:** 501.

Ginzburg, I., A. De Baetselier, M.D. Walker, L. Behar, H. Lehrach, A.M. Frischauf, and U.Z. Littauer. 1980. Brain tubulin and actin cDNA sequences: Isolation of recombinant plasmids. *Nucleic Acids Res.* **8:** 3553.

Hunter, T. and J.I. Garrels. 1977. Characterization of the mRNAs for α-, β- and γ-actin. *Cell* **12:** 761.

Katcoff, D., U. Nudel, D. Zevin-Sonkin, Y. Carmon, M. Shani, H. Lehrach, A.M. Frischauf, and D. Yaffe. 1980. Construction of recombinant plasmids containing rat muscle actin and myosin light chain DNA sequences. *Proc. Natl. Acad. Sci.* **77:** 960.

Long, E.O. and I.B. Dawid. 1980. Repeated genes in eukaryotes. *Annu. Rev. Biochem.* **49:** 727.

Maniatis, T., E.T. Fritsch, J. Lauer, and B. Lawn. 1980. The molecular genetics of human hemoglobins. *Annu. Rev. Genet.* **14:** 145.

Ng, R. and J. Abelson. 1980. Isolation and sequence of the gene for actin in *Saccharomyces cerevisiae. Proc. Natl. Acad. Sci.* **77:** 3912.

Nudel, U., D. Katcoff, R. Zakut, M. Shani, Y. Carmon, M. Finer, H. Czosnek, I. Ginzburg, and D. Yaffe. 1982. Isolation and characterization of rat skeletal muscle and cytoplasmic actin genes. *Proc. Natl. Acad. Sci.* (in press).

Scheller, R.H., L.B. McAllister, W.R. Crain, D.S. Durica, J.W. Posakony, T.L. Thomas, R.J. Britten, and E.H. Davidson. 1981. Organization and expression of multiple actin genes in the sea urchin. *Mol. Cell. Biol.* **1:** 609.

Shah, D.M., R.C. Hightower, and R.M. Meagher. 1982. Complete nucleotide sequence of a soybean actin gene. *Proc. Natl. Acad. Sci.* **79:** 1022.

Shani, M., D. Zevin-Sonkin, O. Saxel, Y. Carmon, D. Katcoff, U. Nudel, and D. Yaffe. 1981a. The correlation between the synthesis of skeletal muscle actin, myosin heavy chain and myosin light chain and the accumulation of the corresponding mRNA sequences during myogenesis. *Dev. Biol.* **86:** 483.

Shani, M., U. Nudel, D. Zevin-Sonkin, R. Zakut, D. Givol, D. Katcoff, Y. Carmon, J. Reiter, A.M. Frischauf, and D. Yaffe. 1981b. Skeletal muscle actin mRNA: Characterization of the 3′ untranslated region. *Nucleic Acids Res.* **9:** 579.

Southern, E.M. 1975. Detection of specific sequences among DNA fragments separated by gel electrophoresis. *J. Mol. Biol.* **98:** 503.

Vandekerckhove, J. and K. Weber. 1978. At least six different actins are expressed in a higher mammal: An analysis based on the amino acid sequence of the amino-terminal tryptic peptide. *J. Mol. Biol.* **126:** 783.

———. 1979. The complete amino acid sequence of actins from bovine aorta, bovine heart, bovine fast skeletal muscle and rabbit slow skeletal muscle. *Differentiation* **14:** 123.

Waring, G.L. and A.P. Mahowald. 1979. Identification and time of synthesis of chorion proteins in *Drosophila melanogaster. Cell* **16:** 599.

Whalen, R.G., S.M. Sell, G.S. Butler-Browne, K. Schwartz, P. Bouveret, and I. Pinset-Härstrom. 1981. Three myosin heavy-chain isozymes appear sequentially in rat muscle development. *Nature* **292:** 805.

Zakut, R., M. Shani, D. Givol, S. Neuman, D. Yaffe, and U. Nudel. 1982. The nucleotide sequence of the rat skeletal muscle actin gene. *Nature* (in press).

Changes in Myosin and Actin Gene Expression and DNase-I Sensitivity Associated with Terminal Differentiation of Myogenic Cultures

MOSHE SHANI
DINA ZEVIN-SONKIN
YORAM CARMON
HENRYK CZOSNEK
URI NUDEL
DAVID YAFFE
Department of Cell Biology
The Weizmann Institute of Science
Rehovot 76100, Israel

A central issue concerning myogenesis, as well as the differentiation of other cell types, is how the changes in gene expression during terminal differentiation are controlled. Are the changes in protein synthesis a result of transcriptional activity of the genome, or are there additional control mechanisms operating at the posttranscriptional level (e.g., processing of nuclear RNA precursors, polyadenylation, or translation control)? Several observations have indicated that mRNAs coding for muscle-specific proteins exist in mononucleated myoblasts, prior to their utilization in protein synthesis. For example, the addition of actinomycin D to rat primary muscle cultures approaching the phase of cell fusion did not prevent cell fusion nor the onset of synthesis of myosin heavy chain or creatine phosphokinase for a period of 6–8 hours (Shainberg et al. 1971; Yaffe and Dym 1972; Dym et al. 1978). In agreement with these results, it was reported that in calf skeletal-muscle cultures, a 26S RNA (presumptive myosin heavy-chain mRNA) was found in

the mononucleated cells, mainly as free messenger ribonucleoprotein (mRNP) (Buckingham et al. 1976). In addition, DNA complementary to this presumptive mRNA was used to measure the accumulation of the mRNA in chick primary muscle cultures (Robbins and Heywood 1978; Dym et al. 1979). Those results indicated that the maximal level of myosin heavy-chain mRNA was found at the onset of cell fusion, when the rate of synthesis of the corresponding protein was relatively low. They also found that this mRNA shifted from a free mRNP fraction to polysomes during differentiation, suggesting that at the onset of cell fusion most of the mRNA was stored in an inactive form. In contrast, analysis of the changes in mRNA populations, by cell-free translation of RNA extracted at various stages of differentiation, has revealed no evidence for the accumulation of significant amounts of inactive mRNA (Yablonka and Yaffe 1977; Devlin and Emerson 1979).

In the first part of this paper, we focus on the question of whether the rate of muscle-specific protein synthesis is controlled primarily by the availability of the corresponding mRNA or by the activation of preexisting mRNA molecules. Using cloned cDNA probes, we measured the relative amounts of mRNA coding for three defined proteins at various stages of differentiation. The results of these experiments show that in differentiating rat muscle cultures there is no evidence for the accumulation of significant amounts of inactive mRNA prior to the phase of rapid synthesis of muscle-specific proteins.

In the second part of this paper, we deal with the question concerning the time of activation of a muscle-specific gene. In particular, we tried to determine whether, in proliferating myoblasts, a gene that is programmed to be expressed later in development is in a different conformation in chromatin than a gene that is not expressed in muscle cells. The DNase-I sensitivity assay (Weintraub and Groudine 1976) was applied in this study.

RESULTS AND DISCUSSION

Characteristics of Recombinant Plasmids Carrying Sequences Homologous to Myosin Heavy-chain, Myosin Light-chain-2, and Actin mRNAs

A prerequisite for a study of the control of muscle-specific gene expression during differentiation is the availability of pure and well-defined probes. We have constructed several recombinant DNA plasmids containing sequences homologous to myosin heavy chain, myosin light chain 2, and actin. Plasmid p82 hybridizes specifically with myosin heavy-chain mRNA from muscle tissue and differentiated culture (Nudel et al. 1980). The insert of another plasmid, p103, contains 213 nucleotides coding for 71 amino acids from the carboxyterminal end of myosin light chain 2 plus 61 nucleotides within the 3′-untranslated region (Katcoff et al. 1980; M. Shani et al., in collaboration with J. Calvo, unpubl.). p749 contains an insert of 567 nucleotides coding for 189 amino acids (171–360) in the carboxyter-

minal half of skeletal-muscle actin (Katcoff et al. 1980; Shani et al. 1981b). This plasmid cross-hybridizes with the cytoplasmic β- and γ-actin mRNAs.

Accumulation of mRNA for Myosin Heavy Chain, Myosin Light Chain 2, and Actin during Differentiation

The three cDNA plasmids were used as probes to measure the accumulation of the corresponding mRNA during differentiation of primary rat muscle cell cultures. Forty-hour-old primary cultures grown in FE medium (i.e., consisting mostly of proliferating cells) were triggered to differentiate by a change to S medium (Yaffe 1973). The phase of rapid cell fusion started at about 16 hours following the change of medium. At several time points after the change of medium, one plate was stained with Giemsa for measurement of cell fusion, and another plate was labeled with [^{35}S]methionine for 1 hour to assay protein synthesis. Polyadenylated RNA was extracted at each time point from additional plates (Auffray and Rougeon 1980). Equal amounts of mRNA were fractionated on denaturing agarose gels (Bailey and Davidson 1976). Each gel also contained a series of dilutions of mRNA from differentiated cultures that served as calibration standards for quantitation of the amounts of mRNA. The RNA was transferred from the gel to DBM-cellulose papers and hybridized to ^{32}P-labeled plasmid DNA (Fig. 1; Alwine et al. 1979). The relative amount of the specific mRNA present at each time point was determined by scanning the autoradiographs. The results, expressed as a percentage of the maximal level, are shown in Figure 2. The accumulation of myosin heavy-chain and myosin light-chain-2 mRNA sequences became detectable about 15 hours after the change of medium (Figs. 1 and 2). During differentiation, there was an increase of about 50-fold in the amount of these mRNAs. The phase of accumulation of α-actin mRNA preceded that of myosin heavy and light chains by 4 hours (Figs. 1 and 2). The amount of this mRNA increased about 13-fold during differentiation. In older cultures (58 hr), the amounts of myosin light-chain and α-actin mRNAs decreased, possibly reflecting deterioration of the cultures due to aging.

Figure 2 also shows that there was a background of muscle-specific RNA at 0–10 hours after the change of medium. This may be due to the small amount of fibers always present in primary cultures prior to the main phase of cell fusion. However, it may also be due to the presence of small amounts of muscle-specific mRNA in the mononucleated cells. To resolve this question, a similar experiment was performed with a cloned myogenic cell line, L8 (Fig. 3), which starts fusing at about 30 hours after the change of medium. In these cultures the muscle-specific mRNA sequences were not detectable during the first 20 hours following the change of medium. The accumulation of α-actin and of myosin heavy-chain and light-chain-2 mRNAs started about 10 hours prior to the onset of cell fusion. However, the main increase in the amount of hybridizable mRNA occurred during the phase of rapid cell fusion.

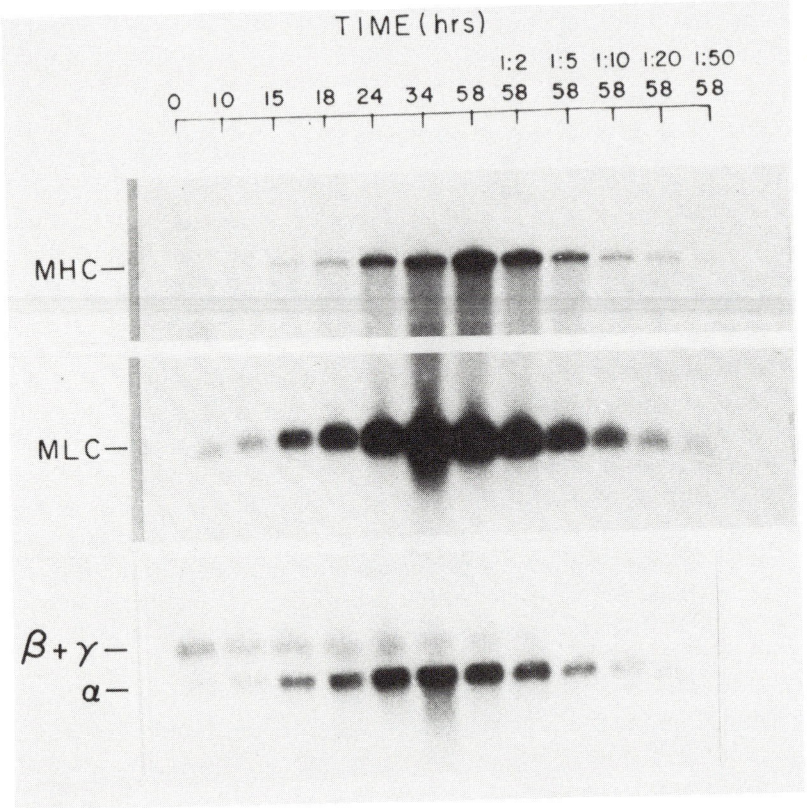

FIGURE 1
Hybridization of cloned myosin heavy-chain (p82), myosin light-chain-2 (p103), and actin (p749) probes to polyadenylated RNA extracted from rat primary muscle cultures harvested at various times after a change to S medium. Polyadenylated RNA (20 μg of RNA extracted at the indicated time points and 10 μg, 4 μg, 2 μg, and 1 μg of RNA extracted at 58 hr) was electrophoresed on 1% agarose gel, transferred to DBM paper, and hybridized to ^{32}P nick-translated inserts of plasmids p82, p103, and p749. (MHC, MLC) The positions of myosin heavy-chain and myosin light-chain-2 mRNAs, respectively. (α, β, γ) The positions of mRNA coding for α-, β-, and γ-actins, respectively. Only the relevant parts of these gels are presented. (Data from Shani et al. 1981a.)

It can also be seen (Fig. 3) that, concomitant with the increase in the amount of α-actin mRNA, there was a decrease in the amount of the cytoplasmic β- and γ-actin mRNAs. This was especially clear when the myogenic cell line was used. Thus, the isoform transition that is observed during terminal differentiation is also reflected at the level of accumulation of the corresponding mRNA.

The above experiments were performed with poly(A)$^+$ RNA. Since myosin heavy-chain mRNA had been reported to exist in the oligo-

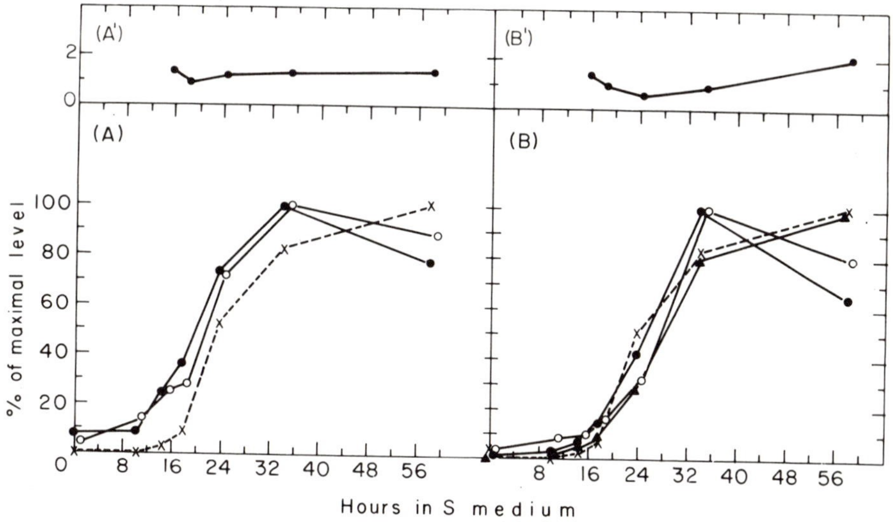

FIGURE 2
Synthesis of muscle proteins and accumulation of corresponding mRNA during differentiation of rat primary skeletal-muscle cultures. The relative concentration of specific mRNA in polyadenylated RNA populations at each time point was determined by scanning an underexposed autoradiograph from the experiment shown in Fig. 1. The relative rate of synthesis of the proteins was determined by cutting the corresponding spots from two-dimensional gels, as described in the text. Cell fusion was measured by counting nuclei in the fibers (Yablonka and Yaffe 1977). Results are expressed as percentages of the maximal level. (*A*) Accumulation of α-actin mRNA (●), rate of synthesis of α-actin (○), and cell fusion (×). (*B*) Accumulation of myosin light-chain-2 mRNA (●), myosin heavy-chain mRNA (▲), rate of synthesis of myosin light chain (○), and cell fusion (×). (*A'*,*B'*) The ratio between the rate of synthesis of α-actin and myosin light chain 2, respectively, to the amount of the corresponding mRNA at each time point (arbitrary units). (Data from Shani et al. 1981a.)

(dT)-unbound RNA fraction (Benoff and Nadal-Ginard 1979), we checked the possible existence, prior to cell fusion, of stored mRNA, which does not bind to oligo(dT) cellulose. The results of this experiment clearly demonstrated that in primary cultures and in L8 cells, both at the onset of cell fusion and in differentiated cultures, the great majority of α-actin and myosin light-chain-2 mRNAs were in the oligo(dT)-bound poly(A)$^+$ RNA fraction. In RNA from differentiated cultures, about 40% of the myosin heavy chain did not bind to oligo(dT). However, no myosin heavy-chain mRNA was detectable in this fraction when RNA from cultures harvested at the onset of cell fusion was tested (Shani et al. 1981a). These results thus rule out the possibility that significant amounts of inactive mRNA coding for myosin heavy chain, myosin light chain 2, and α-actin accumulate in a non-polyadenylated form prior to cell fusion.

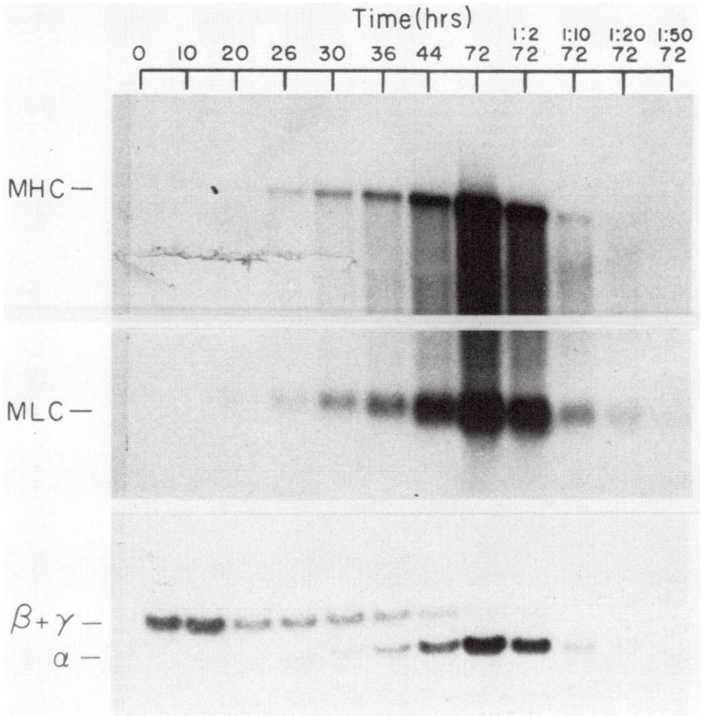

FIGURE 3
Hybridization of cloned myosin heavy-chain (p82), myosin light-chain-2 (p103), and actin (p749) probes to polyadenylated RNA extracted from L8 cultures at various times after a change to 2HI medium. (For details, see legend to Fig. 1). The phase of rapid cell fusion started 30 hr after change to 2HI medium. (Data from Shani et al. 1981a.)

Synthesis of Myosin Light Chain and α-Actin during Myogenesis

The accumulation of the mRNA for myosin light chain 2 and α-actin was compared to the rate of synthesis of the corresponding proteins. At each time point, one plate was labeled with [^{35}S]methionine for 1 hour, and the cell lysates containing an equal number of counts were fractionated by two-dimensional gel electrophoresis (Hunter and Garrels 1977). Quantitation was done by measuring radioactivity in spots corresponding to myosin light chain 2 and α-actin. The results (Fig. 2) show that the rates of synthesis of myosin light chain 2 and α-actin follow very closely the rate of accumulation of these mRNAs. Furthermore, as was the case for the corresponding mRNAs, the synthesis of α-actin preceded that of myosin light chain 2.

Myosin heavy-chain synthesis had to be analyzed by one-dimensional SDS-polyacrylamide gel electrophoresis, since this large protein does not

enter the isoelectric focusing gel. Although we were unable to obtain exact quantitation in this experiment, the increase in the rate of synthesis of myosin heavy chain appeared to parallel the accumulation of myosin heavy-chain mRNA (Shani et al. 1981a).

Distribution of Actin and Myosin Light-chain-2 mRNAs on Polysomes and Free mRNP Fractions

It had been reported that, in primary chick muscle cultures at the beginning of cell fusion (when ~20% of the nuclei are in multinucleated fibers), myosin heavy-chain mRNA was found mostly in the 80S–100S (free mRNP) fraction. In well-differentiated cultures (when ~80% of the nuclei are in fibers), it was found mostly in heavy polysomes. This suggested that prior to cell fusion, myosin heavy-chain mRNA was stored as an inactive free mRNP (Dym et al. 1979).

We investigated the distribution of actin and myosin light-chain-2 mRNAs between polysomes and free mRNPs in primary cultures and in the myogenic cell line. Cytoplasmic fractions obtained from primary cultures harvested 19 hours after the change to S medium (~10% maximal cell fusion) and from differentiated cultures at 54 hours after the change to S medium (80% maximal fusion) were subjected to sucrose gradient centrifugation. Fractions corresponding to polysomes and heavy mRNP and light mRNP were pooled. Equal amounts of total RNA extracted from each fraction were loaded on 1% agarose gels, electrophoresed, transferred to DBM paper, hybridized to ^{32}P-labeled plasmids p749 and p103, and autoradiographed. The results (Fig. 4) showed that more than 85% of α-actin and myosin light-chain-2 mRNA sequences were found on polysomes, both at the beginning of cell fusion and in differentiated cultures.

The results presented here, showing a close correlation between accumulation of mRNA coding for three major muscle proteins and the rate of synthesis of these proteins, are in agreement with the results of experiments in which accumulation of mRNA coding for myosin light chain was assayed by translation in a cell-free system (Yablonka and Yaffe 1977). Similarly, the accumulation of translatable mRNA for a number of muscle proteins in quail muscle cells was found to correlate well with the synthesis of these proteins (Devlin and Emerson 1971). The fact that hybridizable RNA, translatable RNA, and protein all accumulate at about the same time suggests that, in this cell system, activation of mRNA does not serve as a major mechanism to control the timing and rate of protein synthesis. This conclusion is also supported by experiments in which the distribution of α-actin and myosin light-chain-2 mRNAs between polysome and free mRNP fractions was determined. The majority of these mRNAs was associated with polysomes both in cultures containing mostly myotubes and in cultures containing mostly myoblasts.

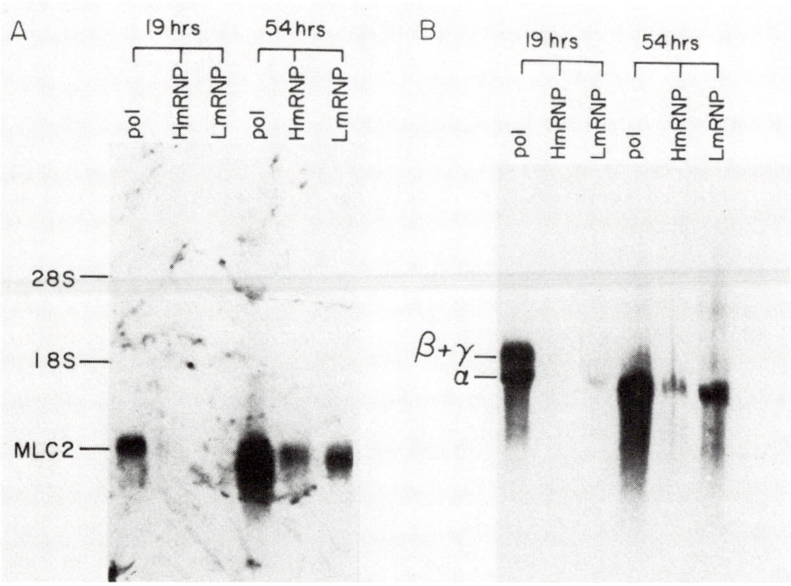

FIGURE 4
Distribution of myosin light-chain-2 mRNA (*A*) and actin mRNA (*B*) on polysomes and free mRNPs. Postmitochondrial supernatants obtained from rat primary muscle cultures were harvested at 19 hr (10% fusion) and 54 hr (80% fusion) after the change to S medium. The supernantants were subjected to sucrose gradient centrifugation. RNA was extracted from fractions designated as pol (>80S), HmRNP (80S–60S), and LmRNP (<60S) and hybridized to inserts of plasmids p103 (*A*) or p749 (*B*).

As mentioned earlier, although the main increase in hybridizable RNA occurred during the phase of rapid cell fusion, small amounts of these mRNAs became detectable a few hours before cell fusion was observed. The small amounts of muscle-specific mRNA present before cell fusion may explain earlier observations on the initial rise in muscle-specific protein synthesis following application of actinomycin D (Shainberg et al. 1971; Yaffe and Dym 1972; Dym et al. 1978).

We do not as yet have an explanation for the difference between our results and those indicating that at the onset of cell fusion most of the mRNA is stored in an inactive form (Robbins and Heywood 1978; Dym et al. 1979). This may stem from the quality of the probes used or the differences in the experimental system.

The available data suggest that during differentiation of rat skeletal-muscle cultures, the rate of muscle-specific protein synthesis is determined by the amount of the corresponding mRNA. However, the mechanism controlling the mRNA levels is unknown. The number of genes, the rate of their

transcription, the processing and transport of the mRNA precursors, and the half-life of these mRNAs in the cytoplasm are all potential steps of control of the amounts of mRNA. Using recombinant DNA techniques it should soon be possible to evaluate which of these mechanisms is operating during terminal differentiation.

Organization of Chromatin of the Myosin Light-chain-2 Gene in Nuclei of Proliferating Myoblasts

In other systems, it has recently been shown that transcriptionally active genes are in an altered chromatin conformation compared to inactive genes (for review, see Mathis et al. 1980). This altered conformation was correlated with the presence of specific nonhistone proteins, such as high mobility group (HMG) 14 and HMG 17, and conferred sensitivity to DNase-I digestion (Weisbrod et al. 1980). A question of interest is whether, in muscle precursor cells, a gene programmed to be expressed during terminal differentiation is in a different conformation in chromatin than a gene that is not expressed in muscle cells.

Nuclei were isolated from cloned populations of proliferating cells of the rat myogenic cell line L8 and from well-differentiated cultures consisting predominantly of multinucleated fibers. The nuclei were digested with increasing concentrations of DNase I. The extent of digestion was monitored chemically by measuring the degree of DNA solubilization in acid. The DNA was isolated and restricted with *Eco*RI. The restriction fragments were separated on agarose gels, blotted onto nitrocellulose filters, and hybridized with nick-translated p103 DNA, carrying myosin light-chain-2 DNA sequences. As an internal control, we chose a plasmid DNA (B1) containing sequences homologous to the constant region of rat immunoglobulin gene (kindly provided by I. Schechter, The Weizmann Institute of Science). This gene is not expressed in muscle cells. Plasmid p103 hybridized to a single 3.0-kb DNA fragment, whereas plasmid B1 hybridized to a 6.8-kb DNA fragment. The autoradiograms of these hybridizations were scanned, and the results are summarized in Figure 5. As expected for an active gene, the myosin light-chain-2 gene was preferentially sensitive to DNase-I digestion in nuclei from differentiated cultures. However, in nuclei from proliferating cells, in which no detectable transcripts for myosin light chain 2 were found, this DNA fragment was as insensitive to DNase-I digestion as the immunoglobulin gene. These results showed that although proliferating myoblasts are programmed to synthesize myosin light chain 2 later during differentiation, the chromatin conformation of this gene resembles that of a gene that is not expressed in muscle cells. Similar results were obtained with a skeletal-muscle actin gene (Carmon et al. 1982).

The conformational change in these genes, reflected in DNase-I sensitivity, indicates a qualitative change in their transcriptional activity. Thus, the change in the amount of muscle-specific mRNAs, which takes place during

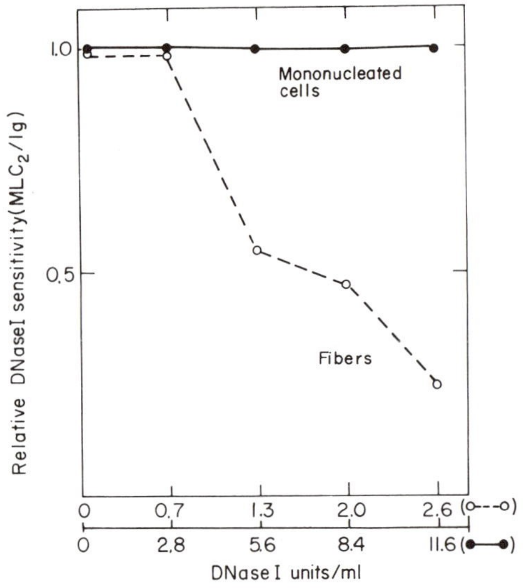

FIGURE 5
Preferential DNase-I sensitivity of myosin light-chain-2 gene in nuclei from differentiated L8 cultures. Nuclei were isolated from proliferating myoblasts (mononucleated cells) and differentiated L8 cultures (fibers) (Jacquet et al. 1974). Nuclei at a final concentration of 1 mg/ml of DNA were digested at 37°C for 10 min with increasing concentrations of DNase I. DNA was extracted from treated and untreated nuclei (Gross-Bellard et al. 1973), restricted with *Eco*RI, electrophoresed on 1% agarose gel, and transferred to nitrocellulose papers (Southern 1975). The DNA-containing filter was hybridized to ^{32}P nick-translated plasmids p103 and B1 (plasmids containing the C_k region [constant region] of rat immunoglobulin $\varkappa$ gene). The autoradiograph was scanned, and the results are expressed as the relative sensitivity of myosin light-chain-2 gene to DNase-I digestion. (MLC2) Myosin light chain 2; (Ig) the C_k immunoglobulin gene. (Data from Carmon et al. 1982.)

terminal differentiation, seems to be primarily a result of a change in transcription rather than a change in the processing or the stability of these mRNAs. Using a cloned population of myogenic cells, we have also demonstrated that the change in chromatin conformation of a developmentally regulated gene occurs within the same cell lineage (Carmon et al. 1982). Thus, the changes that render such genes sensitive to DNase I apparently take place close to the time of their phenotypic expression.

SUMMARY

The accumulation of myosin heavy-chain, myosin light-chain-2, and α-actin mRNAs during differentiation of rat skeletal-muscle cell cultures was measured using cloned cDNA probes. This was compared with the rate

of synthesis of the corresponding proteins. Although small amounts of these mRNA sequences were detectable a few hours before cell fusion, the main increase in hybridizable RNA occurred during the phase of rapid cell fusion. There was a very close correlation between the levels of these three mRNAs and the rates of synthesis of the corresponding proteins. The majority of myosin light-chain-2 and α-actin mRNAs were found associated with polysomes both at the beginning of cell fusion and in well-differentiated cultures. These results therefore suggest that in this cell system, activation of stored mRNA does not play a significant role in the control of the timing and rate of muscle protein synthesis.

Nuclei isolated from proliferating myoblasts and differentiated cultures were treated with DNase I. The preferential sensitivity was measured by blot hybridization with probes for myosin light-chain-2 and actin genes. The results showed that genes coding for the skeletal-muscle actin and myosin light chain 2 are DNase I sensitive in differentiated cultures but not in proliferating myoblasts.

ACKNOWLEDGMENTS

We thank Ora Saxel, Zehava Levi, and Sara Neuman for their excellent assistance in this work. These investigations were supported by a grant from the Muscular Dystrophy Association of America and by the National Institutes of Health (grant RO1-GM-22767).

REFERENCES

Alwine, J.C., D.J. Kemp, B.A. Parker, J. Reiser, J. Renart, G.R. Stark, and G.M. Wahl. 1979. Detection of specific RNAs or specific fragments of DNA by fractionation in gels and transfer to diazobenzyloxymethyl paper. *Methods Enzymol.* **68:** 220.

Auffray, C. and F. Rougeon. 1980. Purification of immunoglobulin heavy chain mRNA from total myeloma tumor RNA. *Eur. J. Biochem.* **107:** 303.

Bailey, J.M. and N. Davidson. 1976. Methylmercury as a reversible denaturing agent for agarose gel electrophoresis. *Anal. Biochem.* **70:** 75.

Benoff, S. and B. Nadal-Ginard. 1979. Most myosin heavy chain mRNA in L_6E_9 rat myotubes has a short poly(A) tail. *Proc. Natl. Acad. Sci.* **76:** 1853.

Buckingham, M.E., A. Cohen, and F. Gros. 1976. Cytoplasmic distribution of pulse-labeled poly(A)-containing RNA, particularly 2GS RNA, during myoblast differentiation. *Dev. Biol.* **69:** 202.

Carmon, Y., H. Czosnek, U. Nudel, M. Shani, and D. Yaffe. 1982. DNase I sensitivity of genes expressed during myogenesis. *Nucleic Acids Res.* **10:** 3085.

Devlin, R.B. and C.P. Emerson. 1979. Coordinate accumulation of contractile protein mRNAs during myoblast differentiation. *Dev. Biol.* **69:** 202.

Dym, H.P., D.S. Kennedy, and S.M. Heywood. 1979. Subcellular distribution of the cytoplasmic myosin heavy chain mRNA during myogenesis. *Differentiation* **12:** 145.

Dym, H., D. Turner, H.M. Eppenberger, and D. Yaffe. 1978. Creatine kinase isoenzyme transition in actinomycin D-treated differentiating muscle cultures. *Exp. Cell Res.* **113:** 15.

Gross-Bellard, M., P. Oudet, and P. Chambon. 1973. Isolation of high molecular weight DNA from mammalian cells. *Eur. J. Biochem.* **36:** 32.

Hunter, T. and J.I. Garrels. 1977. Characterization of the mRNAs for α, β and γ actin. *Cell* **12:** 761.

Katcoff, D., U. Nudel, D. Zevin-Sonkin, Y. Carmon, M. Shani, H. Lehrach, A.M. Frischauf, and D. Yaffe. 1980. Construction of recombinant plasmids containing rat muscle actin and myosin light chain DNA sequences. *Proc. Natl. Acad. Sci.* **77:** 960.

Mathis, D., P. Oudet, and P. Chambon. 1980. Structure of transcribing chromatin. *Prog. Nucleic Acids Res.* **7:** 2105.

Nudel, U., D. Katcoff, Y. Carmon, D. Zevin-Sonkin, Z. Levi, Y. Shaul, M. Shani, and D. Yaffe. 1980. Identification of recombinant phages containing sequences from different rat myosin heavy chain genes. *Nucleic Acids Res.* **8:** 2133.

Robbins, J. and S.M. Heywood. 1978. Quantification of myosin heavy chain mRNA during myogenesis. *Eur. J. Biochem.* **82:** 601.

Shainberg, A., G. Yagil, and D. Yaffe. 1971. Alteration of enzymatic activities during muscle differentiation in vitro. *Dev. Biol.* **25:** 1.

Shani, M., D. Zevin-Sonkin, O. Saxel, Y. Carmon, D. Katcoff, Y. Nudel, and D. Yaffe. 1981a. The correlation between the synthesis of skeletal muscle actin, myosin heavy chain and myosin light chain and the accumulation of corresponding mRNA sequences during myogenesis. *Dev. Biol.* **86:** 483.

Shani, M., U. Nudel, D. Zevin-Sonkin, R. Zakut, D. Givol, D. Katcoff, Y. Carmon, J. Reiter, A.M. Frischauf, and D. Yaffe. 1981b. Skeletal muscle actin mRNA characterization of the 3′ untranslated region. *Nucleic Acids Res.* **9:** 579.

Southern, E.M. 1975. Detection of specific sequences among DNA fragments separated by gel electrophoresis. *J. Mol. Biol.* **98:** 503.

Weintraub, H. and M. Groudine. 1976. Chromosomal subunits in active genes have an altered conformation. *Science* **195:** 848.

Weisbrod, S.M., M. Groudine, and H. Weintraub. 1980. Interaction of HMG 14 and 17 with actively transcribed genes. *Cell* **19:** 289.

Yablonka, Z. and D. Yaffe. 1977. Synthesis of myosin light chains and accumulation of translatable mRNA coding for light chain-like polypeptides in differentiating muscle cultures. *Differentiation* **8:** 133.

Yaffe, D. 1973. Rat skeletal muscle cells. In *Tissue culture: Methods and applications* (ed. P.F. Kruse and M.K. Patterson), p. 106. Academic Press, New York.

Yaffe, D. and H. Dym. 1982. Gene expression during differentiation of contractile muscle fibers. *Cold Spring Harbor Symp. Quant. Biol.* **37:** 543.

Messengers Coding for Myosins and Actins: Their Accumulation during Terminal Differentiation of a Mouse Muscle Cell Line

MARGARET E. BUCKINGHAM
ADRIAN J. MINTY
BENÔIT ROBERT
SERGE ALONSO
ARLETTE COHEN
PHILIPPE DAUBAS
ANDRÉ WEYDERT
Department of Molecular Biology
Pasteur Institute
75015 Paris, France

MARIO CARAVATTI
Friedrich Miescher Institute
Basel CH-4002, Switzerland

The appearance, accumulation, and sarcomeric assembly of muscle-specific isoforms of the contractile proteins characterize the terminal differentiation of muscle cells (for review, see Buckingham 1977). For some of these proteins, different isoforms are also present at different stages of muscle tissue development. For example, in mammalian skeletal muscle, an embryonic form of myosin heavy chain (MHC_{emb}) (Whalen et al. 1981), a newborn form of myosin heavy chain (MHC_{NB}) (Whalen et al. 1981), and an embryonic form of an alkali myosin light chain (MLC_{1emb}) (Whalen et al. 1978) have been described. The contractile-protein genes are thus present as multigene families; different genes of a family may be expressed in different

nonmuscle or muscle tissues and at different stages of development in any one muscle. To understand how a muscle phenotype is established, it is essential to determine at what level of cellular regulation changes in contractile gene expression take place during myogenesis. A number of indirect approaches have been employed, which include experiments with metabolic inhibitors (e.g., Yaffe and Dym 1973), studies of mRNA metabolism (e.g., Buckingham et al. 1976), comparisons of in vivo protein synthesis and the products of mRNA translation in vitro (Strohman et al. 1977; Yablonka and Yaffe 1977; Devlin and Emerson 1979; Daubas et al. 1981), and molecular hybridization experiments using total mRNA populations (Paterson and Bishop 1977; Leibovitch et al. 1979; Affara et al. 1980; Zevin-Sonkin and Yaffe 1980) or highly enriched mRNA fractions (John et al. 1978; Benoff and Nadal-Ginard 1979; Dym et al. 1979). As a result of these attempts, both transcriptional and posttranscriptional control mechanisms have been advocated. The advent of recombinant DNA technology has made it possible to obtain nucleotide sequences coding for specific muscle proteins and to use these as probes of the mRNA population at different stages of myogenesis. Such recombinant probes also make it possible to look at the topological organization, structure, and conformational states of the muscle genes.

We have chosen to work with coding sequences for the contractile proteins in the mouse, both because of the possibility of exploiting current knowledge of mouse genetics and embryology and the potential utility of the mouse teratocarcinoma system (for review, see Jacob 1978) for looking at earlier stages of myogenesis. For part of the work reported here, we have used a permanent skeletal-muscle cell line that was derived from a mouse teratocarcinoma (Jakob et al. 1978). In these cells and in different mouse tissues, cloned myosin and actin probes have been used to look at accumulation of the different mRNAs of these multigene families.

RESULTS AND DISCUSSION

Recombinant plasmids containing DNA sequences complementary to different size fractions of poly(A)$^+$ RNA prepared from the skeletal muscle of newborn (6–12-day-old) mice were obtained by G-C tailing and insertion at the *Pst*I site of pBR322. Recombinants cloned in *Escherichia coli* C600 were identified by gel analysis of the translation products after hybridization with mRNA and then by DNA sequence analysis (for details, see Minty et al. 1981). Plasmids containing DNA insertions coding for actins and myosins are described in Table 1. The sizes of the mRNAs with which they hybridize are also shown. In most of the experiments described, the Northern blot technique (Alwine et al. 1979) was used to obtain information on the size, relative sequence homology, and accumulation of the corresponding mRNAs. Information on the corresponding genes was obtained by Southern blotting (Southern 1975) of mouse DNA.

TABLE 1

Actin and Myosin Recombinant Plasmids Isolated from Newborn Mouse Skeletal Muscle

Clone	Identification	Coding/noncoding		Size of insertion (nucleotides)	Size of homologous RNA (nucleotides)
p32	MHC (F or NB)	+	+	1230	6900
p161	LC_{1F}, LC_{3F}	−	+	400	1050, 900
p91	SM α-actin	+	+	1350	1600
p91-200	SM α-actin	−	+	200	1600
p81	CM α-actin	+	(+)	1080	1600
p41	nonmuscle actin (cloned from lymphoma)	+	(+)	1150	2000

MHC, myosin heavy chain; F, adult fast skeletal muscle; NB, newborn; SM, skeletal muscle; CM, cardiac muscle; (+), a few residues.

MHC mRNAs

The MHC plasmid p32 hybridizes strongly with mRNA from adult and newborn skeletal muscle and from heart muscle. It hybridizes less strongly with mRNA from embryonic tissue and from differentiated muscle cell cultures. These results would suggest that the MHC_{emb} isoform has diverged from the other MHC sequences, at least in the region represented by the probe shown by DNA sequencing to be near the COOH-terminal of the proteins. All of the mRNAs tested are of the same size, 6900 nucleotides. In addition to showing homology with RNA from different muscle tissues, this sequence also cross-hybridizes with MHC mRNA from different species (rat, man, chick). With mouse genomic DNA digested by *Eco*RI, seven to ten bands are seen on hybridization with different restriction fragments of the MHC probe.

Alkali MLC mRNAs

The MLC probe p161 has been shown by DBM filter hybridization and translation to hybridize with the fast alkali light chains LC_1 and LC_3 from adult skeletal muscle. With RNA preparations from cells expressing the embryonic isoform LC_{1emb}, this translation product is not detectable. By Northern blot analysis (Fig. 1), hybridization with two mRNAs from fast skeletal muscle is seen. When the blots are washed under more stringent conditions (50–65°C), hybridization with the two mRNAs diminishes in parallel. Either is sufficiently large to contain the coding sequence of LC_1 (570 nucleotides) or LC_3 (447 nucleotides). Plasmid 161 cross-hybridizes with rat mRNAs but is otherwise species-specific. This cross-hybridization permits us to look at the level of LC_{1emb} mRNA, since LC_{1emb} is the predominant alkali light chain in rat L6 cell lines (Whalen et al. 1978). Under the Northern blot conditions, a faint cross-hybridization with an intermediate size mRNA is seen (Fig. 1). We therefore conclude that the mRNA coding for LC_{1emb} is a rather distinct species differing in size and sequence homology from the adult forms. A similar RNA species is also detected in embryonic and adult heart tissue. This is not unexpected in view of the correspondence between LC_{1emb} and the adult cardiac atrial form of LC_1 (Whalen et al. 1982), which is also the predominant alkali light chain in embryonic hearts (Whalen and Sell 1980).

The DNA sequence of the insertion in p161 shows that it is a noncoding region of the mRNA and includes an 80–100-nucleotide-long tract of poly(A). The alkali light chains LC_1 and LC_3 share a common COOH-terminal sequence of 141 amino acids (Frank and Weeds 1974). The fact, however, that the noncoding sequence is very similar if not identical is surprising. Even in the case of the actins, which have highly conserved amino acid sequences, the 3′-noncoding sequence has diverged (see Fig. 2). The possibility therefore arises that the two sequences share a common 3′-

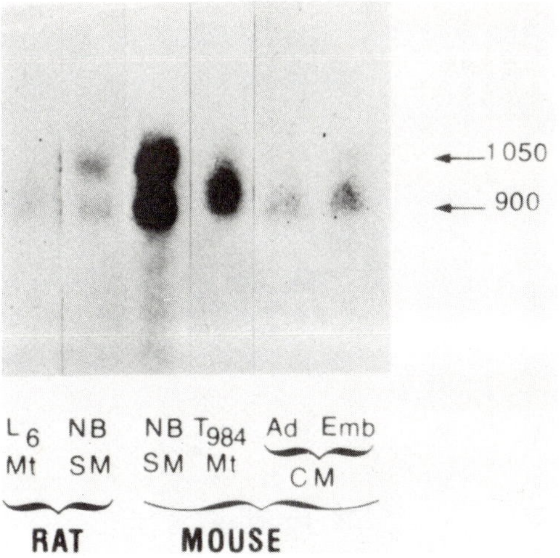

FIGURE 1
A Northern blot showing mRNAs cross-hybridizing with MLC plasmid p161. (*Left to right*) Total RNA prepared from myotubes (Mt) of the rat muscle cell line L6 (44 hr), poly(A)+ RNA from newborn rat skeletal muscle (NB, SM) (14 hr), from myotubes (Mt) of the mouse muscle cell line T984 (44 hr), from adult mouse heart (Ad, CM) (44 hr), and from embryonic mouse heart (Emb, CM) (44 hr). 1 μg of poly(A)+ RNA of each sample was denatured and electrophoresed on the same 1.5% agarose gel (McMaster and Carmichael 1977). After transfer to DBM paper, the blots were hybridized with 5×10^6 cpm of p161 DNA, ^{32}P-labeled by nick translation (Minty et al. 1981), and treated as described by Alwine et al. (1979). The exposure times for autoradiography of the different samples on the gel are shown in parentheses above. Size markers (pBR322 and phage λ DNA cut by various restriction enzymes and denatured as for the RNA) were used as standards to estimate mRNA size.

genomic fragment and that a differential splicing-type mechanism generates the two mRNAs, as suggested from the protein data by Matsuda et al. (1981). Southern blotting analysis of the genomic DNA has proved difficult because of the poly(A) sequence in the probe. We are currently analyzing recombinant phages containing MLC sequences from the mouse genome.

Actin mRNAs

As expected, given the conservation of the actin proteins, the coding sequence of plasmid p91 cross-hybridizes with actin mRNAs from different tissues and species. The muscle actin mRNAs fall into a size class of about 1600 nucleotides, whereas the nonmuscle β- and γ -actin mRNAs are about 2000 nucleotides long (see Fig. 2). Since the proteins are of similar size, this

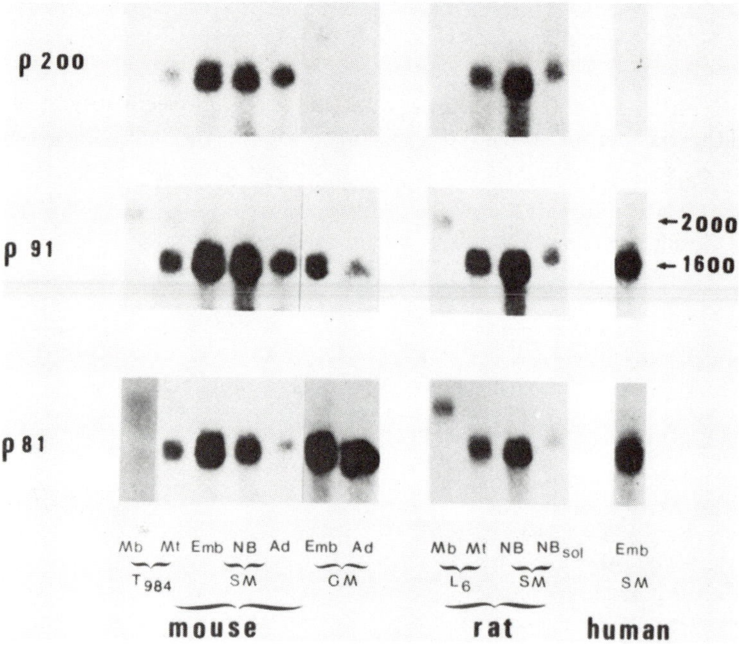

FIGURE 2

A Northern blot showing the relative homologies of actin mRNAs from different tissues, with the actin probes p81 and p91. Samples of poly(A)$^+$ RNA were separated on 1.25% agarose gels, transferred to DBM paper, and hybridized with the nick-translated probes essentially as described for the actins in the legend to Fig. 3. The same region of the blot was reused after loss of counts from the previous probe. The degree of hybridization with the different probes p81, p91, and p91-200 (p200) is therefore directly comparable. The samples shown are (*left to right*) RNA from myoblasts (Mb) and myotubes (Mt) of the mouse muscle cell line T984; from embryonic (Emb), newborn (NB), and adult (Ad) skeletal muscle (SM); and embryonic (Emb), and adult (Ad) cardiac muscle (CM) of the mouse; from myoblasts (Mb) and myotubes (Mt) of the rat muscle cell line L6; from newborn fast (NB) and newborn slow (NB_{sol}) skeletal muscle of the rat; and from human fetal (3 months) skeletal muscle (Emb, SM).

implies that the noncoding sequence of the nonmuscle actin mRNAs is considerably longer than that of the muscle mRNAs. The characterization of several plasmids with insertions corresponding to different actin-coding sequences has proved very useful, since the differing homologies of these probes permit us under appropriate hybridization conditions to distinguish the different genes and mRNAs. Thus, the differences in hybridization of the nonmuscle mRNA band with p41 and p91 suggest the presence of two mRNAs of about 2100 and 1950 nucleotides in length. Similarly, p81, which

contains a cardiaclike sequence, has permitted us to demonstrate the expression of a cardiac actin mRNA during skeletal-muscle myogenesis (see below).

The insertion of p91 extends into the noncoding region (see Minty et al. 1981), and this has permitted us to define and reclone p91-200, a noncoding 200-nucleotide sequence flanked by two *Pst*I sites in this region. Under our standard Northern blot conditions, hybridization of this probe is only seen to the α-actin mRNA from mouse skeletal muscle, and this therefore permits us to look at the expression of the mRNA for this isoform alone (see Figs. 2 and 3).

Expression of Myosin and Actin mRNAs during Myogenesis in a Mouse Muscle Cell Line

We have used the myosin and actin plasmids as probes to look at accumulation of the corresponding mRNAs during myogenesis in the T984 mouse muscle cell line (Jakob et al. 1978). Under our conditions, these cells grow as a population of dividing myoblasts for the first 80 hours in culture. At this time the cultures are confluent, and the serum content of the medium is reduced from 15% to 2% fetal calf serum in order to reduce overgrowth of the cultures. The first myotubes are detectable from about 120 hours, and large spontaneously contracting fibers are seen toward the end of the culture period (144–168 hr). Synthesis of MHC and accumulation of creatine phosphokinase activity are clearly increasing by 120 hours. LC_{1emb} is detectable on a two-dimensional gel at 96 hours and is a major spot by 120 hours. By 168 hours, its synthesis is decreasing. In contrast, LC_1 and LC_3 appear later. LC_3 synthesis is detectable from 144 hours and is still a minor component at 168 hours.

Analysis by the Northern blot technique of RNA isolated from the cultures at different times shows that the accumulation of muscle mRNAs is closely correlated with the synthesis of the muscle proteins. Thus, MHC mRNA and skeletal-muscle α-actin mRNA are detectable from 96 hours and accumulate rapidly, reaching a maximum at 144 hours (Fig. 3). Hybridization with the light-chain probe p161 follows a different time course. At 120 hours, it hybridizes weakly to a rather lighter mRNA band, as expected if this is principally the LC_{1emb} mRNA. By 144 hours, hybridization is stronger and the mRNA front now corresponds to that seen with adult LC_1/LC_3 mRNAs. We interpret these results to indicate that in this case, too, the timing of accumulation of light-chain mRNAs parallels the different kinetics of embryonic and adult MLC synthesis.

The question of whether muscle genes are transcribed in dividing myoblasts is not resolved by this approach. With the homologous probe p91-200, specific for skeletal-muscle α-actin, it is possible to calculate that

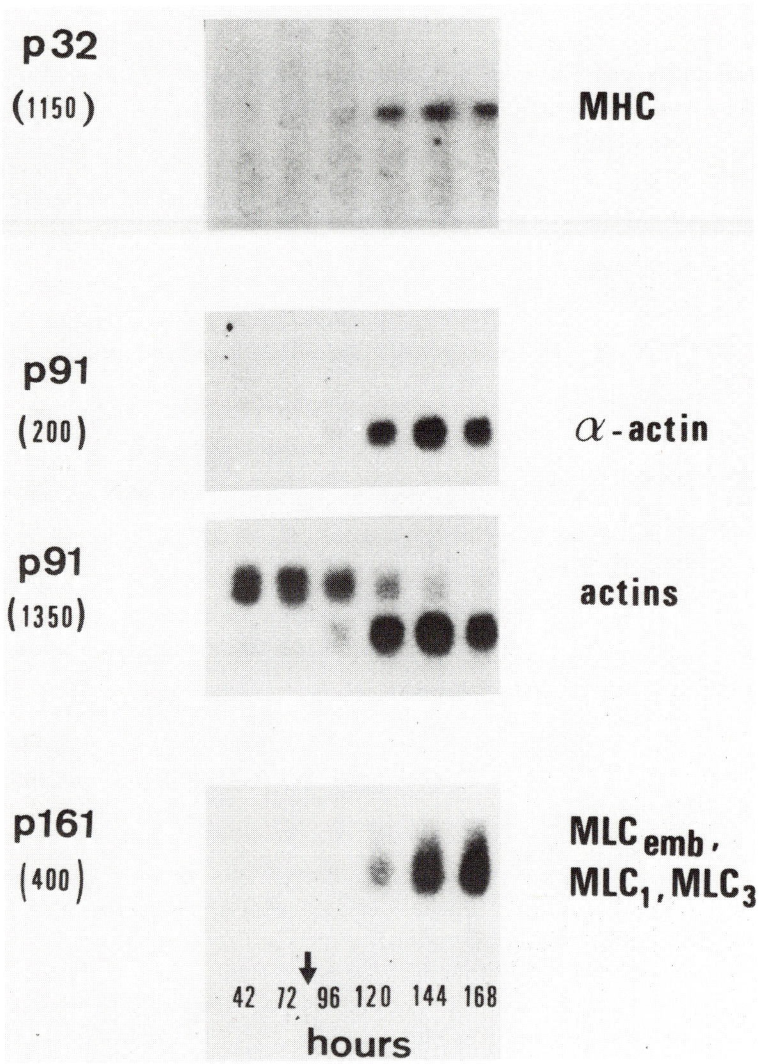

FIGURE 3
Northern blot analysis of accumulated actin and myosin mRNAs. Poly(A)$^+$ RNA (2 μg/preparation), from T984 muscle cells at different days after plating, was fractionated on 1.25% agarose gels and transferred to DBM paper as described in the legend to Fig. 1. The paper was cut into three pieces containing different RNA size classes and hybridized with nick-translated plasmid probe (see Table 1) as indicated in the legend to Fig. 1. The exposure times for the different parts of the blot were as follows: MHC, 82 hr; α-actin, 22 hr; the actins, 72 hr; MLC, 60 hr.

the lower limit of detection under our conditions is about 40 mRNA molecules per nucleus (for details, see Caravatti et al. 1982), and this therefore gives a maximum estimate of the possible concentration of muscle mRNA in dividing myoblasts. It also gives a lower limit for the accumulation of this mRNA of 130-fold. A second quantitative point concerns the stoichiometry of actin and myosin mRNA accumulation. An analysis of this kind is only meaningful if the probes are perfectly homologous to the muscle-coding sequences expressed. This is not the case for the myosin probes used in these experiments, and it is therefore not possible to assess accurately the relative amounts of myosin and actin mRNAs nor any minor differences in the timing of their appearance.

Plasmid p91 hybridizes with other actin mRNAs. Thus, the nonmuscle mRNAs can be seen to decrease at fusion in parallel with the increase in α-actin mRNA (Fig. 3). With this probe, a muscle-type actin mRNA, which is smaller ($\sim$1550 nucleotides) than the skeletal-muscle mRNA, is detectable in dividing myoblasts. We suspect that this is a smooth-muscle form (see Minty et al. 1981).

Expression of a Cardiac Actin Sequence during Skeletal-muscle Development

Plasmid p81, which was also cloned from newborn skeletal muscle, hybridizes very strongly to cardiac mRNA (Fig. 2) and much less strongly to adult skeletal-muscle RNA, compared with the skeletal-muscle α-actin plasmid p91. When RNA from embryonic skeletal muscle or differentiated cultures of T984 is hybridized with the two probes, this difference is no longer evident. With newborn skeletal muscle, an intermediate situation is seen. Similar results are seen with rat and human samples. We therefore conclude that this plasmid contains an actin sequence that is either identical or very similar to cardiac actin and that this is expressed during skeletal-muscle development. DNA sequencing of p81 shows that it codes for cardiac actin. The key nucleotide positions present in the plasmid insertion code for the amino acids that distinguish the heart isoform (Vandekerckhove and Weber 1979). It remains possible that there is a second "embryonic" actin gene that codes for a cardiaclike actin. We have attempted to clarify this by Southern blotting experiments on mouse genomic DNA cut with *Eco*RI, using p91 and p81 to reveal the actin genes most homologous to these probes. At low washing stringencies (0.1$\times$ SSC, 60°C), multiple actin bands are seen with both plasmids. However, after a 75°C wash, in the case of the blot hybridized with p91, a single band is seen at about 7.5 kb (Fig. 4). We therefore conclude that this fragment contains the skeletal-muscle α-actin gene. In the case of p91, the skeletal-muscle actin gene is clearly distinguished as a band at 7.5 kb after a 75°C wash. Plasmid p81, which appears to be less homologous to the other actin sequences, already gives one

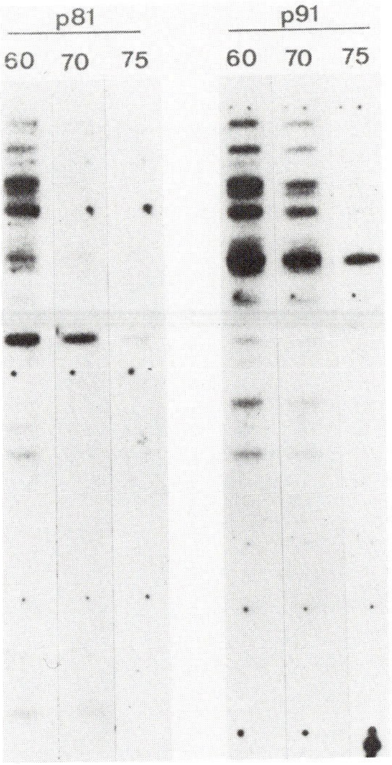

FIGURE 4
Southern blot analysis of actin genes. 5 μg of DNA prepared from mouse (129) embryos was digested with *Eco*RI and separated on 0.7% agarose gels. The DNA was transferred to nitrocellulose (Southern 1975) and hybridized with nick-translated p81 or p91 DNA. The blots were washed successively in 0.1× SSC under increasingly stringent conditions, 60°C, 70°C, and 75°C, as indicated.

major band (5.5 kb) and two minor bands (7.5 kb, 1 kb) after a 70°C wash (Fig. 4). The largest band represents slight cross-hybridization with the skeletal-muscle gene. The major 5.5-kb fragment corresponds to the 5′ end of the cardiac actin gene. The small 1-kb fragment is the 3′ end of the same gene that is cut by *Eco*RI (Minty et al. 1982). We interpret these results to indicate that p81 is homologous with a single gene coding for cardiac actin and that this gene is expressed during skeletal-muscle development. This is rather analogous to the LC_1 situation where LC_{1emb} is also found in the atria of adult hearts.

CONCLUSIONS

Experiments based on the hybridization between different cloned actin and myosin DNA sequences and RNAs from different sources demonstrated certain characteristics of the different mRNAs of these multigene families expressed in different tissues.

Within the alkali MLC group, the LC_{1emb} sequence is distinguished from that for adult LC_1 by the divergence of its noncoding sequence and by its size. A similar RNA is seen in heart RNA preparations and probably corresponds to the atrial LC_{1emb} isoform. In comparision, the adult LC_1 and LC_3 mRNAs of skeletal muscle have a highly conserved 3′-noncoding se-

quence and may indeed be transcribed from the same fragment of genomic DNA, coding for the common COOH-terminal region of these molecules.

From experiments using cloned probes to different actin-coding sequences, we are led to conclude that a cardiac actin gene is expressed during the development of skeletal muscle. Southern blotting experiments with the actin probes show at least 15–20 mouse genomic actin sequences. When washed at high stringency after hybridization with the cardiac- or with the skeletal-muscle actin probe, only the corresponding homologous gene is seen.

Examination of the accumulation of actin and myosin mRNAs during the differentiation of a mouse muscle cell line shows a close temporal correlation between this and the synthesis of the corresponding muscle proteins. No detectable accumulation of muscle mRNAs prior to differentiation is seen.

ACKNOWLEDGMENTS

The laboratory is supported by grants from the Délégation Générale à la Recherche Scientifique et Technique (DGRST), the Centre National de la Recherche Scientifique (CNRS), the Institut National de la Santé et de la Recherche Medicale (INSERM), the Commissariat à l'Energie Atomique, the Ligue Nationale Française contre le Cancer, the Fondation pour la Recherche Médicale Française, and the Muscular Dystrophy Association of America. M.C. was the recipient of a fellowship from the Swiss National Research Foundation and from INSERM; A.M., from the Muscular Dystrophy Association of America; P.D., from the Ligue Française contre le Cancer; and S.A., from the DGRST. We are grateful to R. Williamson for providing the sample of human fetal muscle RNA.

REFERENCES

Affara, N.A., B. Robert, M. Jacquet, M. Buckingham, and F. Gros. 1980. Changes in gene expression during myogenic differentiation. I. Regulation of mRNA sequences expressed during myotube formation. *J. Mol. Biol.* **140:** 441.

Alwine, J.C., D.J. Kemp, B.A. Parker, J. Reiser, J. Renart, G.R. Stark, and G.M. Wahl. 1979. Detection of specific RNAs or specific fragments of DNA by fractionation in gels and transfer to diazobenzyloxymethyl paper. *Methods Enzymol.* **68:** 220.

Benoff, S. and B. Nadal-Ginard. 1979. Most myosin heavy chain mRNA in L_6E_9 rat myotubes has a short poly(A) tail. *Proc. Natl. Acad. Sci.* **76:** 1853.

Buckingham, M.E. 1977. Muscle protein synthesis and its control during the differentiation of skeletal muscles *in vitro*. *Int. Rev. Biochem.* **15:** 269.

Buckingham, M.E., A. Cohen, and F. Gros. 1976. Cytoplasmic distribution of pulse-labeled poly(A)-containing RNA, particularly 26S RNA, during myoblast growth and differentiation. *J. Mol. Biol.* **103:** 611.

Caravatti, M., A.J. Minty, B. Robert, D. Montarras, A. Weydert, A. Cohen, P. Daubas, and M.E. Buckingham. 1982. Regulation of muscle gene expression: The accumulation of mRNAs coding for muscle specific proteins during myogenesis in a mouse cell line. *J. Mol. Biol.* (in press).

Daubas, P., D. Caput, M.E. Buckingham, and F. Gros. 1981. A comparison between the synthesis of contractile proteins and the accumulation of their translatable mRNAs during calf myoblast differentiation. *Dev. Biol.* **84:** 133.

Devlin, R.B. and C.P. Emerson. 1979. Co-ordinate accumulation of contractile protein mRNA during myoblast differentiation. *Dev. Biol.* **69:** 202.

Dym, H.O., D.S. Kennedy, and S.M. Heywood. 1979. Subcellular distribution of the cytoplasmic myosin heavy chain mRNA during myogenesis. *Differentiation* **12:** 145.

Frank, G. and A.G. Weeds. 1974. The amino acid sequence of the alkali light chains of rabbit skeletal muscle myosin. *Eur. J. Biochem.* **44:** 317.

Jacob, F. 1978. Mouse teratocarcinoma and mouse embryo. *Proc. R. Soc. Lond. B.* **201:** 249.

Jakob, H., M.E. Buckingham, A. Cohen, L. Dupont, M. Fiszman, and F. Jacob. 1978. A skeletal muscle cell line isolated from a mouse teratocarcinoma undergoes apparently normal terminal differentiation *in vitro*. *Exp. Cell. Res.* **114:** 403.

John, H.A., M. Patrinou-Georgoulas, and K.W. Jones. 1977. Detection of myosin heavy chain mRNA during myogenesis in tissue culture by *in vitro* and *in situ* hybridization. *Cell* **12:** 501.

Leibovitch, M.P., S.A. Leibovitch, J. Harel, and J. Kruh. 1979. Changes in the frequency and diversity of mRNA populations in the course of myogenic differentiation. *Eur. J. Biochem.* **97:** 321.

Matsuda, G., T. Maita, and T. Umegane. 1981. The primary structure of L-1 light chain of chicken fast skeletal muscle myosin and its genetic implication. *FEBS Lett.* **126:** 111.

McMaster, G.K. and G.G. Carmichael. 1977. Analysis of single and double stranded nucleic acids on polyacrylamide and agarose gels by using glyoxal and acridine orange. *Proc. Natl. Acad. Sci.* **74:** 4835.

Minty, A.J., S. Alonso, M. Caravatti, and M.E. Buckingham. 1982. A fetal skeletal muscle actin mRNA in the mouse, and its identity with cardiac actin mRNA. *Cell* (in press).

Minty, A.J., M. Caravatti, B. Robert, A. Cohen, P. Daubas, A. Weydert, F. Gros, and M.E. Buckingham. 1981. Mouse actin messenger RNAs: Construction and characterization of a recombinant plasmid molecule containing a complementary DNA transcript of mouse α-actin mRNA. *J. Biol. Chem.* **256:** 1008.

Paterson, B.M. and J.O. Bishop. 1977. Changes in the mRNA population of chick myoblasts during myogenesis *in vitro*. *Cell* **12:** 751.

Southern, E.M. 1975. Detection of specific sequences among DNA fragments separated by gel electrophoresis. *J. Mol. Biol.* **98:** 503.

Strohman, R.C., P.S. Moss, J. Micou-Eastwood, D. Spectro, A. Przbla, and B. Paterson. 1977. Messenger RNA for myosin polypeptides: Isolation from single myogenic cell cultures. *Cell* **10:** 265.

Vandekerckhove, J. and K. Weber. 1979. The complete amino acid sequence of actins from bovine aorta, bovine heart, bovine fast skeletal muscle and rabbit slow skeletal muscle. *Differentiation* **14:** 123.

Whalen, R.G. and S.M. Sell. 1980. Myosin from foetal hearts contains the skeletal muscle embryonic light chain. *Nature* **286:** 731.

Whalen, R.G., G.S. Butler-Browne, and F. Gros. 1978. Identification of a novel form of myosin light chain present in embryonic muscle tissue and cultured muscle cells. *J. Mol. Biol.* **126:** 415.

Whalen, R.G., S.M. Sell, L.E. Thornell, and A. Ericksson. 1982. The myosin subunits in foetal and adult ventricles, atria, and Purkinje fibers. *Devel. Biol.* (in press).

Whalen, R.G., S.M. Sell, G.S. Butler-Browne, K. Schwartz, P. Bouveret, and I. Pinset. 1981. Three myosin heavy chain isozymes appear sequentially in developing rat muscle. *Nature* **292:** 805.

Yablonka, Z. and D. Yaffe. 1977. Synthesis of myosin light chains and accumulation of translatable mRNA coding for light chain-like polypeptides in differentiating muscle cultures. *Differentiation* **8:** 133.

Yaffe, D. and H. Dym. 1973. Gene expression during differentiation of contractile muscle fibers. *Cold Spring Harbor Symp. Quant. Biol.* **37:** 543.

Zevin-Sonkin, D. and D. Yaffe. 1980. Accumulation of muscle-specific RNA sequences during myogenesis. *Dev. Biol.* **74:** 326.

Gene Sets in Muscle Development

KENNETH E.M. HASTINGS
CHARLES P. EMERSON, JR.
Department of Biology
University of Virginia
Charlottesville, Virginia 22901

The concept of coordinately regulated gene sets greatly simplifies the differential gene expression model of embryonic development and cellular differentiation. However, it is not at all clear just what constitutes a developmentally significant gene set or how the members of such gene sets are regulated in a coordinate fashion. Two aspects of skeletal-muscle development make this tissue an attractive experimental system in which to identify and analyze developmentally significant gene sets and to study the mechanisms of coordinate gene regulation.

First, we can consider the initial establishment of muscle fibers by the differentiation of myoblasts in the embryo or in culture. The fusion of myoblasts to form myofibers in culture is accompanied by the activation of synthesis of a variety of proteins, including proteins of the muscle contractile apparatus (Devlin and Emerson 1978; Garrels 1979; Affara et al. 1980). This muscle culture system provides an opportunity to study the coordinate activation of a set of genes encoding a structurally diverse group of proteins that form a functional set responsible for the major activity of this specialized cell type, i.e., contraction. The simultaneous activation of a diverse set of genes in one cell type raises the question of how this coordinate gene regulation is achieved. How are these genes recognized and activated as a functional set in these cells? How is the activation of this gene set coupled to the process of terminal muscle differentiation?

A second, complementary aspect of the recognition of gene sets in muscle development concerns the fast versus slow fiber distribution of specific

contractile-protein isotypes in mature skeletal muscle in the animal. Several contractile proteins, e.g., troponin I (Wilkinson and Grand 1978) and troponin C (Wilkinson 1980), are encoded in gene families whose members are differentially expressed in fast and slow muscle fibers. Clearly, the developing organism can sort out, as members of different functional sets, genes as similar as those encoding the fast and slow isotypes of troponin I. At the same time, the organism must also recognize and activate, during myogenesis, genes with unrelated coding sequences such as those encoding α-actin and α-tropomyosin. What is the relationship between the gene set activated during myogenesis and the gene sets expressed in mature muscle fibers? Can a gene belong to more than one coregulated set? If so, are the different sets recognized and regulated by different mechanisms? Here we describe the initial progress we have made using a cDNA cloning approach to study the gene set that is activated during myoblast differentiation.

RESULTS AND DISCUSSION

cDNA Cloning Strategy

The activation of specific protein synthesis during myogenesis appears to be regulated by mechanisms controlling specific mRNA abundance (Paterson and Bishop 1977; Devlin and Emerson 1979; Affara et al. 1980). There is a set of mRNAs whose abundances increase dramatically during myoblast differentiation, and these mRNAs code for the proteins whose synthesis is activated during that time. This suggested to us a strategy for the isolation of cloned cDNA probes corresponding to this coordinately regulated muscle gene set. The strategy is to clone the whole population of poly(A)$^+$ RNA sequences present in cultures of differentiated myofibers and then to identify developmentally regulated sequences by screening the clone library with probes representing myofiber RNA and undifferentiated myoblast RNA.

Secondary cultures of myoblasts from the breast muscle of embryonic quail *Coturnix coturnix* (Buckley and Konigsberg 1974) were used as the source of mRNA for our cDNA cloning. On the second day of culture, almost all of the cells are rapidly proliferating, mononucleated myoblasts. By the fifth day, the myoblasts have undergone extensive fusion, forming multinucleated myofibers, which contain most of the nuclei (> 80%) in the culture and actively synthesize muscle-specific contractile proteins.

To construct a cDNA clone library of myofiber mRNA sequences, whole cell poly(A)$^+$ RNA extracted from day-5 (myofiber) cultures was transcribed into double-stranded cDNA and introduced by G-C tailing into the *Pst*I site of the plasmid vector pBR322 (Chang et al. 1978). After introducing the chimeric DNA into *Escherichia coli* cells, we obtained about 10^4 tetracycline-resistant transformants, of which we estimate (from a control transformation carried out in parallel) at least 95% to be the result of transformation by recombinant molecules.

A set of 890 transformants selected randomly from the myofiber cDNA clone library was screened for developmentally regulated sequences by colony hybridization (Grunstein and Hogness 1975), using as probes day-5 (myofiber) and day-2 (myoblast) poly(A)$^+$ RNA preparations reverse transcribed into highly ^{32}P-labeled cDNAs. Of the transformants screened, 190 gave a detectable colony hybridization signal with the myofiber probe and, of these, 28 gave a substantially reduced or undetectable signal when hybridized with the myoblast probe. In this way, we identified 28 cDNA clones containing what we call myofiber-specific sequences. From analyses by DNA dot hybridization, DNA sequencing, and Northern RNA gel hybridization, we believe that the set of 28 myofiber-specific cDNA clones represent 12–18 different RNA species and, hence, 12–18 coordinately regulated genes. Regardless of whether 12 or 18 is closer to the true number of different genes represented in our sample, we have shown that at least several genes are represented by two or three cDNA clones in this set of 28. Thus, the collection of 28 cDNAs must represent a relatively complete sampling of all the genes whose transcripts become abundant during myogenesis. This conclusion is consistent with the number of proteins whose synthesis is activated during fusion and detected by two-dimensional gel electrophoresis; these estimates range from 8 (Garrels 1979) to about 20–30 (Devlin and Emerson 1978; Affara et al. 1980).

Identifying the Coordinately Regulated Genes

The next important question was which specific genes are members of the coordinately regulated gene set represented by the myofiber-specific cDNA clones? By the technique of hybridization-translation (Fig. 1), we were able to identify cDNA clones complementary to mRNAs encoding α-tropomyosin (Fig. 1C), α-actin (Fig. 1D), myosin light chain 2 (Fig. 1E), and troponin C (Fig. 1F). These identifications were confirmed by sequencing regions of these cDNA clones. By this means, we also discovered that two sets of cDNA clones that had not yielded detectable hybridization-translation products correspond to mRNAs encoding troponin I and myosin heavy chain (Hastings and Emerson 1982). The results of DNA sequencing also enabled us to determine the register and extent of the cDNA clones with respect to the corresponding proteins. These relationships are shown in Figure 2.

The six identified members of the coordinately regulated gene set encode one or more subunits of actin, myosin, tropomyosin, and troponin. Since they are the main structural elements of the thick and thin filaments, there is a clear biological rationale for their coordinate regulation. The situation with the remaining 6–12 genes is less clear. Clones may represent other known contractile proteins (e.g., troponin T, β-tropomyosin, myosin light chain 1) or, perhaps, muscle enzymes such as creatine phosphokinase. Alternatively, some may represent unknown proteins that perhaps play im-

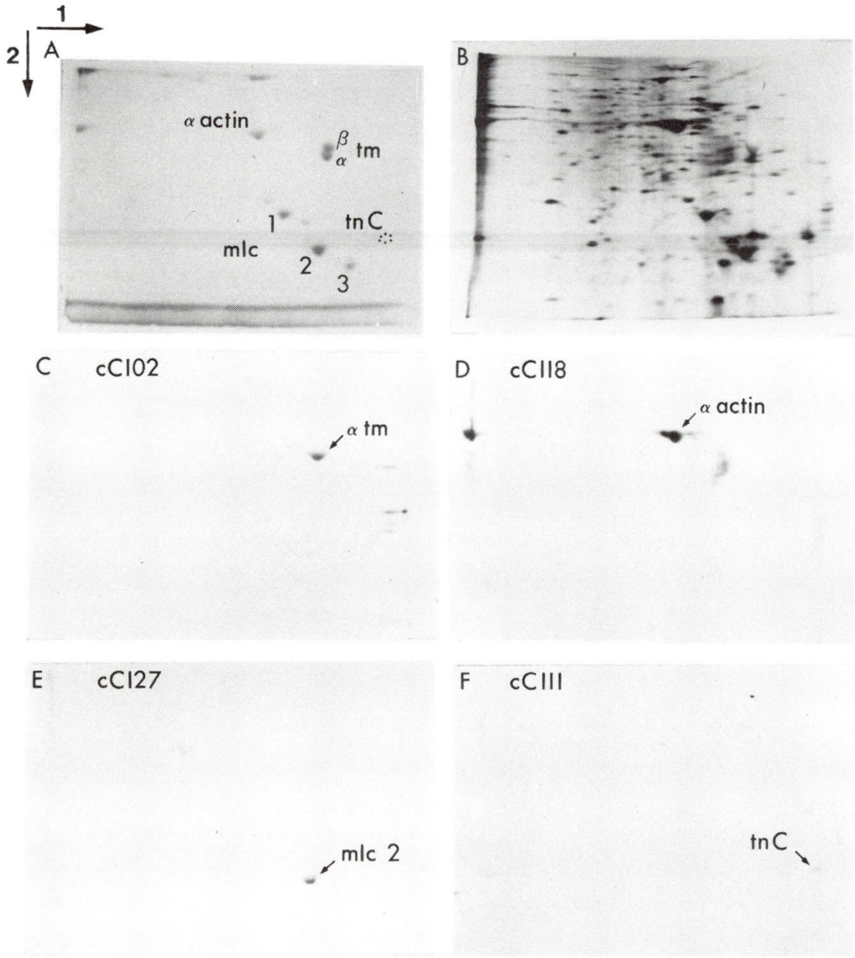

FIGURE 1
Identification of cDNA clones by hybridization-translation. Nitrocellulose-immobilized cDNA clones were hybridized with myofiber RNA, and the hybridized RNA was recovered and translated in a rabbit reticulocyte lysate containing [^{35}S]methionine (Ricciardi et al. 1979). Translation products were analyzed by two-dimensional gel electrophoresis (O'Farrel 1975) and fluorography (Bonner and Laskey 1974). Dimension 1 is isoelectric focusing with the basic end at the left, and dimension 2 is electrophoresis in an SDS-12% polyacrylamide gel. (*A*) Coomassie-blue-stained marker polypeptides from adult quail leg muscle actomyosin. (tm) Tropomyosin; (mlc) myosin light chain; (tnC) troponin C (visible before preparation for fluorography). (*B*) The translation products of the myofiber RNA preparation used in the hybridization-translation analysis. (*C–F*) The hybridization-translation products of the four cDNA clones indicated.

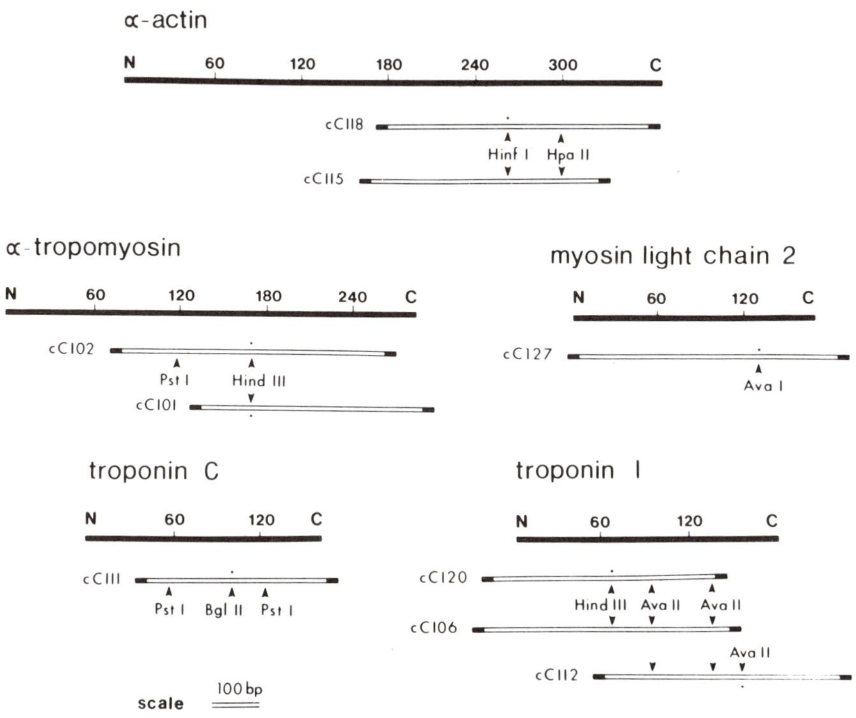

FIGURE 2
Relationships between proteins and cloned cDNAs. Proteins (black) and cloned cDNA inserts (white) are represented by bars whose lengths are proportional to molecular weight. The relative scales are such that 1 amino acid = 3 bp; bar represents 100 bp. Residues are numbered from the amino terminus on the protein symbols. Alignment of cloned cDNAs with proteins is based on DNA sequence information and restriction mapping of overlapping clones. The black ends of the cDNA symbols represent the G-C tail joints with pBR322 DNA, and each junction contains a *Pst*I site. Other restriction sites are indicated by arrowheads.

portant roles in the morphological and physiological changes that constitute myogenesis.

DNA sequencing also provided important information concerning the relationship between the coordinately activated genes of myogenesis and the genes expressed in adult muscle. With one exception, we found that our cDNA-derived amino acid sequences show excellent to perfect matching with amino acid sequences of known "adult" contractile proteins. The one exception was myosin heavy chain. cDNA clone cC128 represents the 3′ end of a 7-kb RNA. A partial DNA sequence determined for cC128 corresponds to an apparently complete 3′-untranslated mRNA region and a sequence of 61 amino acids corresponding to the carboxyl terminus of the encoded protein. This decoded sequence matches at 51 positions the amino

acid sequence determined for an adult rabbit skeletal-muscle myosin heavy chain (Elzinga and Trus 1980). The sequence differences were concentrated near the carboxyl terminus with six differences in the last 25 amino acids (Fig. 3). It is our impression (on the basis of contractile-protein sequence data) that this degree of divergence is rather too great to be considered bird/mammal differences in the "same" gene. We think that the rabbit gene (whose product was sequenced by Elzinga and Trus [1980]) and the quail gene (which we can refer to as the cC128 gene) are most probably two different members of a myosin heavy-chain gene family. However, we think the cC128 gene does correspond exactly with a chicken gene—the clone 251 gene of Umeda et al. (this volume). The last 25 amino acids of the chicken clone 251 gene product are identical with those of the quail cC128 gene product (Fig. 3). Moreover, there is a very high degree of homology (~85%) in the 3′-untranslated mRNA sequences of the two genes (Fig. 3). The expression of this gene in chickens has been studied by Umeda et al. (this volume), who concluded that it is the predominant myosin heavy-chain gene expressed in embryonic leg and breast muscle. They could not detect expression of this gene in adult breast, anterior tibialis, or adult heart and thus concluded that this was an embryonic myosin heavy-chain gene. Our results indicate that this gene is a member of the set of genes that are activated during myoblast fusion. Umeda et al. (this volume) also detected expression of four additional myosin heavy-chain genes in embryonic leg muscle. Whether these additional forms are also part of the coordinate muscle gene set activated during myogenesis is an important question that remains to be determined.

In the cases of the other five contractile-protein cDNA clones, identification of the specific gene family members was somewhat simpler because the cDNA-derived partial amino acid sequences showed excellent to perfect homology with known adult contractile-protein sequences (Hastings and Emerson 1982). Thus, in contrast to the cC128 myosin heavy-chain gene, the coordinately activated genes encoding α-actin, α-tropomyosin, myosin light chain 2, troponin I, and troponin C do not appear to be a special embryonic set but are probably the same genes expressed in adult muscle. The decoded troponin-C sequence corresponds to the adult skeletal-muscle slow-fiber isotype, whereas the troponin-I and myosin light-chain-2 sequences correspond to the skeletal-muscle fast-fiber isotypes.

Relationships among Muscle Gene Sets

Considering that the major breast muscle of adult birds consists almost exclusively of fast fibers, we can recognize from cDNA sequence analysis two functional classes among the contractile-protein genes activated during myogenesis in embryonic breast muscle myoblasts. One class consists of genes that, once activated, appear to remain active throughout subsequent development of the breast muscle. These are genes encoding α-actin,

```
protein sequences

rabbit protein   ...asp ile ala glu ser gln val asn lys leu arg val lys ser arg glu val his thr lys val ile ser glu glu(COOH)

chicken 251 gene
         protein ...asp ile ala glu ser gln val asn lys leu arg ala lys ser arg glu ile gly lys lys ala glu ser glu glu(COOH)
            cDNA ...GAC ATT GCA GAG TCG CAA GTC AAC AAG CTC CGA GCA AAG AGC CGT GAA ATA GGC AAG AAG GCA GAA AGT GAA GAG

quail cC128 gene
         protein ...asp ile ala glu ser gln val asn lys leu arg ala lys ser arg glu ile gly lys lys ala glu ser glu glu(COOH)
            cDNA ...GAC ATT GCA GAG TCA CAG GTC AAC AAG CTT CGA GCA AAA AGC CGT GAA ATA GGC AAG AAG GCA GAA AGT GAA GAG


3'-untranslated sequences

chicken 251 cDNA    TAGATGCCTCAAGCGGTGCAAAGTGAAATAGAATTGCACAAAATGTGAAATTCTAT CACTTTGATTTGTAATTACTGCTTAGTTCT
quail cC128 cDNA    TAGATGCCTCCAGTGGTGCAAAGTGAAAGAGAATTGCACAAAATGTGAAATTCTATTCACTTTGAT_TGTGATTAC_GCT_AGTTCT

chicken 251 cDNA    TCAACTATC??GATAAT TAA?ATTTAGATAATAAAAATTGTAGAGATTTTCCCATGGpolyA
quail cC128 cDNA    TCATCAATC______AATGTAATGTTT_GATAATAAA_ATTGTAGAGATTTTC_CATGpolyA
```

FIGURE 3

Comparison of myosin heavy-chain sequences. The rabbit protein sequence is that determined chemically by Elzinga and Trus (1980). The six underlined residues differ from the corresponding amino acids in the avian proteins whose sequences were deduced from cDNA sequences. Sequence data for the chicken clone 251 gene are from Umeda et al. (this volume). The 3′-untranslated sequences begin with the translation stop codon. Gaps were introduced to maximize homology. Underscores indicate differences in the cDNA sequences.

α-tropomyosin, fast myosin light chain 2, and fast troponin I. The second class consists of genes that are apparently inactivated later in development, i.e., the "embryonic" myosin heavy-chain gene and the slow troponin-C gene. The phenomenon of activation followed by apparent inactivation is actually not unusual among contractile-protein genes for, although adult avian breast muscle contains neither β-tropomyosin nor the slow isotypes of myosin light chains 1 and 2, the synthesis of these proteins is activated during breast myoblast fusion (Devlin and Emerson 1978; Keller and Emerson 1980).

Keller and Emerson (1980) have shown that all of the known skeletal-muscle myosin light-chain genes are activated during fusion of avian myoblasts, whether the myoblasts are obtained from presumptive fast or presumptive slow embryonic muscles and regardless of the exact age of the embryo. Moreover, their clonal studies indicate that the activation of slow and fast forms during fusion is not due to the presence of "slow" and "fast" myoblast subpopulations but seems to be occurring in all of the cells. These observations suggested a total activation–selective-repression model of contractile-protein gene expression. According to this model, there is only one genetic program of skeletal-muscle myoblast differentiation, and this program includes activation of all of the skeletal-muscle contractile-protein genes in the genome. Later in development, different subsets of these genes are selectively repressed in different muscles, leaving active only those genes appropriate to the adult muscle fiber type. Thus, the contractile-protein gene sets expressed in adult skeletal muscle are subsets of the genes activated during myogenesis.

The chief distinctive feature of the total activation–selective-repression model is that no skeletal-muscle contractile-protein gene "ignores" myoblast differentiation; all of them are coordinately activated at this time. To be different, other models must have the feature that certain skeletal-muscle contractile-protein genes are outside of the main coordinate set and ignore myoblast differentiation, and are either never activated in certain myoblast lineages or are activated later in development by a different mechanism. To determine whether such "maverick" contractile-protein genes do exist is important, not only to rule out the total activation–selective-repression model but to indicate what kind of latitude the developing organism has in recognizing and regulating gene sets.

The literature does suggest some possible candidates for maverick contractile-protein genes, namely those encoding neonatal and adult forms of myosin heavy chain (Whalen et al. 1981) and the fast isotype of troponin T (Matsuda et al. 1981). These proteins accumulate relatively late in development, apparently displacing other forms that had accumulated earlier. It is important to determine whether these genes are in fact not activated during myoblast differentiation (as one would guess from the pattern of protein accumulation) or whether they are activated along with the

other contractile-protein genes, but their products are unable to accumulate until synthesis of the earlier forms has been repressed. This should be a straightforward matter to determine when the appropriate cloned cDNA probes become available.

FUTURE PROSPECTS

The cDNA cloning approach can provide new information on muscle gene expression in three main areas. The first, which we have discussed in some detail, is in determining exactly which genes are members of which gene sets. This information is fundamental because, obviously, the enormous power of the molecular cloning approach will be mitigated if there are any misunderstandings about when and where any particular gene is expressed. The second is in determining the mechanisms that regulate the expression of muscle genes. The success of our cDNA cloning strategy indicates that our assumptions regarding contractile-protein mRNA accumulation during myogenesis were correct. We have also verified the dramatic myogenic increase in contractile-protein mRNA levels by RNA gel transfer hybridization analysis using cloned cDNAs as specific probes. Further analysis using cloned cDNAs should reveal whether transcriptional or posttranscriptional mechanisms determine contractile-protein mRNA levels during myoblast differentiation and also the mechanisms of gene regulation operating during fiber-type differentiation. Third, identified cDNA clones can be used to isolate cloned genomic DNA fragments carrying muscle genes and their adjacent chromosomal elements. Comparative structural analysis of diverse coordinately regulated muscle genes may reveal common features that can explain their pattern of coordinate regulation. A similar analysis of gene families encoding differentially expressed contractile-protein isotypes may reveal unique features correlating with their differential expression. These two comparative approaches will generate complementary kinds of information regarding the recognition and regulation of functionally significant gene sets during development.

ACKNOWLEDGMENTS

We thank Ann Glenn, Margaret Ober, Linda Schuman, Druen Robinson, and Ruth Ashforth for excellent technical assistance. We also thank Patrick Umeda and Clifford Kavinsky for allowing us access to the chicken clone 251 cDNA sequence prior to publication. This work was supported by grants from the National Institutes of Health and the Muscular Dystrophy Association of America (to C.P.E.), and a postdoctoral fellowship from the Muscular Dystrophy Association of America (to K.E.M.H.).

REFERENCES

Affara, N.A., P. Daubas, A. Weydert, and F. Gros. 1980. Changes in gene expression during myogenic differentiation. II. Identification of the proteins encoded by myotube-specific complementary DNA sequences. *J. Mol. Biol.* **140:** 459.

Bonner, W.M. and R.A. Laskey. 1974. A film detection method for tritium-labeled proteins and nucleic acids in polyacrylamide gels. *Eur. J. Biochem.* **46:** 83.

Buckley, P.A. and I.R. Konigsberg. 1974. Myogenic fusion and the duration of the mitotic gap. *Dev. Biol.* **37:** 193.

Chang, A.C.Y., J.H. Nunberg, R.J. Kaufman, H.A. Erlich, R.T. Schimke, and S.N. Cohen. 1978. Phenotype expression in *E.coli* of a DNA sequence coding for mouse dihydrofolate reductase. *Nature* **275:** 617.

Devlin, R.B. and C.P. Emerson, Jr. 1978. Coordinate regulation of contractile protein synthesis during myoblast differentiation. *Cell* **13:** 599.

———. 1979. Coordinate accumulation of contractile protein mRNAs during myoblast differentiation. *Dev. Biol.* **69:** 202.

Elzinga, M. and B. Trus. 1980. Sequence and proposed structure of a 17,000 dalton fragment of myosin. In *Methods in peptide and protein sequence analysis* (ed. C. Birr), p. 213. Elsevier/North Holland Biomedical Press, Amsterdam.

Garrels, J.I. 1979. Changes in protein synthesis during myogenesis in a clonal cell line. *Dev. Biol.* **73:** 134.

Grunstein, M. and D.S. Hogness. 1975. Colony hybridization: A method for the isolation of cloned DNAs that contain a specific gene. *Proc. Natl. Acad. Sci.* **72:** 3961.

Hastings, K.E.M. and C.P. Emerson, Jr. 1982. cDNA clone analysis of six coregulated mRNAs encoding skeletal muscle contractile proteins. *Proc. Natl. Acad. Sci.* **79:** 1553.

Keller, L.R. and C.P. Emerson, Jr. 1980. Synthesis of adult myosin light chains by embryonic muscle cultures. *Proc. Natl. Acad. Sci.* **77:** 1020.

Matsuda, R., T. Obinata, and Y. Shimada. 1981. Types of troponin components during development of chicken skeletal muscle. *Dev. Biol.* **82:** 11.

O'Farrel, P.H. 1975. High resolution two-dimensional electrophoresis of proteins. *J. Biol. Chem.* **250:** 4007.

Paterson B.M. and J.O. Bishop. 1977. Changes in the mRNA population of chick myoblasts during myogenesis in vitro. *Cell* **12:** 751.

Ricciardi, R.P., J.S. Miller, and B.E. Roberts. 1979. Purification and mapping of specific mRNAs by hybridization-selection and cell-free translation. *Proc. Natl. Acad. Sci.* **76:** 4927.

Whalen, R.G., S.M. Sell, G.S. Butler-Browne, K. Schwartz, P. Bouveret, and I. Pinset-Harstrom. 1981. Three myosin heavy-chain isozymes appear sequentially in rat muscle development. *Nature* **292:** 805.

Wilkinson, J.M. 1980. Troponin C from rabbit slow skeletal and cardiac muscle is the product of a single gene. *Eur. J. Biochem.* **103:** 179.

Wilkinson, J.M. and R.J.A. Grand. 1978. Comparison of amino acid sequence of troponin I from different striated muscles. *Nature* **271:** 31.

Molecular Cloning and Analysis of Genes Regulated during Embryonic Muscle Differentiation

*CHARLES P. ORDAHL **
*THOMAS COOPER **
CATHERINE E. OVITT
JAMES A. FORNWALD
ANDREW F. CALMAN
Department of Anatomy
Temple University School of Medicine
Philadelphia, Pennsylvania 19140

ARNOLD I. CAPLAN
Department of Biology
Case Western Reserve University
Cleveland, Ohio 44106

INTRODUCTION AND BACKGROUND

The fusion of embryonic myoblasts to form multinucleate myotubes entails a wide variety of cellular and biochemical changes. Many of these changes result from changes in the levels of expression of different genes encoding proteins that are important at different stages of muscle differentiation. To investigate these gene-expression changes, we analyzed the transcript population present in embryonic chick myoblasts and the changes in this population that occur during fusion. The overall complexity of chick myoblast poly(A) RNA, as estimated by nonrepetitive DNA and complementary DNA (cDNA) hybridization, is sufficient to encode over 100,000 diverse transcripts 1.5 kb in length (Ordahl and Caplan 1978). The majority of the diverse transcript species (97%) occur at low frequency, less than an average of 1 copy/cell, and comprise approximately 45% of the poly(A)

*Present address: Department of Anatomy, School of Medicine, University of California, San Francisco, California 94143.

RNA mass. The remaining 60% of the poly(A) RNA mass represents 3000–4000 diverse species present at a range of frequencies between ten and several hundred copies per cell (Ordahl and Caplan 1978).

The transcript population of the postfusion myotube differs from that of the myoblast in two major respects. First, as indicated below and by the work of others (Paterson and Bishop 1977; Devlin and Emerson 1978, 1979; Leibovitch et al. 1979; Zevin-Sonkin and Yaffe 1980), the mRNAs for the muscle-specific proteins are highly abundant in myotubes (hundreds to thousands of copies per nucleus) and, as a group, comprise a substantial fraction of the total poly(A) RNA mass (20–30%; Zevin-Sonkin and Yaffe 1980; C. Ordahl, unpubl.). The second difference is that the complexity of the myotube transcript population is lower, by about half, than that of the myoblast as assayed by nonrepetitive DNA hybridization (Ordahl and Caplan 1976). Thus, in addition to turning on (or up-regulating) the transcription of genes encoding muscle-specific proteins, myotubes also turn off (or down-regulate) a complex set of genes that had previously been active in the myoblast and in the premyoblast mesodermal cells within the embryonic mesenchyme (Ordahl and Caplan 1976).

HOMOLOGY BETWEEN MYOBLAST AND NONMYOBLAST TRANSCRIPT POPULATIONS

Although nonrepetitive DNA hybridization demonstrated that a complex class of low-abundance transcripts is down-regulated during fusion, this type of experiment would not detect changes in high- or moderately high-abundance transcripts. To assay for changes in these transcript classes, we used cDNA hybridization that preferentially detects the high- and middle-abundance transcripts, because the cDNAs are synthesized in proportion to the relative abundance of the RNA template. A radiolabeled cDNA was prepared using myoblast total poly(A) RNA as a template (Ordahl and Caplan 1978). This myoblast cDNA was shown to represent most, and probably all, of the diverse poly(A) RNAs in the myoblast and hybridized to completion with myoblast RNA.

When myoblast cDNA is hybridized to total cell RNA from premyogenic mesenchyme, fused embryonic muscle, or adult muscle (Fig. 1, left), a minimum of 80% of the cDNA hybridizes (as assayed by resistance to digestion with nuclease S1; Ordahl and Caplan 1978). This indicates that at least 80% of the mass of poly(A) RNA present in the myoblast is transcribed from genes that continue to be active in the myotube. When a similar analysis is made using RNA from cartilage and brain (Fig. 1, right), as well as from liver and oviduct (not shown), a hybridization level of greater than or equal to 80% is again obtained. These results indicate, therefore, that at least 80% of the myoblast poly(A) RNA mass is transcribed from genes that are transcribed in all cells and probably represent "housekeeping" genes, which encode structures and functions common to virtually all cells.

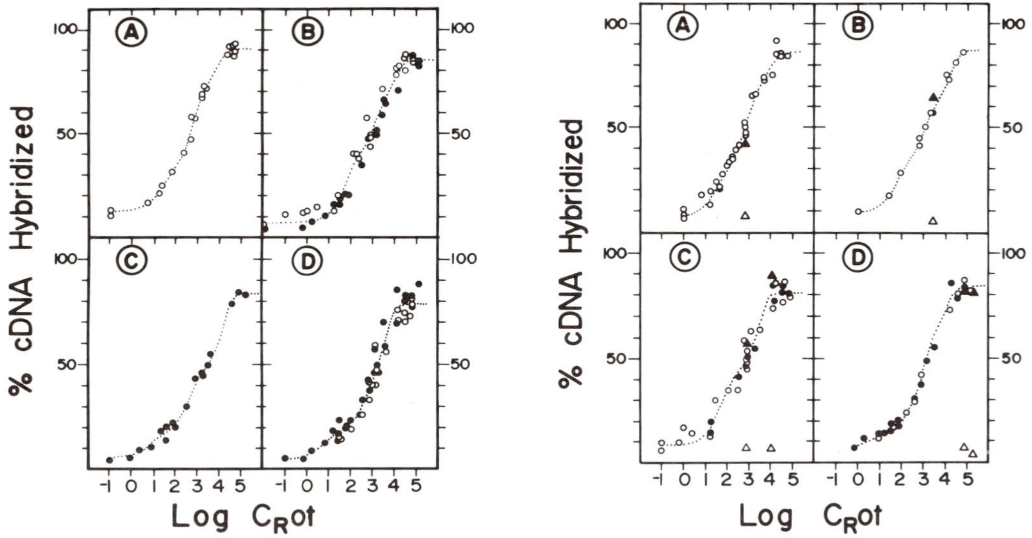

FIGURE 1
Hybridization of myoblast cDNA to total cell RNA from muscle and nonmuscle sources. Synthesis and hybridization of cDNA as described by Ordahl and Caplan (1978). (○) Hybridization of cDNA from day-9 embryo leg-muscle poly(A) RNA; (●) cDNA from day-14 embryo leg-muscle poly(A) RNA. (.....) Least squares fit to data. Binding of hybrids to hydroxylapatite before (▲) and after (Δ) RNase release. (Details as in Ordahl and Caplan 1978.) (*Left*) Total cell RNA driving the hybridization in each panel is stage-20–22 limb buds (*A*), stage-24 limb buds (*B*), day-18 embryo leg muscle (*C*), and adult leg muscle (*D*). Approximate final hybridization values determined by least squares: 90% (*A*); 85% (*B*); 83% (*C*); 79% (*D*). (*Right*) Total cell RNA driving the hybridization in each panel: cartilage cells after 3 days in culture (*A*); cartilage cells after 10 days in culture (*B*); day-18 embryonic brain (*C*); adult brain (*D*). Approximate final hybridization values determined by least squares: 87% (*A*); 88% (*B*); 81% (*C*); 84% (*D*). Cartilage culture cells were prepared as described by Caplan (1970).

Several conclusions can be drawn from the above experiments. First, when embryonic muscle undergoes fusion, some genes are up-regulated and other genes are down-regulated. For simplicity we call any gene that is up-regulated class A and any gene that is down-regulated class B, without implications as to the molecular mechanisms underlying these regulatory patterns. The results discussed above further indicate that in general, class-A gene transcripts are highly abundant in myotubes but not in myoblasts, whereas transcripts of class-B genes are present at low-abundance levels in myoblasts and virtually absent from myotubes. Obviously, these general conclusions, summarized in Figure 2, apply only to populations of RNA in the myoblast, and individual species of RNA will go undetected by these experiments. Thus, there may be high-abundance transcripts that are regulated according to the class-B pattern and class-A transcripts that are present at relatively low-abundance levels in the myotube.

MOLECULAR CLONING OF REPRESENTATIVE CLASS-A AND CLASS-B TRANSCRIPTS

To analyze in greater detail the gene regulatory patterns seen during muscle development, we undertook to clone representative class-A and class-B transcripts using recombinant DNA techniques. The methods for cloning

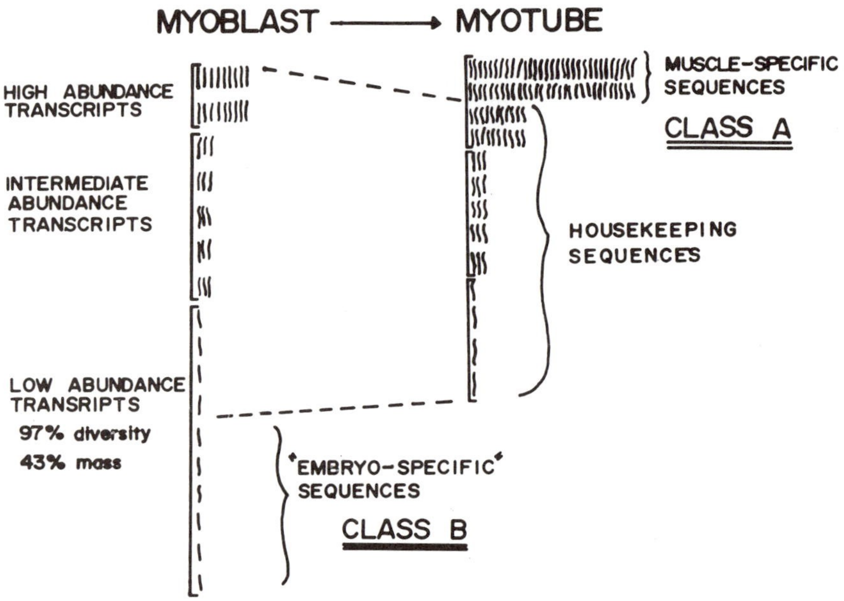

FIGURE 2
Diagrammatic summary of the transcriptional changes occurring when myoblasts fuse into multinucleate myotubes.

abundant mRNAs are relatively straightforward (e.g., see Goodman and MacDonald 1980), and we used this general approach to obtain cDNA clones of several class-A transcripts from myotube RNA where these transcripts are abundant. However, because the class-B transcripts described above are present at low-abundance levels, it was necessary to develop a novel approach to cloning representative cDNAs of these, as no general strategy was available for cloning such low-abundance transcripts.

On the basis of the results presented above, we would expect that (1) the majority of class-B transcripts are present at relatively low-abundance levels in myoblast total poly(A) RNA, and (2) class-B transcripts constitute as much as 50% of the complex class of myoblast poly(A) RNA and therefore as much as 20% of myoblast cDNA (50% of the complex class, which is ~45% of the cDNA mass). On the basis of these estimates, we predicted that as many as 20% of the cDNA clones in a myoblast cDNA library should be derived from class-B transcripts. We therefore decided it was feasible to conduct a blind search for clones of class-B sequences (for details, see Ordahl et al. 1980a). It is important to point out that such a blind search would be unlikely to be successful unless a substantial fraction of the clones contain sequences derived from transcripts regulated according to the class-B pattern. The first phase of this search entailed identifying low-abundance clones by virtue of their failure to detectably hybridize in a standard colony hybridization assay (Grunstein and Hogness 1975). The second phase involved analyzing each low-abundance cDNA clone individually, by blot hybridization, to identify those that hybridize to myoblast RNA but not to myotube RNA. Of the 30 low-abundance clones selected for screening, 3 showed the class-B pattern of hybridization. Whether this reflects the true percentage of class-B clones is not yet clear, because many clones failed to hybridize in the blot hybridization assay, indicating that they may represent transcripts too low in abundance to be analyzed even by this method.

CHARACTERIZATION OF CLONES OF CLASS-A AND CLASS-B TRANSCRIPTS

Five cDNA clones of class-A genes and three cDNA clones of class-B genes were selected for further characterization. To determine the timing of appearance of the corresponding transcripts during development, RNA blot hybridizations were performed using RNA from various stages of limb muscle, brain, and liver development. Figure 3 shows the results of RNA blot hybridization experiments using one cDNA clone of a housekeeping transcript, four cDNA clones of class-A transcripts, and three cDNA clones of class-B transcripts. The housekeeping clone hybridizes to a 1.3-kb transcript that is present throughout the development of all three tissues. The class-A and -B clones hybridize to transcripts that have two distinctly

different patterns of appearance during limb-muscle development; i.e., class-A transcripts increase in abundance during muscle differentiation (Fig. 3B,D,E,F), whereas class-B transcripts decrease in abundance during the same period (Fig 3C,G,H). Below, we discuss further the characterization of the cDNA clones for class-A and class-B transcripts.

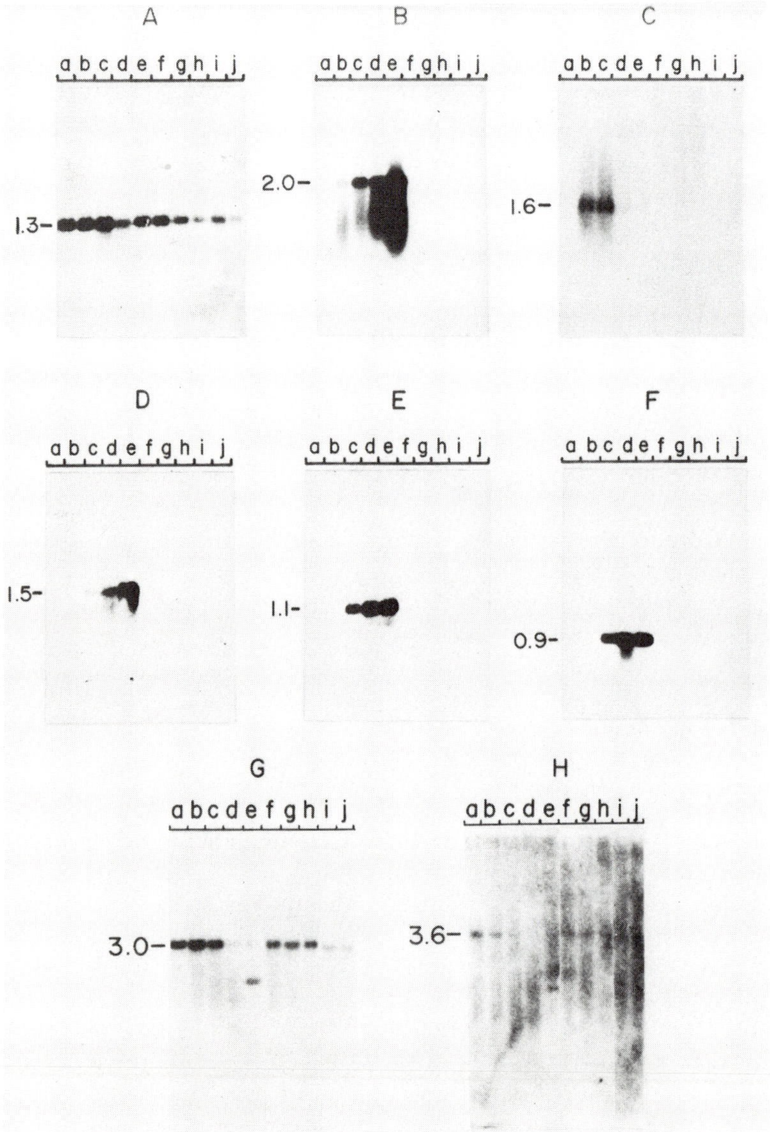

Class-A Transcripts

The class-A transcripts are first detectable as low-abundance sequences at the earliest stages of myogenesis when mononucleate myoblasts predominate. As development proceeds and the proportion of fused, terminally differentiated myotubes increases, the relative abundance of the class-A sequences increases correspondingly. In fully differentiated muscle, the class-A sequences are of high abundance. It has been shown that muscle-specific proteins accumulate only in myotubes (for review, see Hauschka 1968). From the data presented here, a similar conclusion can be drawn for the accumulation of class-A mRNAs.

Within the class-A sequence set as a whole, the pattern of accumulation is similar. Devlin and Emerson (1978, 1979) have shown that the synthesis of contractile proteins is coordinate and that this is probably due to the coordinate accumulation of their respective mRNAs during myogenesis. Our results suggest that it is now possible to extend this conclusion to include some noncontractile muscle-specific proteins as well. The mRNAs for α-actin (a contractile protein) and muscle creatine kinase (a muscle-specific metabolic enzyme) both accumulate in a similar pattern (Fig. 3B,D). The identification of α-actin cDNA clone and gene is published (Ordahl et al. 1980b); the identification of the muscle creatine kinase cDNA clone is shown in Figure 4A. In addition, identification of the polypeptides encoded by two other class-A sequence clones is also shown in Figure 4.

FIGURE 3 (see facing page)
RNA blot hybridization screening of muscle-regulated sequence clones. Total cell poly(A) RNA from the embryonic and adult tissues indicated were electrophoresed, blotted, and hybridized as described previously (Ordahl et al. 1980a,b). Cloned cDNA inserts were isolated, labeled with ^{32}P by nick translation, and hybridized overnight to the blots. After washing to remove unhybridized probe, the blots were exposed to X-ray film (Kodak XR-Omat) for 1–120 hr with an intensifying screen. RNA size is shown in kilobases. (*A*) Housekeeping sequence clone; (*B*) α-actin mRNA; (*C*) class-B sequence; (*D*) class-A sequence—in this case it is muscle creatine kinase mRNA (for characterization, see legend to Fig. 4); (*E*) class-A sequence—encodes a polypeptide 17,000 daltons in size (see Fig. 4); (*F*) class-A sequence—coding function unknown; (*G,H*) class-B sequences discussed more extensively in Ordahl et al. (1980a). The stages of muscle development employed were chosen to span the developmental range from a mesoderm progenitor cell population (*a*) through to adult muscle (*e*). In general terms, the muscle developmental stages can be characterized as follows: stage-23 limb buds, which contain a mesodermal core of progenitor cells for muscle, cartilage, and possibly other phenotypes (*a*); day-10 leg muscle, which consists predominately of myoblasts that proliferate but do not accumulate terminal differentiation marker molecules (*b*); day-14 leg muscle, which consists of a mixture of myoblasts and fused, terminally differentiating multinucleate myotubes (*c*); day-18 leg muscle composed almost entirely of fused, multinucleate myotubes, which are contractile and synthesize and accumulate muscle marker molecules in quantity (*d*); adult leg muscle (*e*); day-10 brain (*f*); day-18 brain (*g*); adult brain (*h*); day-18 liver (*i*); adult liver (*j*).

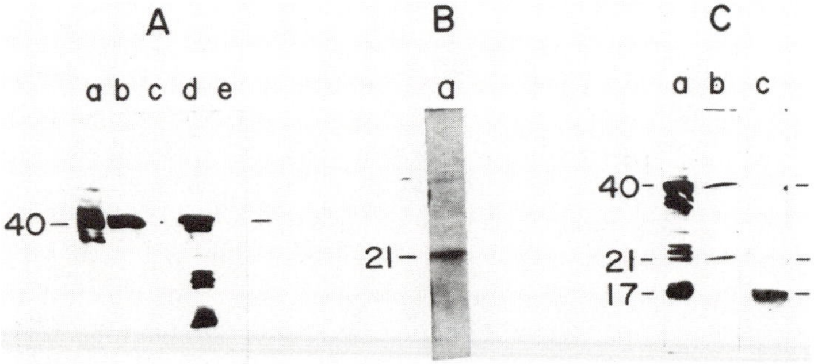

FIGURE 4
Identification of polypeptides encoded by muscle-regulated mRNAs. mRNAs for various muscle-regulated genes were isolated by selective hybridization to cloned cDNAs affixed to nitrocellulose (Ordahl et al. 1980b). Polypeptides were synthesized in vitro with [^{35}S]methionine using nuclease-treated reticulocyte lysate (NEN), fractionated on 10% SDS-polyacrylamide gels, and autoradiographed. (*A*) Immunoprecipitation of A-2-encoded polypeptide with antimuscle creatine kinase antibody. (*a*) Immunoprecipate from translation of day-18 leg muscle total poly(A) RNA; (*b*) immunoprecipitate from translation of A-2-selected mRNA; (*c*) marker lane—dot shows position of authentic chick muscle creatine kinase; (*d*) supernatant from immunoprecipitate in *a*; (*e*) supernatant from immunoprecipitate in *b*. All lanes are deliberately overexposed to show extent of background contaminants. Identification of A-2-encoded protein as creatine kinase was confirmed by Cleveland partial proteolysis (Cleveland et al. 1978; data not shown). (*B*) Polypeptide encoded by B-1 hybridization is shown in Fig. 3C. 100 μg of B-1 cloned DNA was affixed to nitrocellulose and used to isolate the B-1 mRNA from 1 mg of day-12 embryo leg muscle poly(A) RNA. The encoded polypeptide is approximately 21,000 daltons in size. (*C*) Polypeptides encoded by A-3 and A-5 mRNAs. (A-3 hybridization is shown in Fig. 3E; A-5 hybridization is not shown.) (*a*) Translation products of total poly(A) RNA from day-18 embryo leg muscle; (*b*) polypeptides encoded by A-5-selected mRNA (two polypeptides are synthesized); (*c*) polypeptide encoded by A-3-selected mRNA.

Class-B Sequences

The class-B transcripts are present at the earliest stages of myogenesis, when myoblasts predominate, but are absent from the terminal stages, when fused, multinucleate myotubes and myofibers predominate. Thus, the class-B sequences are regulated in reverse phase compared with class-A sequences, because they are turned off concomitant with the fusion of myotubes.

There are interesting differences in the pattern of regulation of the three low-abundance class-B sequences shown in Figure 3. One class-B sequence

(B-1 [Fig. 3C]) resembles a class-A sequence in two respects: (1) It is muscle-specific, and (2) its presence cannot be detected in limb-bud RNA. Thus, its tissue specificity and its first appearance are regulated similarly to class-A sequences. However, the fact that the B-1 sequence disappears in fused muscle demonstrates that it is regulated quite differently from a class-A transcript at this stage of myogenesis. Garrels (1979) has shown that a heart form of tropomyosin appears transiently, at the myoblast stage of skeletal-muscle differentiation. Thus, the pattern of appearance of the heart form of tropomyosin is similar to the pattern of appearance of the B-1 sequence. Interestingly, we have detected the B-1 sequence in heart-muscle RNA (data not shown). However, the B-1 sequence does not encode a tropomyosin, because the molecular weight of its encoded polypeptide is too small (Fig. 4B).

The regulation of the other two class-B sequences (Fig. 3G,H) differs from the one discussed above in two major respects. First, both are present in progenitor limb-bud RNA prior to the onset of muscle differentiation (Fig. 3G,H, lane a), indicating that their genes are active prior to myogenic differentiation. Second, although both are regulated in muscle development, they are not developmentally regulated in brain (Fig. 3G,H, lanes f–h) or in liver (lanes i,j). Thus, these two class-B genes may be housekeeping genes that simply are not active in differentiated muscle. There are genes known to fit this description. For example, the brain form of the creatine kinase gene is a housekeeping gene because its expression is virtually ubiquitous. It too is active in limb progenitor cells and is repressed in myotubes coincident with the muscle creatine kinase gene becoming active.

The most important consideration at present is not the specific function of the individual class-B genes but rather the fact that the components of embryonic muscle cell recognize and regulate the various B genes according to at least two different patterns, and that, collectively, the class-B genes are regulated according to a pattern that is distinct from that of class-A gene regulation. Since we have probes for these various muscle-regulated sequences, it is now possible to begin to study the molecular mechanism underlying these developmentally regulated patterns of gene activity.

CURRENT AND FUTURE GOALS: ISOLATION AND CHARACTERIZATION OF GENOMIC SEGMENTS CONTAINING CLASS-A AND CLASS-B GENES

Our future goals are to determine the molecular mechanisms by which muscle cells differentially regulate class-A and -B gene expression. Our working hypothesis is that the nucleotide sequence of each of these genes, or the DNA flanking the genes, carries information that permits the embryonic muscle cell to recognize and thereby regulate its expression according to one pattern or another. To test this hypothesis we are analyzing the genomic segments containing the five class-A and three class-B genes for which we

have cDNA clones. Each is being isolated from a chicken genomic library cloned in λ Charon 4A phage (Dodgeson et al. 1979) using the hybridization screening procedure of Benton and Davis (1977). The position of the genes within these DNA segments is then mapped by standard procedures involving restriction endonuclease digestion, Southern blotting (Southern 1975) and electron microscopy (Casey and Davidson 1977; White and Hogness 1977), and determination of the nucleotide sequence of important regions (Maxam and Gilbert 1980). Ultimately, we expect that a detailed structural analysis will demonstrate structural features among class-A and class-B genes, which might account for their differential pattern of regulation during development.

Our most extensive progress to date has been in the analysis of the chick α-actin gene, on which we have reported its isolation and gross structural analysis (Ordahl et al. 1980b). Recently, we have determined the nucleotide sequence of the entire gene (Fornwald et al. 1982). At present we cannot compare the α-actin gene structure to other chick-muscle genes, because no other muscle gene has yet been studied in such detail. We can, however, make a comparison of the chick α-actin gene to the actin genes that have been studied in other organisms. Six intervening sequences interrupt the chick α-actin gene (Table 1). The position of the chick α-actin-gene intervening sequences differs substantially from those reported for nonvertebrate actin genes (for review, see Fornwald et al. 1982). Thus, inter-

TABLE 1
Summary of α-Actin Gene Organization

Exon/IVS	Position	Size	Encoded region of mRNA
Exon I	1–61	61	61 of 73 NT 5′ UT
IVS A	62–181	111	—
Exon II	182–322	141	aa 1–41 + 12 NT of 5′ UT
IVS B	323–434	112	—
Exon III	435–759	325	aa 42–150
IVS C	760–893	134	—
Exon IV	894–1055	162	aa 150–204
IVS D	1056–1280	225	—
Exon V	1281–1472	192	aa 204–267
IVS E	1473–1577	105	—
Exon VI	1578–1759	182	aa 267–327
IVS F	1760–1908	149	—
Exon VII	1909–2324	416	aa 328–374 + 272 NT 3′ UT

The size of each exon and intervening sequence (IVS) is given in nucleotides. The position of each is numbered in nucleotides with nucleotide number 1 being the first transcribed nucleotide and nucleotide number 2324 being the poly(A) addition site. The amino acid (aa) residues of actin are numbered according to Elzinga et al. (1973). NT, nucleotide; UT, untranslated region of mRNA.

vening sequence positions are not necessarily as good indicators of the evolutionary relatedness of genes over long evolutionary distances (Fornwald et al. 1982) as they have been shown to be over much shorter evolutionary distances, as in the case of globin genes (Leder et al. 1978; Dodgeson et al. 1979; Lauer et al. 1980).

Structural analysis of the class-A and class-B genes may indicate possible regulation sequences. Ultimately, however, it will be necessary to test the role of such sequences experimentally. At present, no assay system exists that is adequate to experimentally determine the mechanism of developmental gene regulation. We are currently using DNA transfection to study the feasibility of reintroducing modified cloned genes into cells under conditions that will allow them to be regulated.

ACKNOWLEDGMENTS

This work was supported by grants from the National Institutes of Health (GM-25400 and CA-12923), the Muscular Dystrophy Association of America, and a Basil O'Connor starter research grant from the March of Dimes Birth Defects Foundation. C.P.O. is a recipient of a National Institutes of Health Research Career Development Award (HD-00290).

REFERENCES

Benton, W.D. and R.W. Davis. 1977. Screening of recombinant clones by hybridization to single plaques in situ. *Science* **196:** 180.

Caplan, A.I. 1970. Effects of the nicotinamide-sensitive teratogen 3-acetyl-pyridine on chick limb cells in culture. *Exp. Cell Res.* **62:** 341.

Casey, J. and N. Davidson. 1977. Rates of formation and thermal stabilities of RNA:DNA and DNA:DNA duplexes at high concentrations of formamide. *Nucleic Acids Res.* **4:** 1539.

Cleveland, D.W., M.W. Kirschner, and N.J. Cown. 1978. Isolation of separate mRNAs for α- and β-tubulin and characterization of the corresponding in vitro translation products. *Cell* **15:** 1021.

Devlin, R.B. and C.P. Emerson. 1978. Coordinate regulation of contractile protein synthesis during myoblast differentiation. *Cell* **13:** 599.

———. 1979. Coordinate accumulation of contractile protein mRNAs during myoblast differentiation. *Dev. Biol.* **69:** 202.

Dodgeson, J.B., J. Strommer, and J.C. Engel. 1979. Isolation of the chicken β-globin gene and a linked embryonic β-like globin gene from a chicken DNA recombinant library. *Cell* **17:** 879.

Elzinga, M., J.H. Collins, W.M. Kuehl, and R.S. Adelstein. 1973. Complete amino-acid sequence of actin of rabbit skeletal muscle. *Proc. Natl. Acad. Sci.* **70:** 2687.

Fornwald, J.A., G. Kuncio, I. Peng, and C.P. Ordahl. 1982. The complete nucleotide sequence of the chick a-actin gene and its evolutionary relationship to the actin gene family. *Nucleic Acids Res.* **10:** 3861.

Garrels, J.I. 1979. Changes in protein synthesis during myogenesis in a clonal cell line. *Dev. Biol.* **73:** 134.

Goodman, H.M. and R.J. MacDonald. 1980. Cloning of hormone genes from a mixture of cDNA molecules. *Methods Enzymol.* **68:** 75.

Grunstein, M. and D.S. Hogness. 1975. Colony hybridization: A method for the isolation of cloned DNAs that contain a specific gene. *Proc. Natl. Acad. Sci.* **72:** 3961.

Hauschka, S. 1969. Clonal aspects of muscle development and the stability of the differentiated state. In *The stability of the differentiated state* (ed. H.P. Ursprung), p. 37. Springer-Verlag, New York.

Lauer, J., C.J. Shen, and T. Maniatis. 1980. The chromosomal arrangement of human α-like globin genes: Sequence homology and α-globin gene deletions. *Cell* **20:** 119.

Leder, A., H. Miller, D. Homer, T.G. Seidman, E. Norman, M. Sullivan, and P. Leder. 1978. Comparison of cloned mouse α- and β-globin genes: Conservation of intervening sequence locations and extragenic homology. *Proc. Natl. Acad. Sci.* **75:** 6187.

Leibovitch, M.-P., S.-A. Leibovitch, J. Harel, and J. Kruh. 1979. Changes in the frequency and diversity of messenger RNA populations in the course of myogenic differentiation. *Eur. J. Biochem.* **97:** 321.

Maxam, A.M. and W. Gilbert. 1980. Sequencing end-labeled DNA with base-specific chemical cleavages. *Methods Enzymol.* **65:** 499.

Ordahl, C.P. and A.I. Caplan. 1976. Transcriptional diversity in myogenesis. *Dev. Biol.* **54:** 61.

——. 1978. High diversity in the polyadenylated RNA populations of embryonic myoblasts. *J. Biol. Chem.* **252:** 7683.

Ordahl, C.P., D. Kioussis, S.M. Tilghman, C. Ovitt, and J. Fornwald. 1980a. Molecular cloning of developmentally regulated, low-abundance mRNA sequences from embryonic muscle. *Proc. Natl. Acad. Sci.* **77:** 4519.

Ordahl, C.P., S.M. Tilghman, C. Ovitt, J. Fornwald, and M.T. Largen. 1980b. Structure and developmental expression of the chick α-actin gene. *Nucleic Acids Res.* **8:** 4989.

Paterson, B.M. and J.O. Bishop. 1977. Changes in the mRNA population of chick myoblasts during myogenesis in vitro. *Cell* **12:** 751.

Southern, E.M. 1975. Detection of specific sequence among DNA fragments separated by gel electrophoresis. *J. Mol. Biol.* **98:** 503.

White, R.L. and D.S. Hogness. 1977. R loop mapping of the 18S and 28S sequences in the long and short repeating units of *Drosophila melanogaster* rDNA. *Cell* **10:** 177.

Zevin-Sonkin, D. and D. Yaffe. 1980. Accumulation of muscle-specific RNA sequences during myogenesis. *Dev. Biol.* **74:** 326.

The Switching of Creatine Kinase Gene Expression during Myogenesis

JEAN-CLAUDE PERRIARD
URS B. ROSENBERG
THEO WALLIMANN
HANS M. EPPENBERGER
Institute for Cell Biology
Swiss Federal Institute of Technology
CH-8093 Zurich, Switzerland

MARIO CARAVATTI
Friedrich Miescher Institute
CH-4002 Basel, Switzerland

In the course of myogenic development, a number of isoprotein switches occur, characterized by the replacement of one isoprotein (generally a ubiquitous component in many tissues) by another isoprotein specific for differentiated muscle. Examples of such isoprotein transitions have been described for the myosin heavy and light chains and for the members of the actin gene family (e.g., Lowey et al.; Whalen et al.; both this volume). We have chosen to investigate the isoprotein transition of creatine kinase during muscle development. During muscle contraction, this enzyme catalyzes the regeneration of ATP from creatine phosphate and ADP. During the early stages of differentiation in muscle cells, the brain creatine kinase (BB-CK), a homodimer, is accumulated, but as differentiation proceeds, isoenzymes containing M-CK subunits appear until, in adult muscle, they have replaced most of the B-CK (Turner et al. 1974, 1976; Perriard et al. 1978b). In striated muscle, the muscle-specific homodimer MM-CK is localized partially in the M line of myofibrils (Wallimann et al. 1977, 1978).

Several lines of evidence indicate that the M-CK and B-CK isoproteins are encoded by two or more genes. Antibodies to either homodimeric isoenzyme do not cross-react with the heterologous homodimeric isoenzyme but do react with the heterodimer MB-CK (Eppenberger et al. 1967; Perriard et al. 1978b). Antibodies made to chicken creatine kinase do not significantly cross-react with creatine kinases of mammalian origin (Eppenberger et al. 1967; J.C. Perriard, unpubl.), which indicates that these proteins may be evolving rather rapidly. Differences between the M-CK and B-CK subunits have also been observed in their peptide maps (Eppenberger et al. 1967; Rosenberg et al. 1981), amino acid composition (Eppenberger et al. 1967), and isoelectric points. Here, we discuss the usefulness of creatine kinase as a marker protein to monitor the progression of myogenesis, the regulation of the genes involved, and the possible functional significance of the creatine kinase isoprotein switch for myofibrillogenesis.

RESULTS AND DISCUSSION

Tissue-specific Distribution of Creatine Kinase Isoenzymes

Most tissues contain low concentrations of BB-CK, ranging from 0.06 units/mg of tissue protein in spleen or liver up to 2 units/mg in brain or chicken gizzard. However, in adult skeletal muscle, MM-CK is by far the major creatine kinase species and is found at a specific activity 250-fold higher than that of BB-CK in spleen or liver (Turner and Eppenberger 1973; M. Specker, pers. comm.). Chicken heart muscle contains only BB-CK, whereas the hearts of mammals accumulate MM-CK and MB-CK in addition to BB-CK (Turner and Eppenberger 1973).

Quite surprisingly, the thymus also contains the muscle-specific form MM-CK, in addition to variable amounts of BB-CK. Immunological analysis of thymus extracts has also revealed the presence of another muscle-specific protein, the 165,000-molecular weight M-line protein, myomesin (Eppenberger et al. 1981). The distribution of cells containing these muscle proteins was studied by immunohistochemical analysis of cryosections of thymus (Fig. 1). Nodules embedded in the thymic reticulum stain positively for M-CK (Fig. 1A) and myomesin (Fig. 1B), whereas only faint staining with anti-BB-CK (Fig. 1C) and nonimmune (Fig. 1D) serum can be seen. In further experiments, thymus tissue, dissociated into single cells and cultured at high density (1.5×10^7 to 3×10^7 cells/ml), gave rise to myotubes that appeared morphologically similar to myotubes from muscle-derived cultures. These myotubes also stained positively with the antibodies to MM-CK and myomesin. Other investigators working with cells derived from mammalian thymus have also observed the development of cross-striated myotubes after in vitro culture under similar conditions (Wekerle et al. 1975). It is reasonable to assume that thymus contains some tissue with characteristics of differentiated muscle and that there are myoid

cells present capable of forming myotubes. This explains the occurrence of muscle-specific proteins in thymus.

Characteristics of Creatine Kinase Subunits in Different Tissues

The study of the regulatory mechanisms governing the creatine kinase switch would be more easily pursued if the same molecular species of M-CK and B-CK subunits were present in different muscle types. To examine this ques-

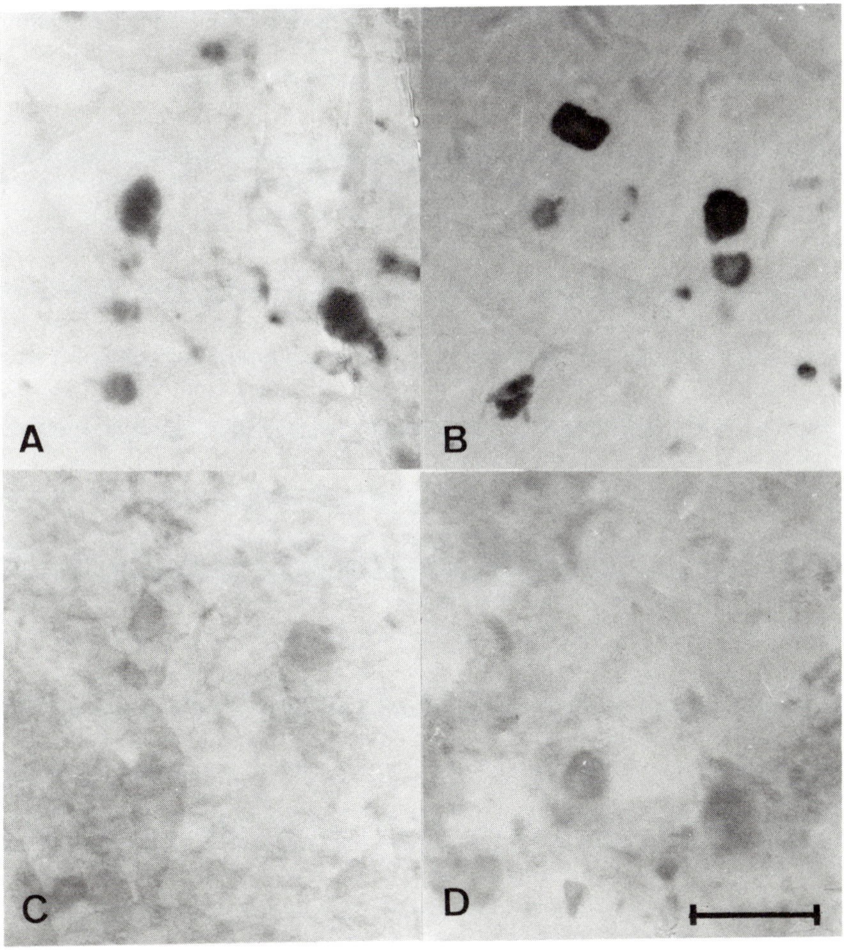

FIGURE 1
Cryosections of thymus explanted from a 4-week-old chicken and stained by the indirect immunoperoxidase technique with anti-MM-CK (*A*), antimyomesin (*B*), anti-BB-CK (*C*), and nonimmune serum (*D*). Bound antibody was revealed by using protein A from *Staphylococcus aureus* conjugated to horse radish peroxidase with diaminobenzidine as a substrate for the peroxidase staining. Bar represents 50 μm.

tion, M-CK subunits were isolated from fast muscle (posterior latissimus dorsi; pectoralis major), slow muscle (anterior latissimus dorsi), and differentiated myogenic cells and subjected to analysis by two-dimensional gel electrophoresis. The resulting patterns of the M-CK peptides were constant for all preparations tested and showed a double spot, indicating the presence of at least two subspecies with similar if not identical molecular weights. The same type of pattern was observed for the B-CK subunits isolated from various tissues, like brain, gizzard, heart, and myogenic cells. These patterns were constant with a more acidic isoelectric point for B-CK than for M-CK and a larger difference of isoelectric points between the two B-CK subspecies. The occurrence of subspecies in both cases does not seem to arise by modification of the proteins after translation or by allelic variation (Rosenberg et al. 1981); their functional significance is unclear at present.

Regulation of Biosynthesis

The accumulation of creatine kinase isoenzymes has been compared in myogenic cell cultures and in embryonic muscle tissue during development and was found to be similar, although the almost complete transition from BB-CK to MM-CK isoenzyme, observed in embryonic muscle, is only partially complete in cultured muscle cells (Perriard et al. 1978b). To investigate the mechanisms regulating this gene switch, measurements of the biosynthesis of M-CK and B-CK subunits were made during the differentiation of myogenic cultures. The incorporation of radioactively labeled amino acids into the M-CK and B-CK subunits was determined using immunoprecipitation methods (Perriard et al. 1978a; Caravatti et al. 1979). In Figure 2, the incorporation of leucine into B-CK and M-CK is shown to be highly regulated in the course of myogenesis. In young cultures (24 hr and 48 hr), when almost no M-CK synthesis occurs, the synthesis of B-CK increased sharply and was maximal at 72 hours. In the later stages of development, the B-CK synthesis dropped to low levels, but synthesis was still significant in 7-day-old cultures. The synthesis of M-CK, however, increased dramatically after the second day of culture, and its synthesis remained at high levels after reaching near maximal values in 96-hour cultures.

In addition to the measurement of total M-CK and B-CK subunit synthesis, we also asked whether the different subspecies of either subunit type were differentially regulated during myogenesis. The radioactivity incorporated into the subspecies of the M-CK and B-CK subunits was determined after separation of the four subspecies on two-dimensional gels. At all stages, the M-CK or B-CK subspecies were synthesized at the same relative proportions, and there was no evidence for differential regulation of any of the subspecies (Rosenberg et al. 1981).

The regulation of the creatine kinases not only could be controlled by the rate of synthesis but also by the rate of differential degradation. Therefore, we determined the turnover kinetics of creatine kinase subunits for both

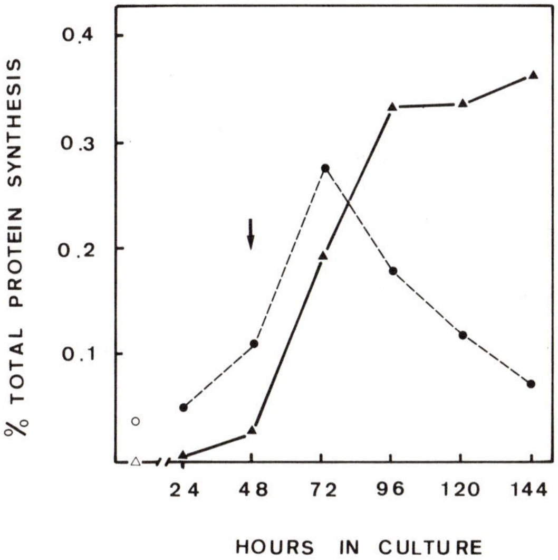

FIGURE 2
Synthesis of creatine kinase subunits during myoblast differentiation. Cultures were pulse-labeled at different times of the culture period, and the incorporation of [^{3}H]leucine was measured as described by Caravatti et al. (1979). (●) Incorporation into B-CK; (Δ) incorporation into M-CK; (○, Δ) values found for B-CK and M-CK, respectively, in subcultured fibroblasts. These cultures were treated with cytosine arabinoside (5 μM) from 48 hr on (→).

M-CK and B-CK in myotubes. The results indicate that M-CK is degraded somewhat more slowly than B-CK, but the difference does not appear to be a major factor in the control of the levels of these isoproteins during their developmental switch (Caravatti and Perriard 1981).

Further attempts to elucidate the mechanism regulating the creatine kinase isoprotein switch were carried out by measuring the relative amounts of translatable mRNA for the creatine kinase subunits using reticulocyte lysates (Perriard et al. 1978a). Here, it could be shown that the availability of mRNA in total cellular RNA or in polysomal RNA parallels the rates of biosynthesis of the creatine kinase subunits. In young cultures up to 48 hours, sizable amounts of mRNA for B-CK could be found, whereas only insignificant traces of M-CK mRNA were detected. During the third day of culture, a dramatic increase in the accumulation of M-CK mRNA was observed, which became relatively abundant. At the same time, there was a further increase in B-CK mRNA. At later stages of differentiation, the level of M-CK mRNA remained high, but the B-CK mRNA dropped to much lower levels. These findings parallel the measurements of the biosynthesis of

the subunits and indicate that the concentration of translatable mRNA regulates the creatine kinase isoprotein transition (Perriard 1979).

The dual control of the creatine kinase genes could be further substantiated using cultures treated with BrdU, a thymidine analog that inhibits the differentiation of myogenic cells (O'Neill and Stockdale 1974; Merill et al. 1980). BrdU treatment was effective in inhibiting the expression of both M-CK isoprotein (Turner et al. 1976; Caravatti et al. 1979) and M-CK mRNA accumulation. The mRNA for B-CK in such BrdU-inhibited cells appears to be accumulated at the relatively high levels characteristic of myogenic cells in early stages of terminal differentiation. A similar peak of BB-CK accumulation was also observed in developing muscle of young embryos prior to the appearance of M-CK-containing isoenzymes (Perriard et al. 1978b). If the cells are released from the BrdU block by subculturing into standard medium, the M-CK gene is expressed as judged by the appearance of translatable M-CK mRNA (Perriard 1979).

The expression of B-CK also appears to be regulated during early phases of terminal differentiation and exhibits a maximum in 3-day-old cultures. In BrdU-blocked cells, the peak of B-CK synthesis was observed concomitant with the one occurring in 3-day-old cultures. It is possible that B-CK expression at an elevated level could be characteristic for an early stage in terminally differentiating muscle cells (Caravatti et al. 1979).

Specific Functional Roles of Newly Acquired MM-CK Isoproteins

A major unresolved question is whether a newly appearing isoprotein has a functional role in differentiating cells that cannot be adequately fulfilled by the isoprotein found in the precursor cells. In the case of the contractile isoproteins, like actin or myosin, this question cannot be answered at present. However, in the case of creatine kinase, we can postulate a unique, functional role for the newly appearing M-CK gene product in terminally differentiating cells, which cannot be fulfilled by the B-CK isoprotein. A portion of the MM-CK isoenzyme is specifically bound to the M line of myofibrils of adult chicken skeletal muscle isolated under physiological conditions. Upon incubation of myofibrils, with low ionic strength buffer or with an excess of Fab fragment of specific anti-MM-CK antibody, MM-CK is specifically extracted and, concomitantly, the electron-dense M line disappears (Wallimann et al. 1978). This structurally bound creatine kinase was recently shown to be active in rephosphorylating ADP during Ca^{++}-stimulated contraction of isolated myofibrils in the presence of creatine phosphate. Furthermore, it is present in sufficient amounts to maintain a maximal actin-activated Mg^{++}-ATPase activity (T. Schlosser and T. Wallimann, in prep.). MM-CK associates specifically with the M-line structure early in development. In somites from 4-day-old chicken embryos, BB-CK is the predominant form of CK, whereas M-CK-containing isoenzymes are barely detectable. Yet MM-CK is already localized at the M line. In con-

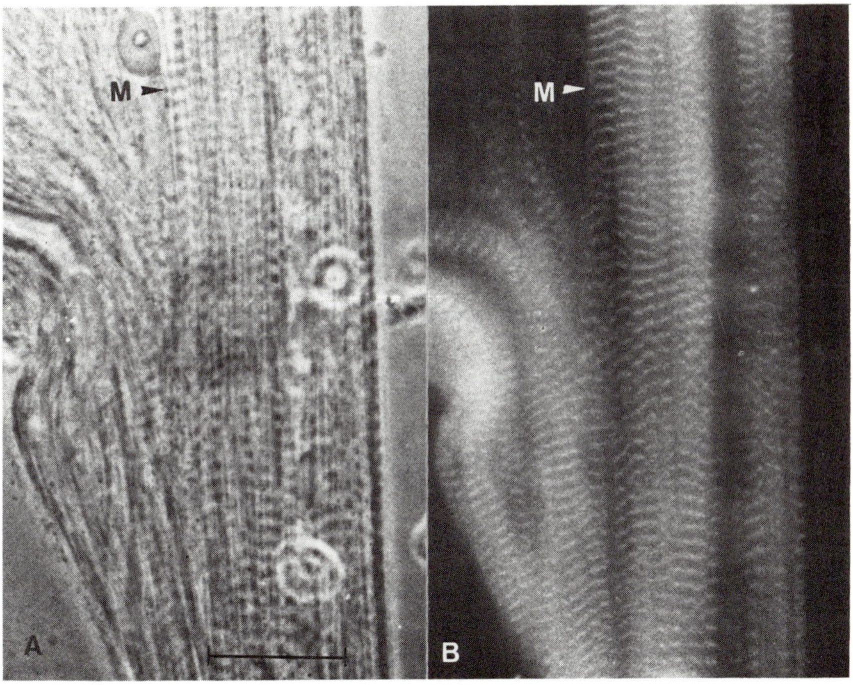

FIGURE 3
Indirect immunofluorescence localization of M-line-bound MM-CK in differentiated skeletal-muscle cells cultured for 5 days. To remove the majority of unbound creatine kinase, cells were permeabilized with Triton X-100, washed extensively, and stained with affinity-purified anti-MM-CK immunoglobulin as described previously (Eppenberger et al. 1981). By superimposing the phase-contrast images (*A*) and immunofluorescence (*B*), staining can be localized within the middle of the A band of sarcomeres. (M) M-line. Bar represents 20 μm.

trast, B-CK is never found in the M-line region (Wallimann et al. 1977). Recently, in this laboratory, a similar isoenzyme-specific association of MM-CK with the M line was demonstrated in differentiated cultured myotubes (H. Moser and T. Wallimann, in prep.) known to contain all three creatine kinase isoenzymes (Turner et al. 1976; Perriard et al. 1978b). The bulk of the soluble creatine kinase molecules was removed selectively by a detergent-containing solution of physiological ionic strength, which did not extract M-line-bound creatine kinase, nor did it destroy the fragile myofibrils. Upon immunofluorescence staining of such cells with antiserum to MM-CK—under these conditions, anti-MM-CK immunoglubulin (IgG) does not extract the M line but binds to it—there was an intense fluorescence staining in the M line (Fig. 3). When the same cell preparations were stained with antibody to BB-CK, only faint staining in the I-band

region of the myofibrils was observed, as reported previously for myofibrils of somites (Wallimann et al. 1977). Thus, only MM-CK, not the other creatine kinase isoproteins present in these cells, is able to associate specifically with the nascent myofibrils of differentiated skeletal-muscle cells.

SUMMARY

This isoprotein switching system appears to be well suited for studying the mechanisms of gene regulation during development. To this end, experiments are under way in our laboratory using recombinant DNA technology to study the structure and expression of the M-CK and B-CK isoprotein genes. We hope to recognize fundamental regulatory mechanisms common to the muscle isoprotein switches by comparing the results of such studies on creatine kinase to those for the other contractile isoproteins. In addition, investigations of how the muscle-specific isoproteins take part in the processes that lead to the highly ordered structures of the contractile elements of the myofibrils should help to elucidate the phenotypic significance of isoprotein switches in muscle development.

ACKNOWLEDGMENTS

We are grateful to Gisela Kunz, Hanni Moser, and Evelyne Perriard for technical assistance. This work was supported by grant 3.187-0.77 from the Swiss National Science Foundation, a grant to H.M.E. from the Muscular Dystrophy Association of America, and predoctoral training grants from the Swiss Federal Institute of Technology to U.B.R. and M.C.

REFERENCES

Caravatti, M. 1979. "Synthese und Degradation von Kreatin-Kinase-Untereinheiten während der Terminaldifferenzierung embryonaler Huehner-Skelettmuskelzellen 'in vitro'." Ph.D thesis, Swiss Federal Institute of Technology, Zurich.

Caravatti, M. and J.C. Perriard. 1981. Turnover of the creatine kinase subunits in chicken myogenic cell cultures and in fibroblasts. *Biochem. J.* **196:** 377.

Caravatti, M., J.C. Perriard, and H.M. Eppenberger. 1979. Developmental regulation of creatine kinase isoenzymes in myogenic cell cultures from chicken. Biosynthesis of creatine kinase subunits M and B. *J. Biol. Chem.* **254:** 1388.

Dawson, D.M., H.M. Eppenberger, and N.O. Kaplan. 1967. The comparative enzymology of creatine kinases. II. Physical and chemical properties. *J. Biol. Chem.* **242:** 210.

Eppenberger, H.M., D.M. Dawson, and N.O. Kaplan. 1967. The comparative enzymology of creatine kinases. I. Isolation and characterization from chicken and rabbit tissues. *J. Biol. Chem.* **242:** 204.

Eppenberger, H.M., J.C. Perriard, U.B. Rosenberg, and E.E. Strehler. 1981. The M_r 165,000 M-protein myomesin: A specific protein of cross-striated muscle cells. *J. Cell. Biol.* **89:** 185.

Merill, G.F., E.B. Wittier, and S.D. Hauschka. 1980. Differentiation of thymidine kinase deficient mouse myoblasts in the presence of 5′-bromodeoxyuridine. *Exp. Cell Res.* **129:** 191.

O'Neill, MC. and F.E. Stockdale. 1974. 5-bromodeoxyuridine inhibition of differentiation. Kinetics of inhibition and reversal in myoblasts. *Dev. Biol.* **37:** 117.

Perriard, J.C. 1979. Developmental regulation of creatine kinase isoenzymes in myogenic cell cultures from chicken. Levels of mRNA for creatine kinase subunits M and B. *J. Biol. Chem.* **254:** 7036.

Perriard, J.C., E.R. Perriard, and H.M. Eppenberger. 1978a. Detection and relative quantitation of mRNA for creatine kinase isoenzymes in RNA from myogenic cell cultures and embryonic chicken tissues. *J. Biol. Chem.* **253:** 6529.

Perriard, J.C., M. Caravatti, E.R. Perriard, and H.M. Eppenberger. 1978b. Quantitation of creatine kinase isoenzyme transitions in differentiating chicken embryonic breast muscle and myogenic cell cultures by immunoadsorption. *Arch. Biochem. Biophys.* **191:** 90.

Rosenberg, U.B., H.M. Eppenberger, and J.C. Perriard. 1981. Occurrence of heterogenous forms of the subunits of creatine kinase in various muscle and non-muscle tissues and their behaviour during myogenesis. *Eur. J. Biochem.* **117:** 87.

Turner, D.C. and H.M. Eppenberger. 1973. Developmental changes in creatine kinase and aldolase isoenzymes and their possible function in association with contractile elements. *Enzyme* **15:** 224.

Turner, D.C., V. Maier, and H.M. Eppenberger. 1974. Creatine kinase and aldolase isoenzyme transitions in cultures of chick skeletal muscle cells. *Dev. Biol.* **37:** 63.

Turner, D.C., R. Gmür, H.G. Lebherz, M. Siegrist, and T. Wallimann. 1976. Differentiation in cultures derived from embryonic chicken muscle. II. Phosphorylase histochemistry and fluorescent antibody staining for creatine kinase and aldolase. *Dev. Biol.* **48:** 284.

Wallimann, T., D.C. Turner, and H.M. Eppenberger. 1977. Localization of creatine kinase isoenzymes in myofibrils. I. Chicken skeletal muscle. *J. Cell Biol.* **75:** 297.

Wallimann, T., G. Pelloni, D.C. Turner, and H.M. Eppenberger. 1978. Monovalent antibodies against MM-creatine kinase remove the M-line from myofibrils. *Proc. Natl. Acad. Sci.* **75:** 4296.

Wekerle, H., B. Paterson, U.P. Ketelsen, and M. Feldman. 1975. Striated muscle fibres differentiate in monolayer cultures of adult thymus reticulum. *Nature* **256:** 493.

Amplification of the Chicken Skeletal α-Actin Gene during Myogenesis

ROBERT J. SCHWARTZ
WARREN E. ZIMMER, JR.
Department of Cell Biology
and Program in Neurosciences
Baylor College of Medicine
Houston, Texas 77030

The changes in gene activity that underlie cell differentiation are generally thought to occur at the RNA level. Most genes are thought to be regulated by transcriptional or messenger-stability mechanisms rather than by changes in the number of genes per genome. In fact, several studies on the induction of contractile-protein mRNA in cultured muscle cells have suggested that myogenesis is regulated both by gene transcription (Paterson and Bishop 1977; Benoff and Nadal-Ginard 1980) and mRNA stability (Buckingham et al. 1974; Singer and Kessler-Icekson 1978). Although control of genes at the RNA level may prove to be the general rule, numerous examples of developmental changes in DNA content and sequence organization have been documented (Brown 1981). The rearrangement of the immunoglobulin genes (Bracke et al. 1978), the amplification of ribosomal (Brown and Dawid 1968) and chorion genes (Spradling and Mahowald 1980; Thireos et al. 1980), the transduction of cellular genetic information by retroviruses (Wang et al. 1978), and the transposition of DNA elements in yeast (Cameron et al. 1979) and in *Drosophila* (Potter et al. 1979) demonstrate the variety of ways the genetic information of eukaryotic cells can be structurally altered at the DNA level to effect the accumulation of specific RNA species.

With the construction of a nearly full-length chicken α-actin cDNA clone, pAC269 (Schwartz et al. 1980), the molecular mechanism underlying the selective induction of the skeletal-muscle α-actin mRNA can be studied in

detail. Previously, it was difficult to use hybridization assays to quantitate the induction of the α-actin gene during muscle development because of a preexistent population of nonmuscle β- and γ-actin mRNAs in prefusion, replicating myoblasts because of their cross-homology to the α-actin cDNA probe. However, characterization of the thermal stability properties of hybrids formed between RNA and ^{3}H-labeled pAC269 DNA has revealed a large sequence divergence ($T_m \geq 10°C$) between cloned α-actin cDNA and nonskeletal-muscle actin mRNA (Schwartz et al. 1980). This difference in the primary nucleic acid sequence has allowed us to develop highly stringent hybridization conditions to eliminate cross-hybridization to nonmuscle actin mRNA species but would allow complete hybridization to α-actin mRNA (Schwartz and Rothblum 1981). We have utilized these specific hybridization conditions to quantitate the level of actin mRNAs during muscle development. One unexpected consequence of these investigations has been the discovery that chicken actin genes can be selectively amplified during normal myogenic development.

RESULTS AND DISCUSSION

Quantitation of Actin mRNA during Myogenesis

In cell culture, replicating myoblasts contained approximately 2000 molecules of actin mRNA per cell, composed predominantly of β- and γ-actin mRNAs. During this same period, the content of muscle-specific α-actin mRNA was estimated at a minimum level of 130 molecules per cell (Fig. 1). At the beginning of fusion, there was no increase in the number of actin mRNA molecules per cell. However, midway through fusion (75 hr in culture), the α-actin mRNA content increased to 14,600 molecules per cell nucleus, and at the completion of myotube formation, the actin mRNA con-

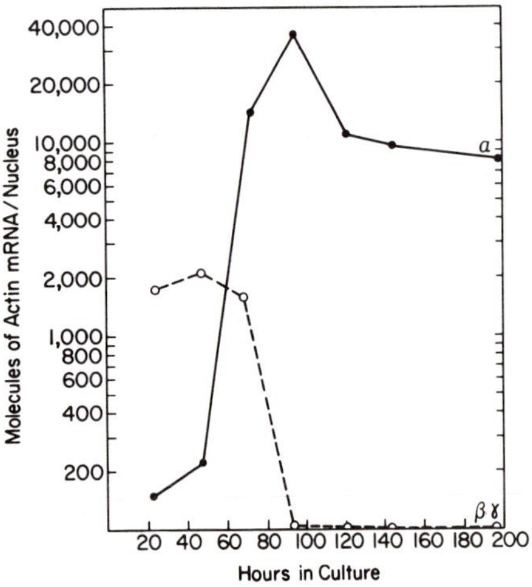

FIGURE 1
Switching of actin mRNA species during chicken myogenesis in culture. The content of α-actin mRNA (●) and β- and γ-actin mRNAs (○) in chicken myoblast cultures was expressed per diploid nucleus, as described by Schwartz and Rothblum (1981).

centration reached a peak level of 36,000 molecules per cell nucleus—a 270-fold increase. Unexpectedly, we found that the amount of α-actin mRNA declined during the next 100 hours in culture, although sarcomeres were functional and there was no loss of myotubes. A maintenance level of 8000 molecules of α-actin per nucleus was reached following the completion of muscle cell differentiation. The reduction of nonmuscle β- and γ-actin mRNA content was quantitated by hybridization assays. At least 94% of the actin mRNA in dividing myoblasts is β- and γ-actin mRNAs, whereas only 6% is the α-isoform. When fusion is complete, all the detectable actin mRNAs are of the α-actin type. These data support a specific developmental program in myogenesis in which the expression of the actin isoform genes is switched.

Identification of the Chicken α-Actin Gene in a Genomic Library

To examine the structure of actin genes in the chicken genome, a genomic λ Charon 4A library composed of 14–22 kb of *Hae*III and *Alu*I partially restricted chicken DNA (Dodgson et al. 1979) was screened with a chicken actin DNA probe. Twenty-six independent clones were isolated. The skeletal-muscle α-actin structural gene was identified because of the large sequence divergence between α-actin and the other actin genes, utilizing pAC23, a probe containing 120 bp of 3′-noncoding sequence of α-actin. Mapping of fragments digested with restriction endonucleases indicated that two of these genomic actin clones, λAC5 and λAC25, were identical.

The transcribed portion of the α-actin gene is approximately 2000 nucleotides in length and contains at least three small introns (Ordahl et al. 1980; Zimmer et al., in prep). A simple restriction endonuclease map of the α-actin gene and its flanking sequences was constructed following the transfer of DNA fragments to nitrocellulose by the Southern (1975) protocol and hybridization to ^{32}P-labeled pAC269 (Fig. 2). Except for the artificial *Eco*RI dodecameric linkers used in the construction of the chicken genomic library, there are no *Eco*RI recognition sites within 9000 nucleotides of the α-actin gene.

Amplification of Actin Genes during Muscle Development

Since approximately 60–70% of the actin cDNA insert in pAC269 is complementary to most nonmuscle actin mRNAs, we used this DNA as a probe to determine the number of actin genes in the genomes of various adult tissues as well as at various stages of developing skeletal muscle. When equal amounts of adult chicken DNA from liver, brain, and muscle were digested with *Eco*RI and subjected to Southern transfer analysis with a pAC269 insert as a probe, an identical pattern of seven major bands of various intensities was seen in each lane (Fig. 3, lanes 1–3). Thus, the actin genes in these tissues have the same position with respect to *Eco*RI sites and the same

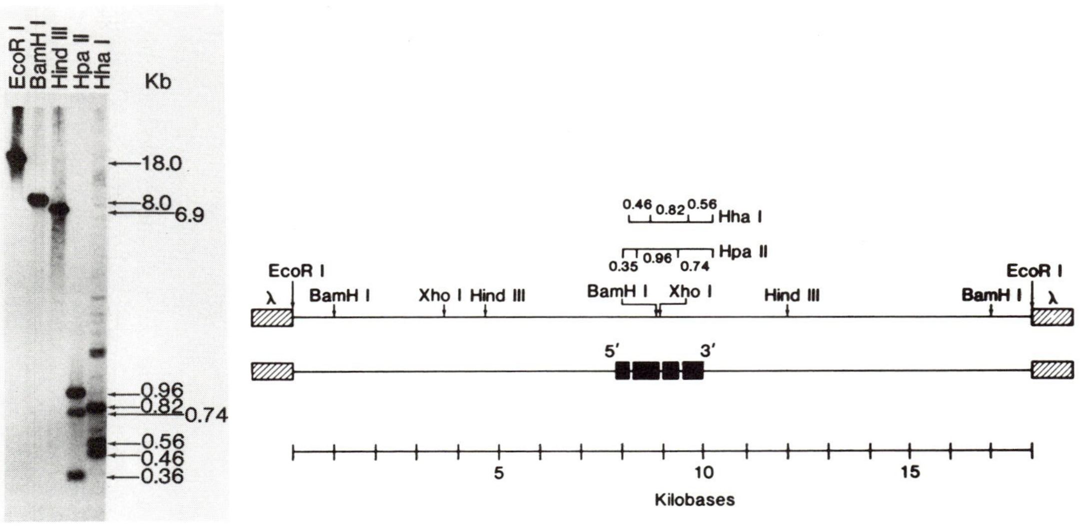

FIGURE 2

Partial restriction map of the chicken chromosomal α-actin gene. The schematic diagram indicates the restriction endonuclease digestion sites of the chicken DNA insert in λAC5. The position and orientation of the transcribed region as determined by a Southern transfer protocol and hybridization with ^{32}P-labeled pAC269 DNA (*left*) is shown by solid blocks (*right*). The orientation of *Hha*I and *Hpa*II fragments was determined by using overlapping actin cDNA clones pAC51 (450 nucleotides of insert) and pAC23 (120 nucleotides of insert). Intervening sequences of ~100 nucleotides were estimated by electron micrographs of R loops and are shown as spaces between blocks (Zimmer et al., in prep.). The hatched blocks represent the arms of λ Charon 4A.

relative copy number per mass unit of total DNA. The most intense autoradiographic band migrated at 20 kb and was identified as the α-actin structural gene by hybridization with the flanking sequences in λAC5 (data not shown). The variability in intensity of the other bands is probably due to sequence divergence between the α-actin probe and the various non-α-actin genes present in the other six fragments. Each of these bands, however, probably contains at least one actin gene or a portion of one gene, since hybridization of actin cDNA to an excess of sheared chicken genomic DNA revealed a copy number on the order of ten actin genes per haploid genome (Schwartz and Rothblum 1980).

When a pAC269 probe was hybridized to Southern blots of genomic DNA cut with *Eco*RI from prefusion (replicating) and late-fusion myoblasts, a striking result was seen (Fig. 3, lanes 4,5). The prefusion DNA yields

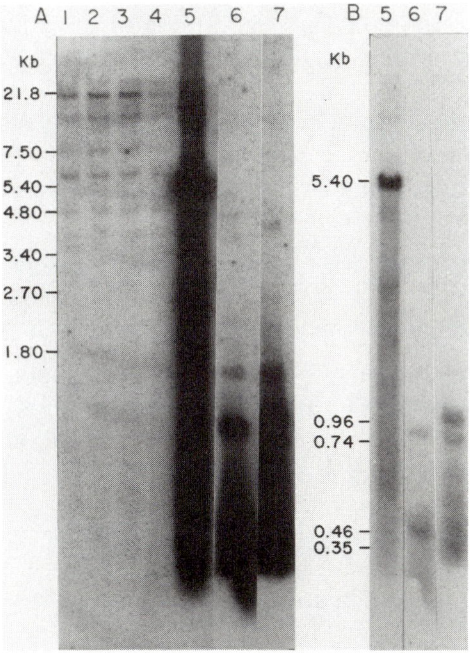

FIGURE 3
Hybridization of ^{32}P-labeled pAC-269 DNA to filter-bound restriction fragments of chicken DNA. Chicken nuclear DNA isolated from 3-week-old brain (*1*), liver (*2*), muscle tissues (*3*), and replicating myoblasts (*4*). Late-fusion myoblasts (*5–7*) were digested separately with restriction endonuclease *Eco*RI (*1–5*), *Hha*I (*6*), and *Hpa*II (*7*) under conditions that ensure a limit digest. Total DNA fragments (10 μg) were loaded onto a 1.0% agarose gel, separated by electrophoresis, transferred to nitrocellulose sheets, and hybridized to ^{32}P-labeled pAC269 (10^8 cpm/μg). The constancy of DNA mass from sample to sample was controlled by sensitive fluorometric measurements of DNA before loading the gel and by ethidium bromide staining of the gel before and after the Southern transfer. The scale (in kilobases) was derived from the locations of restriction fragments of known lengths detected in a separate track of the same gel. Autoradiographs were exposed for 5 days (*A*) and 1 day (*B*).

a pattern similar to those of the adult tissues shown in the first three lanes, whereas the late-fusion sample has an extraordinary increase in intensity of a novel 5.4-kb fragment and a smear of smaller fragments. The 5.4-kb band showed an approximately 85-fold increase in intensity when compared to the 20-kb band from adult and prefusion tissues (Fig. 3). Thus, during fusion of myoblasts in culture, actin DNA molecules appear that have new *Eco*RI sites, different from those in prefusion and adult genomic DNA. The appearance of this DNA results in an approximately 85-fold increase in actin gene dosage per unit of mass DNA.

The DNA from late-fusion myoblasts was also digested with *Hha*I and *Hpa*II because these enzymes produce a series of bands that are characteristic of the α-actin structural gene (Fig. 2; Fig. 3, lanes 6,7). *Hha*I produced three fragments, 820, 560, and 460 nucleotides in length, which hybridized to pAC269 DNA. Digestion with *Hpa*II showed three discrete fragments, 960, 740, and 350 nucleotides in length, complementary to actin cDNA. Restriction digests also revealed a population of intermediately sized DNA fragments, which was probably caused by random nicking of the amplified DNA. We assume that since *Hha*I and *Hpa*II produced cuts that are identical to those observed for the α-actin structural gene, the major portion of the amplified DNA was copied from an α-actin template. Since *Hha*I and *Hpa*II are enzymes that are inhibited by methylation of cytosines in CpG pairs, the digestion patterns observed indicate the absence of methylated cytosines within the transcribable portion of the amplified α-actin gene.

Timing of Actin Gene Amplification in Myoblast Cultures

After finding amplified actin genes in postfusion myoblasts, we repeated the experiment using DNAs from five stages of myoblast development to investigate the time course of the amplification (Fig. 4). Since, in the prior experiment (Fig. 3), amplified DNA obscured the resolution of the actin genomic bands, we reduced the DNA sample concentration by half (5 μg). DNA from prefusion replicating myoblasts yielded a less intense but identical banding pattern on the Southern transfer blots as did DNA from adult tissues (Fig. 4; see also Fig. 3, lanes 1–4). At this stage of culture, myoblasts have a very low level of α-actin mRNA (130 molecules/cell). DNA from the early fusion stage (75 hr in culture) contained a new 7.45-kb band complementary to the actin cDNA. Densitometry of the autoradiographs showed that the 7.45-kb band contained roughly nine times the actin sequences found in the 20-kb (α-actin gene) band. Interestingly, the appearance of the 7.45-kb band remained unchanged during fusion stages (95 hr and 120 hr in culture); but, at completion of myotube formation (120 hr), the 5.4-kb band appeared and was about 80 times more dense than the 20-kb band (Table 1). This approximately 90-fold amplification of the actin gene occurs in concert with the peak of α-actin mRNA level (36,000 molecules/nucleus), found near the end of fusion. During the next 40 hours of culture, the α-actin

FIGURE 4
Hybridization of ^{32}P-labeled pAC269 DNA to filter-bound DNA from cultured myogenic stages digested with *Eco*RI. DNA was isolated from chicken muscle cell cultures and digested with restriction endonuclease *Eco*RI. Total DNA fragments (5 μg) were processed as described in the legend to Fig. 2.

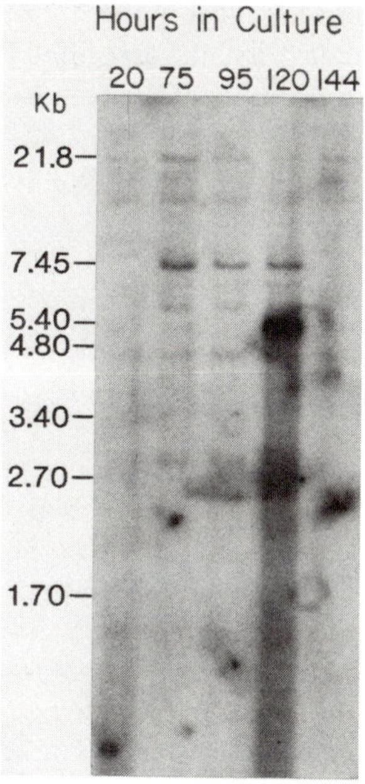

TABLE 1
Relative Content of Actin Genes during Myogenesis in Culture

Molecular weight (kb)	Control[a]	Hours in culture[b] 20	75	95	120	144
20.0	1.00	1.00	1.00	1.00	1.00	1.00
11.5	0.28	0.59	0.15	0.28	0.27	0.33
10.1	0.64	0.79	0.79	0.30	0.27	0.53
8.6	0.27	0.26	0.21	0.24	0.21	0.16
7.8	0.40	0.43	0.16	0.23	0.25	0.52
7.45[c]	—	—	8.9	5.2	9.8	—
6.3	0.38	0.31	0.36	0.25	obs[d]	0.25
5.4[c]	—	—	—	—	84.4	—
4.5	0.31	0.52	0.25	0.22	obs	0.46

[a]Relative densitometric measurements of autoradiographic scans from *Eco*RI-digested genomic DNA isolated from 3-week-old chicken brain, liver, and muscle tissues, which were transferred to nitrocellulose and hybridized to ^{32}P-labeled pAC269, as described in the text and the legend to Fig. 3. The autoradiographic intensity of the 20-kb band was normalized to a value of one. Eighteen separate samples were included in these measurements.
[b]Densitometric measurements of autoradiographic scans of culture myogenic stages as shown in Fig. 4.
[c]The 7.45-kb and 5.4-kb *Eco*RI bands were not detected in control DNA samples.
[d]obs indicates autoradiographic bands that were obscured by the amplified DNA.

mRNA level in the myotubes declined to 20% of the peak value. More dramatically, the amplified DNA species disappeared within 24 hours of reaching its greatest concentration. It should be noted that in late-stage myotubes (Fig. 3, lane 5), the 7.45-kb band was not observed in the autoradiographs. It could be that the 7.45-kb band disappeared before the 5.4-kb band, or that the appearance of the 7.45-kb band is highly variable.

Amplification of the α-Actin Gene in the Intact Embryonic Muscle

When culturing the myoblasts for the experiments shown in Figures 3 and 4, cytosine arabinoside (Ara C), a potent DNA synthesis inhibitor, was added to the cultures at 45 hours and removed at 88 hours to inhibit the growth of contaminating fibroblasts that remained after preplating (Moss et al. 1979). To demonstrate that the amplification of actin DNA was not an artifact caused by the addition of this drug, we repeated the experiment using DNA from thigh muscles removed from untreated embryos at various embryonic stages of development. Figure 5 is an autoradiograph of a Southern transfer blot of *Eco*RI-digested DNA. Samples from 12- to 16-day-old embryos have digestion patterns characteristic of adult tissues and prefusion myoblasts (see Fig. 5A, lanes 12–16). In these stages, the embryonic thigh muscle contains a mixture of mononucleated myoblasts and myoblasts undergoing fusion. The actin mRNA level is about 4000 molecules per cell and consists mostly of the β- and γ-isoforms (Fig. 5B). Even though muscle development in the animal is less synchronous than in culture, on day 18 of embryogenesis, thigh muscle is predominantly composed of multinucleated myotubes. DNA from this stage contains the amplified actin DNA as manifested by the appearance of the 7.45-kb and 5.4-kb bands, which appear to be identical in size to those found in cultured myoblasts. As a result of asynchronous development, a low level of both the 7.45-kb and 5.4-kb amplified DNA species is first apparent in 17-day embryonic muscle. However, the maximal appearance of both amplified DNA species occurs in day-18 thigh muscle, and this coincides with the accumulation of high levels of actin mRNA (20,000 molecules/cell nucleus), as it does in the culture system (Fig. 5B). Furthermore, the amplified actin DNA bands begin to disappear at 9 days, and the 5.4-kb band is absent from the blot of 20-day embryonic thigh DNA and is not seen in any of the older developmental stages. The only differences apparent in actin DNA amplification between cultured myoblasts and embryonic thigh muscles are the noncoordinate appearance of the 7.45-kb and 5.4-kb *Eco*RI bands in cultured myoblasts and their coordinate temporal appearance in embryonic muscle.

During development, these results therefore indicate that α-actin gene dosage changes at least 90-fold, and the amplified actin DNA sequences contain new *Eco*RI sites. The amplified DNA disappears from myotubes within 24 hours of its greatest concentration. The gain and loss of the extra

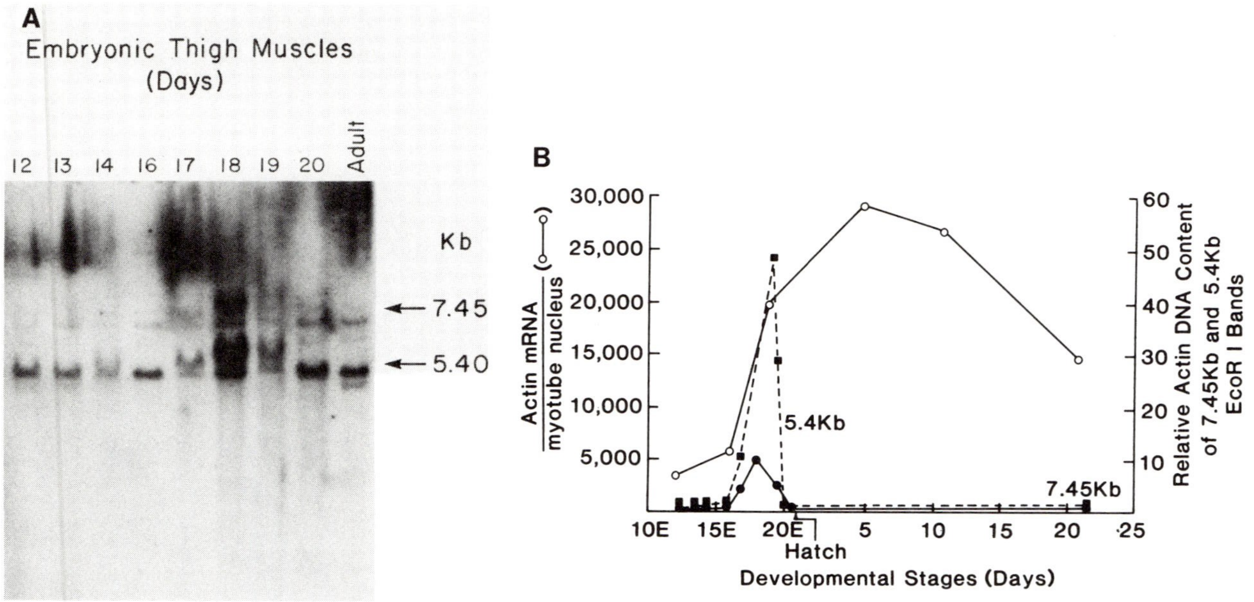

FIGURE 5

Amplification of actin genes and induction of actin mRNA during embryonic muscle development. (*A*) DNA was isolated from staged chicken thighs and digested with restriction endonuclease *Eco*RI. Total DNA fragments (10 μg) were processed as described in the legend to Fig. 3. The autoradiograph was exposed for 5 days. (*B*) The induction of actin mRNA (○) and the relative genomic contents of the 7.45-kb actin DNA band (●) and the 5.4-kb actin DNA band (■) were graphed together to show the correlation between increase in mRNA content and DNA content.

α-actin gene closely parallel the rise and fall of the α-actin mRNA concentration. Thus, amplified DNA may be an intermediate in the developmental expression of α-actin.

Although our experiments do not directly probe the mechanism of the amplification, a few of the findings merit comment in this context. First, the 7.45-kb *Eco*RI band appears before or, perhaps, during the presence of Ara C in culture. Only after removal of the Ara C does the 85-fold-amplified 5.4-kb band appear. This suggests that the synthesis of the two bands of actin DNA may differ. Second, within the domain of the α-actin gene, prior to amplification, the *Eco*RI sites are located thousands of nucleotides from the transcribed region. How were the new *Eco*RI sites placed in the amplified actin DNA? One possibility is that new DNA elements were introduced into the amplified DNA during its replication, and another possibility is that DNA was rearranged during gene amplification. We are trying to isolate and clone the amplified actin DNA in order to verify one of these alternatives directly by nucleic acid sequencing.

Is there any relationship between the amplified DNA and the macrochromosomes found in chicken? It is possible that the amplified DNA exists extrachromosomally as does the ribosomal DNA in *Xenopus* oocytes (Brown and Dawid 1968), or it may be endoreduplicated in the macrochromosomes as is the chorion gene in *Drosophila* follicle cells (Spradling and Mahowald 1980). In *Xenopus* oocytes, ribosomal DNA is degraded after the necessary RNA abundance is attained, whereas in *Drosophila* follicle cells, the amplified chorion DNA disappears via cell death. In muscle, there appears to be no gross cell death or decrease in bulk DNA associated with the disappearance of the amplified sequences. Thus, we tentatively favor the extrachromosomal case because of the speed with which the amplified actin DNA disappears.

ACKNOWLEDGMENTS

We thank Mr. Edwin Stone for his thoughtful commentary on this study and in preparation of this manuscript. This research was supported by U.S. Public Health Service grant NS-15050 and by the Jerry Lewis Neuromuscular Disease Research Center of the Muscular Dystrophy Association of America. R.J.S. is a recipient of a U.S. Public Health Service Research Career Development Award.

NOTE ADDED IN PROOF

During the past year, we examined a number of chicken flocks and found actin gene amplification to be quite variable and, in some cases, not apparent. We assume that amplification of actin genes is an interesting phenomenon but not the major gene regulatory event in the expression of actin during myogenesis.

REFERENCES

Benoff, S. and B. Nadal-Ginard. 1980. Transient induction of poly(A)-short myosin heavy chain messenger DNA during terminal differentiation of L_6E_9 myoblasts. *J. Mol. Biol.* **140:** 283.

Brack, C., M. Horama, R. Lenhard-Schullar, and S. Tonegawa. 1978. Immunoglobulin gene is created by somatic recombination. *Cell* **15:** 1.

Brown, D.D. 1981. Gene expression in eukaryotes. *Science* **211:** 667.

Brown, D.D. and I.B. Dawid. 1968. Specific gene amplification in ooctyes. *Science* **160:** 272.

Buckingham, M.E., D. Caput, A. Cohen, R.G. Whalen, and F. Gros. 1974. The synthesis and stability of cytoplasmic messenger RNA during myoblast differentiation in culture. *Proc. Natl. Acad. Sci.* **71:** 1466.

Cameron, J.R., E.Y. Loh, and R.W. Davis. 1979. Evidence for transposition of dispersed repetitive DNA families in yeast. *Cell* **16:** 739.

Dodgson, J.B., J. Strommer, and J.D. Engel. 1979. Isolation of the chicken β-globin gene and a linked embryonic β-like globin gene from a chicken DNA recombinant library. *Cell* **17:** 879.

Moss, M., B. Asch, and R.J. Schwartz. 1979. Differentiation of actin containing filaments during chick skeletal myogenesis. *Exp. Cell Res.* **121:** 167.

Ordahl, C.P., S.M. Tilgham, C. Ovitt, J. Fornwald, and T. Largen. 1980. Structure and developmental expression of the chicken α-actin gene. *Nucleic Acids Res.* **8:** 4989.

Paterson, B.M. and J.O. Bishop. 1977. Changes in the mRNA population of chick myoblasts during myogenesis *in vitro. Cell* **12:** 751.

Potter, S.S., W.J. Borein, P. Dunsmier, and G.M. Rubin. 1979. Transposition of elements of the 412, copia, and 297 dispersed gene families in *Drosophila. Cell* **17:** 415.

Schwartz, R.J. and K.N. Rothblum. 1980. Regulation of muscle differentiation: Isolation and purification of chick actin messenger ribonucleic acid and quantitation with complementary deoxyribonucleic acid probes. *Biochemistry* **19:** 2506.

———. 1981. Gene switching in myogenesis: Differential expression of the chicken actin multigene family. *Biochemistry* **20:** 4122.

Schwartz, R.J., J.A. Haron, K.N. Rothblum, and A. Dugaiczyk. 1980. Regulation of muscle differentiation: Cloning of sequences from α-actin messenger ribonucleic acid. *Biochemistry* **19:** 5883.

Singer, R.H. and G. Kessler-Icekson. 1978. Stability of polyadenylated RNA in differentiating myogenic cells. *Eur. J. Biochem.* **88:** 395.

Southern, E.M. 1975. Detection of specific sequences among DNA fragments separated by gel electrophoresis. *J. Mol. Biol.* **98:** 503.

Spradling, A.L. and A.F. Mohowald. 1980. Amplification of genes for chorion proteins during oogenesis in *Drosophila melanogaster. Proc. Natl. Acad. Sci.* **77:** 1096.

Thireos, G., R. Griffin-Shea, and F.C. Kafatos. 1980. Untranslated mRNA for a chorion protein of *Drosophila melanogaster* accumulates transiently at the onset of specific gene amplification. *Proc. Natl. Acad. Sci.* **77:** 5789.

Wang, L.H., C.C. Halpern, M. Nadel, and H. Hanafusa. 1978. Recombination between viral and cellular sequences generates transforming sarcoma viruses. *Proc. Natl. Acad. Sci.* **75:** 5812.

RNA Polymerase-II Mutants Defective in Myogenesis

MARK L. PEARSON
Basic Research Program-LBI
Cancer Biology Program
Frederick Cancer Research Facility
Frederick, Maryland 21701

MICHAEL M. CRERAR
Department of Biology
York University
Downsview, Ontario, Canada M3J 1P3

The regulatory proteins that control the expression of muscle-characteristic genes during myogenesis have not yet been identified, let alone characterized at the molecular level. By analogy with the mechanisms known to regulate gene activation and inactivation in prokaryotic cells, gene expression during development in eukaryotic cells probably functions by means of an interlocking series of interactions between RNA polymerase and regulatory sites on the DNA template on the one hand and transcriptional activators and inhibitors on the other. This hypothesis assumes that the primary control of gene expression in such systems is exerted at the level of transcription, an assumption now supported by a large body of experimental data on muscle (see Buckingham et al.; Hastings and Emerson; Nadal-Ginard et al.; Ordahl et al.; Shani et al.; all this volume). Two general experimental strategies for the analysis of genetic control mechanisms have been useful in studies of differential gene expression in prokaryotes. One strategy involves the isolation and characterization of mutants defective in gene regulation, and the other involves the biochemical analysis of gene expression using recombinant DNA techniques. We have applied the genetic strategy to the analysis of the control of gene expression during myogenesis. In this paper we summarize the results of our investigations into the role of RNA polymerase II in myogenic differentiation and the effects of mutations to α-amanitin resistance on this process in the L6 rat myoblast cell line.

α-Amanitin is a mushroom toxin that binds specifically to RNA polymerase II and inhibits both the initiation and elongation of mRNA transcripts in vitro (Cochet-Meilhac and Chambon 1974). In cultured cells, the acquisition of resistance to the cytotoxic action of this toxin is accompanied by an alteration in the amanitin-binding affinity of RNA polymerase II, thus allowing the direct selection of mutant cells with structurally altered RNA polymerase II (Chan et al. 1972; Somers et al. 1975a; Ingles et al. 1976; Crerar et al. 1977). Such mutations breed true; they are stable in the absence of the selective agent. They are also expressed in a codominant fashion in cell hybrids (Lobban and Siminovitch 1975); equal amounts of the wild-type and the α-amanitin-resistant (Ama^R) mutant forms of RNA polymerase II are also found in mutant diploid cells (Crerar et al. 1977), indicating that both alleles of the amanitin-binding subunit structural gene are simultaneously expressed.

In such Ama^R mutants, the relative concentration of the two forms of the RNA polymerase II can be perturbed by culturing the mutant cells in the presence of amanitin (Somers et al. 1975b; Crerar and Pearson 1977). Under these conditions, the wild-type sensitive form of the enzyme is inactivated. Surprisingly, the cell compensates for this rapid drop in the overall intracellular activity level of RNA polymerase II by synthesizing more of the mutant form of the enzyme, thus maintaining the steady-state activity level of polymerase in the cell (Crerar and Pearson 1977; Guialis et al. 1977, 1979). This "autoregulatory" phenomenon may be similar to that studied in bacterial cells in which RNA polymerase probably acts as both an activator and repressor of its own synthesis (Yura and Ishihama 1979). (There is no evidence to date that this regulation of RNA polymerase-II activity in mammalian cells in culture is caused by a gene-amplification mechanism, although this question needs to be studied using cloned RNA polymerase-II structural gene probes.) Experimental manipulation of the ratio of wild-type to mutant RNA polymerase-II activities in Ama^R mutant myoblasts in vivo by simply growing the mutants in the presence or absence of amanitin thus provides a useful tool for studying the influence of different Ama^R mutations in RNA polymerase II on myogenic differentiation.

In bacterial systems, some rifampicin-resistant RNA polymerase mutations exhibit defects in sporulation in *Bacillus subtilis* (Sonenshein et al. 1974) or in phage development in *Escherichia coli* (Georgopoulos 1971; Ghysen and Pironio 1972). These mutations appear to affect the transcriptional control of these "developmental" processes without impairing cell growth. The existence of these pleiotropic mutants suggests that conformational changes in the structure of RNA polymerase can subtly influence the specificity of the altered enzyme in some transcription regulatory interactions and not others. With this in mind, we wondered whether a similar genetic approach, focusing on muscle development, would be useful in the

analysis of the molecular determinants of transcription specificity in mammalian cells. The hypothesis and the experimental plan are as follows:

Hypothesis:	Some RNA polymerase-II mutations are specifically defective in transcription of genes essential for myogenesis.
Model:	Defect is at the level of promoter or regulatory protein interactions with mutant RNA polymerase II.
Methods:	Measure frequency of Myo^- phenotypes in Ama^R mutants. Search for Myo(ama) mutant with conditional myogenic defect.

RESULTS AND DISCUSSION

Isolation and Characterization of the Ama^R Mutants

Over 40 independent, ethylmethanesulfonate (EMS)-induced Ama^R mutants of the L6 D_0 pseudodiploid rat myoblast cell line (Loomis et al. 1973) were isolated and characterized as having an RNA polymerase-II activity resistant to the inhibitory action of α-amanitin in vitro (M.M. Crerar et al., in prep.). The level of resistance relative to wild-type enzyme ($K_i \cong 1$ nM amanitin) varied from 5-fold to 4000-fold in independent mutant clones, indicating that a wide range of resistance phenotypes was possible. (No mutants were found that were completely resistant to amanitin, suggesting that loss of the amanitin-binding site might be lethal with respect to enzyme activity.) In a comparison of the myogenic properties of a parallel population of thioguanine-resistant mutants and EMS "survivors," we noted that the frequency of the impaired myogenic phenotypes was higher in the Ama^R mutants. This led us to study these mutants in more detail.

Myogenic Phenotypes

The myogenic phenotypes of these mutants were examined by assaying the formation of myotubes and the change in muscle creatine kinase (M-CK) specific activity after reaching confluence. The mutants could be divided into three classes on the basis of their myogenic phenotypes (Table 1). Both the myoblast fusion and M-CK assays gave similar results. The frequency distribution of clones in each phenotypic class was approximately the same.

One class, Myo^+, was indistinguishable from the wild-type parent cells with respect to the onset, rate, and extent of fusion or specific activity of M-CK. In these cases, the RNA polymerase-II mutations conferring

TABLE 1
Myogenic Phenotypes of α-Amanitin-resistant L6 Mutants

Phenotype[a]	Fusion; M-CK[b]	Frequency (%)[c]
Myo^{+}	+	35
Myo^{-}	−	32
Myo(ama)		
− ama	+	33
+ ama	−	

[a]There are differences in myogenic phenotypes for cells cultured in the presence or absence of α-amanitin.

[b]The kinetics of myotube formation and the increase in M-CK specific activities are generally similar for wild-type and mutant L6 rat myoblast cultures.

[c]The frequencies of clones exhibiting each of the different myogenic phenotypes after treatment of L6 wild-type cultures with 500 μg/ml of EMS for 24 hr (50% survival) and selected for resistance to 2 μM α-amanitin.

α-amanitin resistance did not appear to interfere with any of the processes required for proper myogenesis in cell culture.

A second class, Myo^{-}, was absolutely defective in myogenesis. In this class, we do not know at present whether the ama^{R} mutation itself is directly responsible for the developmental defect. Other unlinked mutations induced by the EMS treatment could conceivably alter the ability of these cells to form myotubes. No fusion was observed in this class of mutants even under shiftdown conditions (1% fetal calf serum [FCS] and 5 μg/ml of insulin), which stimulate the fusion of wild-type L6 cells (Mandel and Pearson 1974).

The third class of mutant, designated as Myo(ama), had the most interesting phenotype. It was conditionally defective with respect to myogenesis. When these mutants were grown in the absence of amanitin, they differentiated normally. In its presence, they did not. In mutants of this third class, the impairment in myogenesis could be correlated with changes in the intracellular activity level of mutant RNA polymerase II—the higher the ratio of altered versus normal polymerase activity, the greater the myogenic defect. The existence of this class provides strong evidence that some RNA polymerase-II mutations can lead to pleiotropic effects on the ability of L6 myoblasts to differentiate.

The conditional phenotype is illustrated in Figure 1, which shows the specific activity of M-CK at various times after confluence for one representative Myo(ama) mutant, Ama27. In the presence of amanitin, which has essentially no effect on growth, the mutant failed to make elevated levels of M-CK. However, in its absence, Ama27 made even higher levels of M-CK,

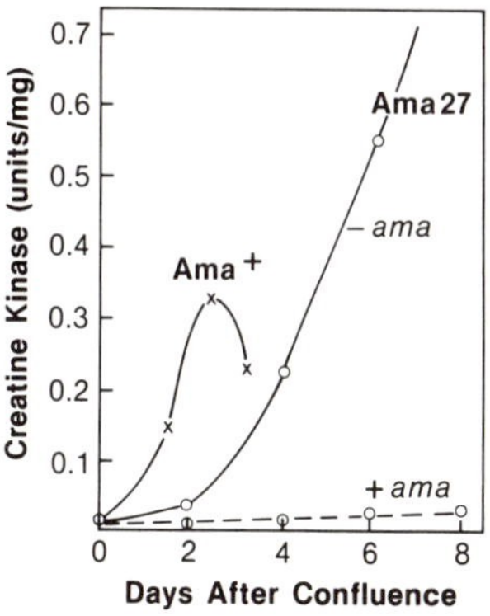

FIGURE 1
M-CK specific activities in wild-type (Ama$^+$) and RNA polymerase-II mutant L6 rat myoblasts (Ama27). Mutant cells were grown for at least 1 week in the absence (−ama) or the presence (+ama) of 1.5 μM α-amanitin prior to plating at high cell density (3 $\times$ 10^4 cells/cm^2) in Dulbecco's modified Eagle's medium containing 10% FCS. M-CK specific activities were measured as described previously (Mandel and Pearson 1974). 1 unit corresponds to 1 μmole of creatine phosphate hydrolyzed per minute at 37°C.

and it exhibited a lag in the rise of M-CK relative to the wild-type parent. This conditional difference in M-CK profile suggests that the *ama27* mutation itself is responsible for the myogenic defect of Ama27 cells.

Patterns of Protein Synthesis in Myo(ama) Mutants

To examine the extent of the myogenic defect in these conditional mutants, we turned to the analysis of the patterns of protein synthesis in mutant cells using two-dimensional isoelectric focusing and polyacrylamide gel electrophoresis (O'Farrell 1975; Garrells 1979). The patterns of protein synthesis in Ama27 mutant cells grown in the presence and absence of amanitin is shown in Figure 2. In the absence of amanitin (when fusion has occurred), proteins labeled with [^{35}S]methionine in Ama27 looked identical to those observed in fused wild-type L6 myotubes (Garrels 1979). However, in the presence of amanitin (when fusion is blocked), the pattern of proteins synthesized looked similar to that found in unfused wild-type L6 myoblasts. Thus, the impairment in myogenesis in the Myo(ama) cell in amanitin is not restricted to fusion and to M-CK synthesis; it is global in nature by this method of analysis. Not only is the synthesis of the major contractile proteins not turned on in the mutant, the synthesis of some of the myoblast characteristic proteins (fibronectin?) is not turned off.

RNA Polymerase-II Pleiotropy and Myogenesis

The observation that the myogenic phenotype of independent clonal isolates of the Myo(ama) mutants can be altered by the addition of amanitin to the medium argues strongly that the inhibition observed is a direct consequence of their RNA polymerase-II mutations, since only the relative level of the mutant and wild-type RNA polymerase II are known to be modulated in vivo by amanitin. Despite intensive searches by us and by others, no other secondary targets of amanitin with binding constants similar to RNA

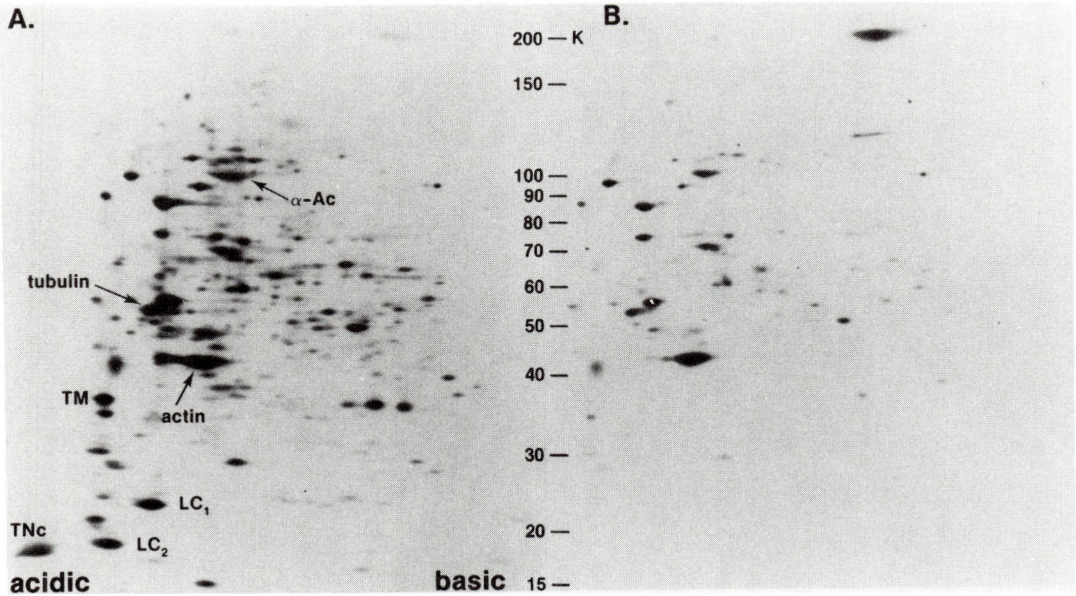

FIGURE 2

Two-dimensional polyacrylamide gel electrophorogram of [^{35}S]methionine-labeled proteins synthesized in Ama27 cells in the absence and presence of α-amanitin. Myotube formation occurred in the absence of 1.5 μM α-amanitin (*A*), but not in its presence (*B*). The first isoelectric dimension ranges over an effective range of pH 5–8; the second dimension utilizes a gradient of 5–20% polyacrylamide. Cultures were labeled with 250 μCi [^{35}S]methionine for 2–3 hr before lysis with SDS-mercaptoethanol and fractionated according to the method of O'Farrell (1975).

polymerase II have been found in mammalian cells. In fact, the existence of the Ama^R class of Myo^+ mutants capable of normal differentiation in amanitin argues that no such targets specific to myogenic development exist in L6. (The likelihood of finding mutations in such a secondary target, as well as in RNA polymerase II, for a large number of independent mutants is extremely low.) Attempts to correlate this Myo(ama) phenotype with the K_i for amanitin resistance of the mutant RNA polymerase II have been fruitless. This suggests that a wide range of conformational alterations in RNA polymerase II can result in the Myo(ama) phenotype. In this regard, the Myo^- class of mutants may represent a more severe structural alteration that absolutely blocks myogenesis, preventing even the wild-type enzyme from successfully participating in myogenic transcription events in these mutants grown in the absence of amanitin.

Our results are consistent with the idea that the altered Ama27 RNA polymerase II of the sort found in the Myo(ama) mutants is unable to promote the synthesis of muscle-characteristic proteins in general. This could involve a defect in some critical early step in the transcription pathway for the initiation of myogenesis. Of course, other possibilities abound. The molecular analysis of the defect clearly requires a more direct demonstration that transcription itself is impaired, an analysis currently under way in our laboratories using appropriate cDNA probes coding for muscle gene mRNAs. At present, we have no way of determining whether the defect involves transcriptional regulatory protein interactions of the polymerase-protein type or promoter recognition interactions of the polymerase-DNA type.

The ama^R mutations in L6 described here give rise to a wide spectrum of RNA polymerase-II phenotypes, some of which exhibit a marked alteration in myogenic development. This may be a specific example of perhaps related pleiotropic effects of similar RNA polymerase-II ama^R mutations in *Drosophila* that exhibit allele-specific variations in fly morphogenesis and affect the expression of several loci (Greenleaf et al. 1979; Greenleaf et al. 1980; Mortin and Lefevre 1981). Thus, further analysis of the influence of such RNA polymerase-II mutations on developmentally regulated gene expression may be useful for unravelling the complexities of the reaction pathways controlling development in general.

SUMMARY

α-Amanitin-resistant L6 rat myoblast mutants were isolated and characterized. These mutants exhibited a wide range of altered RNA polymerase-II phenotypes with respect to the ability of the enzyme to bind amanitin properly. Analysis of the myogenic phenotypes of these mutants suggested that some of these mutations result in impaired myogenesis. One class had a conditional defect in myogenesis; its members did not appear to be able to

develop normally in the presence of amanitin. Such Myo(ama) mutants failed to form myotubes, did not make elevated levels of M-CK, and were defective both in the turn on of myotube proteins and the turn off of myoblast proteins. The differentiation-defective phenotype was correlated with an increase in the ratio of the mutant to wild-type RNA polymerase-II activity when these ama^R/ama^S diploids were grown in amanitin. Such mutations presumably result in a structurally altered RNA polymerase-II enzyme unable to recognize transcriptional control elements, either regulatory proteins or DNA sequences, which are important to the transcriptional activation of myogenesis.

ACKNOWLEDGMENTS

We are indebted to Rick Leather and Emina David for excellent technical assistance. This work, initiated in the Department of Medical Genetics at the University of Toronto, was supported by grants from the National Cancer Institute of Canada, the Medical Research Council of Canada, and a contract from the National Cancer Institute, the National Institutes of Health. It was completed under contract N01-C0-75380 to Litton Bionetics from the National Cancer Institute. The contents of this publication do not necessarily reflect the views or policies of the Department of Health and Human Services, nor does mention of trade names, commercial products, or organizations imply endorsement by the U.S. Government.

REFERENCES

Chan, V.L., G.F. Whitmore, and L. Siminovitch. 1972. Mammalian cells with altered forms of RNA polymerase II. *Proc. Natl. Acad. Sci.* **69:** 3119.

Cochet-Meilhac, M. and P. Chambon. 1974. Animal DNA-dependent RNA polymerases. II. Mechanism of the inhibition of RNA polymerase B by amatoxins. *Biochem. Biophys. Acta* **353:** 160.

Crerar, M.M. and M.L. Pearson. 1977. RNA polymerase II regulation in α-amanitin-resistant rat myoblast mutants. Changes in wild-type and mutant enzyme levels during growth in α-amanitin. *J. Mol. Biol.* **112:** 331.

Crerar, M.M., S.J. Andrews, E.S. David, D.G. Somers, J.-L. Mandel, and M.L. Pearson. 1977. Amanitin binding to RNA polymerase II in α-amanitin-resistant rat myoblast mutants. *J. Mol. Biol.* **112:** 317.

Garrels, J.I. 1979. Changes in protein synthesis during myogenesis in a clonal cell line. *Dev. Biol.* **73:** 134.

Georgopoulos, C.P. 1971. Bacterial mutants in which the gene *N* function of bacteriophage lambda is blocked have an altered RNA polymerase. *Proc. Natl. Acad. Sci.* **68:** 2977.

Ghysen, A. and M. Pironio. 1972. Relationship between the *N* function of bacteriophage λ and host RNA polymerase. *J. Mol. Biol.* **65:** 259.

Greenleaf, A.L., J.R. Weeks, R.A. Voelker, S. Onishi, and B. Dickson. 1980.

Genetic and biochemical characterization of mutants at an RNA polymerase II locus in *D. melanogaster*. *Cell* **21:** 785.

Greenleaf, A.L., L.M. Borsett, P.F. Jiamachello, and D.E. Coulter. 1979. α-Amanitin-resistant *D. melanogaster* with an altered RNA polymerase II. *Cell* **18:** 613.

Guialis, A., K.E. Morrison, and C.J. Ingles. 1979. Regulated synthesis of RNA polymerase II polypeptides in Chinese hamster ovary cell lines. *J. Biol. Chem.* **254:** 4171.

Guialis, A., B.G. Beatty, C.J. Ingles, and M.M. Crerar. 1977. Regulation of RNA polymerase II activity in α-amanitin-resistant CHO hybrid cells. *Cell* **10:** 53.

Ingles, C.J., B.G. Beatty, A. Guialis, M.L. Pearson, M.M. Crerar, P.E. Lobban, L. Siminovitch, D.C. Somers, and M. Buchwald. 1976. α-Amanitin-resistant mutants of mammalian cells and the regulation of RNA polymerase II activity. In *RNA polymerase* (ed. R. Losick and M. Chamberlin), p. 835. Cold Spring Harbor Laboratory, Cold Spring Harbor, New York.

Lobban, P.E. and L. Siminovitch. 1975. α-Amanitin-resistance: A dominant mutation in CHO cells. *Cell* **4:** 167.

Loomis, W.F., Jr., J.P. Wahrmann, and D. Luzzati. 1973. Temperature-sensitive variants of an established myoblast line. *Proc. Natl. Acad. Sci.* **70:** 425.

Mandel, J.-L. and M.L. Pearson. 1974. Insulin stimulates myogenesis in a rat myoblast line. *Nature* **251:** 618.

Mortin, M.A. and G. Lefevre, Jr. 1981. An RNA polymerase II mutation in *Drosophila melanogaster* that mimics ultrabithorax. *Chromosoma* **82:** 237.

O'Farrell, P.H. 1975. High resolution two-dimensional electrophoresis of proteins. *J. Biol. Chem.* **250:** 4007.

Somers, D.C., M.L. Pearson, and C.J. Ingles. 1975a. Isolation and characterization of an α-amanitin-resistant rat myoblast cell line possessing α-amanitin-resistant RNA polymerase II. *J. Biol. Chem.* **250:** 4825.

———. 1975b. Regulation of RNA polymerase II activity in a mutant rat myoblast cell line resistant to α-amanitin. *Nature* **253:** 372.

Sonenshein, A.L., B. Cami, J. Brevet, and R. Cote. 1974. Isolation and characterization of rifampin-resistant and streptolydigin-resistant mutants of *Bacillus subtilis* with altered sporulation properties. *J. Bacteriol.* **120:** 253.

Yura, T. and A. Ishihama. 1979. Genetics of bacterial RNA polymerases. *Annu. Rev. Genet.* **13:** 59.

MEMBRANE EVENTS

Introduction: Membrane Events during Myogenesis

STEPHEN J. KAUFMAN
Department of Microbiology
and School of Basic Medical Sciences
University of Illinois
Urbana, Illinois 61801

During the development of skeletal muscle, myoblasts proliferate, interact, and then fuse to form multinucleate myotubes, which further develop into functional muscle (for reviews, see Bischoff 1978; Pearson 1980). These cells must therefore be specialized to receive and transmit incoming signals (e.g., hormones, acetylcholine, and so on) and to engage in specific cell-cell interactions. As all of these functions are to some extent mediated by the cell membrane, it is reasonable to predict that, as skeletal myoblasts differentiate from single cells into multinucleate, electrically excitable fibers, changes in the myoblast cell surface evolve to promote and/or mediate the changes in the functional capacities of these cells. Thus, three specialized stages of the myoblast membrane presumably exist and reflect the missions of these cells at their respective stages of differentiation: (1) replication, (2) fusion, and (3) excitability and contraction.

Myoblast fusion is a striking morphologic event, obligatory and central to the formation of functional skeletal muscle. Although myoblast fusion is not a prerequisite for the expression of many parameters of the differentiated phenotype, morphologic and biochemical differentiation of skeletal muscle are usually tightly coupled. A great deal of what is understood about myoblast interactions and fusion has come from microcinematographic analyses and from studies in which myogenesis in vitro was perturbed by a variety of chemicals, viruses, and by manipulations of an empirical environment. These observations have led to the description of several important phenomena; however, the mechanisms that mediate the interactions and fusion of myoblasts remain to be resolved. What is clear from the reports presented in this volume is that major progress is under way toward defining and understanding these mechanisms.

In this paper I have tried to integrate some previous observations on myoblast fusion and some of the questions these studies have raised with the new information in the papers presented in this volume. As with most attempts at integration, the reader is cautioned against the pitfall of over-simplification and errors of omission that may follow.

CALCIUM AND FUSION

The requirement for calcium in myoblast fusion has been well documented and often utilized to manipulate and synchronize myogenesis in vitro, yet the basis for this cation requirement is not well understood. M.J.O. Wakelam and D. Pette (pers. comm.)[1] have found that a calcium-induced breakdown of membrane phosphatidylinositol, which renders the membrane more adaptable to fusion, takes place only in myoblasts that are competent to fuse. Selective proteolysis may also have an important role in myoblast fusion. Strittmatter et al. (this volume) have shown that inhibition of a carboxypeptidase and a zinc-dependent endoprotease inhibits calcium-induced myoblast fusion. Release or activation of these proteases is dependent on the presence of calcium. Rapid degradation of a low-molecular-weight, iodinatable cell-surface component also parallels myoblast fusion and only occurs in fusion-competent myoblasts (P.G. Pauw and J.D. David, pers. comm.). Thus, the calcium requirement for myoblast fusion may be related to conformational or degradative alterations of the myoblast membrane, which subsequently lead to fusion.

FIBRONECTIN

Studies using immunofluorescence have shown changes in the pattern and a decrease in fibrillar fibronectin upon myotube formation (Chen 1977; Frucht et al. 1978). In contrast, no significant changes in this glycoprotein were found when it was labeled by iodination and purified by electrophoresis (Hynes et al. 1976) or upon biosynthetic labeling (J. Bohn and S.J. Kaufman, unpubl.). J. Gardner and D. Fambrough (pers. comm.) have confirmed these findings using monoclonal antibodies to localize and immunoprecipitate chick myoblast fibronectin. In addition, they report the assembly of fibronectin into a fibrous matrix upon prolonged cultivation of myotubes. Specific, distinct functional domains have been identified on peptide fragments of fibronectin; one fragment binds to the myoblast, whereas another peptide binds to the collagen matrix (Wagner and Hynes 1980; Ehrisman et al. 1981). Myogenesis in culture often results in branched fibers; however, branching is not usually found in situ. This apparant ar-

[1] In keeping with policies regarding citations of abstracts and unpublished data presented at the meeting from which this volume is derived, this information is being cited as personal communications with the consent of the authors.

tifact of the cell-culture environment may be overcome by precoating dishes with fibronectin (Chiquet et al. 1981). Thus, the ordered deposition of fibronectin may dictate the orientation of cells and guide the formation of unbranched fibers. H. Vandenburgh (pers. comm.) reported that alignment and orientation of myoblasts may also be achieved in vitro via physical stretching of the substratum. The effect of stress-induced orientation may be analogous or directly due to fibronectin-induced orientation. The potential importance of fibronectin to myoblast interactions and development is illustrated in the work of H.F. Elson, who found that myoblasts from some patients with muscular dystrophy fail to form polymerized fibronectin as part of their cell-surface matrix. The lack of filamentous fibronectin may be causally related to the altered cell anchorage and aberrant development of these cells (H.F. Elson, pers. comm.). Since normal myoblasts both synthesize and polymerize fibronectin, in contrast to newly formed myotubes where it is synthesized but not readily polymerized, it is tempting to suggest that the defect in the human dystrophic cells may represent an early onset of this apparently normal switch. A decrease of cell-surface filamentous fibronectin also accompanies malignant transformation (Hynes 1976; Yamada and Olden 1978) and may be a consequence of increased phosphorylation (Ali and Hunter 1981). It is not presently known whether changes in fibronectin on myotubes and transformed cells are mediated by similar mechanisms.

LECTINS

The inhibitory activities of various lectins on myoblast fusion suggest that fusion may be mediated by cell-surface carbohydrates or glycoproteins (Den et al. 1975; Sandra et al. 1977). The finding that thiodigalactoside inhibits myoblast fusion supports the concept that a developmentally regulated lectin may be involved in this event (Gartner and Podleski 1975, 1976). These experiments, however, have been questioned by conflicting experimental data (Den et al. 1976; Kaufman and Lawless 1980), and it remains unclear whether the effects of carbohydrates and lectins with or on myoblast membranes preclude fusion via mechanisms directly associated with the normal processes of myoblast interaction and fusion. One approach to resolve this controversy is to purify cell-surface lectins and more directly assess their role in myogenesis (MacBride and Przybylski 1980; Przybylski and MacBride, unpubl.).

Another approach to examining the role of surface lectins is to isolate appropriate mutants. Gilfix and Sanwal (this volume) have obtained myoblasts defective in either the binding or internalization of several lectins. One of these mutants of L6 rat myoblasts lacks both cell-surface sialic acid and galactose residues, yet it does fuse. This raises the question of

whether cell-surface sialic acid and/or galactose moieties have any significant role in myoblast development.

MEMBRANE COMPOSITION, FLUIDITY, AND CELL CYTOSKELETON

Analyses of the membrane composition of developing myoblasts have not revealed major changes in lipid composition, in lipid-protein ratios, or in the fatty acid or phospholipid composition of myoblast membranes. Likewise, the activities of phospholipase A or phosphatidic acid phosphatase do not markedly change during myoblast development (Kent et al. 1974; Kent and Vagelos 1976). In contrast, the glycolipid content of myoblasts has been reported to increase when cells associate before fusion (Whatley et al. 1976), and an increase in membrane fluidity occurs at, or closely associated with, myotube formation (Prives and Shinitzky 1977; Elson and Yguerabide 1979). Using microscopic fluorescence relaxation, Herman and Fernandez (1978) confirmed this latter finding and further showed that the increased rotational freedom of the fluorescent probe is at the areas of contact between fusing myoblasts.

How do these transient changes in fluidity arise without apparent concomitant alterations in membrane composition? Two possibilities, which are not mutually exclusive, are suggested by several experimental observations. First, these regions of membrane fluidity could result from localized exchanges between internal and surface membranes. Kalderon and Gilula (1979) observed by freeze-fracture scanning electron microscopy and transmission electron microscopy that formation of particle-free domains in the plasma membrane may originate from the fusion of the plasma membrane with submembranous vesicles. Apposition of these particle-free regions on adjacent interacting cells could then result in cell fusion. Alternatively, particle-free domains might arise from localized clearing of the myoblast membrane.

An additional problem that myoblasts face is related to the close opposition of the plasma membrane and the cytoskeleton. What happens to the cytoskeleton when myoblasts fuse? What are the associations of the outer membrane and cytoskeleton and how are these disrupted or realigned during the interaction and fusion of myoblasts? Clearly, there are changes in cell-surface topography, such as the redistribution, loss, or turnover of specific membrane components, which may be associated with changes in membrane fluidity and, perhaps, mediated by changes in the association of the membrane with the cytoskeleton. Studies on the aggregation, mobility, and Triton-X-100-extractability of acetylcholine and concanavalin A (Con A) receptors indicate that localized changes in membrane fluidity may arise from transient alterations in the association of plasma membrane and cytoskeleton (Prives et al.; Fernandez and Herman; both this volume). In earlier studies, inhibition of myoblast fusion by Con A was dependent on

the extent of receptor aggregation and on prior alteration of the myoblast membrane either with trypsin or via calcium deprivation (Sandra et al. 1977). Lateral mobility of the lectin receptors was decreased by treatment with cytochalasin B, suggesting that the association of receptors with the microfilament substructure is involved in the translational mobility of myoblast-surface components (Schlessinger et al. 1976). Using monoclonal antibodies, we found that marked diminutions of several membrane antigens paralleled myoblast fusion (Lee and Kaufman 1981). Some of these cell-surface determinants are associated with a Triton-insoluble matrix, and the distribution of selected membrane components can be altered by disrupting microfilaments (Kaufman et al., this volume). The association of some cell-surface components with microfilaments can also be promoted or stabilized by the binding of antibodies (Flanagan and Koch 1978; Koch and Smith 1978; Akeson and Lessard 1980), and actin filaments have been reported in microprocesses that may be involved in myoblast fusion (Huang et al. 1978). These observations suggest that cell-cell interactions may be mediated by components on the cell surface, which are either directly linked to the cytoskeleton or become associated with it as a consequence of cell interactions. Such associations could alter the mobility, translocation, aggregation, and/or capping of surface components resulting in the formation of relatively particle-free domains of lipid bilayer that could serve as nucleation sites for membrane fusion.

IMMUNOLOGIC APPROACHES

Immunologic analyses of the myoblast and myotube cell surfaces are under way in several laboratories. Using heterologous rabbit antisera, Friedlander and Fischman (1979) showed that myoblasts have unique membrane antigenic determinants and that the immune reactivity of these determinants may vary with development. In analogous experiments, we found that myoblasts have both unique surface determinants and altered distributions of common determinants when compared with developmentally defective cells (Lee and Kaufman 1981). Monoclonal antibodies reactive with the myoblast surface have been prepared (Gottlieb and Greve 1978; Grove and Stockdale 1979; Lee and Kaufman 1979, 1981; Walsh and Ritter 1981) and can be employed to evaluate the role of specific membrane components. These antibodies have been used to discern antigenic alterations on the myoblast cell surface, and they are being applied to study the genetic regulation of specific myoblast antigens. Generation of these monoclonal antibodies has been relatively straightforward compared with the application of these probes to sort out the functions of the respective determinants. Immunization procedures that result in production of monoclonal antibodies reactive with muscle-specific antigens may simplify this approach (Horwitz et al., this volume).

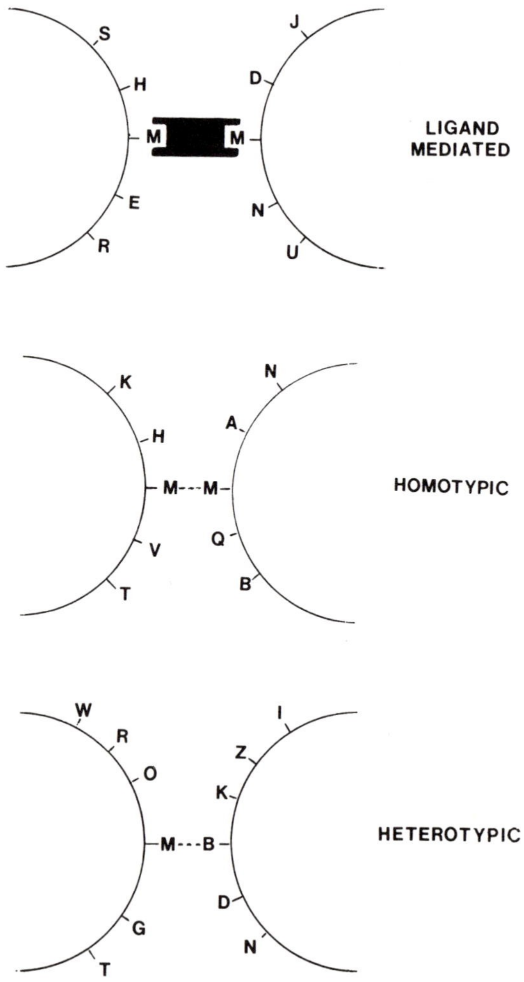

FIGURE 1
Muscle-specific components. Membrane components unique to skeletal myoblasts may mediate cell interactions via ligand bridges or, more directly, via homotypic or heterotypic associations.

SPECIFICITY

A brief consideration of the specificity of the interaction between myoblasts is in order, especially when considering the immunologic approaches currently being taken. Some antigenic determinants are highly representative of or specific for myoblasts (Grove and Stockdale 1979; Walsh and Ritter 1981; Horwitz et al., this volume). The interaction of myoblasts may be mediated by muscle-specific components via ligand, homotypic, or heterotypic interactions (Fig. 1). In the latter case, one might expect to find phenotypic complementation between mutants deficient in only one of the specific components. There are also membrane components that are not unique to myoblasts, e.g., fibronectin (Chen 1977; Frucht et al. 1978), Thy-1 antigen (Lesley and Lennon 1977), and other membrane antigens (Lee and Kaufman 1981), but that do undergo stage-specific changes that accom-

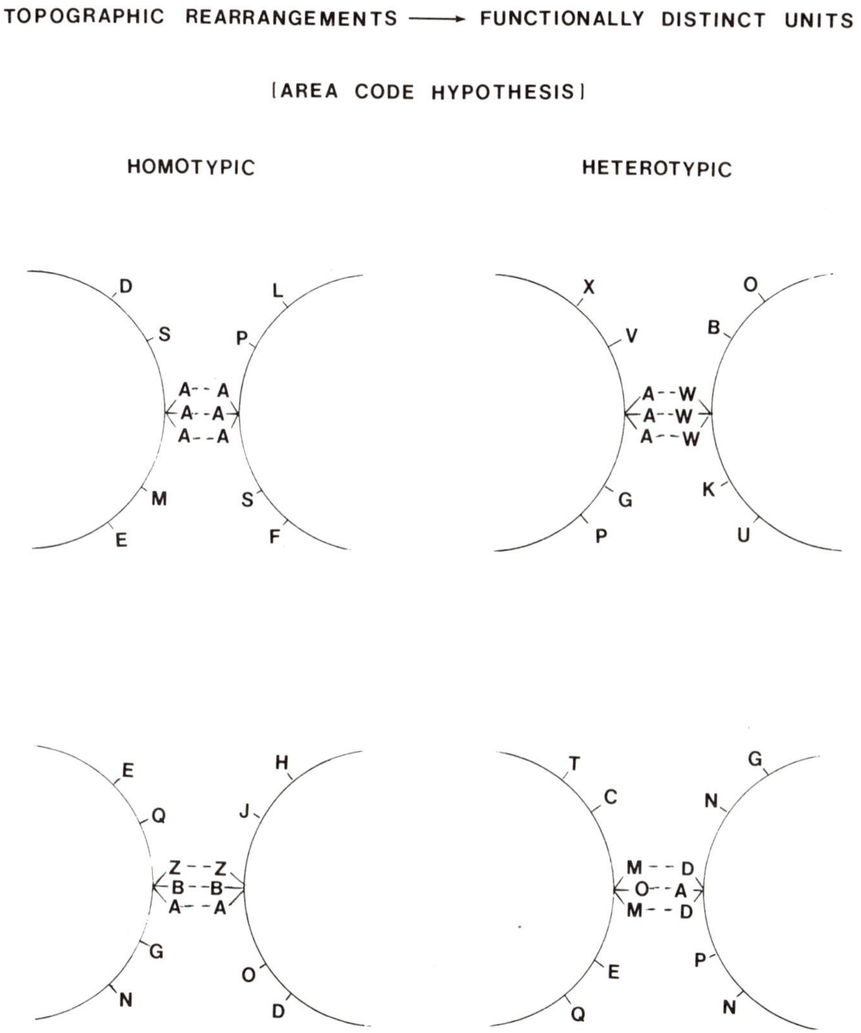

FIGURE 2
Associations of membrane species may result in functionally distinct units. Just as ten digits provide directive area codes, unique associations of cell-surface molecules may also result in functionally and morphologically distinct units for cell interactions (based on information from Hood et al. 1977). Specificity could thus arise not solely from molecular species unique to myoblasts but from molecules shared among several tissues (e.g., histocompatibility molecules, and so on).

pany myoblast development. Topographic rearrangements or associations of molecules common to many cell types may result in unique functional units and thereby provide specificity (Fig. 2). Generation of the molecular complementarity that underlies specific cell-cell associations by common molecules has been discussed previously (Hood et al. 1977; Frazier and Glaser 1979). According to the "area-code" hypothesis (Hood et al. 1977),

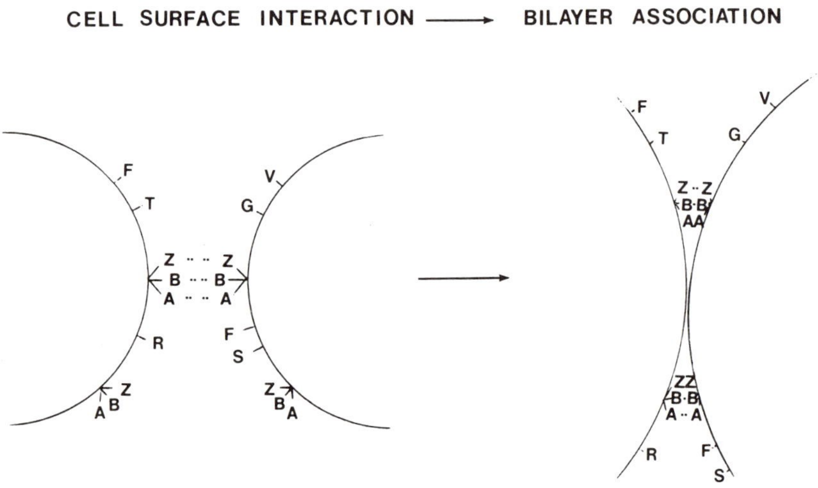

FIGURE 3
Cell-surface interactions may promote lateral movement of membrane components, resulting in relatively particle-free domains of closely apposed lipid bilayers that subsequently fuse. Just as antibodies and lectins may promote aggregation and capping of cell-surface molecules, association of cell membranes might likewise lead to localized clearing of the myoblast surface. Lateral movement of membrane components may result from reversible associations with the cell cytoskeleton.

the association of three membrane species into unique functional groups could provide the great diversity in specificity needed for different cell interactions. As depicted in Figure 3, such interactions might lead to the formation of relatively particle-free domains and myoblast fusion.

DIRECTIONS

Use of antibodies to define the myoblast membrane and to isolate mutants deficient in particular determinants will lead to an understanding of the components and events on the myoblast membrane germane to their interaction with the environment and substrate, to fusion, and to subsequent neuromuscular functioning. Recombinant DNA technology and monoclonal antibodies may also be used to clone, identify, and study the expression of those genes that mediate the membrane events unique to myogenesis. Likewise, these immunologic reagents are the best tools currently available to aid in defining and isolating the lineages of myoblasts that arise during normal and aberrant development.

ACKNOWLEDGMENTS

This work was supported by grants from the Muscular Dystrophy Association of America and U.S. Public Health Service grant GM-28842.

REFERENCES

Akeson, A. and J.L. Lessard. 1980. Actin associated proteins in P3X63 Ag8 myeloma cells during immunoglobulin patching. *J. Cell Biol.* **87:** 100a.

Ali, I.U. and T. Hunter. 1981. Structural comparison of fibronectins from normal and transformed cells. *J. Biol. Chem.* **256:** 7671.

Bischoff, R. 1978. Myoblast fusion. *Cell Surf. Rev.* **5:** 138.

Chen, L.B. 1977. Alteration in cell surface LETS protein during myogenesis. *Cell* **10:** 393.

Chiquet, M., H.M. Eppenberger, and D.C. Turner. 1981. Muscle morphogenesis: Evidence for an organizing function of exogenous fibronectin. *Dev. Biol.* **88:** 220.

Den, H., D. Malinzak, and A. Rosenberg. 1976. Lack of evidence for the involvement of a β-D-galactosyl-specific lectin in the fusion of chick myoblasts. *Biochem. Biophys. Res. Commun.* **69:** 621.

Den, H., D. Malinzak, H.J. Keating, and A. Rosenberg. 1975. Influence of concanavalin A, wheat germ agglutinin, and soybean agglutinin on the fusion of myoblasts in vitro. *J. Cell Biol.* **67:** 826.

Ehrismann, R., M. Chiquet, and D.C. Turner. 1981. Mode of action of fibronectin in promoting chicken myoblast attachment. *J. Biol. Chem.* **256:** 4056.

Elson, H.F. and J. Yguerabide. 1979. Membrane dynamics of differentiating cultured embryonic chick skeletal muscle cells by fluorescence microscopy techniques. *J. Supramol. Struct.* **12:** 47.

Flanagan, J. and G.L.E. Koch. 1978. Cross-linked surface Ig attaches to actin. *Nature* **273:** 278.

Frazier, W. and L. Glaser. 1979. Surface components and cell recognition. *Annu. Rev. Biochem.* **48:** 491.

Friedlander, M. and D.A. Fischman. 1979. Immunological studies of the embryonic muscle surface. *J. Cell. Biol.* **81:** 193.

Frucht, L.T., D.F. Mosher, and G. Wendelshafer-Crabb. 1978. Immunocytochemical localization of fibronectin (LETS protein) on the surface of L_6 myoblasts. *Cell* **13:** 263.

Gartner, T.K. and T.R. Podleski. 1975. Evidence that a membrane bound lectin mediates fusion of L_6 myoblasts. *Biochem. Biophys. Res. Commun.* **67:** 972.

———. 1976. Evidence that the types and specific activity of lectins control fusion of L_6 myoblasts. *Biochem. Biophys. Res. Commun.* **70:** 1142.

Gottlieb, D. and J. Greve. 1978. Monoclonal antibodies against developing chick brain and muscle. *Curr. Top. Microbiol. Immunol.* **81:** 40.

Grove, B.K. and F.E. Stockdale. 1979. Monoclonal antibodies to cell surfaces of differentiating skeletal muscle. *J. Cell Biol.* **83:** 28a.

Herman, B.A. and S.M. Fernandez. 1978. Changes in membrane dynamics associated with myogenic cell fusion. *J. Cell. Physiol.* **94:** 253.

Hood, L., H.V. Huang, and W.J. Dreyer. 1977. The area code hypothesis: The immune system provides clues to understanding the genetic and molecular basis of cell recognition during development. *J. Supramol. Struct.* **7:** 531.

Huang, H.L., R.H. Singer, and E. Lazarides. 1978. Actin-containing microprocesses in the fusion of cultured chick myoblasts. *Muscle and Nerve* **1:** 219.

Hynes, R.O. 1976. Cell surface proteins and malignant transformation. *Biochim. Biophys. Acta* **458:** 73.

Hynes, R.O., G.S. Martin, M. Shearer, D.R. Critchley, and C.J. Epstein. 1976. Viral

transformation of rat myoblasts: Effects on fusion and surface properties. *Dev. Biol.* **48:** 35.

Kalderon, N. and N.B. Gilula. 1979. Membrane events involved in myoblast fusion. *J. Cell Biol.* **81:** 411.

Kaufman, S.J. and M.L. Lawless. 1980. Thiodigalactoside binding lectin and skeletal myogenesis. *Differentiation* **16:** 41.

Kent, C. and P.R. Vagelos. 1976. Phosphatidic acid phosphatase and phospholipase A activities in plasma membranes from fusing muscle cells. *Biochim. Biophys. Acta* **436:** 377.

Kent, C., S.O. Schimmel, and P.R. Vagelos. 1974. Lipid composition of plasma membranes from developing chick muscle cells in culture. *Biochim. Biophys. Acta* **360:** 312.

Koch, G.L.E. and M.J. Smith. 1978. An association between actin and the major histocompatibility antigen H-2. *Nature* **273:** 274.

Lee, H.U. and S.J. Kaufman. 1979. Monoclonal antibodies against rat E63 myoblasts. *J. Cell Biol.* **83:** 386a.

Lee, H.U. and S.J. Kaufman. 1981. Use of monoclonal antibodies in the analysis of myoblast development. *Devel. Biol.* **81:** 81.

Lesley, J.F. and V.A. Lennon. 1977. Transitory expression of Thy-1 antigen in skeletal muscle development. *Nature* **268:** 163.

MacBride, R.G. and R.J. Przybylski. 1980. Purified lectin from skeletal muscle inhibits myotube formation in vitro. *J. Cell Biol.* **85:** 617.

Pearson, M.L. 1980. Muscle differentiation in cell culture: A problem in somatic cell genetics. In *The Molecular Genetics of Development* (ed. T. Leighton and W.F. Loomis), p.361. Academic Press, New York.

Prives, J. and M. Shinitzky. 1977. Increased membrane fluidity precedes fusion of muscle cells. *Nature* **268:** 761.

Sandra, A., M.A. Leon, and R.J. Przybylski. 1977. Suppression of myoblast fusion by concanavalin A: Possible involvement of membrane fluidity. *J. Cell Sci.* **28:** 251.

Schlessinger, J., D.E. Koppel, D. Axelrod, K. Jacobson, W.W. Webb, and E.L. Elson. 1976. Lateral transport on cell membranes: Mobility of concanavalin A receptors on myoblasts. *Proc. Natl. Acad. Sci.* **73:** 2409.

Wagner, D.D. and R.O. Hynes. 1980. Topological arrangement of the major structural features of fibronectin. *J. Biol. Chem.* **255:** 4304.

Walsh, F.S. and M.A. Ritter. 1981. Surface antigen differentiation during human myogenesis in culture. *Nature* **289:** 60.

Whatley, R., S.K. Ng, J. Rogers, W.C. Murray, and B.D. Sanwal. 1976. Developmental changes in gangliosides during myogenesis of a rat myoblast cell line and its drug resistant variants. *Biochem. Biophys. Res. Commun.* **70:** 180.

Yamada, K.M. and K. Olden. 1978. Fibronectin-adhesive glycoproteins of cell surface and blood. *Nature* **271:** 179.

Immunochemical Localization of Myoblast Cell-surface Antigens with Submembranous Components

STEPHEN J. KAUFMAN
DIANE M. EHRBAR
THOMAS I. DORAN
Department of Microbiology
and School of Basic Medical Sciences
University of Illinois
Urbana, Illinois 61801

Monoclonal antibodies have been used to identify cell-surface antigens on myogenic L8E63 rat cells. Several of these determinants on myoblasts are markedly reduced upon fusion into myotubes (Lee and Kaufman 1981). As our goal is to identify and characterize those components on the myoblast surface that are germane to their differentiation, we sought to use these antibodies to isolate some of these membrane components. Since these antibodies react with determinants on the cells in situ, rather than prepare cell or membrane extracts and subsequently attempt to isolate the determinants, the approach taken was to form the antigen-antibody complexes on the intact cell and then to extract the cells and isolate these complexes. However, treatment of myoblasts with 1% Nonidet P-40 or 1% Triton X-100 did not extract the antibody–cell-surface components that undergo stage-specific alterations, indicating their possible association with submembranous components. Disruption of microfilaments and microtubules with cytochalasin B and colchicine, respectively, led to pronounced redistributions and/or quantitative changes that were unique to each determinant.

TABLE 1
Developmental Specificity, Localization, and Triton Extraction of Myoblast Determinants Defined with Monoclonal Antibodies

Determinant	Developmental specificity: myoblast	Developmental specificity: myotube	Cell localization	Persistence after extraction with Triton X-100
A5	+	↓	surface	+
A27	+	↓	surface	+
A80	+	↓	surface	+
B88	+	+	surface and intracellular	↓
B58	+	–	intracellular	–

Developmental specificity and localization were determined using indirect immunofluorescence and radioimmunoassay (Lee and Kaufman 1981). Myoblasts were extracted with Triton X-100 subsequent to binding of the antibodies (see legend to Fig. 1). Determinants on the myoblast surface that undergo stage-specific quantitative changes remained in the Triton-insoluble fraction.

RESULTS AND DISCUSSION

Indirect immunofluorescence and radioimmunoassay were used to establish the developmental specificity and cell localization of the determinants reacting with several monoclonal antibodies (Table 1; Lee and Kaufman 1981). Extraction of cells with Triton X-100 (Fulton et al. 1980) did not release the majority of fluorescein-conjugated secondary antibody (rabbit anti-mouse immunoglobulin [RAMIg]) from cells reacted with A5, A27, and A80 monoclonal antibodies (Table 1; Figs. 1–4). Likewise, purified, iodinated A5 and A80 monoclonal antibodies remained bound after extraction of the cells with Triton X-100 (Table 2). Most bound ^{125}I-labeled A27 antibody was solubilized in Triton; however, upon incubation with RAMIg, the proportion of bound antibody remaining in the Triton residue increased. This shift in Triton-extractability was true to a lesser degree for A5 antibody, and it results from both selective extraction of Triton-soluble determinants and stabilization of the antigen-antibody complex (T.I. Doran and S.J. Kaufman, unpubl.). When cells were first extracted with Triton and then reacted with the ^{125}I-labeled A5 or A80 monoclonal antibodies, the ^{125}I-labeled antibody also bound, indicating that these determinants are in situ, part of a Triton-insoluble complex.

Changes in the localization of determinants resulting from treatment of myoblasts with colchicine or cytochalasin B, either before or after binding

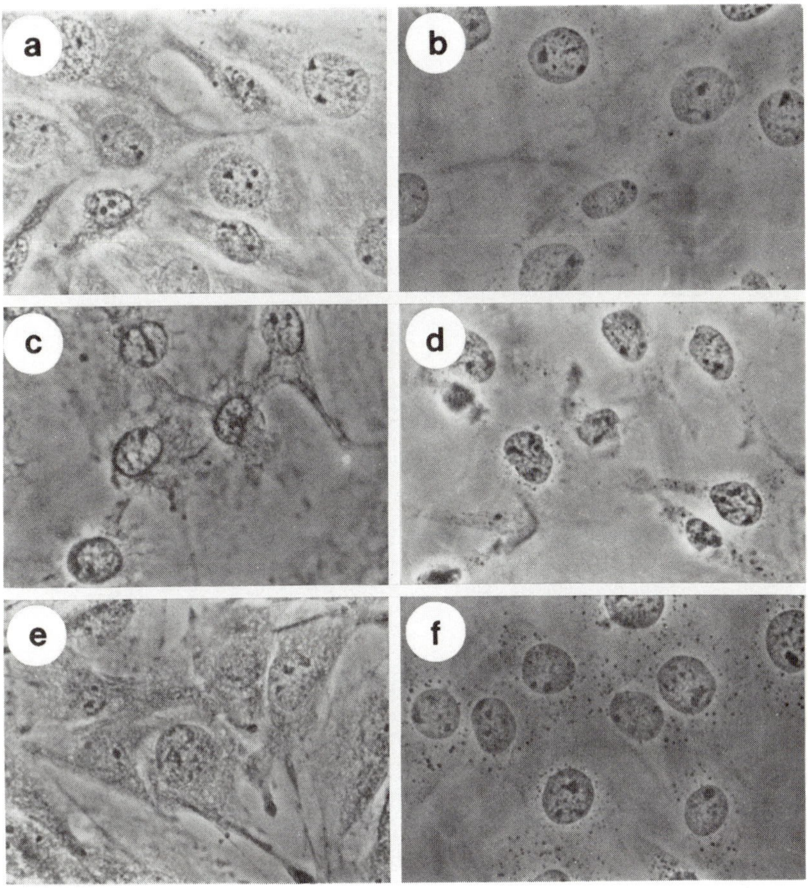

FIGURE 1
Phase-contrast microscopy of L8E63 rat myoblasts treated with cytochalasin B or colchicine and extracted with Triton X-100. Cells grown in Dulbecco's medium + 10% horse serum on glass coverslips (Lee and Kaufman 1981) were washed twice in Dulbecco's phosphate-buffered saline (PBS) (pH 7.4) and either fixed at 20°C for 10 min with 95% ethanol (*a,c,e*) or washed twice in Buffer A (100 mM KCl, 3 mM $MgCl_2$, 1 mM $CaCl_2$, 300 mM sucrose, 10 mM HEPES at pH 6.8) and then extracted with Buffer A + 1% Triton X-100 + 1 mM phenyl-methyl-sulfonyl fluoride for 4 min at 4°C, washed once with Buffer A (*b,d,f*) (Fulton et al. 1980), and mounted in glycerol-PBS and sealed with Flo-Texx. (*a,b*) Controls; (*c,d*) cells pretreated with 8 μg/ml of cytochalasin B in medium containing 0.1% dimethyl sulfoxide at 37°C for 15 min; (*e,f*) cells pretreated with 10 μg/ml of colchicine in medium at 37°C for 1 hr. 0.1% DMSO had no discernible effect on morphology or immunofluorescence.

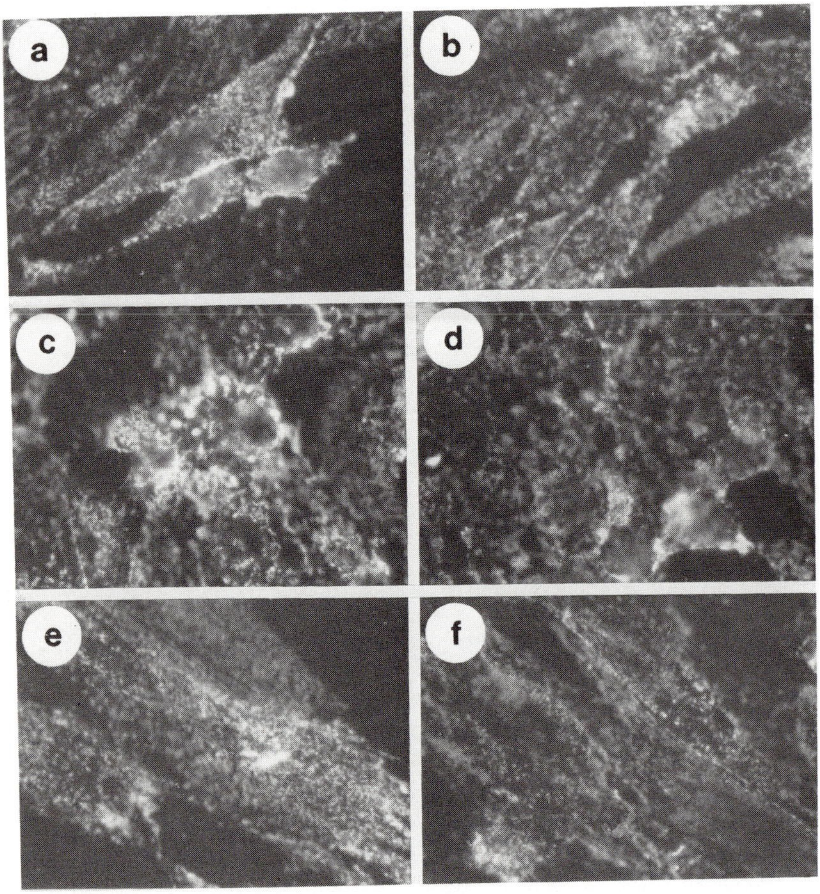

FIGURE 2
Localization of A5 determinants on myoblasts treated with cytochalasin or colchicine. Indirect immunofluoresence was slightly modified from Lee and Kaufman (1981). Cells were grown and processed as indicated in the legend to Fig. 1. Prior to treatment with cytochalasin, colchicine, ethanol, or Triton, cells were reacted at 37°C for 45 min with clone A5 antibody in PBS containing 0.02% sodium azide (PBS-A), washed in PBS-A, reacted with fluorescein-isothiocyanate–RAMIg, and washed in PBS-A. (*a*) Control; (*b*) Triton-extracted control; (*c*) cytochalasin-treated; (*d*) cytochalasin-treated, Triton-extracted; (*e*) colchicine-treated; (*f*) colchicine-treated, Triton-extracted.

of the antibodies, are summarized in Table 3. Disruption of microfilaments with cytochalasin led to aggregation of the A5 determinant and loss of its normal distribution in parallel arrays. A80 antibody reacts with fibronectin

TABLE 2
^{125}I-labeled Monoclonal Antibody in Triton-insoluble Fraction

Determinant	Protocol I[a]	Protocol II[a]	Determinant in Triton-insoluble fraction (%)[b]
A5	60.1	93.5	66.4
A27	16.0	78.1	n.d.
A80	78.7	81.5	111.0

Hybridoma antibodies were purified from ascitic fluid by $(NH_4)_2SO_4$ precipitation and DEAE-Sephacel chromatography, and iodinated (Greenwood and Hunter 1963). Myoblasts were extracted with Triton X-100 before or after reaction with primary antibodies or primary and secondary antibodies (see legend to Fig. 1 for details).

[a]Protocol I: Primary ^{125}I-labeled antibody followed by Triton extraction. Protocol II: Primary ^{125}I-labeled antibody, then secondary antibody, followed by Triton extraction. The percentages of cpm in triplicates of the Triton-insoluble extracts compared to the total cpm bound are given. Total cpm bound/10^4 cells was 8511 (A5), 1842 (A27), and 903 (A80). n.d., not determined.

[b]cpm in the Triton-insoluble residues obtained when cells were first extracted with Triton and then reacted with antibody divided by cpm obtained when cells were initially reacted with antibody and subsequently extracted with Triton X-100.

(I. Sagiv et al., unpubl.). Treatment with cytochalasin led to the loss of the A80 determinant. The effects of cytochalasin on the A5 and A80 determinants were the same regardless of when the cells were reacted with the respective antibodies. In contrast, the distribution of the A27 determinant was only altered when the antibodies were first bound to the cells. As found in the radioimmunoassay, reaction with the antibodies stabilizes the A27 antigen-antibody complexes and promotes their association with the Triton-insoluble material. Treatment with colchicine had no effect on the localization of the A5 and A80 antibodies; however, the distribution of A27 was modified by colchicine. Immunofluorescence analyses showed that the Triton-insoluble status of these antigen-antibody complexes was not altered by disruption of the microfilaments or microtubules after formation of the complexes (Figs. 2–4).

SUMMARY

Several molecular species on the myoblast cell surface, defined by their interaction with monoclonal antibodies, are not extracted by 1% Triton

TABLE 3
Changes in Antigen Distribution upon Treatment of Cells with Cytochalasin or Colchicine

	Prior to reaction with ^{125}I-labeled antibody		After reaction with ^{125}I-labeled antibody	
Determinant	cytochalasin B	colchicine	cytochalsin B	colchicine
A5	aggregation of determinant; loss of parallel arrays	no change	aggregation of determinant; loss of parallel arrays	no change
A27	no change	some areas void of determinant	loss of parallel arrays	continuous distribution over cytoplasm and nucleus
A80	loss of determinant; less fibrous distribution	no change	loss of determinant; less fibrous distribution	no change

X-100 and thus may be associated with internal cytoskeletal components. The association of selected cell-surface molecules with the cell cytoskeleton has been reported in other systems (Ash and Singer 1976; Condeelis 1979; Lehto et al. 1980). The Triton-insolubility of the A80 determinant on fibronectin and the sensitivity of this determinant to cytochalasin are consistent with previous reports on the association of fibronectin with actin filaments (Ali and Hynes 1977; Ali et al. 1977; Willingham et al. 1977; Hynes and Destree 1978; Singer 1979; Burridge and Feramisco 1980). Formation or stabilization of antigen-antibody complexes on cell surfaces can promote the association of cell-surface determinants with actin microfilaments (Flanagan and Koch 1978; Koch and Smith 1978); this also appears to be true for the A27 determinant. We suggest that the interaction of determinants on adjacent myoblasts may also promote analogous associations with internal filaments. The myoblast membrane is in close apposition with the submembranous cytoskeleton (e.g., see Prives et al., this volume), and actin filaments have been found in microprocesses that were proposed

to be involved in myoblast interaction and fusion (Huang et al. 1978). The association of membrane components with the cytoskeleton might regulate the turnover, aggregation, and mobility of cell-surface components (see Prives et al.; Fernandez and Herman; both this volume); this may be true for the A5, A27, and A80 determinants, which are lost from the myoblast surface upon myotube formation (Lee and Kaufman 1981). Membrane-cytoskeletal interactions, modulated by contact between neighboring myoblasts, could alter the topography of the myoblast surface. The altered mobility, aggregation, or loss of membrane components might result in

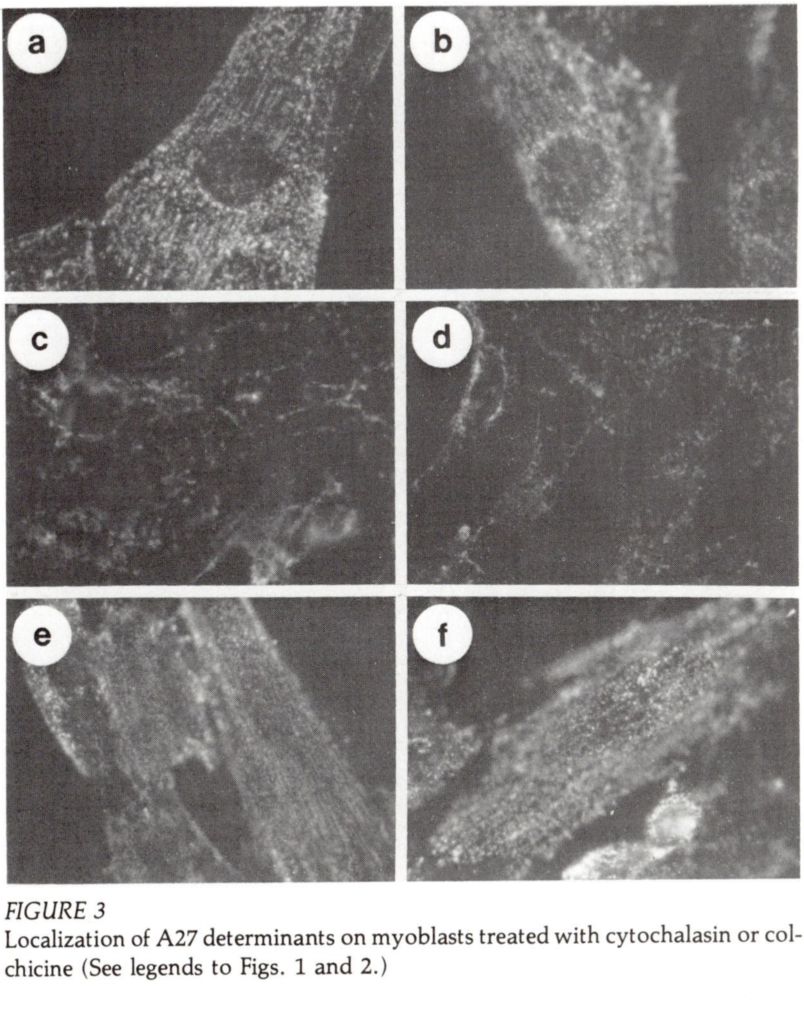

FIGURE 3
Localization of A27 determinants on myoblasts treated with cytochalasin or colchicine (See legends to Figs. 1 and 2.)

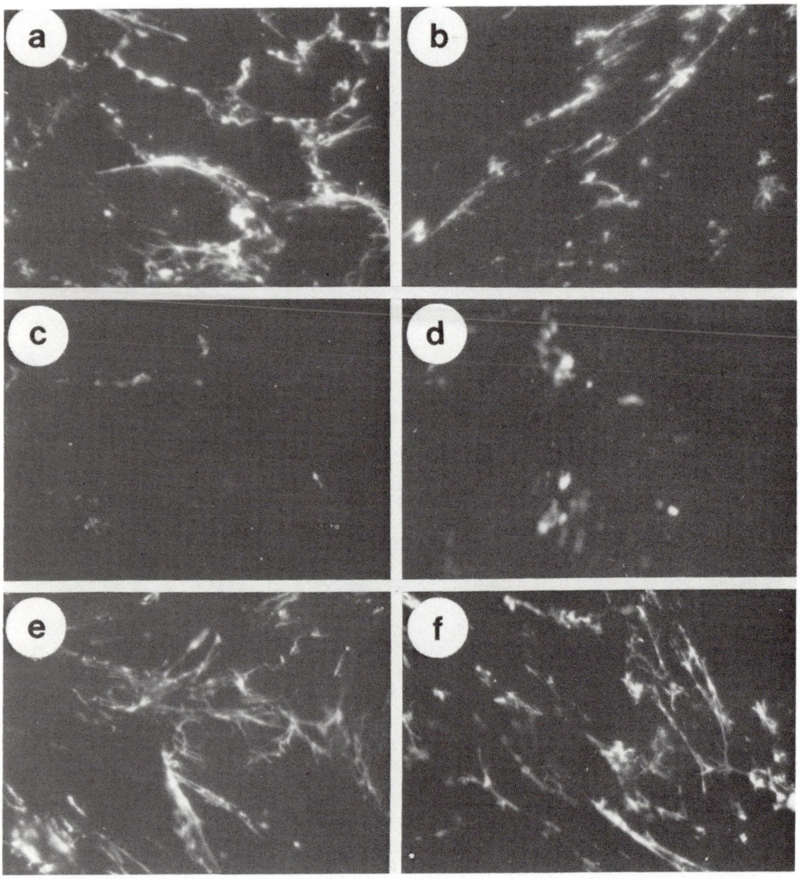

FIGURE 4
Localization of A80 determinants on myoblasts treated with cytochalasin or colchicine (see legends to Figs. 1 and 2). A80 antibody reacts with fibronectin and can be adsorbed by purified human cold insoluble globulin and cell-surface protein (I. Sagiv and S.J. Kaufman, unpubl.).

relatively particle-free, highly fluid domains of lipid bilayer (Herman and Fernandez 1978; Kalderon and Gilula 1979) and lead to myoblast fusion (Kaufman, this volume).

ACKNOWLEDGMENTS

This work was supported by grants from the Muscular Dystrophy Association of America and a grant from the U.S. Public Health Service (GM-28842).

REFERENCES

Ali, I.U. and R.O. Hynes. 1977. Effects of cytochalasin B and colchicine on attachment of a major surface protein of fibroblasts. *Biochim. Biophys. Acta* **471:** 16.

Ali, I.U., V. Mautner, R. Lanza, and R.O. Hynes. 1977. Restoration of normal morphology: Adhesion and cytoskeleton in transformed cells by addition of a transformation-sensitive surface protein. *Cell* **11:** 115.

Ash, J.F. and S.J. Singer. 1976. Concanavalin A-induced transmembrane linkage of concanavalin A surface receptors to intracellular myosin-containing filaments. *Proc. Natl. Acad. Sci.* **73:** 4575.

Burridge, K. and J.R. Feramisco. 1980. Microinjection and localization of 130K protein in living fibroblasts: A relationship to actin and fibronectin. *Cell* **19:** 587.

Condeelis, J. 1979. Isolation of concanavalin A caps during various stages of formation and their association with actin and myosin. *J. Cell Biol.* **80:** 751.

Flanagan, J. and G.L.E. Koch. 1978. Cross-linked surface Ig attaches to actin. *Nature* **273:** 278.

Fulton, A.B., K.M. Wan, and S. Penman. 1980. The spatial distribution of polyribosomes in 3T3 cells and the associated assembly of proteins into the skeletal framework. *Cell* **20:** 849.

Greenwood, F.C. and W.M. Hunter. 1963. The preparation of ^{131}I-labelled human growth hormone of high specific radioactivity. *Biochem. J.* **89:** 114.

Herman, B.A. and S.M. Fernandez. 1978. Changes in membrane dynamics associated with myogenic cell fusion. *J. Cell. Physiol.* **94:** 253.

Huang, H.L., R.H. Singer, and E. Lazarides. 1978. Actin-containing microprocesses in the fusion of cultured chick myoblasts. *Muscle and Nerve* **1:** 219.

Hynes, R.O. and A.T. Destree. 1978. Relationship between fibronectin (LETS protein) and actin. *Cell* **15:** 875.

Kalderon, N. and N.B. Gilula. 1979. Membrane events involved in myoblast fusion. *J. Cell. Biol.* **81:** 411.

Koch, G.L.E. and M.J. Smith. 1978. An association between actin and the major histocompatibility antigen H-2. *Nature* **273:** 274.

Lee, H.U. and S.J. Kaufman. 1981. Use of monoclonal antibodies in the analysis of myoblast development. *Dev. Biol.* **81:** 81.

Lehto, V.-P., T. Vartio, and I. Virtanen. 1980. Enrichment of a 140 KD surface glycoprotein in adherent, detergent-resistant cytoskeletons of cultured human fibroblasts. *Biochem. Biophys. Res. Comm.* **95:** 909.

Singer, I.I. 1979. The fibronexus: A transmembrane association of fibronectin-containing fibers and bundles of 5 nm microfilaments in hamster and human fibroblasts. *Cell* **16:** 675.

Willingham, M.C., K.M. Yamada, S.S. Yamada, J. Pouyssegur, and I. Pastan. 1977. Microfilament bundles and cell shape are related to adhesiveness to substratum and are dissociable from growth control in cultured fibroblasts. *Cell* **10:** 375.

Cellular Interactions in Myogenesis

ALAN HORWITZ
NICOLA NEFF
ALICE SESSIONS
CINDI DECKER
Department of Biochemistry and Biophysics
University of Pennsylvania Medical School
Philadelphia, Pennsylvania 19104

Embryonic myogenic cells derived from avian pectoral explants display several interesting social phenomena in vitro. They adhere to a collagen-coated substrate, a proxy for cell-matrix interactions; they fuse with one another to form multinucleate myotubes; and they form synapses when cocultured with neuronal explants (Bischoff 1978; Patrick et al. 1978). These phenomena are readily studied in tissue culture and have been at the focus of our interest for the past few years. In this paper we outline some of our observations on fusion and myoblast-substrate interactions.

RESULTS AND DISCUSSION

Fusion as a Sequence of Events

Until recently, fusion was scored simply by estimating the fraction of nuclei in multinucleate cells growing on tissue-culture dishes. This "dish assay" is ambiguous for several reasons (Bischoff 1978). First, contributions due to proliferation, differentiation, and migration, all of which are prerequisite to fusion, cannot be distinguished. Second, the extent of fusion is difficult to score reproducibly, especially during the early stages of fusion. Third, the time at which membrane continuity is established cannot be determined with certainty using morphological criteria.

To minimize some of the objections to the traditional methods of measuring fusion, we have studied the aggregation and fusion of chick embryo myoblast suspensions (Knudsen and Horwitz 1977). Myogenic cells are

grown in a low calcium medium for 52 hours to produce fusion-competent myoblasts. The cells are then harvested with EDTA. Upon addition of calcium, the suspended cells aggregate within 10–30 minutes; whereas, in the absence of calcium, there is little aggregation. A number of experiments indicate that this aggregation is not artifactual but, rather, a reflection of the normal fusion process itself. Cells that do not normally fuse, i.e., young myoblasts, old myoblasts, and fibroblasts, also show no calcium-mediated aggregation and are excluded from the myoblast aggregates. Futhermore, all metabolic inhibitors and structural perturbing agents that inhibit this rapid calcium-mediated aggregation in suspension also inhibit fusion on tissue-culture dishes and at similar concentration optima. However, at early times (15–30 min), the aggregated cells do not appear fused; they are readily dispersed into single cells by pipetting, treatment with EDTA, or treatment with trypsin. When the aggregates are allowed to incubate for longer periods of time ($\sim$30 min; the precise time depends on the composition of the incubation medium), they become resistant to dispersion by EDTA, and increasingly harsh treatments are required to disperse them. Electron micrographs of sections from EDTA-resistant aggregates do not reveal significant fusion in these aggregates (H. Barsky and A. Horwitz, unpubl.). Finally, after 1–2 hours, the aggregates become resistant to dispersion with trypsin, and one observes multinucleate cells in suspension. We interpret these observations as supporting the concept of a complex, sequential fusion mechanism. In this view, the different treatments required to disperse the aggregates afford an operational dissection of the fusion process into sequential events. Inhibitions of fusion by treatments that perturb cellular metabolism and structure support this notion of a sequential mechanism (Knudsen and Horwitz 1978). The essence of this sequential mechanism is that the nature of the cellular components and their interactions change as fusion proceeds (Knudsen and Horwitz 1977).

Suspension Assays of Membrane Union

The aggregation assay is particularly useful for studying the initial events in fusion, i.e., the calcium-mediated aggregation (recognition). However, the later events, like membrane union (the establishment of membrane continuity), are not easily observed or quantified. We have recently developed three further assays of fusion in suspension, i.e., myoball formation (Fig. 1). One, a dye-transfer assay, measures the establishment of membrane continuity. The other two, a nuclear-staining and a Coulter-counter-based assay, measure the morphologic endpoint of fusion in suspension, the myoball. These latter two assays are analogous to the traditional assay of fusion scored on tissue-culture dishes.

In the dye-transfer assay, cells are grown and harvested after 50 hours as described above for the aggregation assay. After harvesting, the myoblasts

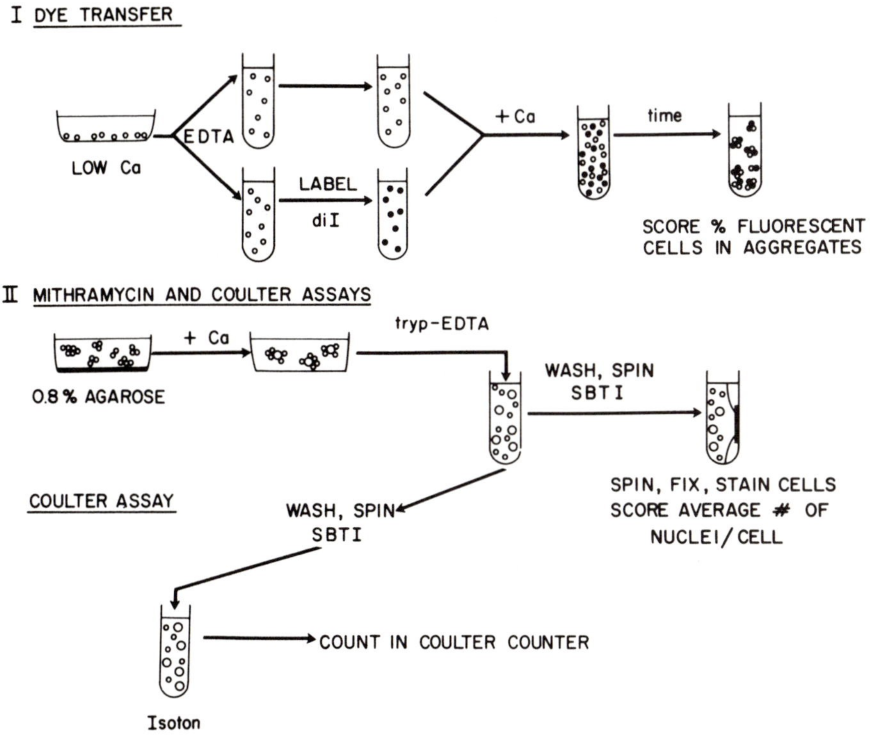

FIGURE 1
Suspension assays of fusion.

are divided into two populations; one is labeled with [diI-C_{18}(3)], a fluorescent lipid. The two populations are then mixed and calcium is added. The fraction of fluorescent cells in aggregates is then scored by microscopy observation. In the presence of calcium, this fraction increases continuously from an initial value of 50% to a plateau value of 70% after 60–90 minutes. Aggregation under conditions that do not promote fusion, such as low calcium concentrations or the addition of concanavalin A (Con A) (either in the presence or absence of calcium), do not produce a significant increase in the fraction of fluorescent cells in aggregates. A simple combinatoric analysis yields a rough estimate of the fractional fusion corresponding to the plateau value. After 60–90 minutes, roughly two cells of every five initially present have fused.

We have also compared the kinetics of dye transfer with that of EDTA and trypsin-resistant aggregate formation. There is no detectable dye transfer under conditions where 30–35% of the cells reside in EDTA-resistant aggregates. In contrast, the onset of trypsin-resistant aggregation is

not apparent until well after the onset of dye transfer. Thus, fusion occurs after the aggregates are resistant to dispersal with EDTA.

The nuclear-staining and Coulter-counter-based assays rely on morphologic criteria of fusion. For these assays we grow cells, using a low calcium medium, in agarose-containing dishes. The agarose prevents cellular attachment to the substrate, and the suspended cells adhere to one another in clumps. After 50 hours in culture, the cells are allowed to incubate in fusogenic calcium levels for the desired length of time followed by trypsinization to disperse the aggregates. They are then either sized using a Coulter counter or spun in a cytocentrifuge, fixed, and their nuclei stained with mithramycin for observation in the fluorescence microscope.

The mithramycin assay provides a quantitative, reproducible estimate of fusion. It is scored as the average number of nuclei per cell. Control cultures of low calcium or BrdU-treated myogenic cells or fibroblasts give values between 1.0 nuclei/cell and 1.2 nuclei/cell, whereas fusing cultures show an increase in the number of nuclei per cell with increased incubation time. After 4 hours, e.g., there are 1.5 nuclei/cell. This corresponds to one fusion for each three cells initially present. Although this assay is performed under conditions somewhat different from those of the dye-transfer assay, it is interesting that fusion measured by dye transfer precedes the morphologic estimate of fusion by at least 1–2 hours.

Phospholipids in Membrane Union

In this section some of the molecular aspects of fusion are discussed: (1) the role of phospholipids in membrane union and (2) an approach to the identification of other fusion-related molecules.

Membrane union is particularly hard to study in fusing myoblasts since it is embedded between the early adhesive and later morphologic phenomena. It has been hypothesized, however, that membrane union is mediated by the fusion of lipid bilayers. The evidence supporting a lipid-mediated fusion for myoblasts derives primarily from the observations of freeze-fracture electron micrographs by Kalderon and Gilula (1979). They report intramembrane-particle-free regions at putative fusion sites. This observation appears common to membrane fusion in a wide variety of biological systems studied (Ahkong et al. 1975). Other observations also support this view. Manipulation of the membrane lipids modulates fusion, and treatment of myoblasts with phospholipase C inhibits fusion (Nameroff et al. 1973; Horwitz et al. 1978, 1979).

Assuming that membrane union results from the fusion of apposed regions of membrane lipid and that this mechanism is analogous to that of lipid vesicles, two particular phospholipid classes, phosphatidylserine (PS) and phosphatidylethanolamine (PE) are likely to play a major role in this process (Papahadjopoulos et al. 1978; Cullis and de Kruijff 1979). Lipid

vesicles comprised of a binary mixture of phosphatidylcholine and high relative concentrations of PS fuse spontaneously in the presence of calcium (Papahadjopoulos et al. 1978). PE, presumably due to its nonlamellar polymorphism, reduces the PS requirement to a more physiologic concentration (Duzgunes et al. 1981; Uster and Deamer 1981). However, the role of the aminophospholipids in fusion raises an interesting problem since, in the animal cells studied thus far, they are believed to reside predominantly on the cytoplasmic side of the membrane (Sandra and Pagano 1978; Op den Kamp 1979).

We have addressed this issue by measuring the asymmetry of the aminophospholipids in myoblasts by using isethionylacetimidate and trinitrobenzenesulfonate, two amidating reagents (Op den Kamp 1979; Sessions and Horwitz 1981). Two plasma-membrane preparations were used. An inside-out preparation was derived from phagocytic vesicles (Sandra and Pagano 1978), and a right-side-out preparation of plasma-membrane blebs was prepared according to the procedure described by Scott (1976). The gross lipid compositions of these two preparations agree well with each other, as expected. The sideness predicted was confirmed by Con-A-mediated agglutination studies.

As a control, we first measured the membrane phospholipid asymmetry in fibroblasts. Membrane labeling under impermeable conditions (4°C) plateaued after 2 hours. Using right-side-out vesicles, the plateau occurred after 35% of the PE and 20% of the PE and PS, respectively, had reacted. Amidation of inside-out membranes displayed reciprocal plateau values, i.e., after 65% and 80% of the PE and PS, respectively, had reacted. Thus, in fibroblasts, 65–70% of the PE and 80% of the PS are internally disposed. These values are in general accord with other data published for fibroblasts (Sandra and Pagano 1978; Op den Kamp 1979).

The labeling of myogenic cells cultured for 52 hours paralled that of the fibroblasts, with one exception. The plateau occurred after 60–65% of the PE and 40–45% of the PS of the right-side-out blebs had reacted. Again, the labeling of the inside-out phagosomes yielded complementary values. Thus, the myoblasts possess a distribution of aminophospholipids between cytoplasmic and outer-membrane faces that differs from that described for other animal cells measured thus far. This eccentric distribution of aminophospholipids is consistent with their participation in fusion. Recently, Deutsch and Kelly (1981) published values for the aminophospholipid asymmetry of another membrane that participates in fusion: the synaptic vesicles isolated from a marine ray. Their values are nearly identical to ours for myoblasts (Table 1).

More recently, we have studied the asymmetry during a developmental window of 24–96 hours in tissue culture. Provided the cultures are free of contaminating fibroblasts, there is no significant change in myoblast asymmetry over this time period. The asymmetry is also present in BrdU-treated

TABLE 1
Aminophospholipid Distribution in Nonfusogenic and Fusogenic Membranes

	PE (%)		PS (%)	
Membrane type	inner monolayer	outer monolayer	inner monolayer	outer monolayer
Chick fibroblast plasma membrane[a]	66	34	83	17
Chick myoblast plasma membrane[a]	37	63	54	46
Marine ray synaptic vesicles[b]	40	60	55–60	40–45

[a]Data from Sessions and Horwitz (1981).
[b]Data from Deutsch and Kelly (1981).

cells. This implies that, whereas the eccentric distribution of aminophospholipids may contribute to fusion, it does not determine fusion competence. (Recall that under our culture conditions, fusion-competent cells do not appear until after ~35 hr in culture, and fusion competency wanes substantially after ~76 hr in culture.) Finally, the asymmetry is unaffected by growth in either a low or high calcium medium.

Membrane Antigens in Fusion

Whereas it seems likely that the membrane lipids participate directly in membrane union, it also seems likely that other molecular constituents participate at other stages.

We have taken an immunologic approach to identifying fusion-related molecules. Our paradigm is to screen for hybridomas secreting monoclonal antibodies (McAbs) that inhibit fusion. The antigens to which these antibodies are directed should be selective for myoblasts and vary with culture age in parallel with the fusion activity. We do, however, recognize that some fusion-related antigenic determinants may not show myogenic specificity, and their membrane concentration may not vary with culture age. These molecules should, however, at least be transiently immobilized and possibly accumulate at fusion sites.

Two immunogens have been used for these studies. One is the muscle plasma-membrane bleb preparation described by Scott (1976) and Sessions and Horwitz (1981). This induces an exceptionally large and diverse repertoire of hybridomas secreting cell-surface-directed antibodies. The other immunogen is the same muscle plasma-membrane bleb preparation precoated with a mouse antifibroblast antisera. In one experiment this induced a hybridoma population secreting antibodies selective for cell surface of mus-

cle; however, this observation has not been reproduced and the appropriate conditions have not been characterized.

The Coulter-counter-based assay of fusion provides a reproducible and sensitive screen for fusion-blocking McAbs (Neff and Horwitz 1982). In this assay, myogenic cells are grown in suspension for 52 hours, McAbs and fusion-permissive levels of calcium are added for 6–16 hours, and the resulting cells are treated with trypsin to disperse any aggregated but unfused cells. The cells are then counted in the Coulter counter at two prechosen threshold values: a low threshold, which yields an accurate estimate of the total number of cells present, and a higher threshold, which distinguishes mononucleate cells (10 μm dia.) and binucleate cells (14 μm dia.). Although this method of scoring does not provide a quantitative estimate of fusion, it does provide a simple and rapid screen. By using these techniques, we are optimistic that fusion-blocking antibodies will be identified and will prove useful in identifying and characterizing fusion-related antigens.

Membrane Antigens in Substrate Adhesion

The immunologic approach is also useful for identifying molecular participants in activities other than fusion. For example, McAbs directed against antigens participating in the interaction of myogenic cells with an extracellular matrix can be identified by their ability to cause cells to round up and detach from a gelatin substrate (Wylie et al. 1979). Using this screen, we have identified a hybridoma secreting an immunglobulin $(IgG)_{2b}(k)$ antibody that selectively rounds and detaches embryonic skeletal muscle in a variety of different states, i.e., myoblasts, myotubes, and myoblasts treated with BrdU and phorbol-12-myristate-13-acetate, but not fibroblasts (Fig. 2). The specificity of the antibody for myoblasts was established morphologically and by measurements of creatine phosphokinase activity. Although this antibody selectively detaches myogenic cells (and also selectively inhibits their attachment), it binds with high affinity to both myoblasts and fibroblasts. Finally, the antibody has no measurable effect on fusion itself, as assayed in suspension cultures.

Although a convincing demonstration that this antibody interferes directly with morphology and adhesion remains to be made, it does seem likely that this is its mode of action. First, the antibody is not cytotoxic, and its effect is reversible. Second, the antibody has no detectable effect on fusion. Third, the antigen, as detected by fluorescent antibody, shows a punctate distribution at the cell periphery and along stress fibers in fibroblasts (C. Damsky, pers. comm.). Fourth, using saturating levels of antibody, the kinetics of binding and cell detachment are rapid and almost superimposable.

Even at this early stage in our studies, the antibody has provided the interesting observation that it distinguishes the substrate adhesion of myoblasts from that of fibroblasts and that involved in myoblast fusion.

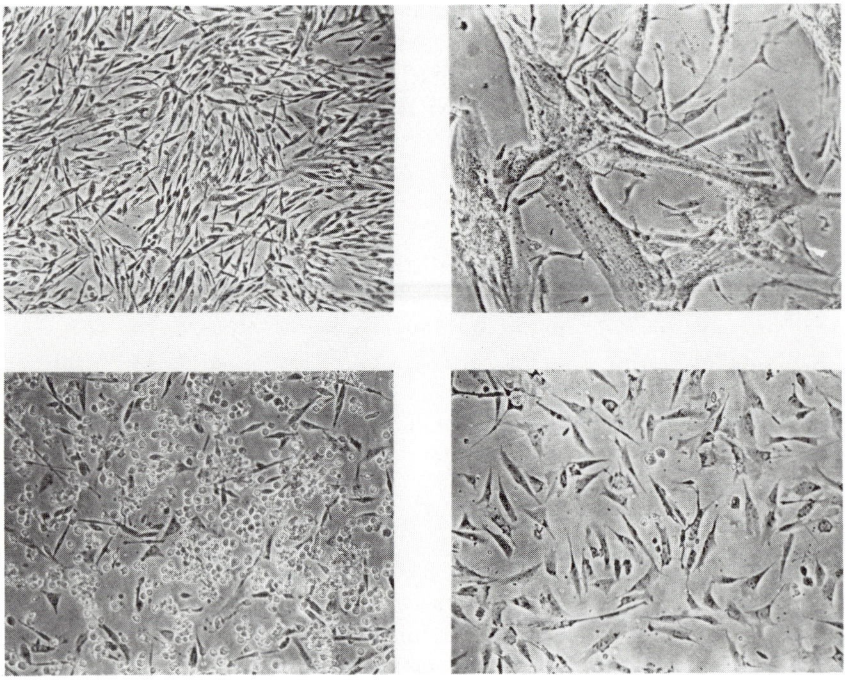

FIGURE 2
Effect of rounding antibody on myogenic cells. *(Top left)* 50-hr culture of myogenic cells grown in a low-calcium medium to produce fusion-competent myoblasts. *(Bottom left)* Same as above, except incubated with antibody. *(Top right)* The cells detached by the antibody were removed, replated, and grown for several days. *(Bottom right)* The cells (grown for several days) remaining after removal of those detached by the antibody.

Membrane Antigens in Myoblast Differentation

To complement the activity assays of the type just described, we have also estimated the concentration of antibody-binding sites on chick myoblasts and fibroblasts for a variety of hybridoma clones. Using an enzyme-linked immunoassay (ELISA), the antibody binding to both myoblasts and fibroblasts is measured, the values are normalized to those obtained using positive- and negative-control supernatants, and the ratio of myoblast to fibroblast binding is then computed.

We have used this screen on over 2000 hybridoma supernatants. Although relative antigenicity and the limited sensitivity of our assay determines the class of useful antibodies, we do find a broad spectrum of ratios of myoblast binding to fibroblast binding. The most frequent ratios observed correspond to those antigens with similar concentrations on the two cell types. Although a minor fraction of the McAbs indicate a signifi-

cant degree of antigen enrichment on myogenic cells, very few appear to be directed toward antigens specific for myogenic cells. A sample of about 30 clones have been purified and their antibodies more fully characterized. Those antibodies with a myoblast-to-fibroblast binding ratio near unity show very little or no change during development in culture. In contrast to these putative housekeeping antigens, those with ratios indicating either myoblast or fibroblast enrichment do not show a general pattern of change during development in culture. Some change substantially (i.e., increase or decrease), whereas others do not change at all.

Assuming that our observations are representative, it thus appears that there are substantially more quantitative than qualitative differences between the surfaces of myogenic and fibroblastic cells. Clearly, much remains to be done to establish the nature of such determinants and their functional role in cell physiology.

ACKNOWLEDGMENTS

This work was supported by the National Institutes of Health (grant GM-23244), the American Heart Association (grant 78-1056), and the H.M. Watts, Jr., Neuromuscular Disease Research Center. N.N. and A.S. are postdoctoral trainees of the National Institutes of Health. We thank H. Bashey and A. Tovar for their assistance.

REFERENCES

Ahkong, Q., D. Fischer, W. Tampion, and P. Lucy. 1975. Mechanisms of cell fusion. *Nature* **253:** 194.

Bischoff, R. 1978. Myoblast fusion. In *Membrane fusion* (ed. G. Post and G. Nicolson), vol. 5, p. 128. Elsevier/North-Holland, Amsterdam.

Cullis, P.R. and B. de Kruijff. 1979. Lipid polymorphism and the functional roles of lipids in biological membranes. *Biochim. Biophys. Acta* **559:** 399.

Deutsch, J.W. and R.B. Kelly. 1981. Lipids of synaptic membranes: Relevance to the mechanism of membrane fusion. *Biochemistry* **20:** 378.

Duzgunes, N., J. Wilscheet, R. Fraley, and D. Papahadjopoulos. 1981. Studies on the mechanism of membrane fusion: Role of head group in calcium- and magnesium-induced fusion of mixed phospholipid vesicles. *Biochim. Biophys. Acta* **642:** 182.

Horwitz, A., A. Wight, and K. Knudsen. 1979. A role for lipid in myoblast fusion. *Biochem. Biophys. Res. Commun.* **86:** 514.

Horwitz, A., A. Wight, R. Cornell, and P. Ludwig. 1978. Interrelated lipid alterations and their influence on the proliferation and fusion of cultured myogenic cells. *J. Cell Biol.* **77:** 334.

Kalderon, N. and N.B. Gilula. 1979. Membrane events involved in myoblast fusion. *J. Cell Biol.* **81:** 411.

Knudsen, K. and A. Horwitz. 1977. Tandem events in myoblast fusion. *Dev. Biol.* **58:** 328.

———. 1978. Differential inhibition of myoblast fusion. *Dev. Biol.* **66:** 294.

Nameroff, M., J.A. Trotter, J.M. Keller, and E. Munar. 1973. Inhibition of cellular differentiation by phospholipase C. I. Effects of the enzyme on myogenesis and chondrogenesis *in vivo*. *J. Cell Biol.* **58:** 107.

Neff, N.J. and A.F. Horwitz. 1982. A rapid assay for fusion of embryonic chick myoblasts. *Exp. Cell Res.* (in press).

Op den Kamp, J.A.F. 1979. Lipid asymmetry in membranes. *Annu. Rev. Biochem.* **48:** 47.

Papahadjopoulos, D., A. Portis, and W. Pangborn. 1978. Calcium-induced lipid phase transitions and membrane fusion. *Ann. N.Y. Acad. Sci.* **308:** 50.

Patrick, J., S. Heinemann, and D. Schubert. 1978. Biology of cultured cells and muscle. *Annu. Rev. Neurosci.* **1:** 417.

Sandra, A. and R.E. Pagano. 1978. Phospholipid asymmetry in LM cell plasma membrane derivatives. I. Polar head group and acyl chain distribution. *Biochemistry.* **17:** 332.

Scott, R.E. 1976. Plasma membrane vesiculation: A new technique for the isolation of plasma membrane. *Science* **194:** 743.

Sessions, A. and A. Horwitz. 1981. Myoblast aminophospholipid asymmetry differs from that of fibroblasts. *FEBS Lett.* **134:** 75.

Uster, P. and D. Deamer. 1981. Fusion competence of phosphatidylserine-containing liposomes quantitatively measured by a fluorescence resonance energy transfer technique. *Arch. Biochem. Biophys.* **209:** 385.

Wylie, D., C. Damsky, and C. Buck. 1979. Studies on the function of cell surface proteins. I. Use of antisera to surface membranes in the identification of membrane components relevant to cell-substrate adhesion. *J. Cell Biol.* **80:** 385.

Interaction of the Acetylcholine Receptor and Acetylcholinesterase of the Muscle Cell Surface with the Cytoskeletal Framework

JOAV PRIVES
LYNNE HOFFMAN
ANTHONY ROSS
NORMA SERAFIN
Department of Anatomical Sciences
Health Sciences Center
State University of New York at Stony Brook
Stony Brook, New York 11794

It is generally accepted that the spatial arrangement of many integral membrane proteins is an important determinant of cell-surface properties. Beyond the expectation that topographical organization must involve the immobilization of proteins in the plasma-membrane lipid bilayer by anchorage to structural components (Nicolson 1976), the structural basis for the regulation of cell-surface topography is still poorly understood. To analyze the role of the peripheral cytoskeleton in plasma-membrane structural organization, we are studying the interaction of the cytoskeletal framework with components on the surface of embryonic muscle cells in culture. The acetylcholine receptor (AChR) is a well-characterized, integral membrane protein that undergoes pronounced changes in topographical distribution during differentiation and innervation of muscle cells. Biosynthesis of AChR in embryonic muscle is initiated before innervation, and these receptors are initially distributed homogeneously on the cell surface (for review, see Patrick et al. 1978; Fambrough 1979). After innervation, the entire AChR population becomes concentrated into a small patch of

membrane adjacent to the nerve ending (Barnard et al. 1971; Hartzell and Fambrough 1972; Fertuck and Salpeter 1976). Even in the absence of innervation, AChRs on cultured embryonic muscle cells form clusters (Fischbach and Cohen 1973; Hartzell and Fambrough 1973; Sytkowski et al. 1973; Prives et al. 1976) that are similar to the junctional AChR of innervated muscle in having an elevated packing density and diminished lateral mobility (Axelrod et al. 1976). Recent observations suggest the colocalization of peripheral cytoskeletal components with AChR at regions of muscle cells containing high densities of surface AChR (Jacob and Lentz 1979; Weldon and Cohen 1979; Burrage and Lentz 1981). Therefore, AChR is a suitable surface marker with which to monitor the role of the cytoskeletal framework in the developmental expression of cell-surface properties in cultured skeletal muscle.

A second major component of the neuromuscular junction is the enzyme acetylcholinesterase (AChE). AChR and AChE appear on the myotube surface at similar rates during myogenesis in vitro (Prives et al. 1976; Rotundo and Fambrough 1980). AChE accumulates at newly formed innervation sites (Weinberg et al. 1981) and can also become localized in patches on cultured embryonic muscle in the absence of innervation (Moody-Corbett and Cohen 1981).

Recent studies utilizing mild extraction with nonionic detergent have indicated that most cell-surface proteins are not extracted but remain attached to the cytoskeletal framework (Yu et al. 1973; Ben Ze'ev et al. 1979; Sheetz 1979). In contrast, the major proportion of cellular proteins is extracted by this procedure (Ben Ze'ev et al. 1979). We have investigated the interaction of AChR and AChE with the cytoskeletal framework of cultured muscle cells, using the nonionic detergent Triton X-100 to achieve a mild selective extraction that preserves cytoskeletal structure. The retention of these cell-surface components, under the extraction conditions detailed below, is used to define their association with the cytoskeleton.

RESULTS AND DISCUSSION

To monitor the interaction of cell-surface components with cytoskeletal structures, cultured muscle cells are incubated for 4 minutes at 22°C with an extraction buffer consisting of 0.5% (v/v) Triton X-100 in 0.3 M sucrose, 50 mM KCl, 1 mM $MgCl_2$, and 10 mM PIPES (pH 6.9). The detergent-resistant structure consists of the filament systems (microfilaments, intermediate filaments, and microtubules), the microtrabeculae, and a peripheral sheet, the surface lamina (Ben Ze'ev et al. 1979). This structure has been termed the cytoskeletal framework (Ben Ze'ev et al. 1979; Fulton et al. 1981). Scanning electron micrographs show intact (Fig. 1a) and extracted (Fig. 1b) 4-day chick myotubes. As can be seen in Figure 1b, the surface lamina covers most of the skeletal framework, and the surface morphology resem-

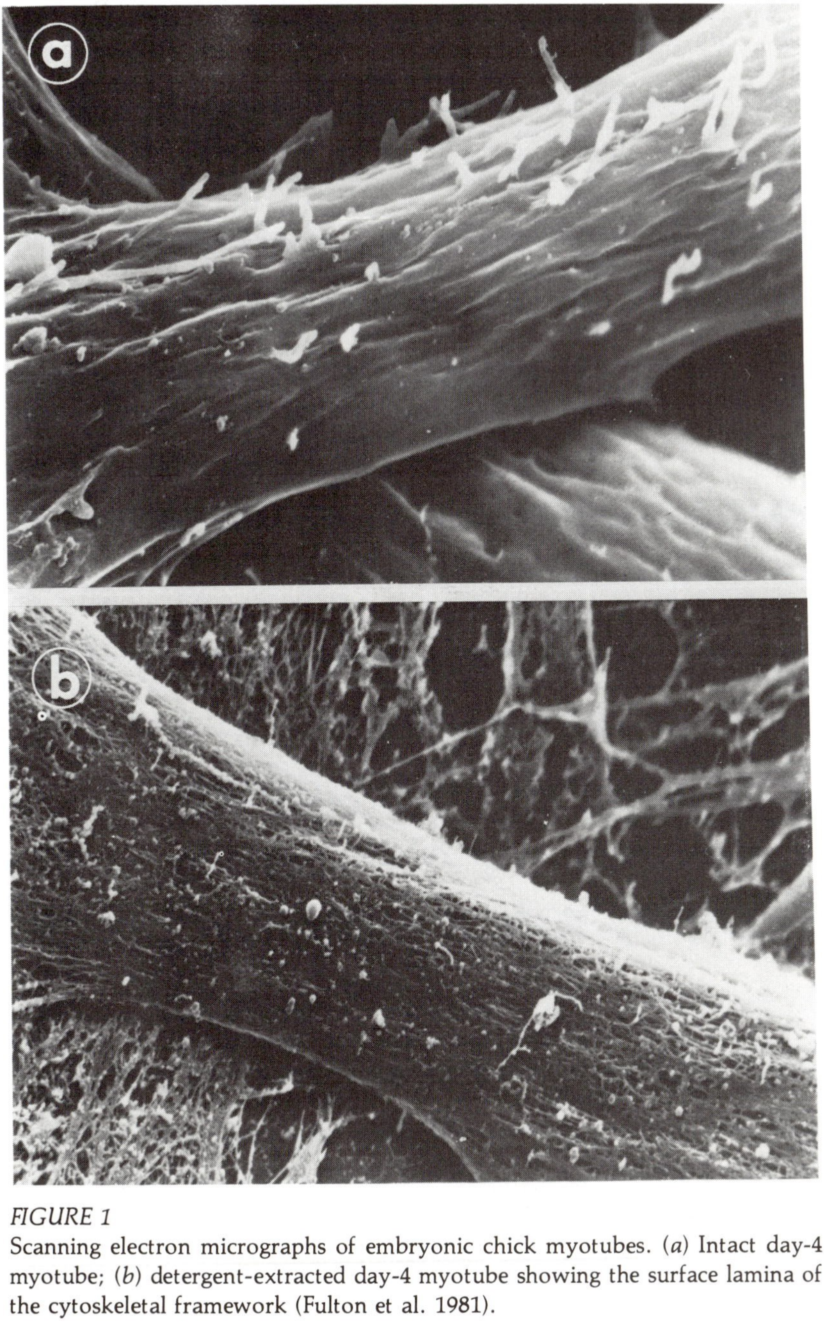

FIGURE 1
Scanning electron micrographs of embryonic chick myotubes. (*a*) Intact day-4 myotube; (*b*) detergent-extracted day-4 myotube showing the surface lamina of the cytoskeletal framework (Fulton et al. 1981).

bles that of the intact cell. Early in differentiation, this framework undergoes an extensive reorganization associated with myoblast fusion, which includes pronounced changes in the surface lamina (Fulton et al. 1981).

We have monitored the developmental changes in the interaction of surface proteins and surface AChR with the cytoskeletal framework. Surface proteins of intact cells are labeled with ^{125}I, using the lactoperoxidase-catalyzed iodination method (Hynes 1973), and surface AChR in replicate cultures is labeled with the specific irreversible ligand ^{125}I-labeled α-bungarotoxin (α-Bgt). Intact cells at various stages of development are labeled and subsequently exposed to the extraction buffer. The partitioning of labeled surface components between the detergent-resistant cytoskeletal framework and the detergent-soluble fractions is measured. As can be seen in Figure 2a, the major proportion of ^{125}I-labeled surface proteins is retained

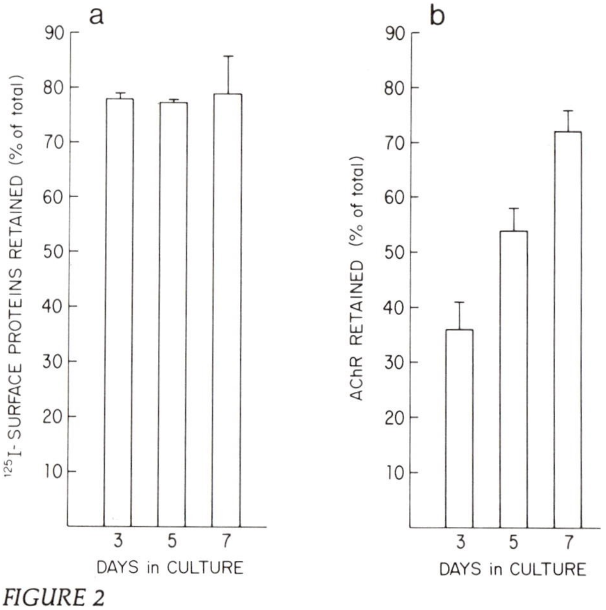

FIGURE 2

Association of iodinated surface proteins (*a*) and AChR (*b*) with the cytoskeletal framework of cultured chick muscle cells at different stages of development. Surface proteins on intact cells were labeled with ^{125}I using lactoperoxidase-catalyzed iodination. AChR in replicate cultures was labeled with ^{125}I-labeled α-Bgt. Cultures were extracted with Triton X-100, and the partitioning of radioactivity to the detergent-soluble and -insoluble fractions was measured. The extent of retention is expressed as the percentage of total labeling.

on the detergent-resistant cytoskeletal framework at all stages of development. In contrast, Figure 2b shows that the association of surface AChR with the cytoskeletal framework changes with culture age. Although in newly fused myotubes, 3 days after plating, a relatively small proportion of surface AChR is retained on detergent-insoluble cytoskeletal framework, this proportion increases markedly with maturation of the muscle cells.

To examine the relationship of cytoskeletal attachment to cell-surface distribution of AChR, intact myotubes are labeled with fluorescent α-Bgt and then extracted with detergent. Figure 3a is a fluorescence photomicrograph of intact embryonic chick muscle cells, previously cultured for 7 days and stained with tetramethyl-rhodamine (TMR)/α-Bgt, which shows AChR aggregates together with regions of diffuse receptor staining. Figure 3b shows the same field after a 5-minute extraction with detergent. The aggregated AChR regions are still prominent, whereas the diffuse receptor fluorescence has diminished below the threshold of visibility. These results suggest that the diffuse AChR is preferentially extracted, whereas the patched receptors are more stably associated with the cytoskeletal framework. These results have been verified by quantitative fluorescence measurements (Prives et al. 1982).

AChE in cultured muscle cells is distributed in the cytoplasm as well as on the cell surface (Rotundo and Fambrough 1980). We have studied the interaction of surface and intracellular AChE with the cytoskeletal framework by the detergent extraction method and have used two pharmacological methods to distinguish cell-surface AChE from intracellular AChE (Emmerling 1980). To selectively measure surface AChE activity, the irreversible inactivator diisopropyl fluorophosphate (2×10^{-4} M) is used to block intracellular AChE activity, whereas the surface enzyme is protected by the reversible inhibitor BW284c51 (10^{-4} M). After reactivation of surface AChE by removal of the reversible inhibitor, intact cells are extracted as described above. Surface AChE is measured in both detergent-soluble and -insoluble fractions according to the method of Johnson and Russell (1975), as modified by Rotundbrough (1980). To selectively measure intracellular AChE, the enzyme on the external surface of intact myotubes is inactivated by ecothiopate iodide (5×10^{-6} M), an irreversible AChE inhibitor that does not penetrate the cell membrane. The muscle cells are then extracted, and intracellular AChE is determined, as described above, in both detergent-soluble and -insoluble fractions. During myotube development the proportion of cell-surface AChE retained on the cytoskeletal framework increases markedly (Fig. 4a). This developmental change is similar to the concomitant increase in cell-surface AChR association with the cytoskeletal framework and clearly different from the partitioning of the radioiodinated surface components. In comparison, Figure 4b shows that only a minor proportion ($\sim 30\%$) of intracellular AChE is retained on the cytoskeletal framework at all stages of development.

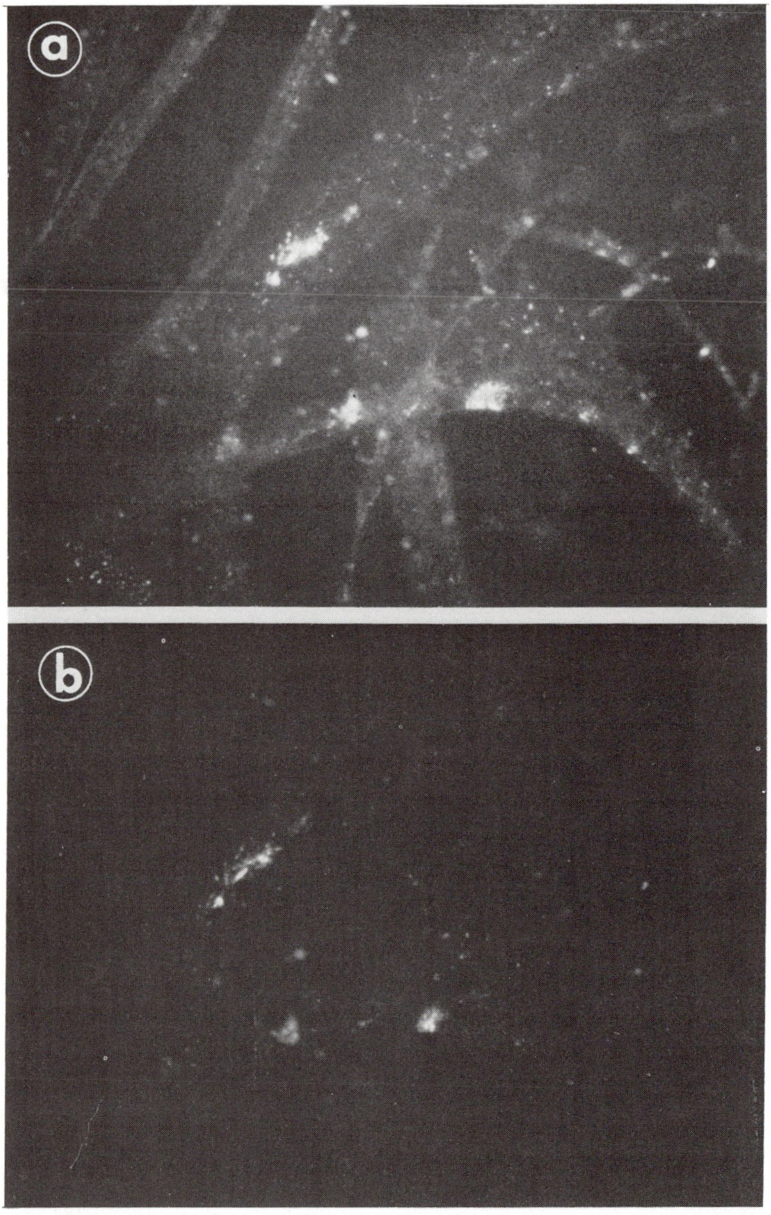

FIGURE 3

The effect of detergent extraction on the distribution pattern of AChR on cultured chick muscle cells. Intact myotubes were labeled with TMR/α-Bgt, and staining was visualized by fluorescence microscopy using a 40× water immersion objective. (*a*) Intact myotubes in a 7-day-old culture; (*b*) the same field 5 min after adding the 0.5% Triton X-100 extraction buffer.

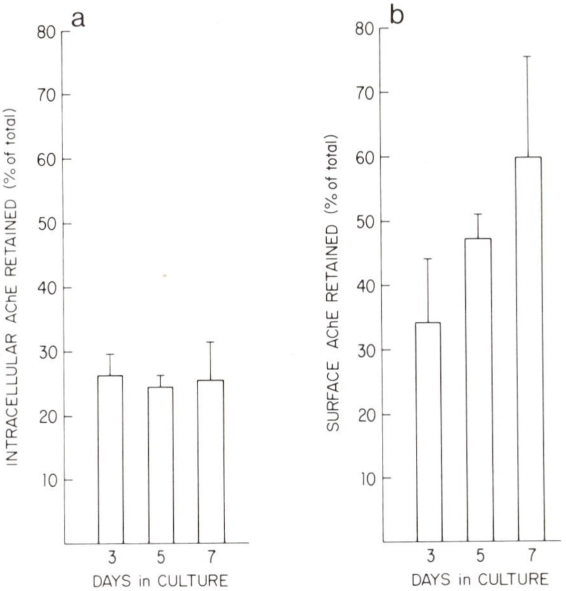

FIGURE 4
Association of intracellular AChE (*a*) and cell-surface AChE (*b*) with the cytoskeletal framework of cultured chick muscle cells at different stages of development. Cultures were extracted with Triton X-100, and AChE was assayed in the detergent-soluble and -insoluble fractions. Before extraction, intact cultures were treated with pharmacological agents to produce selective inhibition of either intracellular AChE or cell-surface AChE. The extent of retention is expressed as the percentage of total surface AChE or of total intracellular AChE.

SUMMARY

Using the Triton X-100 nonionic-detergent-extraction method to detect the association of surface components with the cytoskeletal framework in cultured muscle cells, we report the following observations:

1. The major proportion (> 80%) of surface proteins labeled by radioiodination of intact myotubes are retained on the cytoskeletal framework after detergent extraction.
2. During the early development of muscle cells, most of the AChR is extracted by detergent. However, during myotube maturation, the proportion of AChR that is attached to the cytoskeletal framework increases markedly.
3. AChR association with the cytoskeletal framework appears to be related to the topographical distribution of these receptors. Aggregated AChR

appears to be tightly bound to the framework, whereas most of the diffusely distributed AChR is not tightly bound and is extracted with detergent.

4. Like AChR, a portion of cell-surface AChE is not extracted by detergent but is retained on the cytoskeletal framework of cultured myotubes. This proportion increases during muscle cell differentiation. In contrast to the cell-surface-associated AChE, most intracellular AChE is extracted by detergent at all stages of development.

ACKNOWLEDGMENTS

This work was supported by grants from the Muscular Dystrophy Association of America and the U.S. Public Health Service (NS-16782). We thank D. Bar Sagi for critical reading of the manuscript. We are grateful to Drs. M. Emmerling and R. Rotundo for advice and comments regarding the measurement of AChE activity.

REFERENCES

Axelrod, D., P. Ravdin, D. Koppel, J. Schlessinger, W. Webb, E. Elson, and T. Podleski. 1976. Lateral motion of fluorescently labeled acetylcholine receptors in membranes of developing muscle fibers. *Proc. Natl. Acad. Sci.* **73:** 4594.

Barnard, E., A. Wieckowski, and T. Chiu. 1971. Cholinergic receptor molecules and cholinesterase molecules at mouse skeletal muscle junctions. *Nature* **234:** 207.

Ben Ze'ev, A., A. Duerr, F. Solomon, and S. Penman. 1979. The outer boundary of the cytoskeleton: A lamina derived from plasma membrane proteins. *Cell* **17:** 859.

Burrage, T. and T. Lentz. 1981. Ultrastructural characterization of surface specializations containing high-density acetylcholine receptors in embryonic myotubes in vivo and in vitro. *Dev. Biol.* **85:** 267.

Emmerling, M. 1980. "Molecular forms of acetylcholinesterase in quail muscle and their interaction with fibronectin." Ph.D. thesis, University of Wisconsin, Madison.

Fambrough, D. 1979. Control of acetylcholine receptors in skeletal muscle. *Physiol. Rev.* **59:** 165.

Fertuck, H. and M. Salpeter. 1976. Quantitation of junctional and extrajunctional acetylcholine receptors by electron microscope autoradiography after ^{125}I-α-bungarotoxin binding at mouse neuromuscular junctions. *J. Cell Biol.* **69:** 144.

Fischbach, G. and S. Cohen. 1973. The distribution of acetylcholine sensitivity over uninnervated and innervated muscle fibers grown in cell culture. *Dev. Biol.* **31:** 147.

Fulton, A., J. Prives, S. Farmer, and S. Penman. 1981. Developmental reorganization of the skeletal framework and its surface lamina in fusing muscle cells. *J. Cell Biol.* **91:** 103.

Hartzell, H. and D. Fambrough. 1972. Acetylcholine receptor distribution and extrajunctional density in rat diaphragm after denervation correlated with acetylcholine sensitivity. *J. Gen. Physiol.* **60:** 248.

——. 1973. Acetylcholine receptor production and incorporation into membranes of developing muscle fibers. *Dev. Biol.* **30:** 153.

Hynes, Richard O. 1973. Alterations of cell surface proteins by viral transformation and by proteolysis. *Proc. Natl. Acad. Sci.* **70:** 3170.

Jacob, M. and T. Lentz. 1979. Localization of acetylcholine receptors by means of horseradish peroxidase-α-bungarotoxin during formation and development of the neuromuscular junction in the chick embryo. *J. Cell Biol.* **82:** 195.

Johnson, C.D. and R.. Russell. 1975. A rapid, simple, radiometric assay for cholinesterase suitable for multiple determinations. *Anal. Biochem.* **64:** 229.

Moody-Corbett, F. and M.W. Cohen. 1981. Localization of cholinesterase at sites of high acetylcholine receptor density on embryonic amphibian muscle cells cultured without nerve. *J. Neurosci.* **1:** 596.

Nicolson, G.L. 1976. Transmembrane control of the receptors on normal and tumour cells. I. Cytoplasmic influence over cell surface components. *Biochim. Biophys. Acta* **457:** 57.

Patrick, J., S. Heinemann, and D. Schubert. 1978. Biology of cultured nerve and muscle. *Annu. Rev. Neurosci.* **1:** 417.

Prives, J., I. Silman, and A. Amsterdam. 1976. Appearance and disappearance of acetylcholine receptor during differentiation of chick skeletal muscle in vitro. *Cell* **7:** 543.

Prives, J., A. Fulton, S. Penman, M.P. Daniels, and C.N. Christian. 1982. Interaction of the cytoskeletal framework with acetylcholine receptor on the surface of embryonic muscle cells in culture. *J. Cell Biol.* **92:** 231.

Rotundo, R. and D. Fambrough. 1980. Synthesis, transport, and fate of acetylcholinesterase in cultured chick embryo muscle cells. *Cell* **22:** 583.

Sheetz, M. 1979. Integral membrane protein interaction with triton cytoskeletons of erythrocytes. *Biochim. Biophys. Acta* **557:** 122.

Sytkowski, A., Z. Vogel, and M. Nirenberg. 1973. Development of acetylcholine receptor clusters on cultured muscle cells. *Proc. Natl. Acad. Sci.* **70:** 270.

Weinberg, C.B., J.R. Sanes, and Z.W. Hall. 1981. Formation of neuromuscular junctions in adult rats: Accumulation of acetylcholine receptors, acetylcholinesterase and components of synaptic basal lamina. *Dev. Biol.* **84:** 255.

Weldon, P. and M. Cohen. 1979. Development of synaptic ultrastructure at neuromuscular contacts in an amphibian cell culture system. *J. Neurocytol.* **8:** 239.

Yu, J., D. Fischman, and T. Steck. 1973. Selective solubilization of proteins from red blood cell membranes by nonionic detergents. *J. Supramol. Struct.* **1:** 233.

Role of Specific Proteases in Rat Myoblast Fusion

WARREN J. STRITTMATTER[*†‡]
CHRISTINE B. COUCH[*]
STANTON B. ELIAS[*†]
**Program in Neuroscience*
Departments of †Neurology and ‡Biochemistry
Baylor College of Medicine
Houston, Texas 77030

During muscle development, myoblasts proliferate, aggregate, align, and fuse to form multinucleate myotubes (Bischoff 1978). Myotubes then differentiate into adult muscle. We have used primary cultures of rat myoblasts to study the fusion process. Fusion of these cells in primary tissue culture can be regulated by calcium in the media (Shainberg et al. 1969). Myoblasts grown in calcium-free medium containing dialyzed horse serum will proliferate, aggregate, and align but will not fuse. Introducing calcium into the medium then initiates fusion, which can be first observed within 2 hours and is completed in 2 days. By controlling calcium, fusion can be dissociated from many other developmental events.

The morphology of myoblast fusion has been extensively studied, but the biochemistry is poorly understood. Gap junctions form between adjacent myoblasts, but their role, if any, in mediating fusion is not known (Rash and Fambrough 1973). Particle-free zones appear between regions of myoblast cell membrane, which ultimately fuse, followed by the formation of 15-Å cytoplasmic bridges, which then enlarge (Kalderon 1980).

During fusion, myoblast proteins undergo many changes in both molecular weight and in their distribution over the cell surface. Cell-surface proteins change during fusion, with a decrease in high-molecular-weight species and an increase in low-molecular-weight species (Moss et al. 1978; Pauw and David 1979). The large external transformation-sensitive (LETS) protein redistributes during fusion. In unfused myoblasts, LETS protein is dispersed over the entire cell surface. Following fusion, the amount of LETS protein decreases, and the protein becomes discretely clustered on some

myotubes (Chen 1977). A 16S glycoprotein complex is released from myoblasts during fusion and may mediate recognition or adhesion (Schubert and LaCorbiere 1980). Myoblast fusion in both rat (Gilfix and Sanwal 1980) and quail (Olden et al. 1981) appears to require glycoproteins based on the observation that tunicamycin, an inhibitor of glycosylation, inhibits myoblast fusion. The carbohydrate on the glycoprotein appears to play an indirect role in fusion by preventing nonspecific proteolysis of the protein core, since the addition of the thiol and serine protease inhibitor leupeptin permits fusion of tunicamycin-treated cells. Thus, proteolysis of myoblast glycoproteins prevents fusion, whereas the carbohydrate moieties of glycoproteins appear to prevent nonspecific proteolysis.

Because of these changes in cell-surface proteins during fusion, we examined the hypothesis that limited proteolysis may be a required step in fusion of myoblasts and that the signal to stop fusion may also be regulated by proteolysis. Fusion of other biological membrane requires limited proteolysis. For example, the fusion of Sendai and influenza virus with host cells requires proteolytic cleavage of a coat protein. The influenza virus coat protein, hemagglutinin, must be hydrolyzed to generate a peptide fragment of 221 amino acids, HA_2, before the virus can fuse with the host cell (Wilson et al. 1981). Nonspecific proteolysis of hemagglutinin occurs if the virus is incubated with tunicamycin (Schwartz et al. 1976), and this results in loss of viral infectivity. Specific proteolytic cleavage of a surface glycoprotein in virus–host-cell fusion may serve as a useful model in examining the process of membrane protein processing in myoblast-myoblast fusion.

RESULTS AND DISCUSSION

Metalloendoprotease Required for Myoblast Fusion

Proteases can be divided into four groups by their response to specific protease inhibitors: (1) metal dependent, (2) carboxyl dependent, (3) thiol dependent, and (4) serine dependent. Proteases can be further characterized by determining substrate specificity with synthetic peptide derivatives (Barret 1977). To test the hypothesis that myoblast fusion requires selective proteolysis, rat myoblasts were first grown in calcium-deficient medium for 5 days. These cells proliferated, aggregated, and aligned but did not fuse. Protease inhibitors were then added to the medium with calcium. Thiol protease inhibitors (4-chloromercuribenzoate and leupeptin) and serine protease inhibitors (phenylmethanesulfonyl fluoride, soybean and Kunitz trypsin inhibitors, L-1-tosylamide-2-phenylethyl-chloromethyl-ketone, and aprotinin) had no effect on fusion. Pepstatin, an inhibitor of carboxyl proteases, delayed the onset of fusion but had no effect on fusion after 24 hours, even after repeated addition.

1,10-Phenanthroline, which inhibits metalloproteases, prevented myoblast fusion (as shown in Fig. 1A, B and Fig. 2). Half-maximal inhibition of fusion

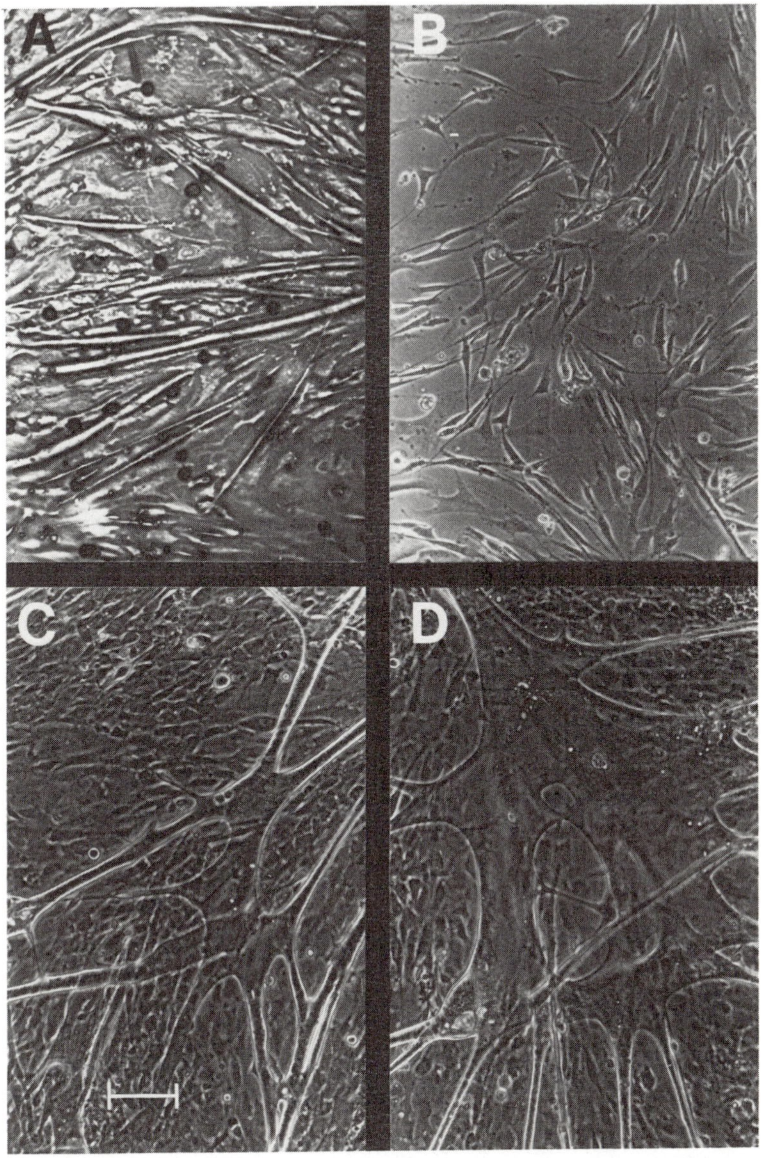

FIGURE 1

Phase-contrast microscopy showing effects of protease inhibitors on myoblast fusion. (*A*,*B*) Effect of the metalloprotease inhibitor 1,10-phenanthroline. Primary cultures of rat myoblasts, grown for 5 days in calcium-free Dulbecco's modified Eagle's medium, containing 10% dialyzed horse serum, were incubated for 24 hr in 1.4 mM $CaCl_2$ (*A*) and in 1.4 mM $CaCl_2$ after a 2-hr incubation in 40 μg/ml of 1,10-phenanthroline (*B*). (*C*,*D*) Effect of the plasmin inhibitor, aprotinin. Myoblast cultures grown in calcium-deficient medium were cultured for 3 days in 1.4 mM $CaCl_2$ (*C*) and in 1.4 mM $CaCl_2$ with aprotinin (1.0 trypsin inhibitor units/ml) (*D*); media were replaced daily. Bar represents 100 μm.

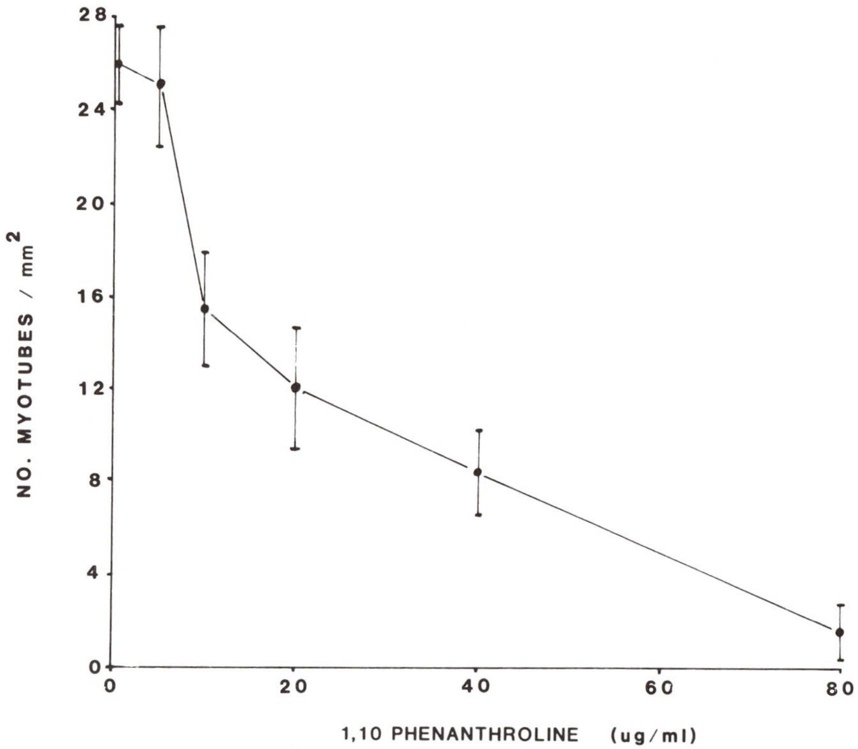

FIGURE 2
Inhibition of myoblast fusion by 1,10-phenanthroline. Rat myoblasts, grown in calcium-deficient medium, were incubated for 2 hr in increasing concentrations (0–80 μg/ml) of 1,10-phenanthroline. The medium was replaced with medium containing 1.4 mM $CaCl_2$. Fusion was quantitated 24 hr later.

required 20 μg/ml of 1,10-phenanthroline. Inhibition of fusion with 1,10-phenanthroline depended on its ability to chelate metal, since prior chelation of 1,10-phenanthroline with equimolar amounts of zinc before addition to myoblast cultures had no effect on fusion.

To further demonstrate that myoblast fusion requires a metal-dependent protease, myoblast cultures were incubated with synthetic amino acid substrates of characterized metal-dependent proteases. Carbobenzoxy (CBZ)-L-phenylalanine, which competitively binds to the active site of zinc-dependent endopeptidase (Burstein et al. 1974), inhibited myoblast fusion with half-maximal inhibition at 1.5 mM. The effect was stereospecific, since CBZ-D-phenylalanine had no effect. Substrates of zinc-dependent exopeptidases, hippuryl-L-arginine and hippuryl-L-phenylalanine, had no effect on fusion.

To determine the substrate specificity of this zinc-dependent endopep-

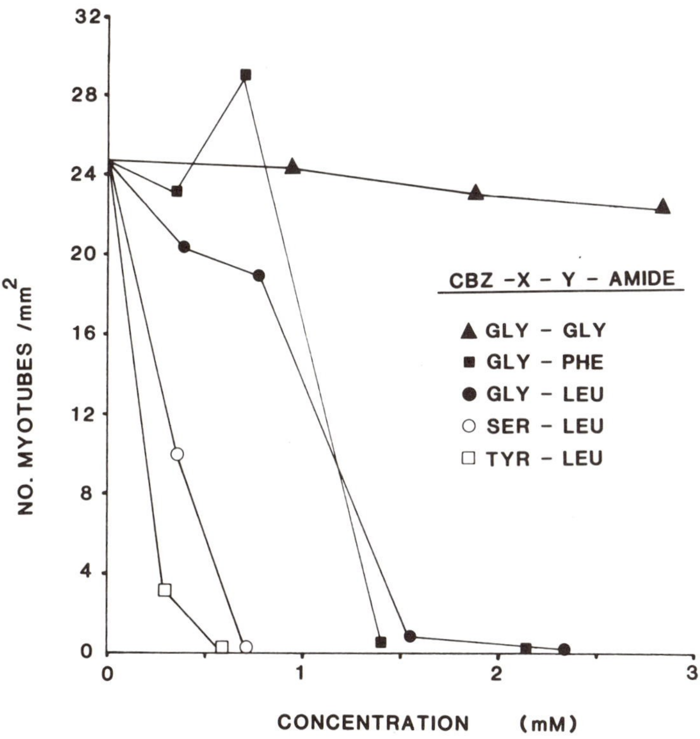

FIGURE 3

Effects of synthetic dipeptide derivatives on myoblast fusion. Rat myoblasts grown in calcium-deficient medium were incubated for 15 min with varying concentrations of CBZ-dipeptide-amide derivative before the introduction of 1.4 mM $CaCl_2$. Cells were incubated for 24 hr before fusion was quantitated.

tidase, a group of CBZ-dipeptide-amide derivatives, with varying specificities for characterized zinc-dependent endopeptidases (Morihara 1974), were tested for their effect on myoblast fusion. High-affinity substrates of zinc-dependent neutral endopeptides, CBZ-tyr-leu-amide and CBZ-ser-leu-amide, inhibited myoblast fusion at low concentrations; intermediate-affinity substrates of these proteases, CBZ-gly-leu-amide and CBZ-gly-phe-amide, required intermediate concentrations to inhibit fusion, whereas CBZ-gly-gly-amide, which is not a substrate for these proteases, had no effect on fusion, as shown in Figure 3. Further evidence that these derivatives inhibit myoblast fusion by competing for an endoprotease is that both the CBZ and amide groups are necessary for both interacting with the endoprotease and for inhibiting fusion, since gly-phe-amide and CBZ-gly-phe had no effect on fusion.

The fusion of myoblasts into multinucleate myotubes therefore requires a metalloendopeptidase. Metal-dependent neutral endoproteases have been characterized in mammalian pancreas (Mumford et al. 1980), pituitary (Orlowski and Wilk 1981), kidney brush border (Kenny 1977), and in bacteria (Morihara 1974). These proteases hydrolyze the amino terminus of aromatic and hydrophobic amino acids. The enzyme in rat myoblasts is now being characterized by determining its cellular localization and possible activation by calcium and by isolating the endogenous substrates and products. Protease activity may be required for fusion by eliminating charge or steric restraints on membrane apposition or by generating a fusogenic peptide, as demonstrated in virus–host-cell fusion.

Plasminlike Protease Terminates Myoblast Fusion

Plasminogen activator is a protease released into the extracellular space by myotubes, which converts proteolytically inactive plasminogen to the active plasmin. Cultured chick myoblasts release plasminogen activator into the media, and the release of this protease increases 2.5-fold as myoblast fusion ends (Miskin et al. 1978). Drugs, such as phorbolesters, increase the release of plasminogen activator and also inhibit myoblast fusion. Since the release of plasminogen activator peaks as myoblast fusion ends and since drugs that block fusion increase plasminogen activator, we examined the possibility that limited proteolysis by a plasminlike protease, regulated by plasminogen activator, may block myoblast fusion. If this hypothesis is true, then the addition of exogenous plasmin to fusing myoblasts should prevent fusion, whereas the addition of the serine protease inhibitor aprotinin should increase fusion.

The addition of the protease plasmin to rat myoblast cultures inhibited fusion, as shown in Figure 4. In contrast, the daily addition of serine protease inhibitor, aprotinin, caused extensive syncytia formation, as shown in Figure 1C and D.

SUMMARY

The fusion of rat myoblasts in primary culture appears to require metal-dependent endopeptidase activity, since 1,10-phenanthroline and dipeptide derivatives, which bind to metalloendoproteases, block fusion. Limited proteolysis during cell fusion may eliminate steric or charge restraints to membrane apposition imposed by large, charged surface glycoprotein; it may permit movement of membrane-bound protein to create particle-free zones of membrane; or it may generate a polypeptide fragment that mediates fusion.

Plasminlike protease activity appears to stop fusion. The localization and substrate specificity of this enzyme is not known. Plasminogen activators

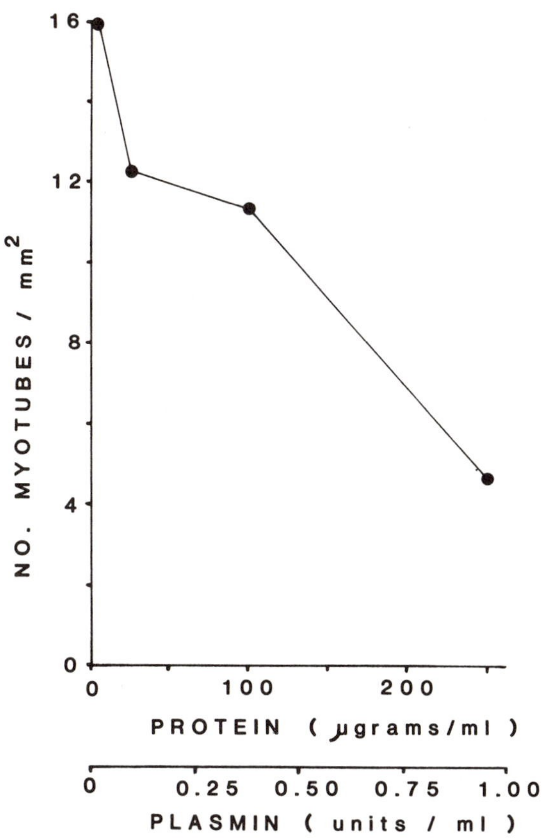

FIGURE 4
Inhibition of myoblast fusion by the protease, plasmin. Rat myoblasts grown in calcium-deficient medium were incubated for 24 hr with 1.4 mM $CaCl_2$ and increasing concentrations of plasmin (Sigma).

released from cultured myotubes may prevent continuing fusion, since exogenous plasmin prevents fusion, and the plasmin inhibitor aprotinin increases fusion. A plasminlike protease may stop fusion by inactivating the metal-dependent endoprotease or a fusogenic peptide.

ACKNOWLEDGMENTS

This work was supported in part by the Jerry Lewis Neuromuscular Disease Center sponsored by the Muscular Dystrophy Association of America, and by Teacher Investigator awards from the National Institute of Neurological and Communicative Disorders and Stroke (NS-00497 to W.J.S. and NS-00483 to S.B.E.).

REFERENCES

Barrett, A.J. 1977. Introduction to proteinases. In *Proteinases in mammalian cells and tissues* (ed. A.J. Barrett), p 1. Elsevier/North Holland, Amsterdam.

Bischoff, R. 1978. Myoblast fusion. In *Membrane fusion* (ed. G. Poste and G.L. Nicolson), p. 127. Elsevier/North Holland, Amsterdam.

Burstein, Y., K.A. Walsh, and H. Neurath. 1974. Evidence of an essential histidine residue in thermolysin. *Biochemistry* **13:** 205.

Chen, L.B. 1977. Alteration in cell surface LETS protein during myogenesis. *Cell* **10:** 393.

Gilfix, B.M. and B.D. Sanwal. 1980. Inhibition of myoblast fusion by tunicamycin and pantomycin. *Biochem. Biophys. Res. Commun.* **96:** 1184.

Kalderon, N. 1980. Muscle cell fusion. In *Membrane—membrane interactions* (ed. N.B. Gilula), p. 99. Raven Press, New York.

Kenny, A.J. 1977. Proteinases associated with cell membranes. In *Proteinases in mammalian cells and tissues* (ed. A.J. Barrett), p. 394. Elsevier/North Holland, Amsterdam.

Miskin, R., T.G. Easton, and E. Reich. 1978. Plasminogen activator in chick embryo muscle cells: Induction of enzyme by RSV, PMA and retinoic acid. *Cell* **15:** 1301.

Morihara, K. 1974. Comparative specificity of microbial proteinases. *Adv. Enzymol.* **41:** 179.

Moss, M., J.S. Norris, E.J. Peck, Jr., and R.J. Schwartz. 1978. Alterations in iodinated cell surface proteins during myogenesis. *Exp. Cell Res.* **113:** 445.

Mumford, R.A., A.W. Strauss, J.C. Powers, P.A. Pierzchala, N. Nishino, and M. Zimmerman. 1980. A zinc metalloendopeptidase associated with dog pancreatic membranes. *J. Biol. Chem.* **255:** 2227.

Olden, K., J. Law, V.A. Hunter, R. Romain, and K. Parent. 1981. Inhibition of fusion of embryonic muscle cells in culture by tunicamycin is prevented by leupeptin. *J. Cell Biol.* **88:** 199.

Orlowski, M. and S. Wilk. 1981. Purification and specificity of a membrane-bound metalloendopeptidase from bovine pituitaries. *Biochemistry* **20:** 4942.

Pauw, P.G. and J.D. David. 1979. Alterations in surface proteins during myogenesis of a rat myoblast cell line. *Dev. Biol.* **70:** 27.

Rash, J.E. and D. Fambrough. 1973. Ultrastructural and electrophysiological correlates of cell coupling and cytoplasmic fusion during myogenesis *in vitro. Dev. Biol.* **30:** 166.

Schubert, D. and M. LaCorbiere. 1980. Role of a 16S glycoprotein complex in cellular adhesion. *Proc. Natl. Acad. Sci.* **77:** 4137.

Schwartz, R.T., J.M. Rohrschneider, and M.F.G. Schmidt. 1976. Suppression of glycoprotein formation of Semliki Forest, influenza and avian sarcoma virus by tunicamycin. *J.Virol.* **19:** 782.

Shainberg, A., G. Yagil, and D. Yaffe. 1969. Control of myogenesis in vitro by Ca^{+2} concentration in nutritional medium. *Exp. Cell Res.* **58:** 163.

Wilson, I.A., J.J. Skehel, and D.C. Wiley. 1981. Structure of the hemagglutinin membrane glycoprotein of influenza virus at 3 Å resolution. *Nature* **289:** 366.

Topography and Mobility of Concanavalin A Receptors during Myoblast Fusion

SALVADOR M. FERNANDEZ
*BRIAN A. HERMAN**
Physiology Department
University of Connecticut Health Center
Farmington, Connecticut 06032

The fusion of mononucleated myoblasts to form multinucleated myotubes is one of the earliest and most overt events in the process of muscle differentiation. The basic mechanisms by which two muscle cells recognize each other and subsequently fuse remain to be elucidated. A number of observations, however, indicate that cell-surface lectins (Den et al. 1975; Teichberg et al. 1975; Gartner and Podleski 1976; Nowak et al. 1976; Sandra et al. 1977) and the lipids of the plasma membrane (Prives and Shinitzky 1977; Herman and Fernandez 1978; Horwitz et al. 1978, 1979; Kalderon and Gilula 1979) play important roles in this process.

On the basis of the evidence just cited and in view of the suspected role for lectins as recognition and/or adhesion molecules (Rosen et al. 1975), we postulate that lectin receptors on the myoblast surface must undergo some form of transient topographical rearrangement during fusion. Furthermore, we propose that the changes in membrane fluidity that occur during development serve to modulate the translational mobility of these and other surface components.

To test these hypotheses, we have examined the topographical distribution and the lateral mobility of concanavalin A (Con A) receptors on the surface of chick pectoralis muscle cells at different times during development in vitro. For this purpose, we employ a resonance energy transfer

*Present address: Anatomy Department, Harvard Medical School, Boston, Massachusetts 02115.

(RET) technique (Fernandez and Berlin 1976) that permits quantitation of the average proximity of these receptors on the surface of single cells.

RET METHOD FOR THE STUDY OF RECEPTOR TOPOGRAPHY

The method employed in this work is essentially the same as that previously described (Fernandez and Berlin 1976), with some minor modifications. Briefly, the RET method is founded on the fact that a fluorophore (donor) in an excited state may transfer its excitation energy to a neighboring chromophore (acceptor) through dipole-dipole interactions (Forster 1965). The efficiency of the transfer process is dependent on, among other factors, the donor-acceptor (D-A) separation distance. Thus, RET can be employed to study proximity relationships of suitably labeled cell-surface components. (We assume that the effects of orientation or changes in orientation upon RET are not significant in these experiments.)

Cells attached to substrate are labeled in a solution containing a suitable ratio (1:6.8) of donor (pyrene)- and acceptor (fluorescein isothiocyanate [FITC])-conjugated Con A and examined with a photon-counting microspectrofluorimeter, which has been described previously (Fernandez and Berlin 1976; Herman and Fernandez 1978).

Since RET results in an increase in acceptor emission (sensitized fluorescence) and a decrease in donor fluorescence, we have taken the ratio of acceptor to donor fluorescence (I_A/I_D) as a convenient experimental measure of RET. The value of this ratio depends on the average distance between D-A pairs and, hence, on the state of aggregation of Con A receptors. As the average distance between D-A pairs increases, this ratio approaches a limiting value that corresponds to the absence of RET. Likewise, an upper limit for I_A/I_D exists corresponding to a maximal packing density of Con A receptors. We assume that the total concentration of Con A receptors, the lateral diffusion rate, and the local environment of the fluorophore–Con A receptor complex all do not vary significantly.

To interpret the values of I_A/I_D obtained from single cell studies, we employed the following procedure to empirically determine the magnitude of the limiting values of I_A/I_D. Solutions of Con A containing donor and acceptor conjugates in the same ratio as used for cellular studies (1:6.8) were serially diluted with native unlabeled Con A in such a way as to vary the relative proportions of labeled and unlabeled species while maintaining the total Con A concentration and the D-A ratio constant. These mixtures of Con A were then precipitated with anti–Con A antibody, and the precipitates were smeared on glass slides for observation under the microscope. I_A/I_D values were then determined for the various samples. The maximum value of I_A/I_D obtained was 1.75; and as the average D-A distance increased with the presence of greater amounts of unlabeled Con A, I_A/I_D decreased and approached a constant value of 0.75. This empirical

calibration thus provides a semiquantitative basis for interpreting the cellular RET data.

RESULTS AND DISCUSSION

Surface Con A Receptors Undergo a Marked Reorganization during the Period of Myoblast Fusion

Results from studies of the inherent Con A receptor topography at sequential times during differentiation are shown in Figure 1. Several findings become readily apparent upon examination of these data. Most salient among these is the marked decrease and subsequent increase in RET that takes place between 30 hours and 70 hours in culture. These results indicate that between 24 hours and 48 hours in culture, a significant increase in average Con A receptor separation takes place on the muscle cell surface.

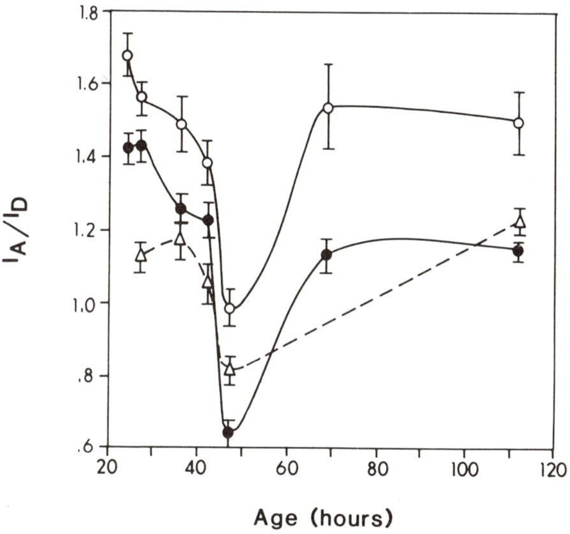

FIGURE 1
RET from myogenic cells at different times during development. Cells of the appropriate age were fixed by incubation for 20 min at 37°C in 2% (w/v) paraformaldehyde and subsequently labeled with Con A (20 μg/ml) for 1 hr at 4°C. The extent of RET is quantitated as the ratio I_A/I_D of acceptor (FITC)-to-donor (pyrene) fluorescence. (○) Values obtained from regions of cell contact or fusion; (●) data obtained from perinuclear regions; (Δ) data obtained from myotube regions. RET values are the mean ± S.E.M. of 48 individual measurements (duplicates on 24 cells).

This interval corresponds to a time period in which these cells have ceased to proliferate and are undergoing fusion or becoming ready to do so. Figure 1 also shows that this increase in average receptor separation is transient since, after 48 hours—a time when fusion is essentially complete—RET increases rapidly to a level somewhat lower than that seen at prefusion times.

To more precisely define the nature of the "changes in average receptor separation" that take place in this time period, it is necessary to take into account the following considerations. First, the R_0 value for the pyrene-FITC pair is 43 Å. (R_0 is the D-A separation for which the transfer efficiency is 50%.) From the inverse sixth-power dependence of RET on D-A separation (Forster 1965), it is clear that transfer will decline rapidly for D-A separations much greater than R_0. Thus, R_0 is a useful parameter that gives information on the range of D-A separations, which can be investigated with a given pair of fluorophores. Since Con A has molecular dimensions that are commensurate with the R_0 for pyrene-FITC (Quiocho et al. 1971), significant RET will take place only when receptors are in very close proximity, a configuration that we designate as clustered. A second consideration to take into account is that I_A/I_D ratios in Figure 1 range from a maximum of about 1.74 at 24 hours to a minimum of about 0.75 at 48 hours, corresponding to the limiting values obtained from the calibration data.

All of this suggests that in the short time span between 24 hours and 48 hours, Con A surface receptor distribution undergoes a drastic reorganization from a state in which receptors are predominantly in microclusters to one in which they are essentially disperse. Furthermore, after 48 hours, once the fusion process is completed, a reversal of this pattern back to a clustered state takes place. During the time period in which the decrease and subsequent increase in RET are observed, the overall density of surface-accessible Con A receptors does not change significantly, as evidenced by autoradiographic analysis (Sandra et al. 1977).

From Figure 1 it can be seen that, qualitatively, the same temporal variation in RET is observed at all three cellular regions examined, although there appears to exist spatial heterogeneity in the actual I_A/I_D ratios. For example, at prefusion times (<36 hr), regions of cell contact exhibit the highest level of RET and thus the highest degree of clustering.

Fusion-arrested cells (grown in the presence of 1.8 mM EGTA) do not exhibit the marked temporal alteration in RET seen in the control cells. As shown in Figure 2, RET in fusion-arrested cells remains relatively constant up to about 68 hours, after which time it begins an upward trend. Addition of calcium at 68 hours to the fusion-arrested cultures results in a period of synchronized fusion activity that leads to 90% fusion in 5 hours. Addition of calcium also results in an oscillation in RET values analogous to that seen in the control cells allowed to develop normally (data not shown). RET values from cells that remain in EGTA do not exhibit this oscillation but, rather, continue to increase at an approximately linear rate. It is noteworthy that although the kinetics of fusion are accelerated in the synchronized cells, the observed changes in RET and the time course of fusion maintain a

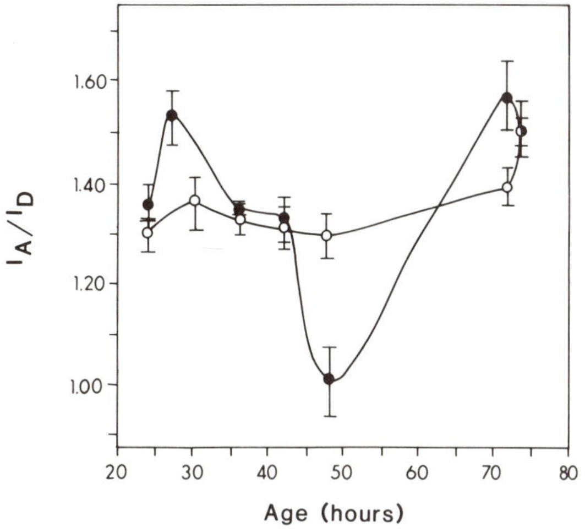

FIGURE 2
RET from perinuclear regions of control cells (●) and fusion-arrested cells grown in media supplemented with 1.8 mM EGTA (○). Cells were fixed and labeled as described in the legend to Fig. 1.

temporal correlation similar to that observed in normal development, i.e., the changes in RET in this case also occur on a proportionately faster time scale.

One additional important aspect of these results stems from our microscopic approach, which enables us to make measurements on single cells. Thus, at any time during the course of fusion activity we can selectively examine isolated single cells that are not in contact with others—cells which, at the light microscope level, still appear as separate entities but are in contact or fusing with another cell—or multinucleated myotubes, which have resulted from the fusion process. We have found that, at any given time in culture, the values of I_A/I_D from perinuclear and myotube regions are similar regardless of whether the cell being examined is an isolated cell or one that has undergone cell contact or fusion. In other words, the transition from a clustered to a disperse distribution of Con A receptors does not appear to be triggered by cell contact and/or fusion. Rather, it seems that these changes are part of a developmental program, i.e., for a certain period of time, the cells develop a surface architecture that is conducive or necessary for recognition, adhesion, and fusion to occur. After this period of time, the duration of which is developmentally regulated, these cell-surface characteristics that render the cells "fusion competent" are realtered. The initial presence of Con A receptor clusters at the beginning of fusion activity may be related to a recognition/adhesion step.

In view of available evidence, which indicates that the cytoskeleton plays an important role in the regulation of the topographic display of surface Con A receptors in a number of cell lines (Albertini et al. 1977), we have examined the effect of the microtubule-disrupting agents colchicine and nocodazol (10^{-7} M) on the topography of Con A receptors during synchronized fusion. The effect of nocodazol and colchicine on RET (data not

shown) are similar. In cultures tested with either drug, the biphasic change in RET normally seen after the addition of calcium is not observed. Instead, a monotonic increase in RET takes place. Thus, disruption of microtubules not only prevents the normally observed clustered-to-disperse transition in Con A receptor topography, but it results in increased receptor clustering. These results suggest that microtubules are involved in the regulation of Con A surface topography during myoblast fusion.

Modulation of Con A Receptor Mobility by Membrane Fluidity

We have shown previously that, coincident with the onset of fusion activity, a significant decrease in the apparent rotational correlation time of membrane-bound 1-anilino-8-naphthalene sulfonate (ANS) occurs in cultured myogenic cells labeled with this probe. This increase in ANS rotational mobility is associated with an increase in membrane fluidity (Herman and Fernandez 1978). Prives and Shinitzky (1977) have also reported similar changes in membrane microviscosity in these cells. The temporal relationships among the changes in ANS rotational mobility, Con A RET, and fusion activity are shown in Figure 3.

To investigate the possibility that the translational diffusional mobility of membrane components may be regulated or at least modulated by such

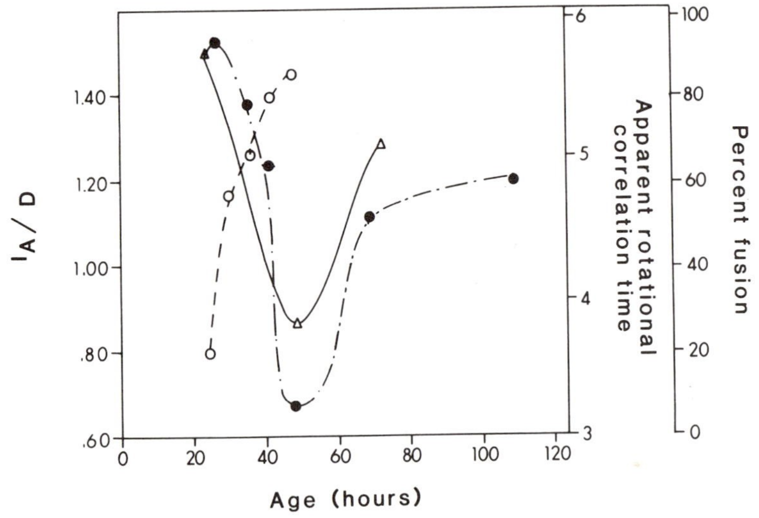

FIGURE 3
Composite graph illustrating the time course of fusion (O), the apparent rotational correlation time of membrane-bound ANS (Δ), and the extent of RET (●) in myogenic cells of different times during development in vitro. Data shown represent average values of several experiments.

alterations in lipid dynamics, we have also studied the short-term kinetics of ligand-induced clustering of Con A receptors by tetravalent Con A at three different times during the cycle of change in fluidity; namely at 24 hours, 48 hours, and 72 hours in culture. Figure 4 illustrates the results of one such experiment, in which the time course of increase in RET is measured. These

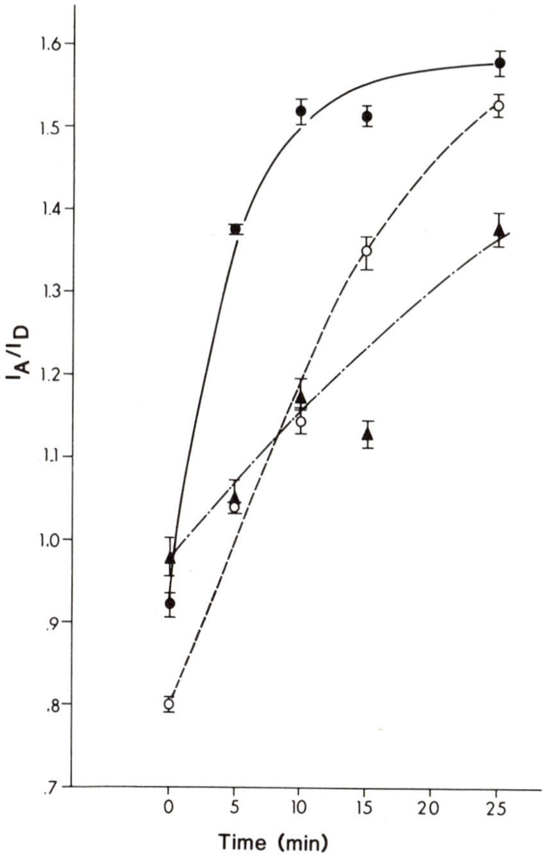

FIGURE 4
Time course of changes in RET (I_A/I_D) during ligand-induced aggregation of Con A on the surface of myogenic cells at 24 hr (○), 48 hr (●), and 72 hr (▲) in culture. Cells were labeled with pyrene- and FITC-conjugated Con A at the usual D-A ratio of 1:6.8 for 1 hr at 4°C. The total concentration of Con A was 20 μg ml^{-1}. At the end of 1 hr incubation, the cells were placed at 37°C for various periods of time and subsequently fixed with 2% paraformaldehyde for 20 min at 37°C. Time zero on the abcissa refers to cells fixed immediately after labeling at 4°C.

data show that ligand-induced aggregation of Con A receptors occurs most rapidly at 48 hours in culture, a time when the apparent rotational diffusion of lipid-bound ANS is also highest (Fig. 3). At 24 hours and at 72 hours, the kinetics of ligand-induced clustering are significantly slower than that at 48 hours, a behavior that correlates with the slower rotational diffusion of ANS at these time points. These results thus provide a direct correlation between the rotational diffusional mobility of a lipid-bound probe and the lateral translation mobility of a heterogeneous class of surface glycoproteins and suggest that the endogenous developmentally regulated change in membrane fluidity associated with myoblast fusion serves to modulate the mobility and display of cell-surface components.

SUMMARY

These results indicate that a dramatic reorganization of the muscle cell surface occurs during the period of myoblast fusion. Prior to fusion activity, Con A receptors exist predominantly in microclusters. During the period of fusion, this pattern changes to one in which receptors are mostly disperse, and after fusion is completed, a transition back to a more clustered distribution takes place. Our data also suggest that microtubules and membrane fluidity play a role in the regulation of these topographical changes.

ACKNOWLEDGMENTS

This research was supported by National Institutes of Health grant GM-23235.

REFERENCES

Albertini, D.F., R.D. Berlin, and J.M. Oliver. 1977. The mechanism of concanavalin A formation in leukocytes. *J. Cell Sci.* **26:** 57.

Den, H., D.A. Malinzak, H.J. Keating, and A. Rosenberg. 1975. Influence of concanavalin A, wheat germ agglutinin and soybean agglutinin on the fusion of myoblasts *in vitro*. *J. Cell Biol.* **67:** 826.

Fernandez, S.M. and R.D. Berlin. 1976. Cell surface distribution of lectin receptors determined by resonance energy transfer. *Nature* **264:** 411.

Forster, T. 1965. Delocalized excitation and excitation transfer. In *Modern quantum chemistry* (ed. O. Sinanoglu), part III, p.93. Academic Press, New York.

Gartner, T.K. and T.R. Podleski. 1976. Evidence that membrane bound lectin mediates fusion of L6 myoblasts. *Biochem. Biophys. Res. Comm.* **67:** 972.

Herman, B.A. and S.M. Fernandez. 1978. Changes in membrane dynamics associated with myogenic cell fusion. *J. Cell Physiol.* **94:** 253.

Horwitz, A.F., A. Wight, K. Knudsen. 1979. A role for lipid in myoblast fusion. *Biochem. Biophys. Res. Commun.* **86:** 514.

Horwitz, A.F., A. Wight, P. Ludwig, and R. Cornell. 1978. Interrelated lipid alter-

ations and their influence on the proliferation and fusion of cultured myogenic cells. *J. Cell Biol.* **77:** 334.

Kalderon, N. and N.B. Gilula. 1979. Membrane events involved in myoblast fusion. *J. Cell Biol.* **81:** 411.

Nowak, T.P., P.L. Haywood, and S.H. Barondes. 1976. Developmentally regulated lectin in embryonic chick muscle and a myogenic cell line. *Biochem. Biophys. Res. Commun.* **68:** 650.

Prives, J. and M. Shinitzky. 1977. Increased membrane fluidity precedes fusion of muscle cells. *Nature* **268:** 761.

Quiocho, F.A., G.N. Reeke, J.W. Becker, W.N. Lipscomb, and G.M. Edelman. 1971. Structure of concanavalin A at 4 Å resolution. *Proc. Natl. Acad. Sci.* **68:** 1853.

Rosen, S.D., R.W. Reitherman, and S.H. Barondes. 1975. Distinct lectin activities from six species of cellular slime molds. *Exp. Cell Res.* **95:** 159.

Sandra, A., M.A. Leon, and R.J. Przybylski. 1977. Suppression of myoblast fusion by concanavalin A: A possible involvement of membrane fluidity. *J. Cell Sci.* **28:** 251.

Teichberg, V.I., I. Silman, D.D. Beitsch, and G. Resheff. 1975. A β-D-galactoside binding protein from electric organ tissue of electrophorus electricus. *Proc. Natl. Acad. Sci.* **75:** 1383.

Lectin-resistant Myoblasts

*BRIAN M. GILFIX**
BISHNU D. SANWAL
Department of Biochemistry
University of Western Ontario
London, Canada N6A 5C1

The plasma membrane of the mammalian cell is considered to be involved in diverse aspects of cellular interactions (Edelman 1976; Baker and Ling 1979). Among the various components of the membrane, the complex carbohydrate-bearing structures (primarily glycoproteins) have been suggested to be involved in the phenomena of cell-cell recognition (Roseman 1970; Hughes 1974), although the molecular basis for such interactions is unclear.

One biological system in which the role of protein-bound carbohydrates in differentiation can be conveniently studied is provided by skeletal myoblasts (Sanwal 1979; Pearson 1980). During differentiation, the myoblasts undergo fusion to form multinucleate myotubes, a process that obviously involves some initial recognition event between individual myoblasts (Knudsen and Horwitz 1977, 1978). That cell-surface carbohydrates are probably involved in myoblast recognition is suggested by reports that plant lectins (Den et al. 1975) and β-galactosides (Gartner and Podleski 1975) inhibit myotube formation. An endogenous galactose-specific lectin has also been reported that increases during myoblast differentiation (Nowak et al. 1976). However, other workers have failed to observe the inhibition of differentiation by β-galactosides (Den et al. 1976; Knudsen and Horwitz 1978; Kaufman and Lawless 1980) or exogenously added purified muscle lectin (MacBride and Przybylski 1980; Den and Chin 1981).

These contradictory results emphasize the difficulty in assigning a function to any component on the basis of the use of inhibitors of uncertain specificity. A more powerful approach is one borrowed from classical microbiology. It involves the isolation of mutants with defined biochemical

*Present address: Department of Physiology, Harvard University Medical School, Cambridge, Massachusetts 02138.

alterations and the observation of the behavior of such mutants (Trowbridge et al. 1978). The use of cytotoxic plant lectins has previously allowed the selection of mammalian cell mutants with structural alterations in the carbohydrate groups of cell-surface glycoproteins and glycolipids (Stanley 1980). Therefore, to define the role of cell-surface carbohydrates in myogenesis, we sought to isolate several lectin-resistant (LEC^R) mutants from a permanent rat skeletal-muscle cell line in the hope that some of them may have alterations in their cell-surface carbohydrate structures, which may affect their behavior in terms of myotube formation.

RESULTS AND DISCUSSION

Isolation and Characterization of LEC^R Myoblasts

By using standard techniques, it proved possible to select LEC^R myoblasts that could be placed into four phenotypic groups named after the lectin used in their selection: (1) concanavalin A resistant (Con A^R), phytohemagglutinin resistant (PHA^R), and wheat germ agglutinin resistant, type I (WGA^{RI}) and type II (WGA^{RII}). The mutants were stable in the absence of the selective agent, and their frequency of appearance increased after mutagenesis. The LEC^R lines, like their parent L6, were near tetraploid (69–72 chromosomes), which may explain the necessity for mutagenesis before their recovery (Siminovitch 1976). Table 1 lists the relative resistance of the LEC^R lines in comparison with the wild-type L6.

The Con A^R mutants from L6 , like those from Chinese hamster ovaries

TABLE 1
Characterization of the Phenotype of Different LEC^R Lines

Cell line	RD_{10} Con A	LCA	PGA	RIC	DT	WGA	Fusion (%)
L6	1 (10.7)[a]	1 (12.1)[a]	1 (22)[a]	1 (1.4)[b]	1 (47)[b]	1 (1)[a]	90
Con A^R	1.7	2.4	0.7	0.0	0.7	0.5	0
PHA^R	0.8	1	5	2.3	1.9	8.5	0
WGA^{RII}	1	1	>3	0.8	1	9	0
WGA^{RI}	<1	<1	<4	29.8	0.4	65	42

A permanent line of rat skeletal-muscle myoblasts, L6 (isolated by Yaffe 1968), able to differentiate in culture, was used for the isolation of WGA, PHA, and Con A mutants. Mutants were selected by recovering surviving colonies, after plating ethylmethanesulfonate-treated cells with sufficient lectin to reduce survival to 10^{-6}, and recloned by limit dilution prior to analysis. Survival curves and fusion indices were determined as described previously (Ng et al. 1977; Gilfix and Sanwal 1980). D_{10} is the concentration of lectin that reduces survival to 10%. RD_{10} is the D_{10} of the mutant divided by the D_{10} of the wild-type L6.

[a] D_{10} in μg/ml.
[b] D_{10} in ng/ml.

(CHO) (Stanley and Siminovitch 1977), were of low resistance to Con A (Table 1) but, unlike the Con A^R mutants from CHO, they were not cross-resistant to sialic acid (WGA) or galactose-binding lectins (ricin [RIC], PHA). The PHA^R mutants from L6 were also distinct from those described previously in terms of cross-resistances and biochemical lesions. The Con A^R and PHA^R mutants from L6, e.g., had wild-type *N*-acetylglucosaminyl-transferase activity.

The WGA^R mutants were found to fall into either a low-resistance or a high-resistance group (Table 1). The low-resistance mutants, WGA^{RII}, although hypersensitive to the galactose-binding lectin RIC, were not deficient in sialic acid content (vide infra). The high-resistance WGA^R mutants, WGA^{RI}, were cross-resistant to galactose-binding lectins (PHA, RIC) but were hypersensitive to *N*-acetylglucosamine (GlcNAc)-binding lectins, such as Con A, *Lens culinaris* agglutinin (LCA), and diphtheria toxin (DT), suggesting that they probably possessed exposed GlcNAc residues. As became apparent, WGA^{RI} lines were similar in lectin sensitivity to a PHA^R lymphoma line (Trowbridge et al. 1978) and a WGA^R CHO line described previously (Briles et al. 1977). Of the LEC^R mutants isolated, only isolates and clones of the WGA^{RI} phenotype (but not WGA^{RII}) were observed to form myotubes. This occurred both in the presence and absence of 10 μg/ml of WGA.

In an effort to understand the molecular basis of the difference in the differentiation properties of the two classes of WGA^R mutants and, thus, hopefully, to identify the glycoproteins responsible for fusion, we undertook three kinds of studies. In one, we measured the reversible and irreversible binding (i.e., endocytosis) of WGA to these lines. In the second study, we analyzed the surface glycopeptides of the wild-type and mutant cells. In the third, we measured the levels of several metabolites in vivo involved in glycoprotein synthesis.

WGA Binding

Binding was characterized in terms of apparent affinity ($X_{1/2}$) and maximum binding ($\overline{B}$). As seen from Table 2, WGA^{RI} bound less WGA but with a higher affinity than the wild type, both reversibly and irreversibly. Presumably, this reflects the replacement of the terminal sialic acid residues by less numerous but higher affinity-terminal GlcNAc residues (Bhavanandan and Katlic 1979). In contrast, WGA^{RII} seemed to have a slight increase in WGA-binding sites, with no significant change in affinity when reversible binding was measured. However, on measuring the irreversible binding of WGA (i.e., endocytosis), WGA^{RII} displayed a marked decrease in affinity for this lectin. This suggests that WGA^{RII} may be an endocytosis-defective type of mutant, much like mutants of A-431 cells (Anderson et al. 1981) and human fibroblasts (Brown and Goldstein 1976), which internalize lipoproteins inefficiently.

TABLE 2
Binding of ^{125}I-labeled WGA to Myoblast Lines

		Binding parameters[b]	
Cell line	Mode of binding[a]	$X_{1/2}$ (μg/ml)	$\overline{B}$ (μg/10^6 cells)
L6	R	96.9 ± 5.7	30.1 ± 2.3
	I	30.6 ± 2.8	3.9 ± 0.57
WGA^{RII}	R	117 ± 17	43.5 ± 10.2
	I	77.3 ± 13.6	3.93 ± 0.42
WGA^{RI}	R	24.6 ± 4.8	9.05 ± 0.95
	I	19.1 ± 5.3	1.83 ± 0.52

[a]R denotes reversible and I, the irreversible binding of WGA. R binding of radioiodinated WGA to whole cells at 4°C and I binding at 37°C were performed by modifications of the procedures described by Meager et al. (1976) and Sandvig et al. (1978), respectively. Reversibly bound WGA was defined as the amount bound that could be eluted by 0.2 M GlcNAc. Irreversibly bound WGA was the amount not eluted by GlcNAc. Controls to measure nonspecific binding included GlcNAc.

[b]$X_{1/2}$ is the lectin concentration that gives one-half maximal binding. $\overline{B}$ is the amount of lectin bound at saturation. Both values were determined from least-square analysis of double-reciprocal plots; the error was determined by the "worst slope" method.

Analysis of Cell-surface-associated Glycopeptides

We have isolated cell-surface-associated glycopeptides that had been labeled metabolically with a variety of radioactive monosaccharides from the L6 and the WGA^R mutants. The glycopeptides were analyzed by sequential glycosidase digestion and Bio-Gel P chromatography (Kobata 1979; Natowicz and Baenziger 1980). Some of these results are given in Tables 3 and 4.

L6, WGA^{RII}, and WGA^{RI} had similar high mannose glycopeptides, $Man_8GlcNAc_2$, $Man_7GlcNAc_2$, and $Man_6GLcNAc_2$, as identified by their sensitivities to α-mannosidase, endogylcosidase H, and endoglycosidase CII, and by their molecular mass on Bio-Gel P-4 (400 mesh). The complex glycopeptides from WGA^{RII} were similar to L6 except that they were resistant to hydrolysis by *Clostridium perfringens* sialidase, unless pretreated with base. This is illustrated by the relative inability of *Aspergillus niger* β-galactosidase to release β-galactose from glycopeptides from WGA^{RII}-6C when *Clostridium* sialidase rather than acid was used to remove terminal sialic acid (Table 3). The sialic acid that appeared to be present in L6 and WGA^{RII} was primarily *N*-acetylneuraminic acid (NeuNAc).

Upon chromatography of the isolated glycopeptides on Bio-Gel P-10, it was found that the complex glycopeptides ran with a K_d of 0.2–0.4, whereas the high mannose glycopeptides ran with a K_d of 0.4–0.5 (Gilfix and Sanwal

TABLE 3

Sequential Exoglycosidase Digestion of Galactose Glyocopeptides

Treatment[a]	Radioactivity released (%) L6	WGA^{RII}
A. niger β-galactosidase	17.5	27.5
C. perfringens sialidase/ *A. niger* β-galactosidase	42.0	31.6
H_2SO_4/*A. niger* β-galactosidase	46.6	48.8

[a][6-^{3}H]galactose-labeled glycopeptides (10^5 dpm) were incubated with one of the following mixtures at 37°C. (1) Acid hydrolysis: 0.1 N H_2SO_4 (45 μl) at 30°C for 1 hr and then neutralized with 5 μl of 1 N NaOH; (2) *C. perfringens* sialidase and *A. niger* β-galactosidase: a mixture of sialidase (0.3 units) and β-galactosidase (0.5 units) in 54.6 mM citric acid/90.9 mM Na_2HPO_4/0.5% (w/v) NaN_3 at pH 4.5 (100 μl) for 72 hr; (3) *A. niger* β-galactosidase: 0.043 units of enzyme in 69.7 mM citric acid/60.6 mM Na_2HPO_4/0.1% (w/v) NaN_3/0.2 M GlcNAc at pH 3.5 (100 μl) for 24 hr. Enzyme digestions were terminated by heating the sample for a few minutes in a boiling water bath. The digest was loaded (along with dextran T500 and mannose to act as void and included volume markers, respectively) onto a Bio-Gel P-6 (200–400 mesh) column (1 × 120 m). 0.75-ml fractions were collected and analyzed for radioactivity. The percentage of total radioactivity eluting with the mannose peak ($K_d=1$) is given. $K_d = (V_{sample} - V_{dextran})/(V_{mannose} - V_{dextran})$, where V is the elution volume of the indicated species.

TABLE 4

Incorporation of Radioactive Monosaccharides into Glycopeptides

	Radioactivity (dpm × $10^{-3}/10^6$ cells)					
	L6[a]		WGA^{RII}[a]		WGA^{RI}[b]	
Label	(dpm)	(%)[c]	(dpm)	(%)[c]	(dpm)	(%)[c]
[2-^{3}H]mannose	9.23	(27.5)	23.1	(19.3)	52.9	(67.9)
[6-^{3}H]galactose	37.6	(50.3)	47.5	(32.6)	14.9	(14.9)
[^{3}H]*N*-acetylmannosamine	1.63	(30.5)	1.92	(23.3)	0.31	(3.3)
[6-^{3}H]glucosamine	74.2	(31.9)	119	(36.0)	382	(42.3)

Glycopeptides were metabolically labeled with radioactive monosaccharides and prepared as described previously (Gilfix and Sanwal 1980). Glycopeptides were labeled with 5 μCi/ml of one of the following: [2-^{3}H]mannose (14.1 Ci/mmole), [6-^{3}H]galactose (9.1 Ci/mmole), [6-^{3}H-ManN]ManNAc (19 Ci/mmole) and [6-^{3}H]GlcN (19 Ci/mmole). The glycopeptides were analyzed on Bio-Gel P-4, P-6, and P-10 columns (1 × 120 cm) (Hubbard and Robbins 1979). K_d is defined in the footnote to Table 3.

[a]In material with K_d of 0.2–0.4 on Bio-Gel P-10.

[b]In material with K_d of 0.4–0.5 on Bio-Gel P-10.

[c]Percentage of total radioactivity recovered.

1980). Although the total yield of glycopeptides varied, each region represented a highly reproducible percentage of the radioactivity loaded onto a column. Although WGA^{RII} showed similar elution profiles to L6, WGA^{RI} was deficient in sialic acid and galactose as measured by labeling with $[^3H]$galactose and $[^3H]$ManNAc (Table 4). The complex glycopeptides from WGA^{RI} were shifted to a lower molecular mass similar to that of the high mannose glycopeptides. Sequential glycosidase digestion of $[^3H]$mannose-labeled glycopeptides and analysis on Bio-Gel P-4 (400 mesh) columns revealed that L6 and WGA^{RII} had glycopeptides of the structure $(NeuNAc\alpha Gal\beta GlcNAc\beta)_\eta Man_3GlcNAc(\pm fucose)GlcNAc$, whereas those of WGA^{RI} were $(GLcNAc\beta)_\eta Man_3GlcNAc(\pm fucose)GlcNAc$.

These results were found to conform with the SDS-polyacrylamide gel electrophoresis patterns obtained with both $[^3H]$mannose-labeled and ^{125}I-labeled glycoproteins (not shown). The patterns from L6 and WGA^{RII} were identical, whereas that from WGA^{RI} revealed a shift of several bands to lower molecular mass reflecting loss of carbohydrate from its surface oligosaccharides.

Biosynthesis of Glycoproteins

To pinpoint the biochemical lesion producing the observed alterations in carbohydrate structure, a number of metabolites involved in glycoprotein synthesis were examined. WGA^{RI} and WGA^{RII} showed no loss or decrease in any of the following as compared to L6: UDP-hexose, UDP-hexosamine and CMP–sialic-acid pools; uptake of UDP-galactose into microsomes; glycosaminoglycan synthesis; GlcNAc transfer to ribonuclease B and desialylated, degalactosylated, dehexosaminylated α-acid glycoprotein and galactose transfer to lactose; ovalbumin; and desialylated, degalactosylated fetuin. WGA^{RI} had a slight decrease, and WGA^{RII}, a slight increase in β-galactosidase activity. Therefore, there was no ready explanation to account for the inability of WGA^{RI} to add sialic acid and galactose to its glycopeptides. Unlike the CHO and mouse fibroblast systems (Stanley 1980), no alteration in *N*-acetylglycosaminyltransferase activity was discernible in L6.

CONCLUSIONS AND SUMMARY

We have demonstrated that it is possible to isolate LEC^R mutants of a permanent line of rat skeletal-muscle myoblasts. Some of these mutants are unable to differentiate (i.e., fuse to form myotubes) and probably, therefore, lack particular macromolecular components involved in the process of fusion. Thus far, we have studied two WGA^R lines: one sharing high resistance (WGA^{RI}), and another, low resistance (WGA^{RII}). As indicated by the pattern of cross-resistance to various lectins and confirmed by sequen-

tial exoglycosidase digestion, WGA^{RI} possessed GlcNAc-β-terminal glycopeptides, unlike the wild type (L6) and WGA^{RII} mutants, which possessed NeuNAcαGalβGlcNAcβ-terminal, cell-surface glycopeptides. Since WGA^{RI} mutants fuse normally, it is clear that the normal levels of sialic acid and galactose residues on surface glycoproteins are not obligatory for the differentiation of myoblasts into myotubes. Our results also suggest (but do not prove) that the endogenous, galactose-binding lectin, shown to be present in differentiating myoblasts (Nowak et al. 1976) and postulated to be involved in the process of differentiation, is probably irrelevant to the mechanism of fusion. Further chemical studies of the several nonfusing LEC^{R} myoblast mutants now available will, no doubt, throw light on the nature of glycoproteins involved in differentiation.

ACKNOWLEDGMENTS

This work was supported by operational grants from the Medical Research Council and Muscular Dystrophy Association of Canada. B.M.G. was supported by scholarship grants from the National Science and Engineering Research Council and the National Cancer Institute of Canada.

REFERENCES

Anderson, R.G.W., M.S. Brown, and J.L. Goldstein. 1981. Inefficient internalization of receptor-bound low density lipoprotein in human carcinoma A-431 cells. *J. Cell Biol.* **88:** 441.

Baker, R.M. and V. Ling. 1979. Membrane mutants of mammalian cells in culture. *Methods Membr. Biol.* **9:** 337.

Bhavanandan, V.P. and A.W. Katlic. 1979. The interaction of wheat germ agglutinin with sialoglycoproteins. The role of sialic acid. *J. Biol. Chem.* **254:** 4000.

Briles, E.B., E. Li, and S. Kornfeld. 1977. Isolation of wheat germ agglutinin resistant clones of Chinese hamster ovary cells deficient in membrane sialic acid and galactose. *J. Biol. Chem.* **252:** 1107.

Brown, M.S. and J.L. Goldstein. 1976. Analysis of a mutant strain of human fibroblasts with a defect in the internalization of receptor-bound low density lipoprotein. *Cell* **9:** 663.

Den, H. and J.H. Chin. 1981. Endogenous lectin from chick embryo skeletal muscle is not involved in myotube formation *in vitro*. *J. Biol. Chem.* **256:** 8069.

Den, H., D.A. Malinzek, and A. Rosenberg. 1976. Lack of evidence for the involvement of a β-D-galactosyl-specific lectin in the fusion of chick myoblasts. *Biochem. Biophys. Res. Commun.* **69:** 621.

Den, H., D.A. Malinzek, H.J. Keating, and A. Rosenberg. 1975. Influence of concanavalin A, wheat germ agglutinin, and soybean agglutinin on the fusion of myoblasts *in vitro*. *J. Cell Biol.* **67:** 826.

Edelman, G.M. 1976. Surface modulation in cell recognition and cell growth. *Science* **192:** 218.

Gartner, T.K. and T.R. Podleski. 1975. Evidence that a membrane bound lectin mediates fusion of L6 myoblasts. *Biochem. Biophys. Res. Commun.* **67:** 972.

Gilfix, B.M. and B.D. Sanwal. 1980. Inhibition of myoblast fusion by tunicamycin and pantomycin. *Biochem. Biophys. Res. Commun.* **96:** 1184.

Hubbard, S.C. and P.W. Robbins. 1979. Synthesis and processing of protein-linked oligosaccharides *in vivo. J. Biol. Chem.* **254:** 4568.

Hughes, R.C. 1974. Membrane glycoproteins. Butterworth, London

Kaufman, S.J. and M.L. Lawless. 1980. Thiogalactoside binding lectin and skeletal myogenesis. *Differentiation* **16:** 41.

Knudsen, K.A. and A.F. Horwitz. 1977. Tandem events in myoblast fusion. *Dev. Biol.* **58:** 328.

——. 1978. Differential inhibition of myoblast fusion. *Dev. Biol.* **66:** 294.

Kobata, A. 1979. Use of endo- and exoglycosidases for structural studies of glycoconjugates. *Anal. Biochem.* **100:** 1.

MacBride, R.G. and R.J. Przybylski. 1980. Purified lectin from skeletal muscle inhibits myotube formation *in vitro. J. Cell Biol.* **85:** 617.

Meager, A., A. Ungkitchanukit, and R.C. Hughes. 1976. Variants of hamster fibroblasts resistant to *Ricinus communis* toxin (ricin). *Biochem. J.* **154:** 113.

Natowicz, M. and J.I. Baenziger. 1980. A rapid method for chromatographic analysis of oligosaccharides on Bio-Gel P-4. *Anal. Biochem.* **103:** 159.

Ng, S., J. Rogers, and B.D. Sanwal. 1977. Alterations in differentiation and pyrimidine pathway enzymes in 5-azacytidine resistant variants of a myoblast line. *J. Cell Physiol.* **90:** 361.

Nowak, T.P., P.L. Haywood, and S.H. Barondes. 1976. Developmentally regulated lectin in embryonic chick muscle and a myogenic cell line. *Biochem. Biophys. Res. Commun.* **68:** 650.

Pearson, M.L. 1980. Muscle differentiation in cell culture. A problem in somatic and molecular genetics. In *The molecular genetics of development: An introduction to recent research on experimental systems* (ed. T. Leighton and W.F. Loomis, Jr.), p. 361. Academic Press, New York.

Roseman, S., 1970. The synthesis of complex carbohydrates by multiglycosyltransferase systems and their potential function in intercellular adhesion. *Chem. Phys. Lipids* **5:** 270.

Sandvig, K., S. Olsnes, and A. Pihl. 1978. Binding, uptake and degradation of the toxic proteins abrin and ricin by toxin-resistant cell variants. *Eur. J. Biochem.* **82:** 13.

Sanwal, B.D. 1979. Myoblast differentiation. *Trends Biochem. Sci.* **4:** 155.

Siminovitch, L. 1976. On the nature of hereditable variation in cultured somatic cells. *Cell* **7:** 1.

Stanley, P. 1980. Surface carbohydrate alterations of mutant mammalian cells selected for resistance to plant lectins. In *The biochemistry of glycoproteins and proteoglycans* (ed. W.J. Lernarz), p. 161. Plenum Press, New York.

Stanley, P. and L. Siminovitch. 1977. Complementation between mutants of CHO cells resistant to a variety of plant lectins. *Somat. Cell Genet.* **3:** 391.

Trowbridge, J.S., R. Hyman, T. Ferson, and C. Mazauskas. 1978. Expression of Thy-l glycoprotein on lectin-resistant lymphoma cell lines. *Eur. J. Immunol.* **8:** 716.

Yaffe, D. 1968. Retention of differentiation potentialities during prolonged cultivation of myogenic cells. *Proc. Natl. Acad. Sci.* **61:** 477.

EMBRYOGENESIS

Introduction: Myoblast Commitment and the Embryogenesis of Skeletal Muscle

FRANK E. STOCKDALE
Stanford University School of Medicine
Department of Medicine
Stanford, California 94305

In a broad sense, the fundamental issue in the cell biology of developing muscle is whether there is a cellular basis for the development of heterogeneous muscle cell types found in the adult. In this section, the cellular basis for commitment of cells to the myogenic lineages during limb development is explored, as is the basis for the expansion in number of cells committed to myogenesis; also discussed are the contractile proteins synthesized by committed cells and the importance of cytoskeletal elements in the assembly of myofibrils.

Implicit in nearly all studies on the mechanisms of differentiation of embryonic skeletal-muscle cells (particularly those performed in cell culture) has been the assumption that there is a single lineage of myogenic cells in the developing embryo. However, is it not possible that the cellular heterogeneity of fully developed muscles is actually based upon diversity of myogenic cell types during embryonic development? It is unknown whether one or more lineages of myogenic cells is committed to produce daughter cells restricted to differentiate into specific muscle fibers of distinctive morphology, physiology, and/or biochemistry. Commitment means that a cell, and presumably its progeny, are restricted in the types of differentiation that they are able to express. It is not known whether the diverse cell types within a muscle emerge from a single or two or more committed cell lineages within the embryo. One can advance two hypotheses that are consistent with the diversity of cell types found in muscle. The first is that a single lineage of cells (myoblasts) emerges during embryonic development; the diver-

sity of fiber types characteristic of the adult arises by extrinsic modulation of this single cell lineage once the cells overtly differentiate into muscle fibers. The second is that the diversity of muscle fiber types, as well as the formation of satellite cells or muscle spindle cells, is based on commitment of these cells (myoblasts) during early development to diverse types of terminal differentiation. The distinction to be made between these hypotheses is that functional diversification occurs following overt differentiation in the former and before it in the latter. These hypotheses are not mutually exclusive, since modulation of cell function could occur following muscle fiber formation whether or not there were one or more myoblast cell lineages.

Holtzer and his associates have written extensively on the developmental lineage of myogenic cells (Holtzer 1978). They have provided a conceptual framework for the diversification of mesenchymal cells into myogenic, chondrogenic, and fibroblastic cell lineages. This framework principally concerns the formation of myoblasts, chondrocytes, and fibroblasts from mesenchymal cells and the delineation of a myoblast hierarchy in terms of proliferative potential. Diversification of cell lineages into myogenic and chondrogenic components, especially in the developing limb, undoubtedly occurs. However, it should be seriously considered that specification of muscle fiber type may be related to differences in myoblast commitment, and modulation of function following differentiation could be rooted in primary structural differences that are established as myoblasts become committed. Recent experimental evidence has been put forward that supports the possibility of commitment of mesenchymal cells to more than a single morphological myogenic cell lineage (Hauschka et al. 1977). These workers have cloned myoblasts from limb buds of both early human and avian fetuses. They find two basic types of myoblasts within the developing limb of these animals. One group, the "early" type of myoblast, requires conditioned medium for differentiation, whereas the "late" type of myoblast does not. These two types of myoblasts differentiate into myotubes (muscle fibers), which differ in morphology. Those requiring conditioned medium for differentiation are short and have very few nuclei. Those differentiating in unconditioned medium are large and branched and contain many nuclei. There is no evidence that these two distinctive myoblast populations differentiate into distinctive muscle fiber types in vivo, nor is it known whether the early type is a precursor to the late type.

Support for the notion that the clonal morphology of muscle fibers may represent different lineages of myoblasts has come from several experiments. Bonner (1978, 1980) has shown that denervation of the limb changes the types of clonable myoblasts that can be isolated from the chick limb. Furthermore, there is a specific period during development when innervation appears to be of importance for the appearance of this myoblast type. Cocultivation of myoblasts with neurons prior to this period can salvage this myoblast type. The observations by Hauschka et al. (1977) and Bonner

and colleagues (Bonner 1978, 1980) suggest that distinct myoblast populations exist and that interactions with other cells may be important in their diversification. However, the observations on myoblast clonal types and the requirements for their formation and differentiation need further refinement before they can be used as evidence to confirm the models proposed by Holtzer and his colleagues (Holtzer 1978) or the hypothesis outlined above.

An important question related to myogenic cell lineages is how proliferation and myotube formation are regulated within a myoblast cell lineage so that there is amplification of the committed cell population. Proliferation of committed myoblasts is very dependent on the ambient nutrient conditions in the cell culture (Yaffe 1973; Konigsburg 1977; Stockdale 1977). The principal ingredients influencing proliferation appear to be serum and embryo extract components (perhaps the somatomedins and/or platelet-derived growth factor) and other specific growth-promoting proteins, such as fibroblast growth factor (FGF) (Linkhart et al. 1980, 1981). The mechanisms by which cells committed to myogenesis either initiate cell fusion and the synthesis of muscle-specific cell products or continue to proliferate is dependent upon medium components and is not rigidly programmed within the myoblast (Konigsberg 1971; O'Neill and Stockdale 1972; Doering and Fischman 1974; Holtzer 1978; Nadal-Ginard 1978). Recent work by Linkhart and colleagues suggests that the presence of specific mitogens underlies the continued expansion of a myoblast population (Linkhart et al. 1981). In the presence of FGF and serum, committed myoblasts continue to proliferate, whereas in their absence, overt differentiation occurs. Recently, Lim and Hauschka (1981) have shown that a mitogen receptor disappears from the cell surface as myogenic cells differentiate. These observations on mitogen regulation of myoblast proliferation suggest mechanisms that may channel cells to proliferate or to express differentiated functions, as well as mechanisms for the uniform or selective expansion of cell lineages during the formation of a muscle.

As cells committed to myogenesis begin to fuse, do they express the functions of only one type of muscle fiber or more than one? If newly formed muscle fibers contain myosin isozymes of both fast- and slow-twitch type, perhaps a single myogenic lineage hypothesis would be more likely. Investigations of this type have become complex, since it has been demonstrated that embryonic forms of many of the myosin isozymes may precede the appearance of those characteristic of the adult (Dhoot and Perry 1979; Rushbrook and Stracher 1979; Whalen et al. 1979). However, the development of sensitive techniques for separation and detection of myosin isozymes has made such investigations feasible (Gauthier et al. 1978; Hoh 1979; Whalen et al. 1979; Cantini et al. 1980; Rubinstein and Kelly 1981).

Under conditions used for most avian muscle cell cultures, "late" myoblast differentiation is fostered and, therefore, some conclusion about this

myoblast type can be reached. Analyses of the types of myosin light chain synthesized in newly formed muscle fibers in cell culture indicate that "late" myoblasts form fibers that synthesize both fast and slow myosin (Cantini et al. 1980; Keller and Emerson 1981; Stockdale et al. 1981a,b). This is true whether the myoblasts are isolated from a region that, in the adult, will be either fast- or slow-type muscle. Furthermore, it is clear that these isozymes exist within the same fiber, whether innervation has occurred or is a possibility (in vivo studies) or whether it is an impossibility (cell-culture studies). Therefore, the "late" myoblast is a cell committed to form muscle fibers that synthesize both slow and fast myosin light-chain isozymes. This coexpression of isozyme types is also independent of innervation. The following questions have not been answered, however. Is the same observation on myosin isozymes true when myoblasts of earlier stages of development differentiate into muscle fibers? Could myogenic cell lineages differ quantitatively in the expression of myosin isozymes synthesized? Could myoblast lineages give rise to muscle fibers that specify different types of cell interactions (innervation) and, thus, control of plasticity or modulation that is seen later in development within the fiber?

One area of investigation that has received little attention is concerned with the mechanisms responsible for the assembly of myofibrillar structures and regulation of this assembly process. If the first myosins synthesized in a muscle fiber from late myoblasts are of the fast or slow type, these proteins are assembled into myofibrils independent of innervation and, perhaps, independent of myosin isozyme type. What is the cellular basis for the assembly of myofibrillar proteins into sarcomeric structures? Is this simply a process of self-assembly or do components of the cytoskeleton regulate this process? This question is discussed at length in another section of this volume, but the work of Holtzer and colleagues presented in this section suggests that two important roles may be played by cytoplasmic structures such as microtubules during the initial phases of overt differentiation of myogenic cells. The first role of this cytoplasmic structure is that it may establish the morphology of the emerging muscle fiber. In the presence of agents that interfere with polymerization of microtubular protein, the distinctive linear muscle fiber shape is lost (Bischoff and Holtzer 1967). Cells treated with colchicine become rounded and amorphous in configuration. Do these cytoplasmic structures have other roles? For example, are there factors in common between the observation that microtubular assembly is involved in maintaining the morphology of the muscle fiber and that the first myofibrils are assembled in the longitudinal axis of the cell? The use of drugs, such as taxol, which interfere with the disassembly of microtubules, offers opportunities for pursuing these questions. The observations reported here show that there is a close association between microtubules and the early assembly of thick filaments in newly formed muscle fibers (Holtzer et al., this volume).

SUMMARY

It is no longer sufficient to view muscle tissue as homogeneous in composition at either the biochemical, structural, or cellular level. The possibility is raised that, in part, the heterogeneity is based upon cells committed to different developmental lineages. The evidence for multiple lineages of myoblasts is scanty at best, and most evidence favors a single lineage with diversification being imposed, not at the myoblast level as required by a multiple-lineage model, but most often after the formation of a muscle fiber. For example, it is clear that both fast and slow myosin are synthesized in newly formed muscle fibers whether or not they are innervated, and only at later times can single types of myosin be found within a fiber. However, there is morphological evidence of diversification at the myoblast level and evidence that cell interactions can change the diversity of myoblast lineages. The role of environmental cues, in particular, mitogens, in fostering the expansion of such committed myogenic cell populations remains to be explored. It is far from clear that there is any basis for the assumption that the diverse cell types found in skeletal muscle emerge from separate myoblast lineages rather than the modulation of differentiated muscle fibers brought on by cell interactions. But the techniques for investigating this possibility are now available.

ACKNOWLEDGMENTS

These investigations were supported by a grant from the National Institutes of Health (AG-02822) and a grant from the Muscular Dystrophy Association of America.

REFERENCES

Bischoff, R. and H. Holtzer. 1967. The effect of mitotic inhibitors on myogenesis. *J. Cell Biol.* **36:** 111.

Bonner, P.H. 1978. Nerve-dependent changes in clonable myoblast populations. *Dev. Biol.* **66:** 207.

———. 1980. Differentiation of chick embryo myoblasts is transiently sensitive to functional denervation. *Dev. Biol.* **76:** 79.

Cantini, M., S. Sartore, and S. Schiaffino. 1980. Myosin types in cultured muscle cells. *J. Cell Biol.* **85:** 903.

Dhoot, G.K. and S.V. Perry. 1979. Distribution of polymorphic forms of troponin components and tropomyosin in skeletal muscle. *Nature* **278:** 714.

Doering, J.L. and D.A. Fischman. 1974. The *in vitro* cell fusion of embryonic chick muscle without DNA synthesis. *Dev. Biol.* **36:** 2251.

Gauthier, G.F., S. Lowey, and A.W. Hobbs. 1978. Fast and slow myosin in developing muscle fibers. *Nature* **274:** 25.

Hauschka, S.D., C. Haney, J.C. Angello, T.A. Linkhart, P.H. Bonner, and N.K. White. 1977. Clonal studies of muscle development: Analogies between human and chicken cells *in vitro* and their possible relation to muscular dystrophies. In *Pathogenesis of human muscular dystrophies* (ed. L. Rowland), p. 835. Excerpta Medica, Amsterdam.

Hoh, J.F.Y. 1979. Developmental changes in chicken skeletal myosin isozymes. *FEBS Lett.* **98:** 267.

Holtzer, H. 1978. Cell lineages, stem cells and the "quantal" cell cycle concept. In *Stem cells and tissue homeostasis* (ed. L. Lord et al.), p. 1. Cambridge University Press, London.

Keller, L.R. and C. P. Emerson 1981. Synthesis of adult myosin light chains by embryonic muscle cultures. *Proc. Natl. Acad. Sci.* **78:** 1020.

Konigsberg, I.R. 1971. Diffusion-mediated control of myoblast fusion. *Dev. Biol.* **26:** 133.

———. 1977. The culture environment and its control of myogenesis. In *Regulation of cell proliferation and differentiation* (ed. W.W. Nichols and D.G. Murphy), p. 105. Plenum Press, New York.

Lim, R.W. and S.D. Hauschka. 1981. Decrease in EGF binding accompanies the commitment and differentiation of a clonal mouse myoblast line MM14. *J. Cell Biol.* **91:** 198a.

Linkhart, T.A., C.H. Clegg, and S.D. Hauschka. 1980. Control of mouse myoblast commitment to terminal differentiation by mitogens. *J. Supramol. Struct.* **14:** 483.

———. 1981. Myogenic differentiation in permanent clonal mouse myoblast cell lines: Regulation by macromolecular growth factors. *Dev. Biol.* **86:** 19.

Nadal-Ginard, B. 1978. Commitment, fusion, and biochemical differentiation of a myogenic cell line in the absence of DNA synthesis. *Cell* **15:** 855.

O'Neill, M.C. and F.E. Stockdale. 1972. Differentiation without cell division in cultured skeletal muscle. *Dev. Biol.* **29:** 410.

Rubinstein, N.A. and A.M. Kelly. 1981. Development of muscle fiber specialization in the rat hind limb. *J. Cell Biol.* **90:** 128.

Rushbrook, J.I. and A. Stracher. 1979. Comparison of adult, embryonic and dystrophic myosin heavy chains from chicken muscle by sodium dodecyl sulphate polyacrylamide gel electrophoresis and peptide mapping. *Proc. Natl. Acad. Sci.* **76:** 4331.

Stockdale, F.E. 1977. Proliferative growth during myogenesisis *in vitro*. In *Regulation of cell proliferation and differentiation* (ed. W.W. Nichols and D.G. Murphy), p. 165. Plenum Press, New York.

Stockdale, F.E., H. Baden, and N. Raman. 1981a. Slow muscle myoblasts differentiating *in vitro* synthesize both slow and fast myosin light chains. *Dev. Biol.* **82:** 168.

Stockdale, F.E., N. Raman, and H. Baden. 1981b. Myosin light chains and the developmental origin of fast muscle. *Proc. Natl. Acad. Sci.* **78:** 931.

Whalen, R.G., K. Schwartz, P. Bouveret, S.M. Sell, and F. Gros. 1979. Contractile protein isozymes in muscle development: Identification of embryonic form of myosin heavy chain. *Proc. Natl. Acad. Sci.* **76:** 5197.

Yaffe, D. 1973. Rat skeletal muscle cells. In *Tissue culture: Methods and applications* (ed. P.F. Krus and M.K. Patterson, Jr.), p. 106. Academic Press, New York.

Myosin Light-chain Isozymes in Developing Avian Skeletal-muscle Cells in Ovo and in Cell Culture

FRANK E. STOCKDALE
MICHAEL T. CROW
PAMELA S. OLSON
Department of Medicine
Stanford University School of Medicine
Stanford, California 94305

Muscle fibers in the early embryos of birds and mammals arise from the fusion of mononucleated myoblasts in anatomic regions destined to become discrete muscle groups (Firket 1958; Capers 1960; Konigsberg et al. 1960; Stockdale and Holtzer 1961; Mintz and Baker 1967). One experimental approach to the question of whether or not muscles of a specific biochemical type in the adult originate from muscles in the embryo of the same biochemical type is to study these muscle groups when they first appear in the embryo. Avian pectoral muscle is a favorable muscle for such studies since, in the adult, it is composed almost exclusively of the fast myosin isozyme type. In addition, the muscle is easily dissected from the embryo and can be isolated cleanly from relatively young embryos. The first fibers begin to appear in the pectoral muscle at about 7–9 days of embryonic development in the chick, and from then on there is a rapid accumulation of myosin and other myofibrillar proteins (Herrmann et al. 1957; Marchok and Herrmann 1967).

A particularly good marker to study fiber specialization in developing muscle is the myosin light chain. Unlike the myosin heavy-chain subunits, which are present in distinct embryonic forms (Rushbrook and Stracher 1979; Whalen et al. 1979; Bandman and Strohman, pers. comm.), the light chains synthesized and accumulated by developing chick muscles appear

similar, by all experimental methods used, to those of the adult (Keller and Emerson 1981; Stockdale et al. 1981a,b). Furthermore, the light chains associated with the various isozyme types present a distinctive pattern that is easily identifiable by either one- or two-dimensional SDS-polyacrylamide gel electrophoresis.

The pectoral muscle of the adult chicken is a fast-twitch muscle that contains myosin characterized by three light chains of distinctive mobilities. Figure 1a is a two-dimensional gel electrophoretogram of extracts from adult pectoralis major showing the position of all three fast myosin light chains ($LC1_f$, $LC2_f$, and $LC3_f$). No myosin light chains typical of slow-twitch myosin could be detected in either crude or purified extracts of this muscle. If one removes tissue from the pectoral region of chick embryos at various ages, a different pattern of myosin light-chain composition is seen. From day 9 of development until at least day 14 (Stockdale et al. 1981b), both slow ($LC1_s$ and $LC2_s$) and fast myosin light chains ($LC1_f$ and $LC2_f$) are present (Fig. 1b). The third fast myosin light chain ($LC3_f$ is not seen until later stages of development (Fig. 3; Roy et al. 1979).

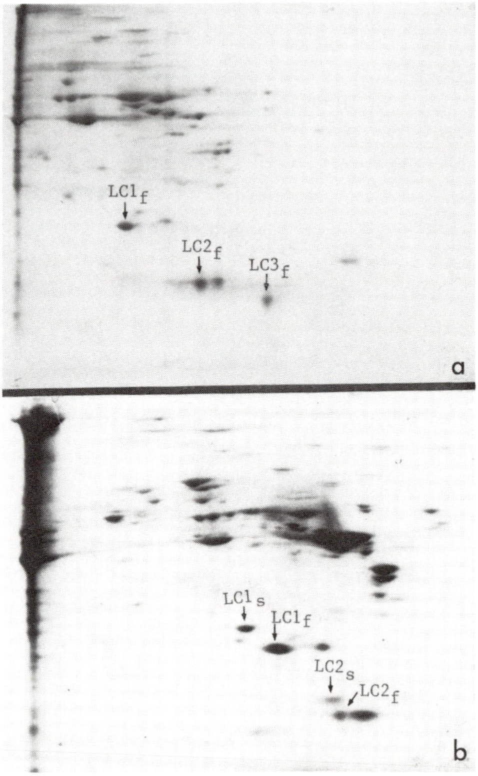

FIGURE 1
(*a*) Two-dimensional gel electrophoretogram of purified myosin (100 μg) from the pectoralis major of the adult chicken. (*b*) Two-dimensional gel electrophoretogram of myosin (80 μg) purified from 11-day pectoral muscle of the chick embryo. Purification and electrophoresis were performed according to the methods of Stockdale et al. (1981b). The gels were stained with Coomassie blue R250. The fast myosin light chains are indicated by arrows.

Hybridomas were developed that secreted monoclonal antibodies to the myosin light chains $LC1_s$ or $LC1_f$ and $LC3_f$. Figure 2 demonstrates the method used to define the specificity of these antibodies. Myosin purified from the medial adductor muscle of the adult chicken thigh was subjected to two-dimensional gel electrophoresis. This muscle is a mixed muscle containing both fast-twitch and slow-twitch muscle fibers (Fig. 2a). Proteins on this gel were then electrophoretically transferred to nitrocellulose (Burnette 1981) and exposed to monoclonal antibodies against either $LC1_f/LC3_f$ (Fig. 2b) or $LC1_s$ (Fig. 2c). The antibodies are specific to each myosin type and show no cross-reaction between types.

These monoclonal antibodies were used to determine the types of myosin present in developing pectoral muscle of the chick embryo. Extracts of myosin from early and late embryonic stages of the chick pectoral muscle were fractionated by one-dimensional gel electrophoresis (Fig. 3) and then transferred electrophoretically to nitrocellulose. The nitrocellulose sheets were exposed to either anti-$LC1_f/LC3_f$ (Fig. 3b) or anti-$LC1_s$ (Fig. 3c) antibodies, as described in the legend to Figure 2. At day 12 of development, both $LC1_f$ and $LC1_s$ were present in the myosin of pectoral muscle. By day 18 of development, however, only small amounts of $LC1_s$ could still be detected, and by day 19, no slow myosin light chains were detected. Myosin light chain $LC3_f$ appeared by day 16 of development. In this study and in other experiments, there was a small amount of $LC1_s$ detected on the day of hatching, which disappeared by the fourth day after hatching.

These observations suggest that muscles that initially form in presumptive fast muscle regions are those that synthesize both slow and fast myosin

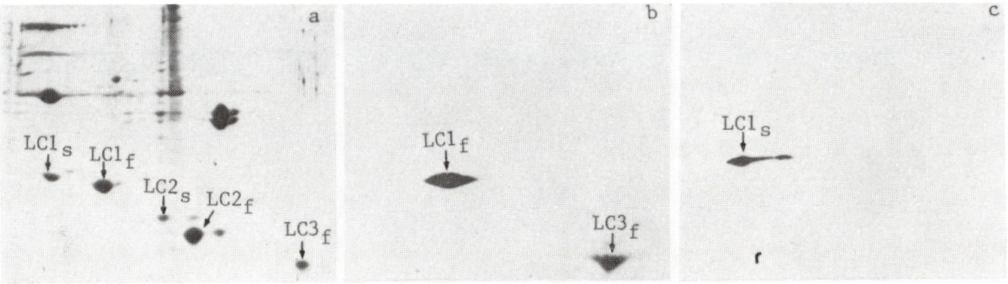

FIGURE 2

A two-dimensional gel electrophoretogram of myosin (10 μg) from the medial adductor muscle of the adult chicken. (*a*) Gel stained with Coomassie blue R250 to reveal both slow ($LC1_s$ and $LC2_s$) and fast ($LC1_f$, $LC2_f$, and $LC3_f$) myosin light chains. (*b*) Reaction of monoclonal antibody $IIIB_{10}$ (anti-$LC1_f/LC3_f$) with a replicate gel electrophoretically transferred to nitrocellulose (Burnette 1981). Binding was visualized by reacting the transfer with horseradish-peroxidase (HRP)-conjugated rabbit anti-mouse immunoglobulin and diaminobenzidine/H_2O_2 (Sternberger 1979). (*c*) Reaction of monoclonal antibody IIA_1 (anti-$LC1_s$) with a replicate gel electrophoretically transferred to nitrocellulose.

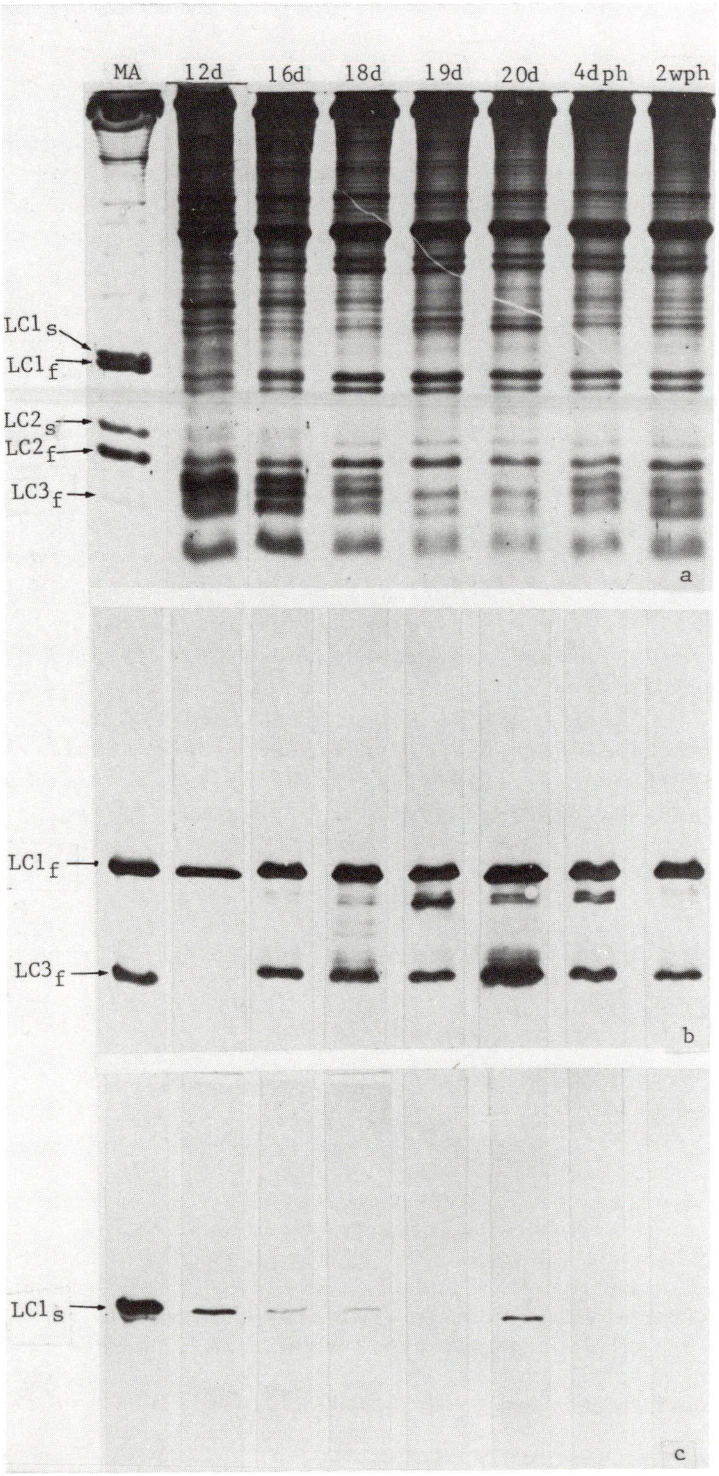

MA
12d
16d
18d
19d
20d
4dph
2wph
LC1s
LC1f
LC2s
LC2f
LC3f
a
LC1f
LC3f
b
LC1s
c

light chains. They further suggest that these light chains are immunologically similar to the adult myosin light chains and that synthesis of slow light chains in fast muscle regions stops in late embryonic development. These studies do not address the issue of whether fiber specialization arises from the repression of myosin synthesis within fibers that synthesize both fast and slow myosin light chains or from the selective cell death of specific fibers. Nor do these studies indicate the intrinsic potential of individual muscle fibers for the synthesis of either or both isozyme types and the modification of this potential by events during embryonic development, such as innervation.

One approach to this latter problem is to study the synthesis of these light chains in muscle fibers formed in cell culture. If one removes myoblasts from slow or fast muscle regions of the chick embryo and incubates them in vitro, they will fuse to form muscle fibers that contain skeletal-muscle myosin. Analysis of such cultures grown in the presence of [^{35}S]methionine using SDS–two-dimensional gel electrophoresis reveals that these fibers synthesize both slow and fast myosin light chains (Keller and Emerson 1981; Stockdale et al. 1981a). Figure 4a is an autoradiograph of a two-dimensional gel of myosin purified from a culture of muscle fibers formed from myoblasts from 14-day embryonic anterior latissimus dorsi (ALD) muscle (a predominantly slow muscle in the adult). Light chains $LC1_s$, $LC1_f$, $LC2_s$, and $LC2_f$ are apparent in myotubes formed in culture. Figure 4b is an autoradiograph of a two-dimensional gel from a culture of fibers formed from myoblasts isolated from 11-day embryonic pectoral muscle regions (a fast muscle in the adult). Light chains $LC1_s$, $LC1_f$, $LC2_s$, and $LC2_f$ are also present in these myotubes. Light chain $LC1_s$ is more difficult to identify in cultures from pectoral regions than in ALD cell cultures. Figure 4b is an autoradiograph of a two-dimensional gel of extracts from 11-day pectoral cultures in which the autoradiographs were exposed for 3 days; Figure 4c is the same gel exposed for 11 days. With longer exposure, $LC1_s$ becomes more apparent. These gels are not appropriate for resolving the third fast myosin light chain.

FIGURE 3 *(see facing page)*
One-dimensional gel electrophoretogram of whole cell extracts (35 μg) from chick embryonic pectoral muscle. The first lane is myosin (10 μg) from the medial adductor (MA) muscle of the adult showing both fast and slow myosin light chains. The other lanes are indicated by the age of the embryonic pectoral muscle used. (*a*) Coomassie-blue-stained gel. (*b*) Replicate gel transferred to nitrocellulose and reacted with monoclonal antibody $LC1_f/LC3_f$, using the methods indicated in the legend to Fig. 2. (*c*) Replicate gel transferred to nitrocellulose and reacted with a monoclonal antibody to $LC1_s$, using the technique indicated in the legend to Fig. 2. Note that both slow and fast myosin light chains are present in early embryonic pectoral muscle (12 days through 18 days) and that the third myosin light chain does not appear until approximately 16 days of development. (4dph) 4 days after hatching; (2wph) 2 weeks after hatching.

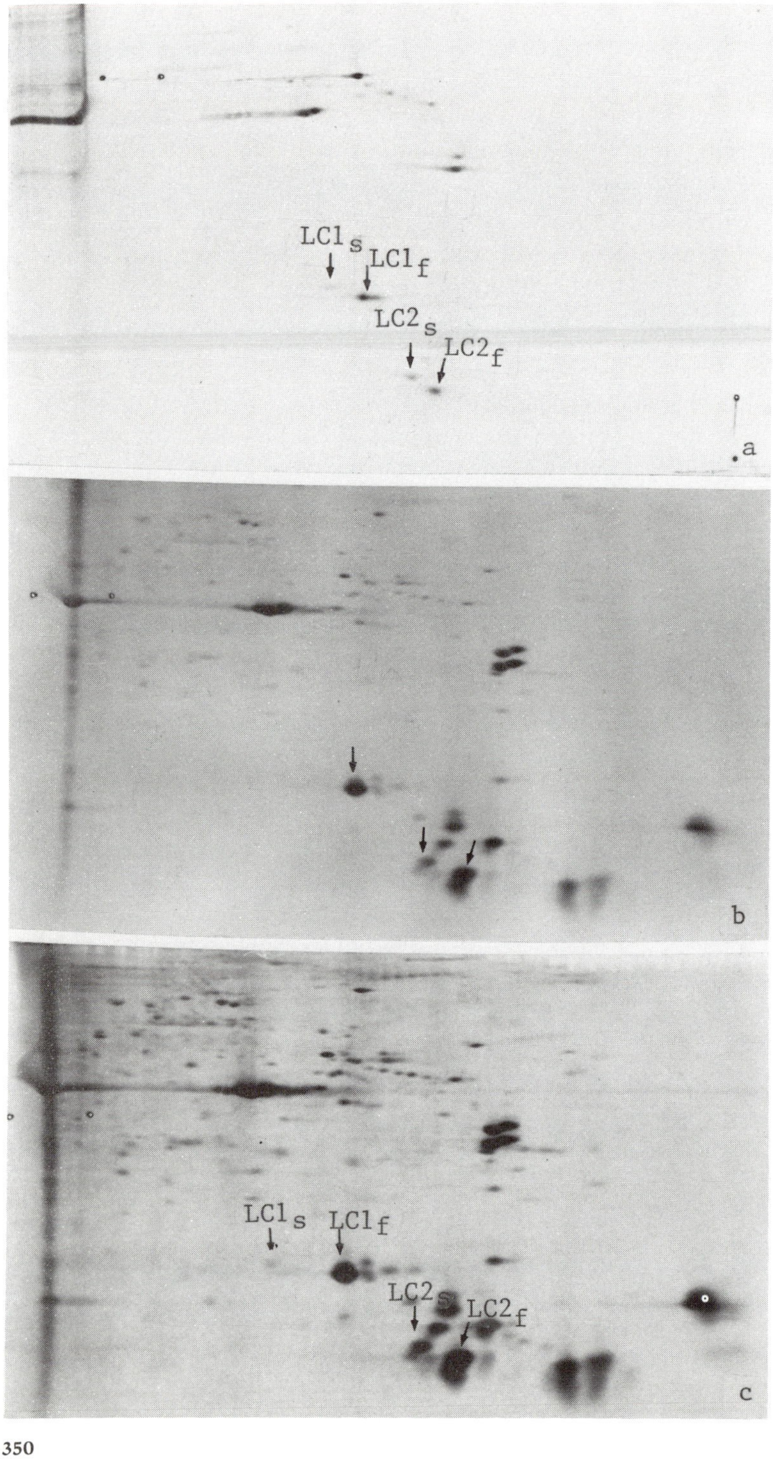
LC1s
LC1f
LC2s
LC2f
a
b
LC1s
LC1f
LC2s
LC2f
c

These observations and those of Keller and Emerson (1981) and Cantini et al. (1980) demonstrate that muscle fibers formed from myoblasts in tissue culture in the absence of innervation synthesize both fast and slow myosin light chains. Therefore, innervation per se is not required for the synthesis of either light-chain group. These observations differ from those reported by Rubinstein et al. (Rubinstein and Holtzer 1979; Rubinstein and Kelly 1980) and Streter et al. (1972), in that slow myosin light chains are found in myotube cultures derived from myoblasts from either slow or fast muscle regions.

With regard to the problem of isozyme expression in individual fibers, the work of Gauthier and Lowey (1979), as well as the recent study of Rubinstein and Kelly (1980), has demonstrated that even in the adult, both myosin types can be expressed within individual muscle fibers. Although our studies on the immunocytochemical localization of myosin in developing muscle fibers are still in progress, the staining patterns observed with monoclonal antibodies already reveal some differences from those obtained with antisera. The monoclonal-antibody staining profile for serial cross sections of adult chicken muscle is shown in Figure 5. Sections were stained with the same anti-$LC1_f/LC3_f$ or anti-$LC1_s$ monoclonal antibodies used in the nitrocellulose transfer experiments. Staining was visualized with HRP-conjugated rabbit anti-mouse secondary antibody. The fast-twitch pectoral muscle of the adult chicken stained only with anti-$LC1_f/LC3_f$ (Fig. 5a). No staining was evident with the anti-slow antibody at any dilution of this antibody (Fig. 5c). On the other hand, most of the fibers in the slow-tonic ALD muscle stained with both anti-fast and anti-slow antibodies (Fig. 5b,d). The fibers in the adult ALD stained more uniformly with anti-$LC1_s$ than with anti-$LC1_f/LC3_f$. Of the muscles examined in the adult, only the larger diameter slow fibers of the hind-limb medial adductor stained exclusively with anti-$LC1_s$ (M.T. Crow et al., unpubl.).

The staining pattern in the adult chicken ALD muscle is consistent with other observations made in this laboratory that fast myosin light chains are

FIGURE 4 (see facing page)
Autoradiographs of two-dimensional gels of myosin extracted from muscle fibers formed in tissue culture. (*a*) Myoblasts isolated from the ALD muscle of 14-day embryos and plated on gelatin-coated 60-mm dishes (O'Neill and Stockdale 1972); 4-day culture. (*b,c*) Myoblasts isolated from pectoralis muscle of 11-day embryos and plated as above; 4-day culture. On day 4, when myotubes were present, [^{35}S]methionine was added to the medium for 6 hr. The method of incubation and myosin extraction is described in Stockdale et al. (1981a). 10^6 cpm of [^{35}]methionine was added to the gels along with 7 μg of purified myosin adult ALD muscle. The gels were run, stained, and autoradiographed as described in Stockdale et al. (1981a,b). The autoradiograph in *b* was exposed for 3 days and then reexposed for 11 days (*c*). Proteins were identified by superimposition of the stained gel containing the carrier adult myosin.

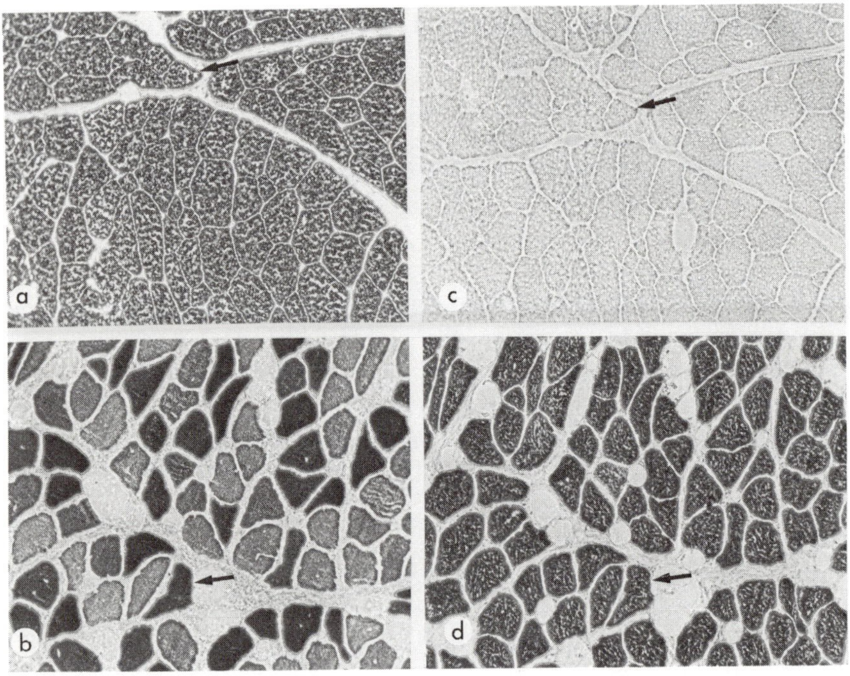

FIGURE 5

Transverse frozen sections of adult chicken pectoralis major (*a,c*) and ALD (*b,d*) muscles reacted with monoclonal antibodies. Serial sections were cut at a thickness of 10 μm and postfixed in ethyl-dimethylaminopropyl carbodiimide. Antibodies from hybridoma clones were applied at a dilution of 1:25. Binding was visualized following reaction with HRP-conjugated rabbit anti-mouse immunoglubulin for 4 hr at room temperature and then incubation with diaminobenzidine/H_2O_2 (Sternberger 1979). Endogeneous peroxidase was blocked with methanol/H_2O_2. (*a,b*) Reaction with anti-$LC1_f$/$LC3_f$ monoclonal antibody. (*c,d*) Reaction with anti-$LC1_s$ monoclonal antibody. (→) The position of the same muscle fiber in cross section. All photomicrographs were done under phase optics so that the fibers in *c* that do not react with anti-$LC1_s$ would be visible.

detectable in SDS-polyacrylamide gel extracts of the ALD both by silver staining and by hybridization with anti-$LC1_f$/$LC3_f$ on nitrocellulose. A similar immunocytochemical profile is also seen in the ALD using monoclonal antibodies to fast and slow myosin heavy chains (M.T. Crow et al., unpubl.), rendering less likely the possibility that these fibers only synthesize the fast light chains but do not assemble them into functional contractile units.

SUMMARY

In conclusion, our observations both in ovo and in cell culture suggest that all muscles, regardless of their adult phenotype, initially synthesize both

myosin light-chain isozyme types. It is only later that the more restricted patterns of myosin synthesis characteristic of the adult appear. These observations suggest a common origin for all muscles, whether of fast or slow type, from fibers that synthesize both fast and slow myosin. These studies, however, do not rule out that the variety of fiber types found within particular muscles of the adult could originate from different myogenic cell lineages. Immunological staining with monoclonal antibodies to myosin light chains reveals that fibers in the adult as in the embryo synthesize both types of myosin.

ACKNOWLEDGMENTS

These investigations were supported by a grant from the National Institutes of Health (AG-02822) and a grant from the Muscular Dystrophy Association of America. M.T.C. is a Muscular Dystrophy Association of America postdoctoral fellow.

REFERENCES

Burnette, W.N. 1981. "Western blotting." Electrophoretic transfer of proteins from sodium dodecyl sulphate-polyacrylamide gels to unmodified nitrocellulose and radiographic detection with antibody and radioiodinated Protein A. *Anal. Biochem.* **112:** 195.

Cantini, M., S. Sartore, and S. Schiaffino. 1980. Myosin types in cultured muscle cells. *J. Cell Biol.* **85:** 903.

Capers, C. 1960. Multinucleation of skeletal muscle *in vitro. J. Biophys. Biochem. Cytol.* **7:** 559.

Firket, H. 1958. Recherches sur la synthese des acides desoxyribonucleiques et la preparation a la mitose dans des cellules. *Arch. Biol.* **69:** 1.

Gauthier, G.F. and S. Lowey. 1979. Distribution of myosin isoenzymes among skeletal muscle fiber types. *J. Cell Biol.* **81:** 10.

Herrmann, H., B.N. White, and M. Cooper. 1957. The accumulation of tissue components in the leg muscle of the developing chick. *J. Cell. Comp. Physiol.* **49:** 227.

Keller, L. and C. Emerson. 1981. Synthesis of adult myosin light chains by embryonic muscle culture. *Proc. Natl. Acad. Sci.* **78:** 1020.

Konigsberg, I., N. McElvain, M. Tootle, and H. Herrman. 1960. The dissociability of deoxyribonucleic acid synthesis from the development of multinuclearity of muscle cells in culture. *J. Biophys. Biochem. Cytol.* **8:** 333.

Marchok, A.C. and H. Herrmann. 1967. Studies of muscle development. I. Change in cell proliferation. *Dev. Biol.* **15:** 129.

Mintz, B. and W.W. Baker. 1967. Normal mammalian muscle differentiation and gene control of isocitrate dehydrogenase synthesis. *Proc. Natl. Acad. Sci.* **58:** 592.

O'Neill, M.C. and F.E. Stockdale. 1972. Differentiation without cell division in cultured skeletal muscle. *Dev. Biol.* **29:** 410.

Roy, R.K., F.A. Sreter, and S. Sarkar. 1979. Changes in tropomyosin subunits and

myosin light chains during development of chicken and rabbit striated muscles. *Dev. Biol.* **69:** 15.

Rubinstein, N.A. and H. Holtzer. 1979. Fast and slow muscle in tissue culture synthesize only fast myosin. *Nature* **280:** 323.

Rubinstein, N.A. and A.M. Kelly. 1980. The sequential appearance of fast and slow myosins during myogenesis. In *Plasticity of muscle* (ed. D. Pette), p. 147. de Gruyter, Berlin.

Rushbrook, J.I. and A. Stracher. 1979. Comparison of adult, embryonic and dystrophic myosin heavy chains from chicken muscle by sodium dodecyl sulphate polyacrylamide gel electrophoresis and peptide mapping. *Proc. Natl. Acad. Sci.* **76:** 4331.

Sternberger, L.A. 1979. *Immunocytochemistry*, second edition. Wiley, New York.

Stockdale, F.E. and H. Holtzer. 1961. DNA synthesis and myogenesis. *Exp. Cell. Res.* **24:** 508.

Stockdale, F.E., H. Baden, and N. Raman. 1981a. Slow muscle myoblasts differentiating *in vitro* synthesize both slow and fast myosin light chains. *Dev. Biol.* **82:** 168.

Stockdale, F.E., N. Raman, and H. Baden. 1981b. Myosin light chains and the developmental origin of fast muscle. *Proc. Natl. Acad. Sci.* **78:** 931.

Streter, F.S., S. Holtzer, J. Gergely, and H. Holtzer. 1972. Some properties of embryonic myosin. *J. Cell Biol.* **55:** 586.

Whalen, R.G., K. Schwartz, P. Bouveret, S.M. Sell, and F. Gros. 1979. Contractile protein isozymes in muscle development. Identification of an embryonic form of myosin heavy chain. *Proc. Natl. Acad. Sci.* **76:** 5197.

The Activation of Myosin Synthesis and Its Reversal in Synchronous Skeletal-muscle Myocytes in Cell Culture

*BLYTHE H. DEVLIN**
PETER A. MERRIFIELD
IRWIN R. KONIGSBERG
Department of Biology
University of Virginia
Charlottesville, Virginia 22903

During the differentiation of multinucleated skeletal-muscle fibers in culture from populations of embryonic myoblasts, three temporally related events occur: (1) the fusion of individual myoblasts, (2) the initiation of synthesis of a battery of muscle-specific contractile proteins and associated enzymes, and (3) the cessation of DNA synthesis in the nuclei of the nascent, syncytial muscle fibers that form. The precise temporal order in which these events normally occur has not been rigorously established. In normal development in culture, they must occur either simultaneously or follow so closely in succession as to appear concomitant (Konigsberg 1977). Whether concomitant or not, it is quite clear that two of these component processes, fusion and the initiation of cell-type-specific protein synthesis, are not obligatorily coupled (Emerson and Beckner 1975; Holtzer et al. 1975; Moss and Strohman 1976; Turner et al. 1976; Vertel and Fischman 1976; Zubrzycka and MacLennan 1976).

When fusion is blocked by decreasing the Ca^{++} concentration in low-growth medium, myosin synthesis is initiated and proceeds at a rate (per nucleus) indistinguishable from the rate observed in cultures maintained under fusion-permissive conditions (Emerson and Beckner 1975). These single, differentiated skeletal myocytes, if they are not withdrawn from the

*Present address: Department of Microbiology and Immunology, Emory University, School of Medicine, Atlanta, Georgia 30322.

cell cycle, are at least cycling very slowly (Emerson 1977). Thus, the only parallel event common to the activation of contractile protein synthesis that is observed in both fusing and fusion-blocked cultures appears to be some restriction (or delay) of reentry into the DNA synthetic phase.

More recently, we have used immunofluorescence techniques to study the accumulation of muscle-specific myosin in individual myogenic cells in the absence of fusion to define more precisely the relationship between cell-cycle withdrawal and the initiation of the differentiative program. Our data are consistent with the hypothesis that, late in G_1, mechanisms exist that switch the myoblast from the pathway leading to DNA synthesis into a shunt, culminating in myogenic differentiation. Furthermore, our results indicate that, for a brief period of time at least, this switch to the differentiative shunt can be reversed.

RESULTS AND DISCUSSION

Initiation of Myosin Synthesis in the "Terminal" G_1

Embryonic quail (*Coturnix coturnix japonica*) myoblasts, collected in mitosis by gentle "shake-off" and plated in low-Ca^{++}, mitogen-poor medium, complete telophase (within 20 min), traverse one complete cycle of normal duration (total generation time $[G_T] \cong 12$ hr) with good synchrony, and enter a second G_1 (see Fig. 1A). If left unperturbed, these cells remain in this second G_1 for as long as 42 hours (the longest period sampled following 15-min pulses with [^{3}H]thymidine). During this protracted G_1, myosin synthesis was scored using an affinity-column-purified, rabbit polyclonal antibody to skeletal-muscle myosin, which does not cross-react with either quail skin fibroblasts or proliferating myoblasts. The antibody localizes to the A band of cross-striated muscle fibers and to myocytes exclusively. Synthesis is initiated after approximately 8 hours in G_1, with the percentage of myosin-positive cells reaching 50% after 14 hours in G_1 and exceeding 95% by 25 hours in G_1 (see Fig. 1B). Thus, virtually all of these myoblasts initiate myosin synthesis during the medium-imposed, protracted G_1.

If the cultures are refed growth-promoting medium after 50% of the myoblasts have initiated myosin synthesis, the population reenters S (Fig. 1A, dashed line). Reentry into S cannot be ascribed to fibroblast overgrowth, however. If refed cultures are maintained for an additional 40 hours, the dividing cells deplete the medium of mitogens, the labeling index declines, and myosin-positive cells begin to accumulate, reaching a maximum in excess of 95%. Since myoblasts in what would be their terminal G_1 can be stimulated to reenter the cell cycle, it is clear that the control of differentiative events resides in G_1 and that differentiation is not the consequence of a qualitatively different mitotic event preceding the terminal G_1 (Bischoff and Holtzer 1969).

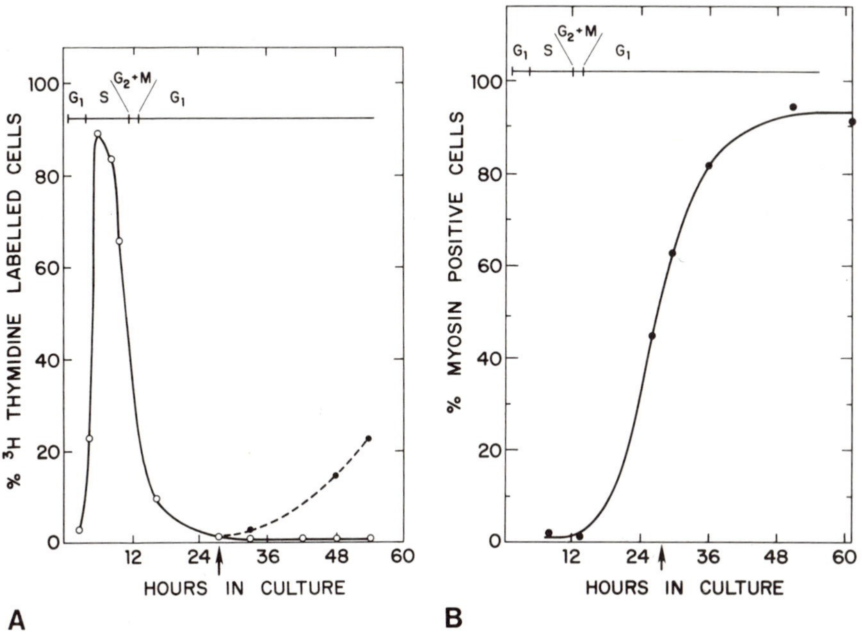

FIGURE 1

The labeling index (*A*) and the accumulation of immunofluorescent myosin-positive cells (*B*) in populations of synchronous myoblasts under growth-restrictive, fusion-impermissive conditions. (*A*) (○) Percentage of cells that incorporate [^{3}H]thymidine during a 15-min pulse at the times indicated. (●---●) Labeling index following a refeeding with growth-promoting, fusion-impermissive medium at the time indicated (→). (*B*) (●) Percentage of immunofluorescently labeled, myosin-positive cells scored at successive times during the terminal G_1 in sister cultures of those described in *A*. Muscle-specific myosin was detected by indirect immunofluorescence using a fluorescein-conjugated goat anti-rabbit immunoglobulin (IgG) to localize affinity-column-purified rabbit antimyosin.

Reentry of Differentiated Myocytes into the Cell Cycle

To determine whether only the undifferentiated cells can reenter S when cultures are refed (at the time when 50% of the cells are myosin-positive), double-labeling experiments were performed, assaying both DNA synthesis ([^{3}H]thymidine) and myosin content (immunofluorescence). In these studies, at three intervals after refeeding (6 hr, 12 hr, and 24 hr), a 15-minute pulse with [^{3}H]thymidine was administered immediately preceding fixation. Such cultures were first processed and scored for myosin-positive cells in randomly selected, marked fields. Following autoradiographic processing and development, these same fields were reexamined to

determine what percentage of the myosin-positive cells were in S (had incorporated [^{3}H]thymidine) during the final 15-minute pulse (Fig. 2). The average percentage of doubly labeled cells at all three time points was approximately 3%. A frequency of this magnitude might simply represent a small number of cells which, in some manner, escape the normal repression of proliferation in differentiated myocytes. Alternatively, it might reflect the low frequency of cells in S following a mitogenic stimulus as cells slowly and asynchronously progress to S. Kinetics of reentry of this kind are observed in stimulated, quiescent fibroblasts (e.g., after serum replacement of serum-deprived cells [Todaro et al. 1965]).

To test the thesis that myosin-positive cells are capable of DNA synthesis (reentry into S), measurements were made to determine the percentage of cells in S at each of the three time intervals at which the frequency of doubly labeled cells was determined. Synchronized myoblasts were cultured under conditions identical to those used in the double-labeling study. These cultures, however, were filmed using time-lapse cinematography and followed continuously for a total of 53 hours (i.e., from the completion of the initial mitosis of the plated, synchronized cohort through refeeding and an additional 32 hr beyond). Of the 74 cells followed, 37 entered and completed mitosis during filming. Each of these 37 cells is plotted in Figure 3, and from this data (see legend to Fig. 3 and Table 1) the percentages of all cells in S (black bar) during the three time points sampled were 5%, 15%, and 7%, respectively. If one assumes an equal probability of reentry into S for myosin-positive cells and undifferentiated cells, one arrives at an expected percentage of double-labeled cells close to the figure actually obtained (Table 1).

Compatible with the assumption of equal probability of reentry into S by myosin-positive and undifferentiated myoblasts, the time-lapse records indicate that irrespective of time spent in G_1 before refeeding (< 9 hr, between 12 hr and 15 hr, or >15 hr), the distribution of times between refeeding and completion of mitosis is the same. The correlation between the time spent in G_1 prior to refeeding and the transit time to mitosis was also tested by analysis of regression. Neither the slope (−0.24) nor the coefficient of determination ($r^2 = 0.04$) indicates any significant correlation. Since a larger proportion of the cells should be myosin-positive with increased time in G_1 (see Fig. 1B), the synthesis of myosin per se does not appear to effect the probability of transition into S. These results suggest that irreversible withdrawal from the cell cycle is neither an obligate precondition nor a direct consequence of the initiation of muscle-specific protein synthesis. The more likely explanation is that some consequence of the fusion event imposes more rigid control of DNA synthesis than is imposed by low-growth medium alone.

Half of the 74 cells that could be followed continuously in our time-lapse records failed to divide before filming was terminated. To extend the time period under observation, we used a clonal analysis to score the retention of

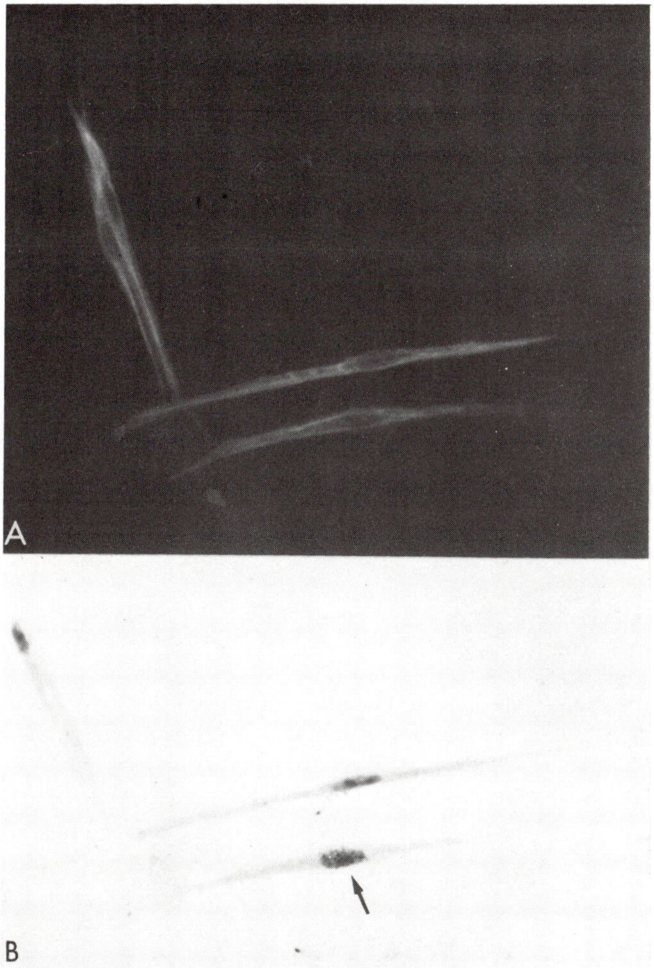

FIGURE 2
Incorporation of [^{3}H]thymidine into myosin-positive cells refed with growth-promoting, fusion-impermissive medium at time indicated in Fig. 1A (→). Cells were fixed at progressively longer times after refeeding (see text), processed, and randomly scored for myosin-positive cells. After autoradiographic processing, the previously scored fields were located and rephotographed. Cells that exhibited both labels (grains over the nucleus and myosin immunofluorescence) comprised ~3% of the myosin-positive cells (total of 359) examined. Of the four myosin-positive cells in *A*, one cell incorporated [^{3}H]thymidine during the 15-min pulse preceding fixation (→, *B*).

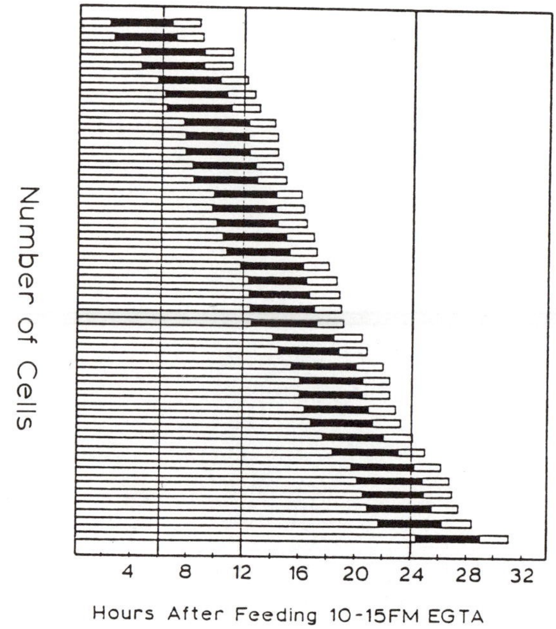

FIGURE 3
Determination of the percentage of cells in S at times corresponding to time points sampled following refeeding with high-growth, low-Ca^{++} medium (10–15 FM [minimum essential medium] EGTA) in the double-labeling experiment. The bar graph represents the interval between refeeding and mitosis of the 37 cells that completed mitosis during filming. Time of entry into S (6½ hr before mitosis) and time in S (4½ hr) were calculated from previous data (Buckley and Konigsberg 1974). Black portion of each bar represents traverse through S phase. Vertical lines at 6 hr, 12 hr, and 24 hr (the times sampled [see text]) intercept each horizontal bar. The ratio of cells in S to the total number of cells progressing to mitosis is determined by the ratio of interception of black and open regions of each bar. Both the percentage of cells in S (which eventually divided) and the percentage of all cells in S (as a function of all cells filmed) at each time sampled are given in Table 1.

proliferative capacity as a function of time in low-growth, low-Ca^{++} medium. If loss of the ability to proliferate were coupled to the initiation of differentiation, one would expect the decline in response to the mitogenic stimulus to occur during the time when the percentage of myosin-positive

TABLE 1
Cells in S at Intervals Sampled after Refeeding

	6 hr	12 hr	24 hr
Dividing cells (37) in S (%)	10	28	14
All cells (74) in S (%)	5	14	7
Double-labeled cells (%)			
expected	2.5	7	3.5
actual	0.9	3	4

cells is rising steeply. What we observe, in fact, is that the drop occurs before the detection of the first myosin-positive cell and remains constant at 50% throughout the period when the accumulation of myosin-positive cells increases to virtually 100% (see Fig. 4). This observation suggests that the loss of clonability is not related to the initiation of myosin synthesis but, perhaps, to marginal concentrations of components of the medium required to maintain physiological integrity. Whatever the nature of the block, these cells must also initiate myosin synthesis, since the percentage of myosin-positive cells approximates 100% (not 50%) under conditions (i.e., time-lapse cinematography) in which cell death, had it occurred, would have been noted.

To examine the kinetics of reentry into the cycle of fusion-blocked myocytes, transitional probability plots were constructed of the lapsed time (Fig. 5) between refeeding and mitosis (for the stimulated myocytes) and between birth and the subsequent mitosis of each of the daughter cells observed during filming. The daughter cell curve is exponential, as one would expect of rapidly dividing, log-phase cells (Konigsberg et al. 1978). The log plot of the original cohort, however, is curvilinear. The actual percentage of stimulated myocytes that have not yet entered mitosis plotted against time fits a straight line extremely well. What these entry kinetics appear to reflect is that mitogen-stimulated myocytes progress into mitosis from a different compartment in G_1 than do their log-phase daughters. These differentiated myocytes are blocked, perhaps, in a state analogous to the G_0 postulated for cultured cell lines deprived of any of a number of nutrients.

Myosin Synthesis in Mitogen-stimulated Myocytes

If we assume that muscle-specific myosin synthesis is initiated within the same compartment from which mitogen-stimulated myocytes reenter the cycle, we can ask whether myosin synthesis is interrupted during reentry and, if so, when. To monitor synthesis, mass cultures of differentiating

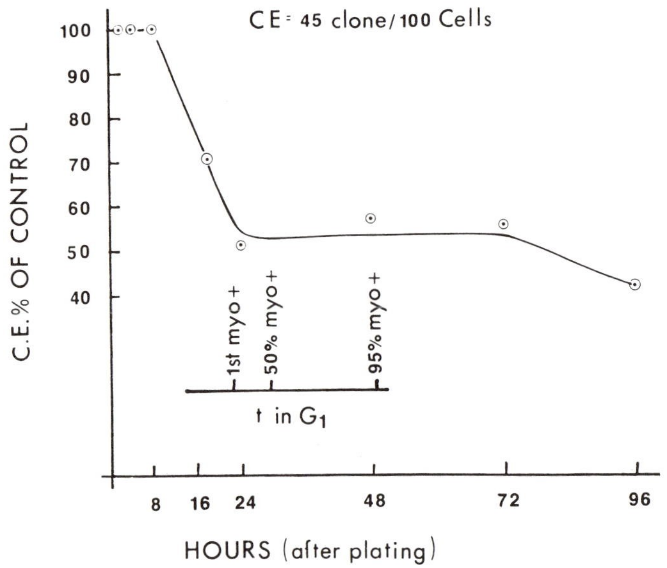

FIGURE 4
Clonal analysis of the retention of the ability to proliferate as a function of time in mitogen-poor, fusion-impermissive medium. Myoblasts synchronized in mitosis were plated at clonal density (200 cells/60-mm petri plate) and switched at the time points indicated to growth-promoting, fusion-impermissive medium. These refed plates were cultured for an additional 8 days to permit clonal development; they were then fixed, stained, and scored for macroscopic clones. Numbers of clones were normalized to cloning efficiency (CE) of controls (i.e., synchronized myoblasts seeded and cultured in high-growth medium for the entire 8-day period).

myocytes were labeled during 2-hour pulses with [^{35}S]methionine at various time points before and after refeeding with high-growth, low-Ca^{++} or low-growth, low-Ca^{++} medium. Incorporation into myosin was determined by immunoprecipitation of cell extracts with rabbit anti-myosin, which had a high affinity for the muscle-specific isoform of quail myosin. Crude immunoprecipitates were subjected to SDS-polyacrylamide gel electrophoresis followed by fluorography, and incorporation into myosin heavy chain exclusively was calculated as a percentage of total incorporation by densitometric tracings from fluorograms. (This rigorous assessment of incorporation was necessitated by the fact that this particular batch of rabbit anti-myosin had not been affinity-column-purified. The crude immunoprecipitates therefore contained a nonmyosin peptide of about 55,000 molecular weight [Fig. 6], which is synthesized in both myoblasts and myocytes. By normalizing myosin heavy-chain synthesis as a percentage of total incorpo-

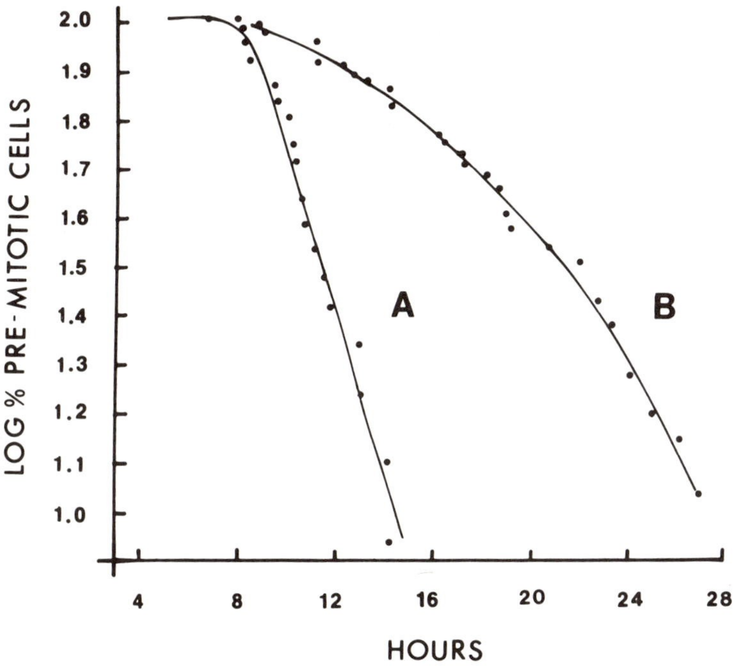

FIGURE 5
Semilog, transitional probability plot of interval from refeeding to mitosis of the original cohort of stimulated myocytes (*B*) and from birth to the next mitosis of the daughters of the original cohort (*A*), derived from time-lapse cinematographic studies.

ration, we also circumvented the problem of changes in amino acid precursor pools, which might be expected to occur consequent to refeeding.) In the low-Ca^{++}, low-mitogen medium used in these experiments, myoblasts differentiate in the absence of fusion. This is shown in Figure 7 by the increase in myosin heavy-chain synthesis from a value of 1.3% of total protein synthesis in cycling myoblasts (1 hr) to approximately 8.0% of total incorporation after 48 hours in culture—the time point at which greater than 95% of the cells are myosin-positive by immunofluorescence criteria and the rate of cell division is negligible (Emerson 1977). When these cultures are refed (at 48 hr) with low-growth, low-Ca^{++} medium, there is a slight decrease in [^{35}S]methionine incorporation into myosin followed by a recovery to levels observed before the medium change. In contrast, the rate of incorporation into myosin heavy chain drops precipitously within 6 hours of refeeding with high-growth, low-Ca^{++} medium, reaching a level similar to that observed in cycling myoblasts. This down regulation of myosin synthesis is not, however, a result of cell division per se. The decrease in incorporation into myosin heavy chain is virtually complete within 6 hours after re-

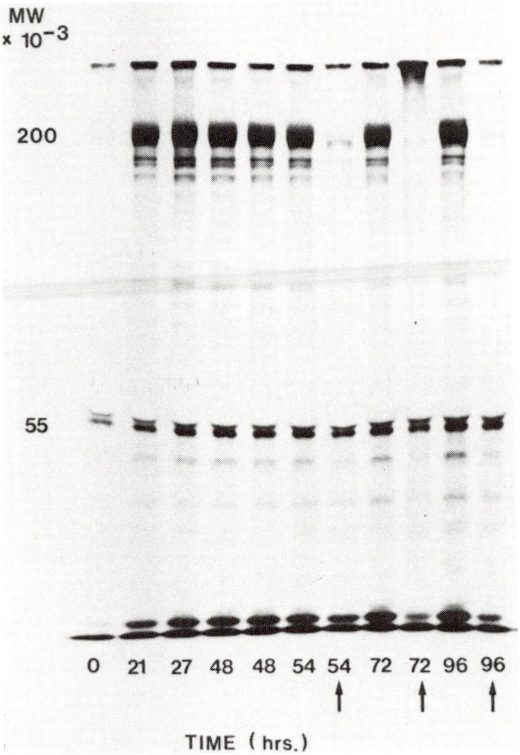

FIGURE 6
Myosin immunoprecipitates from [^{35}S]methionine-labeled myocytes, cultured for various periods of time (0–96 hr), were analyzed by electrophoresis on 8% acrylamide-SDS gels and fluorography. Myoblasts were cultured in low-growth, fusion-impermissive medium for 0 hr, 21 hr, 27 hr, and 48 hr and then refed at 48 hr with either low-growth, fusion-impermissive or high-growth, fusion-impermissive medium. Refed cultures were analyzed for myosin synthesis after 6 hr, 24 hr, and 48 hr (i.e., 54 hr, 72 hr, and 96 hr total time in culture). The myosin heavy-chain band (i.e., 200K) virtually disappears in cultures refed with growth-promoting medium (→).

feeding, during the period before the first mitotic event is observed in our time-lapse cinematographic study. The dramatic change in muscle-specific myosin synthesis occurs, moreover, before a significant number (no more than 7% according to the time-lapse data) of the stimulated cells enter S. Taken together, these data suggest that the down regulation of myosin synthesis occurs as myocytes leave a compartment in late G_1 in which differentiation is initiated and enter a prereplicative G_1 phase of the cell cycle (as has been described for other mitogen-stimulated cells [Pledger and Wharton 1980]). Thus, the earliest biochemical response to the mitogenic stimulus suggests that differentiative reprogramming may be a concomitant process of reentry into the cycle.

SUMMARY

These studies indicate that in quail myoblasts synchronized by mitotic shake-off and cultured under low-growth, fusion-impermissive conditions, myosin synthesis is initiated after 8 hours into a medium-protracted G_1. Thus, both myoblast fusion and the detection of skeletal-muscle myosin in

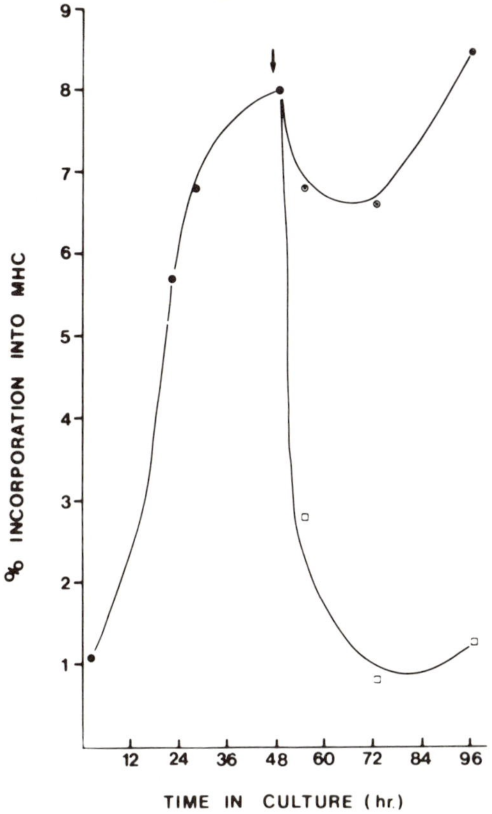

FIGURE 7
The percentage of total trichloroacetic acid (TCA)-precipitable counts incorporated into myosin heavy chain by quail myocytes cultured in low-growth, fusion-impermissive medium (●) for 48 hr prior to refeeding (→) with the same medium (⊙) or high-growth, low-Ca^{++} medium (□). Fluorograms of myosin immunoprecipitates were scanned at 495 nm, and the myosin heavy-chain peak area was determined using an electronic planimeter. Each point represents the mean of duplicate cultures.

fusion-blocked myocytes occur late during an extended G_1 period, two to four times longer than the average G_1 in log-phase myoblasts. These observations are consistent with the hypothesis that a temporal window exists in late G_1 that permits the entrainment of the differentiative program.

In addition, these studies indicate that, for some period of time at least, differentiated mononucleated myocytes can reenter the cell cycle in response to a mitogenic stimulus. In response to this stimulus, the rate of myosin synthesis decreases to the levels observed in cycling myoblasts. Thus, the initiation of contractile protein synthesis in itself does not commit the myoblast to withdraw irreversibly from the cell cycle. Earlier studies using established cell lines of other species have suggested that the commitment to differentiate is associated with irreversible withdrawal from the cell cycle (Bayne and Simpson 1977; Nadal-Ginard 1978; Linkhart et al. 1980). At this time, it is unclear whether the differences with these other studies are a result of cell-culture conditions or whether they represent some species difference in the timing of two consecutive but independent events.

ACKNOWLEDGMENTS

This work was supported by a postdoctoral fellowship from the National Sciences and Engineering Research Council of Canada to P.A.M. and grants from the National Institutes of Health (HD-07083) and the Muscular Dystrophy Association of America (subgrant from the University of Virginia, Jerry Lewis Neuromuscular Disease Research Center) to I.R.K.

REFERENCES

Bayne, E.K. and S.B. Simpson. 1977. Detection of myosin in prefusion G_0 lizard myoblasts in vitro. *Dev. Biol.* **55:** 306.

Bischoff, R. and H. Holtzer. 1969. Mitosis and the process of differentiation of myogenic cells in vitro. *J. Cell Biol.* **41:** 188.

Buckley, P.A. and I.R. Konigsberg. 1974. Myogenic fusion and the duration of the post-mitotic gap (G_1). *Dev. Biol.* **37:** 193.

Emerson, C.P., Jr. 1977. Control of myosin synthesis during myoblast differentiation. In *Pathogenesis of the human muscular dystrophies* (ed. L.P. Rowland), p. 799. Excerpta Medica, Amsterdam.

Emerson, C.P., Jr. and S. Beckner. 1975. Activation of myosin synthesis in fusing and mononucleated myoblasts. *J. Mol. Biol.* **93:** 431.

Holtzer, H., K. Strahs, J. Biehl, A.P. Somlyo, and H. Ishikawa. 1975. Thick and thin filaments in postmitotic, mononucleated myoblasts. *Science* **188:** 943.

Konigsberg, I.R. 1977. The role of the environment in the control of myogenesis *in vitro*. In *Pathogenesis of the human muscular dystrophies* (ed. L.P. Rowland), p. 779. Excerpta Medica, Amsterdam.

Konigsberg, I.R., P.A. Sollman, and L.O. Mixter. 1978. The duration of the terminal G_1 of fusing myoblasts. *Dev. Biol.* **63:** 11.

Linkhart, T.A., C.H. Clegg, and S.D. Hauschka. 1980. Control of mouse myoblast commitment to terminal differentiation by mitogens. *J. Supramol. Struct.* **14:** 483.

Moss, P.C. and R.C. Strohman. 1976. Myosin synthesis by fusion-arrested chick embryo myoblasts in cell culture. *Dev. Biol.* **48:** 431.

Nadal-Ginard, B. 1978. Commitment, fusion and biochemical differentiation of a myogenic cell line in the absence of DNA synthesis. *Cell* **15:** 855.

Pledger, W.J. and W. Wharton. 1980. Regulation of early cell cycle events by serum components. In *Cell mechanisms in animal cells* (ed. L. de Asua Jimenez), p. 165. Raven Press, New York.

Todaro, G.J., J. Lazar, and H. Green. 1965. The initiation of cell division in a contact inhibited mammalian cell line. *J. Cell. Comp. Physiol.* **66:** 325.

Turner, D.C., R. Gmür, M. Siegrist, E. Burckhardt, and H.M. Eppenberger. 1976. Differentiation in cultures derived from embryonic chicken muscle. 1. Muscle-specific enzyme changes before fusion in EGTA-synchronized cultures. *Dev. Biol.* **48:** 258.

Vertel, B.M. and D.A. Fischman. 1976. Myosin accumulation in mononucleated cells of chick muscle cultures. *Dev. Biol.* **48:** 438.

Zubrzycka, E. and D.M. MacLennan. 1976. Assembly of the sarcoplasmic reticulum; biosynthesis of calsequestrin in rat skeletal muscle cell cultures. *J. Biol. Chem.* **251:** 7733.

Regional Distribution and Cell Lineage States of Myogenic Cells during Early Stages of Vertebrate Limb Development

STEPHEN HAUSCHKA
RICHARD RUTZ
CLAIRE HANEY
Department of Biochemistry
University of Washington
Seattle, Washington 98195

Most studies reported in this volume concern developmental changes that occur during terminal phases of muscle differentiation. This interval begins when proliferating myoblasts withdraw from the cell cycle, continues through the period of contractile protein synthesis and myofibril assembly, and culminates in the modulations regulated by nerve-muscle interactions. Although encompassing the most dramatic aspects of muscle differentiation, this period is preceded by an equally crucial interval involving changes at the single-cell level within muscle precursor cells.

The early period is thus far devoid of biochemical markers of developmental change; nevertheless, changes can be detected at the level of single myogenic cells by in vitro clonal analysis. This approach measures the number of cells with the capacity for muscle colony formation (MCF cells) during successive embryonic stages. Previous chick and human studies indicated that MCF cells first appear during comparable limb stages (~day 3 and day 35, respectively) and that the subsequent kinetics of MCF cell increase are similar (Bonner and Hauschka 1974; Hauschka 1974; Hauschka et al. 1977). In addition, these studies revealed that MCF cells derived from early versus late embryonic stages differ with respect to media

requirements and clonal phenotypes (White et al. 1975; Linkhart and Hauschka 1979). More recently, clonal analysis has been used to determine the location of early and late MCF cells during limb development (Rutz et al. 1982) and to investigate whether early MCF cells are the cell lineage precursors of late MCF cells (Rutz and Hauschka 1982). Results from these studies together with a combined explant and clonal analysis of earlier stages are summarized below.

RESULTS AND DISCUSSION

Location of MCF Cells during Limb Development

Spatial analysis of MCF cells during chick leg development has been facilitated by the development of a technique for embedding and serially sectioning living limb buds (Hauschka and Haney 1978). With this procedure, gelatin-embedded limbs can be sliced in a proximal-distal series to thicknesses of 30–300 μm. The resulting flat sections may then be photographed, manually dissected along prospective tissue boundaries, and the dissected pieces subjected to clonal analysis.

Techniques employed in these studies are provided by Hauschka and Haney (1978), Rutz and Hauschka (1982), and Rutz et al. (1982). Clonal assays for MCF cells were performed in conditioned medium (White et al. 1975), and explant assays were performed in fresh medium (81% Ham's F-10 nutrients, 1% antibiotics, 15% preselected horse serum, 3% day-12 chick embryo extract). Clonal assays for cartilage colony-forming cells were performed in a medium containing 84% Ham's F-10 nutrients, 1% antibiotics, and 15% preselected fetal calf serum. Explant cultures were grown for the first 24 hours in 1 ml of fresh medium to promote adhesion and were then fed with an additional 2 ml of fresh medium. All culture dishes were coated with 0.7% gelatin.

Initial experiments involving 3–5½-day chick leg buds (stages 21–26) determined that MCF cells are located in two peripheral tissue compartments (dorsal and ventral) that extend distally in a declining gradient from the proximal limb boundary to the prospective ankle region. In contrast, cartilage colony-forming cells are located exclusively within a central core compartment that extends to within 50–100 μm of the limb's distal tip (Hauschka et al. 1982).

PROXIMO-DISTAL GRADIENTS OF EARLY AND LATE MCF CELLS. MCF cells derived from early embryos can be distinguished (retrospectively) from those derived from late embryos on the basis of clonal morphology. Clones from early MCF cells exhibit small numbers of short myotubes, whereas those from late MCF cells contain large numbers of long myotubes; the average percentages of multinuclearity are about 30% and 60%, respectively. The spatial distribution of early and late MCF cells within developing chick

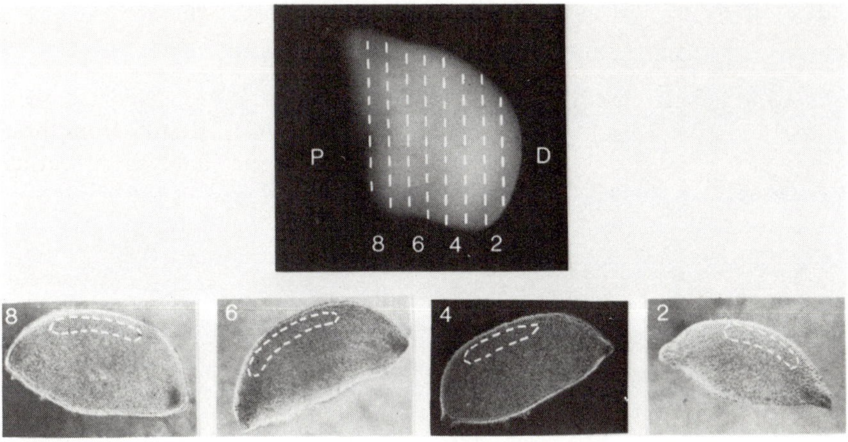

FIGURE 1
Transverse, 100-μm sections from stage-23 (day 3½–4) chick leg bud with sections numbered sequentially from the distal end. Prospective dorsal (dashed lines) and ventral myogenic regions are dissected from each section and subjected to clonal analysis (see Table 1 for clonal analysis).

limbs was determined for stage-23–31 (day 3½–7½) embryos (Figs. 1–3; Tables 1–3).

Several conclusions emerge from this data. (1) A distinct proximo-distal difference in total MCF cells is discernible at all stages. (2) As development proceeds, the proximal myogenic region exhibits an increasingly high plateau level of MCF cells, and the "boundary" between regions containing high and low percentages of MCF cells "moves" to progressively more distal locations. (3) Early-type MCF cells are the only MCF cell type detected

TABLE 1
Proximo-distal Distribution of MCF Cells in Stage-23 Chick Leg Buds

	Section no. (100-μm sections)[a]								
(proximal)	8	7	6	5	4	3	2	1	(distal)
MCF cells[b]	19[c]	35	30	30	27	11	0	0	
Late MCF cells[d]	0	0	0	0	0	0	—	—	

[a]Refer to Fig. 1 for section location and depiction of regions dissected for clonal analysis.
[b]Percentage of total colonies that form myotubes. Clonal plating efficiencies ranged from 0.4% in the most distal sections to 10% in the most proximal sections.
[c]Means of four experiments. S.E.M. for percentage of MCF cells, sections 8–1, are 0.6, 3.2, 3.9, 3.5, 3.3, 1.4, 0, and 0, respectively.
[d]Percentage of the total muscle colonies that exhibit the charcteristic late muscle clone phenotype (i.e., extensive multinuclearity—see text); stage-23 MCF cells all show the early phenotype.

TABLE 2
Proximo-distal Distribution of MCF Cells in Stage-27 Chick Leg Buds

	Section no. (200-μm sections)[a]											
(proximal)	11	10	9	8	7	6	5	4	3	2	1	(distal)
MCF cells[b]	75[c]	87	81	88	85	84	76	28	0	0	0	
Late MCF cells[d]	65	59	57	45	40	31	14	1	—	—	—	

[a]Refer to Fig. 2.
[b]See footnote b in Table 1. Clonal plating efficiencies ranged from 0.3% in the most distal sections to 21% in the most proximal sections.
[c]Means of five experiments. S.E.M. for percentage of MCF cells, sections 11–1, are 10.9, 3.7, 4.0, 2.1, 4.8, 5.0, 6.0, 9.8, 0, 0, and 0, respectively; S.E.M. for percentage of late MCF cells are 9.5, 5.8, 4.6, 6.0, 7.7, 4.4, 4.0, 0.7, —, —, and —, respectively.
[d]See footnote d in Table 1.

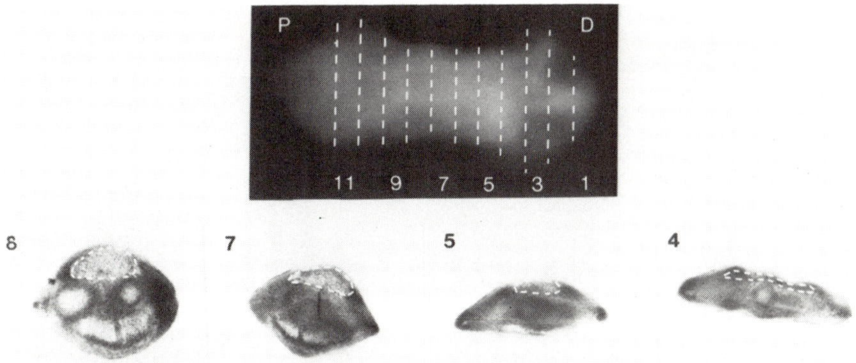

FIGURE 2
(*6–8*) Transverse, 200-μm sections from stage-27 (day 5) chick leg bud showing initial splitting of ventral myogenic mass (see Table 2 for clonal analysis).

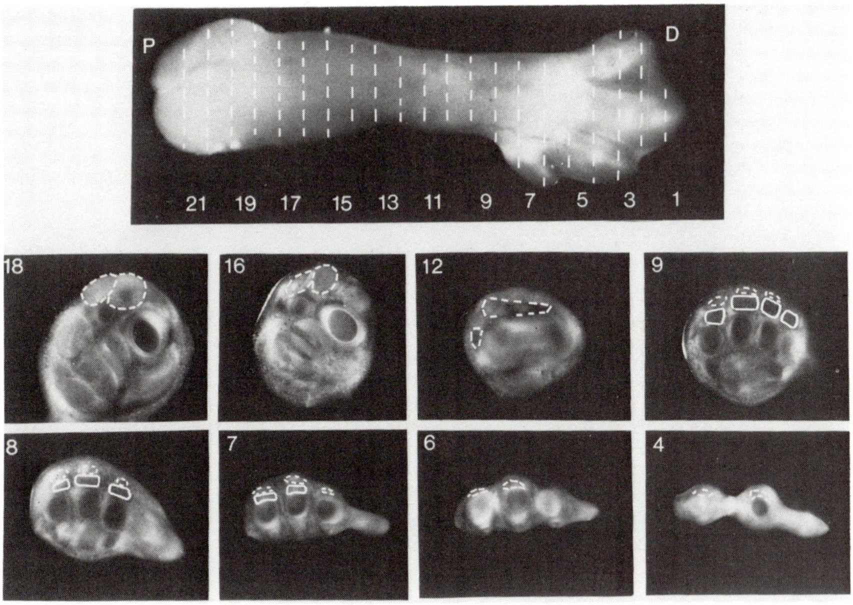

FIGURE 3
(*16,18*) Transverse, 300-μm sections from stage-31 (day 7) lower leg bud showing presence of all adult lower leg muscles. Myogenic elements (*7–9*) that belong to the dorsal toe muscle group are outlined by solid lines; dashed lines in these sections surround tendinous extensions of leg muscles (see Table 3 for clonal analysis).

TABLE 3

Proximo-distal Distribution of MCF Cells in Stage-31 Chick Leg Buds

	Section no. (300-μm sections)[a]															
(proximal)	21	20	19	18	17	16	15	14	13	12	11	10	9	8	7	6 to 1 (distal)
MCF cells[b]	95[c]	93	97	95	89	85	84	72	23	0	0	0	0	0	0	0
(toe muscles)[d]												94	91	76	30	0
Late MCF cells[e]	97	97	97	98	98	97	97	95	93	—	—	—	—	—	—	—
(toe muscles)[d]												97	94	89	46	—

[a]Refer to Fig. 3.

[b]See footnote b in Table 1. Clonal plating efficiencies ranged from 0.1% in the most distal sections to 36% in the most proximal leg muscle sections and from 0.4% to 23% for the toe muscles.

[c]Means of seven experiments ($n = 3$ for the toe muscles). S.E.M. for percentage of MCF cells, sections 21–1, are 1.4, 1.3, 1.3, 1.2, 4.0, 3.3, 2.5, 7.3, 8.3, 0, 0, 0, 0, 0, 0, and 0, respectively; sections 10–7 for the toe muscles are 0.4, 3.5, 4.6, 9.8; S.E.M. for percentage of late MCF cells are 0.6, 0.7, 0.9, 0.8, 0.9, 0.9, 1.1, 1.4, 2.5, —, —, —, —, —, —, and —, respectively; sections 10–7 for the toe muscles are 1.0, 1.6, 1.4, 8.5.

[d]Sections 7–10 contain both the tendinous extensions of the lower leg muscles (which contain no MCF cells) and the toe muscles.

[e]See footnote d in Table 1.

through stage 23 (day 3½), and they remain the predominant MCF cell type until stage 27 (day 5). (4) The superseding of early MCF cells by late MCF cells occurs in a proximo-distal gradient in which proximal myogenic regions of older embryos contain increasingly higher percentages of the late MCF cell type. (5) Similar proximo-distal gradients in total MCF cells and early versus late MCF cells occur within the toe muscles, as the foot and leg muscle regions become separated by the intervening ankle tendons (see Table 3). (6) Early MCF cells persist as a minor MCF cell type throughout all stages. An additional finding, from analysis of well-delineated myogenic regions dissected from sequential sections, is that early and late MCF cells appear to be located within the same muscle groups (e.g., along the entire length of the developing gastrocnemius). Finally, studies of early human limb buds indicate similar temporal and spatial distributions of MCF cell types (Hauschka et al. 1982).

LOCATION OF MUSCLE-FORMING REGIONS DURING EARLIER DEVELOPMENTAL STAGES. Detection of MCF cells within discrete regions of limb sections defines specific myogenic domains as early as stage 21. To determine whether cells with myogenic capacities are located in comparable regions of earlier stages, wing and leg buds from stage-16–20 embryos were sectioned (at 40–50 μm), dissected into regions, and analyzed for myotube formation in explant cultures.

Results from these experiments can be summarized as follows. (1) Myotube-forming cells are detected in explants from stages as early as 17 in the leg bud and 16 in the wing bud. (2) During stages 18–20 (when regional dissections are most feasible), myotube-forming cells are located primarily within the peripheral dorsal and ventral regions; there are, however, small numbers of myotube-forming cells within the central core region. (This may be due to the difficulty in preparing totally "clean" regional dissections at these early stages.) (3) As estimated from myotube density and the timing of their appearance, myotube-forming cells are distributed in a proximo-distal gradient similar to that observed for MCF cells at later stages—the only difference being that at earlier stages myotube-forming cells are detected as close as 100 μm and 150 μm from the distal wing and leg-bud tips.

IDENTIFICATION OF MCF CELL PRECURSORS IN EXPLANT CULTURES FROM EARLY LIMB BUDS. With present techniques, MCF cells are not detectable in leg buds prior to stage 21 (Bonner and Hauschka 1974). Since, however, the location of MCF cells in stage-21 limbs is essentially identical to that of myotube-forming cells in stage-17–20 limbs, it seemed likely that MCF cell precursors would also be localized in these regions. This possibility was examined by establishing explant cultures from myogenic and nonmyogenic regions of stage-18–20 leg buds, which were subsequently passaged and subjected to clonal analysis. Results from these experiments indicate that MCF cells are indeed detectable within myogenic explants after 2–4 days in vitro

and are absent from nonmyogenic explants. These observations suggest that the stage-18–20 precursors to MCF cells require a more complex multicellular environment before they or their progeny acquire the capacity for MCF cell expression.

Heritability of MCF Cell Type

Since early MCF cells precede late MCF cells both temporally and spatially during development, it seemed possible that the cells were related by a precursor-product relationship. If this were so, late MCF cells might be derived from early MCF cells in vitro. This possibility was examined by selecting primary colonies from stage-23–31 embryos, which expressed the early and late MCF cell phenotypes, and then serially subcloning these through four to five passages (> 40 cell doublings) without further selection (for details, see Rutz and Hauschka 1982). Representative data from stage 25 (Fig. 4) indicate that MCF cell type (i.e., clonal morphology) is inherited without deviation over this period of 40 cell doublings. Even in subclones derived from

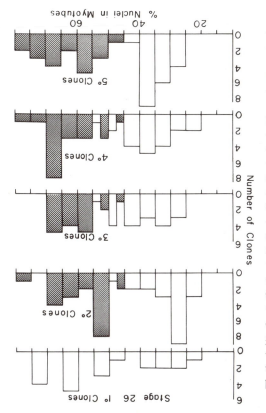

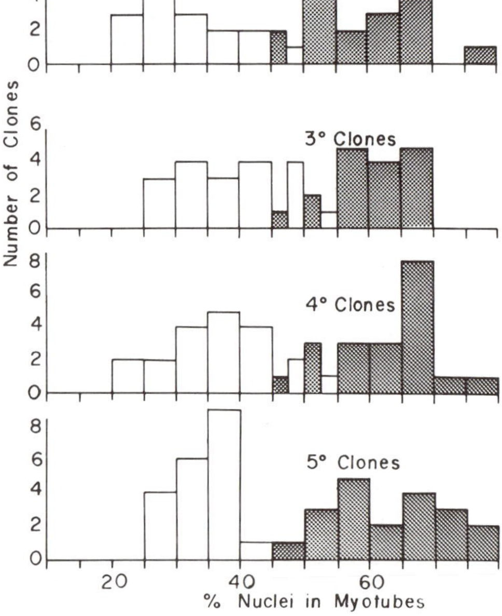

FIGURE 4
Heritability in culture of early- and late-type muscle colony phenotypes derived from stage-26 embryo. Primary (1°) clones representative of the two types were subcloned into separate secondary (2°) cultures, and subsequent (3°–5°) subclonings were made at 5–6-day intervals without selection of colony type. At the time of maximal myotube formation (day 10–14), companion cultures were fixed, 20 random colonies were scored for the percentage of nuclei incorporated into myotubes, and the data were plotted as a function of the percentage of colonies with different levels of multinuclearity. The 1° clone data represent the heterogeneity of the original unselected MCF cells, whereas 2°–5° data depict the heterogeneity of subclonal descendents from individual early- and late-type colonies. (□) Descendants from early-type colony; (▒) descendants from late-type colony.

stage-31 embryos, there is no indication of an early-to-late-type MCF cell transition. Similar results have been obtained with early-type human MCF cells.

Although the culture conditions used for these experiments permit expression of both clonal phenotypes, it is possible that the conditions required for an MCF cell transition have not been duplicated. With this in mind, Womble and Bonner (1980) attempted to demonstrate an MCF cell conversion within pooled chick muscle clones that had been implanted into quail limb buds. As with the subclonal analysis reported above, however, they found no data supporting the existence of a transition. Neither of these approaches categorically rules out a direct precursor-product relationship between early and late MCF cells, but they do suggest that the cell lineage relationship may be more complex. One alternative is that early and late MCF cells may be the products of separate precursors (see Identification of MCF Cell Precursors in Explant Cultures from Early Limb Buds), which are themselves sequential "steps" in a muscle cell lineage. Demonstration of this possibility will require perfecting in vitro conditions that permit myoblast precursors to replicate as well as to differentiate into their respective MCF cell types.

SUMMARY

The location of myogenic precursor cells in developing limb buds has been determined utilizing a combination of clonal and explant analyses together with techniques permitting reproducible spatial analysis. Myogenic cells are located in discrete dorsal and ventral regions and, within these domains, they are distributed in proximo-distal gradients. Although several different types of myogenic precursor cells have been identified, evidence supporting a direct cell lineage relationship among the precursors is lacking. It also remains to be determined what developmental roles the different myogenic cell types may play.

ACKNOWLEDGMENTS

We thank our co-workers Christy Lin and Jennifer Seed for their contributions to this work, Robert Lim and Tom Linkhart for their helpful criticisms, and Heidi Muller for her typing.

REFERENCES

Bonner, P.H. and S.D. Hauschka. 1974. Clonal analysis of vertebrate myogenesis. I. Early developmental events in the chick limb. *Dev. Biol.* **37:** 317.

Hauschka, S.D. 1974. Clonal analysis of vertebrate myogenesis. III. Developmental changes in the muscle colony-forming cells of the human fetal limb. *Dev. Biol.* **37:** 345.

Hauschka, S.D. and C. Haney. 1978. Use of living tissue sections for analysis of positional information during development. *J. Cell Biol.* **79:** 24a.

Hauschka, S.D., C. Haney, J.C. Angello, T. Linkhart, P.H. Bonner, and N.K. White. 1977. Clonal studies of muscle development: Analogies between human and chicken cells *in vitro* and their possible relevance to muscle diseases. In *Pathogenesis of human muscular dystrophies* (ed. L.P. Rowland), p. 835. Excerpta Medica, Amsterdam.

Hauschka, S.D., R. Rutz, T.A. Linkhart, C.H. Clegg, G.F. Merrill, C.M. Haney, and R.W. Lim. 1982. Skeletal muscle development. 1. Developmental changes in muscle colony-forming cell type and location during vertebrate limb development. 2. Mitogen-regulated myoblast commitment to terminal differentiation. In *Disorders of the motor unit* (ed. D.L. Schotland), p. 903. Houghton Mifflin, New York.

Linkhart, T.A. and S.D. Hauschka. 1979. Clonal analysis of vertebrate myogenesis. VI. Acetylcholinesterase and acetylcholine receptor in myogenic and non-myogenic clones from chick embryo leg cells. *Dev. Biol.* **69:** 529.

Rutz, R. and S. Hauschka. 1982. Clonal analysis of vertebrate myogenesis. VII. Heretability of muscle colony type through sequential subclonal passages *in vitro. Dev. Biol.* **91:** 103.

Rutz, R., C. Haney, and S.D. Hauschka. 1982. Spatial analysis of limb bud myogenesis: A proximo-distal gradient of muscle colony-forming cells in chick embryo leg buds. *Dev. Biol.* **90:** 399.

White, N.K., P.H. Bonner, D.R. Nelson, and S.D. Hauschka. 1975. Clonal analysis of vertebrate myogenesis. IV. Medium-dependent classification of colony-forming cells. *Dev. Biol.* **44:** 346.

Womble, M.D. and P.H. Bonner. 1980. Developmental fate of a distinct class of chick myoblasts after transplantation of cloned cells into quail embryos. *J. Embryol. Exp. Morphol.* **58:** 119.

Control of Mouse Myoblast Commitment to Terminal Differentiation by Mitogens

*THOMAS A. LINKHART**
CHRISTOPHER H. CLEGG
ROBERT W. LIM
GARY F. MERRILL
JEFFREY S. CHAMBERLAIN
STEPHEN D. HAUSCHKA
Department of Biochemistry
University of Washington
Seattle, Washington 98195

Studies of myoblast differentiation in vitro have suggested that growth-promoting or mitogenic activity in the culture medium is inhibitory to differentiation and that the mitogenic activity is eliminated before terminal differentiation begins (reviewed in Linkhart et al. 1981). Although many laboratories manipulate culture-medium composition to trigger the onset of myoblast differentiation, the mechanisms by which myoblast proliferation and differentiation are regulated by mitogenic activity are poorly understood.

Studies of a permanent clonal mouse myoblast cell line, MM14, derived in our laboratory have shown that proliferation is stimulated and differentiation prevented by nanomolar concentrations of a purified polypeptide growth factor, bovine pituitary fibroblast growth factor (FGF), added to mitogen-depleted medium (Linkhart et al. 1980, 1981). The removal of mitogens from mouse myoblasts triggers (after a 2–3-hr delay) irreversible withdrawal from the cell cycle and commitment to terminal differentiation (Linkhart et al. 1980). This paper briefly describes the time course of MM14 myoblast responses to mitogen removal; the evidence from mitotically syn-

*Present address: Mineral Metabolism Laboratory, Jerry Pettis Memorial Veterans Hospital and Loma Linda University Medical School, Loma Linda, California 92357.

chronized cultures that the cells commit in G_1; and the evidence that FGF stimulation directly, rather than proliferation per se, prevents terminal differentiation.

RESULTS AND DISCUSSION

Commitment to Terminal Differentiation

Data compiled from several experiments (Fig. 1) illustrate the kinetics of the responses of MM14 myoblasts to mitogen removal. (1) Cells cease entering S phase (as indicated by the [^{3}H]thymidine pulse-labeling index) and accumulate in G_1 or G_0 2 hours after mitogen removal; all cells have cycled out of S by 10 hours. (2) After 2 hours of mitogen removal, cells first commit to terminal differentiation (as indicated by their failure to incorporate [^{3}H]thymidine after mitogen refeeding), and all are committed by 12 hours. In similar experiments, committed cells were shown to elaborate acetylcholine receptors (AChRs) and to form myotubes after mitogen refeeding; thus, commitment of mouse myoblasts is both the irreversible withdrawal from the cell cycle and initiation of terminal differentiation. (3) AChR-positive cells in autoradiographs and creatine kinase activity in cell extracts are detectable 2–4 hours after MM14 cells begin committing. (4) Myoblast fusion commences 8–10 hours after the first cells commit.

Commitment Occurs in G_1

When mitotically synchronized MM14 myoblasts are plated into mitogen-rich media, all cells incorporate [^{3}H]thymidine over the next 24 hours and none become AChR-positive. When plated into mitogen-depleted medium, no cells incorporate [^{3}H]thymidine and all become AChR-positive (Linkhart et al. 1980). This observation and the kinetics studies with asynchronous cultures (Fig. 1; Linkhart et al. 1980) suggested that mouse myoblasts commit as they cycle into G_1 in the absence of mitogens. This was further tested with mitotically synchronized cells (Fig. 2). When synchronized cells were exposed to 6-hour periods of mitogen withdrawal beginning at successively later times after plating, the percentage of cells committing in response to mitogen withdrawal decreased as more and more cells had left G_1 and entered S phase. Commitment was determined by a colony-forming assay and by the appearance of myosin-positive cells after mitogen refeeding; entry into S phase was followed by continuous [^{3}H]thymidine labeling and autoradiography. The results suggest that the presence or absence of mitogens during G_1 determines whether mouse myoblasts enter S phase and continue through another cell cycle or stop in G_1 and commit to differentiation.

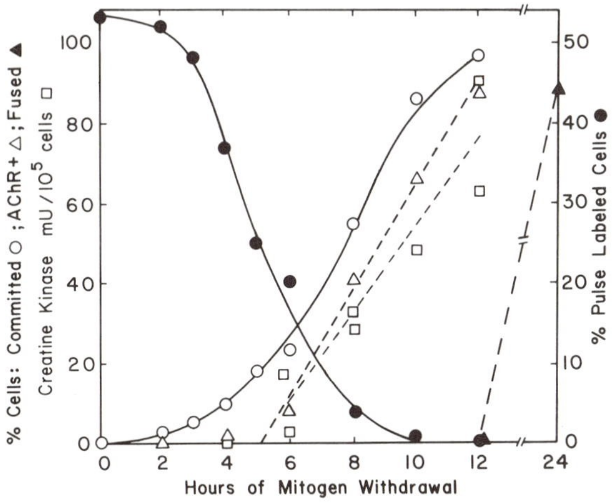

FIGURE 1

Kinetics of commitment and terminal differentiation. Proliferating MM14 cells were rinsed and fed mitogen-depleted medium (conditioned growth medium diluted 1:1 with Hams F-10 medium). At each time point, cells were pulse-labeled with [^{3}H]thymidine and fixed to determine the percentage of cells in S, fed growth medium (15% horse serum + 3% chick embryo extract) + [^{3}H]thymidine and fixed at 30 hr to determine the percentage of cells committed (not incorporating label), or incubated with ^{125}I-labeled α-bungarotoxin and fixed to determine the percentage of AChR-positive cells. All three parameters were analyzed (by autoradiography) in the same experiment, as described by Linkhart et al. (1980), at cell densities of 5×10^3–10×10^3/60-mm plate. Creatine kinase activity was determined in extracts of ten pooled 100-mm plates (1×10^5 - 2×10^5 cells/plate) as described by Hørder et al. (1979). Fusion (percentage of nuclei in myotubes) was observed in clonal density cultures switched to mitogen-depleted medium when clone sizes were ~80 cells (Linkhart et al. 1981).

Myoblast Proliferation and Commitment

Experiments presented elsewhere (Linkhart et al. 1982) indicate that FGF stimulation of MM14 myoblast proliferation is serum dependent and that virtually no proliferation occurs at optimal FGF concentrations in less than 3% horse serum. However, under the same conditions, commitment is prevented by FGF. When MM14 cells were exposed to 20 ng/ml of FGF and 2% horse serum for 12 hours (Fig. 3), the pulse-labeling index decreased to

FIGURE 2
Commitment of mitotically synchronized mouse myoblasts. Cells collected by mitotic shake-off from proliferating cultures (Linkhart et al. 1980) were plated in mitogen-rich (growth medium) or mitogen-depleted medium (conditioned 15% horse serum). The percentage of cells having entered S phase (●) in growth medium was followed by continuous labeling with [^{3}H]thymidine and fixing at each time point (average ± range from two experiments). The percentage of cells committing during 6-hr periods of mitogen withdrawal (ending at each time point) was determined by loss of colony-forming ability (□) of cells plated (at 0 hr) at 200/ 60-mm plate (average ± S.E.M. of four plates each from two experiments). In one experiment, commitment was also determined by the number of myosin-positive cells (Δ) 30 hr after plating (by indirect immunofluoresence with monoclonal MF-20 anti-skeletal-muscle myosin heavy chain, kindly supplied by Dr. D. Fischman, SUNY, Downstate Medical Center, Brooklyn).

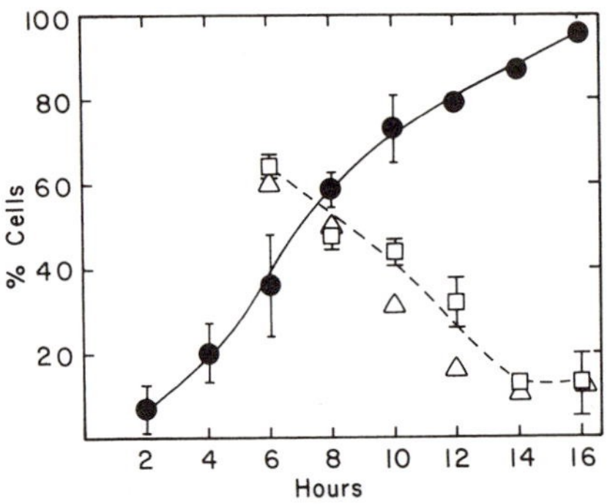

FIGURE 3
Regulation of myoblast proliferation and commitment by horse serum and FGF. Proliferating MM14 myoblasts (5 × 10^3/60-mm plate) were rinsed and fed 2% horse serum (*A*), 15% horse serum (*B*), 2% horse serum + 20 ng/ml of FGF (*C*), or 15% horse serum + 20 ng/ml of FGF (*D*). Open bars represent cells incorporating [^{3}H]-thymidine (30-min pulse) at 12 hr. Stippled bars represent committed cells—cells not incorporating [^{3}H]-thymidine in continuous labeling between refeeding growth medium at 12 hr and fixing at 30 hr.

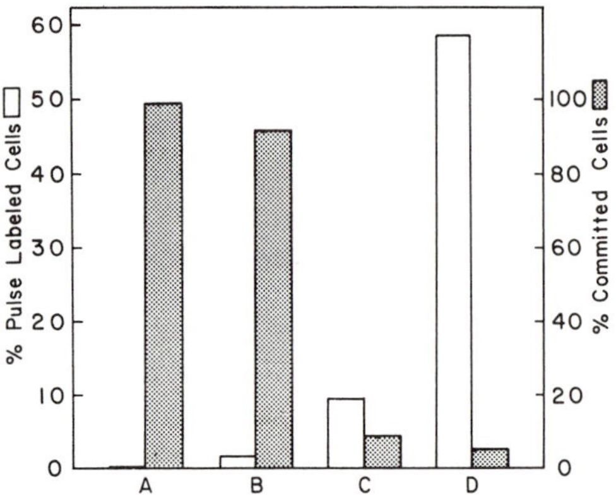

10%, indicating that the cells withdrew from the cell cycle (or were proliferating very slowly) but did not withdraw irreversibly, since 90% incorporated [^{3}H]thymidine upon refeeding growth medium. These results suggest that mouse myoblast commitment to terminal differentiation is repressed by FGF even under conditions in which the cells are not cycling. In a related series of experiments, MM14 myoblasts were observed to withdraw from the cell cycle without irreversibly committing when the cells were allowed to incorporate bromodeoxyuridine before mitogen removal (Merrill et al. 1980).

SUMMARY

MM14 myoblast commitment begins within 2–3 hours after mitogen withdrawal. All cells commit within one cell cycle, and the commitment "decision" appears to be made in G_1. Commitment occurs 2–4 hours before muscle-specific gene products are detectable, but the causal relationship between commitment and the initiation of muscle-specific gene expression is not known. It is clear that mouse myoblast commitment does not require fusion, as cells commit and elaborate muscle-specific gene products before myotube formation. In similar experiments, we found that the amount of epidermal growth factor binding to MM14 myoblasts decreases tenfold as the cells commit (Lim and Hauschka 1982). This suggests that mitogen receptor modulation may be involved in the rapid loss of mitogen responsiveness.

Myoblast commitment to differentiation appears to be restricted to G_1 in species other than mouse (Lavie and Yaffe 1976; Bayne and Simpson 1977; Konigsberg et al. 1978; Nadal-Ginard 1978) and might be regulated at the level of G_1-phase "restriction points," which are thought to control cell proliferation (Pardee et al. 1978). However, even though both are regulated in G_1, commitment is not regulated as a secondary effect of proliferation. We are currently investigating whether the reversible inhibition of proliferation in low-serum, high-FGF conditions is sufficient to stimulate reversible expression of muscle-specific gene products, as has been observed (under different conditions) in quail myoblasts and L6 rat myoblast temperature-sensitive variants (I.R. Konigsberg; B. Nadal-Ginard; both pers. comm.).

REFERENCES

Bayne, E.K. and S.B. Simpson. 1977. Detection of myosin in prefusion G_0 lizard myoblasts *in vitro*. *Dev. Biol.* **55:** 306.

Hørder, M., E. Magid, E. Pitkänen, M. Härkönen, J.H. Strömme, L. Theodorsen, W. Gerhardt, and J. Waldenström. 1979. Recommended method for the determination of creatine kinase in blood modified by the inclusion of EDTA. *Scand. J. Clin. Invest.* **39:** 1.

Konigsberg, I.R., P.S. Sollmann, and L.O. Mixter. 1978. The duration of the terminal G_1 of fusing myoblasts. *Dev. Biol.* **63:** 11.

Lavie, G . and D. Yaffe. 1976. Studies on the nature of the last cell division of rat skeletal muscle cells. In *Progress in differentiation research* (ed. N. Muller-Bérat et al.), p. 25. Elsevier/North Holland, Amsterdam.

Lim, R.W. and S.D. Hauschka. 1982. Differential EGF responsiveness and EGF receptor modulation in a clonal line of mouse myoblasts and a differentiation defective variant. In "Growth of cells in hormonally defined media" (ed. D.A. Sirbasku et al.). *Cold Spring Harbor Conf. Cell Proliferation* **9:** 877.

Linkhart, T.A., C.H. Clegg, and S.D. Hauschka. 1980. Control of mouse myoblast commitment to terminal differentiation by mitogens. *J. Supramol. Struct.* **14:** 483.

——. 1981. Myogenic differentiation in permanent clonal mouse myoblast cell lines: Regulation by macromolecular growth factors. *Dev. Biol.* **86:** 19.

Linkhart, T.A., R.W. Lim, and S.D. Hauschka. 1982. Regulation of normal and variant mouse myoblast proliferation and differentiation by specific growth factors. In "Growth of cells in hormonally defined media" (ed. D.A. Sirbasku et al.). *Cold Spring Harbor Conf. Cell Proliferation* **9:** 867.

Merrill, G.F., C.H. Clegg, T.A. Linkhart, and S.D. Hauschka. 1980. Bromodeoxyuridine inhibits the commitment decision of mouse myoblasts to terminally differentiate. *Eur. J. Cell Biol.* **22:** 402.

Nadal-Ginard, B. 1978. Commitment, fusion and biochemical differentiation of a myogenic cell line in the absence of DNA synthesis. *Cell* **15:** 855.

Pardee, A.B., R. Dubrow, J.L. Hamlin, and R.F. Kletzien. 1978. Animal cell cycle. *Annu. Rev. Biochem.* **47:** 715.

Effects of Taxol, a Microtubule-stabilizing Agent, on Myogenic Cultures

HOWARD HOLTZER
SUSAN FORRY-SCHAUDIES
PARKER ANTIN
YOSHIRA TOYAMA
Department of Anatomy, School of Medicine
University of Pennsylvania
Philadelphia, Pennsylvania 19104

MARGARET A. GOLDSTEIN
DAVID L. MURPHY
Department of Medicine
Baylor College of Medicine
Houston, Texas 77030

Treatment of myogenic cultures of chick breast muscle with 10 μM taxol reveals differences between the cell types in such cultures. In addition to stabilizing microtubules, taxol is known to arrest and/or kill cells in mitosis (Schiff and Horwitz 1980; De Brabander et al. 1981; Holtzer et al. 1982b). Presumptive myoblasts and fibroblasts, the replicating cells in primary myogenic cultures, undergo a 75% decrease in number when treated with taxol for 4 days. In contrast, postmitotic myoblasts remain constant in number over 4 days of treatment, thus demonstrating their stable phenotype. Although they are reared at low density in full growth medium, these postmitotic myoblasts do not reenter the cell cycle nor do they fuse during their 4 days of treatment. Instead, they form mononucleated, star-shaped cells that assemble unprecedented striated "myofibrils" of interdigitating thick filaments and microtubules from which thin filaments are absent (Antin et al. 1981). Presumptive myoblasts and fibroblasts also exhibit increased microtubule numbers but are distinguished from postmitotic myoblasts in the distribution of this excess of microtubules.

MATERIALS AND METHODS

Primary chick breast muscle cultures were prepared from 11-day chick embryos as described in Antin et al. (1981). A final concentration of 10 μM taxol (Natural Products Branch, Division of Cancer Treatment, National Cancer Institute, National Institutes of Health) was added to the cells (1.5 $\times$ 10^6/ml) at the time of plating and kept on for 4 days. Alternatively, 3-day muscle cultures (initial inoculum 0.5 $\times$ 10^6/ml) were treated with 10 μM taxol for 3 days. Cells were grown on collagen-coated, 35-mm dishes and were fed with 1.5 ml of medium per dish. Immunofluorescence staining with antidesmin and antivimentin was performed as described by Bennett et al. (1978, 1979) and Holtzer et al. (1981, 1982). Fluorescein-conjugated rabbit anti-light meromyosin (LMM) was used for visualization of striated myofibrils (Chi et al. 1975). Nuclear staining was performed through the use of propidium iodide at 30 μg/ml. Electron microscopy was performed as described by Toyama et al. (1979).

RESULTS

Effects of Taxol on Replicating Cells in Primary Myogenic Cultures

The debris and moribund cells present in taxol-treated primary myogenic cultures of embryonic chick breast muscle make it difficult to determine when round, mitotic-arrested cells first appear after plating. However, after 24 hours in taxol, approximately 50% of all cells in culture are rounded and arrested in mitosis (Fig. 1). There is no suggestion of chromosomes aligning along a metaphase plate; however, chromatin material is very condensed in such rounded cells and there is no nuclear membrane. Although some round chick cells may lyse in mitosis, many "break through" the mitotic arrest and survive for varying periods with varying numbers of micronuclei (Holtzer et al. 1982b). Most chick cells with micronuclei lyse within 10–15 hours (Fig. 1).

In primary chick myogenic cultures in the absence of taxol, the total cell population increases eightfold during the first 4 days (Bischoff and Holtzer 1970; Dienstman and Holtzer 1977). Of the cells that replicate during this period, over 60% become postmitotic, definitive, mononucleated myoblasts. Fusion of the in vitro newly born postmitotic myoblasts and of those present in the original inoculum begins during day 2. The remaining 40% in day-4 cultures are largely replicating presumptive myoblasts and fibroblasts (Abbott et al. 1974; Yeoh and Holtzer 1977). The cytotoxic effect of taxol is demonstrated by the observation that there are ten times as many nuclei in control day-4 cultures (Fig. 2a,b) as in toxol-treated cultures (Fig. 1; Antin et al. 1981). Taxol has no obvious cytotoxic effect on postmitotic, definitive myoblasts and/or myotubes (see below).

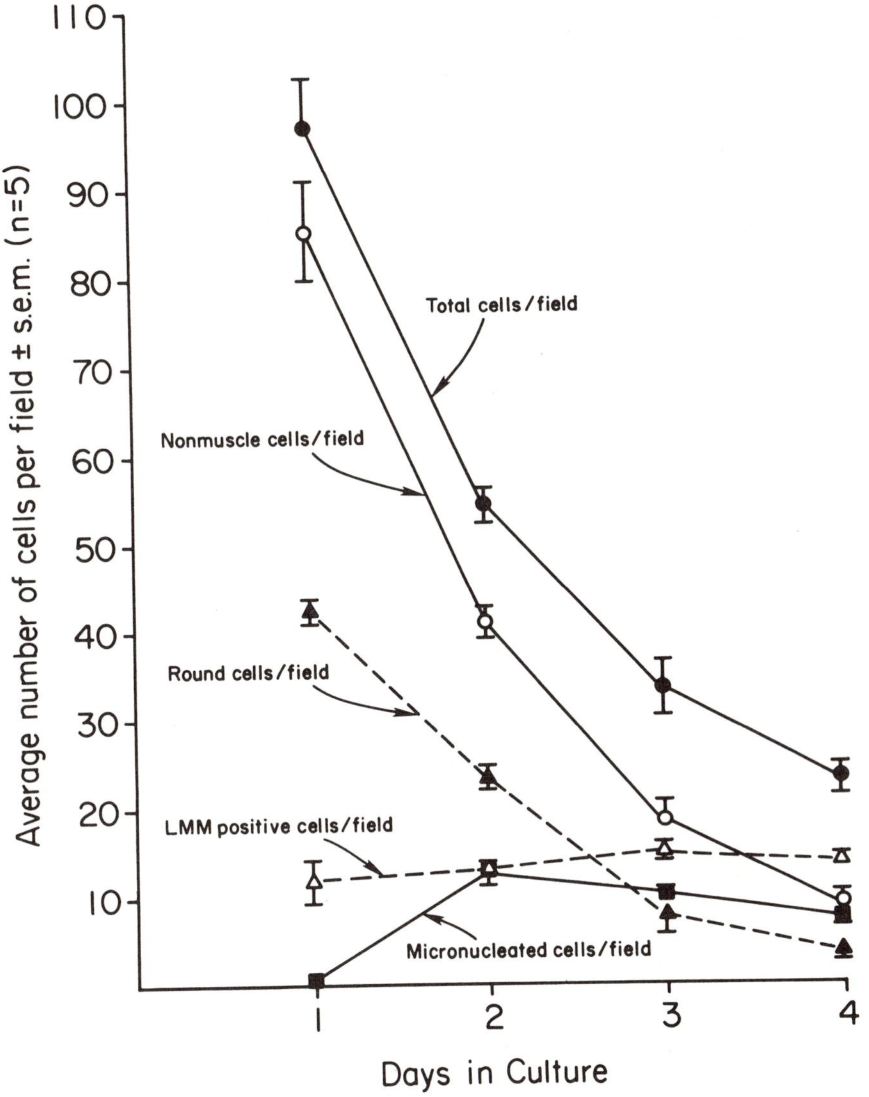

FIGURE 1
Effects of taxol treatment of primary muscle cells. Primary muscle was cultured in the presence of taxol. Plates were sacrificed and stained with anti-LMM, antidesmin, antivimentin, and propidium iodide on each of 4 successive days after culture. The total number of cells in each category of the graph were tabulated in each of five fields. Averages and S.E.M.s were calculated for each category. Nonmuscle cells are defined as presumptive myoblasts and fibroblasts that are antivimentin-positive and anti-LMM- and antidesmin-negative (Holtzer et al. 1981).

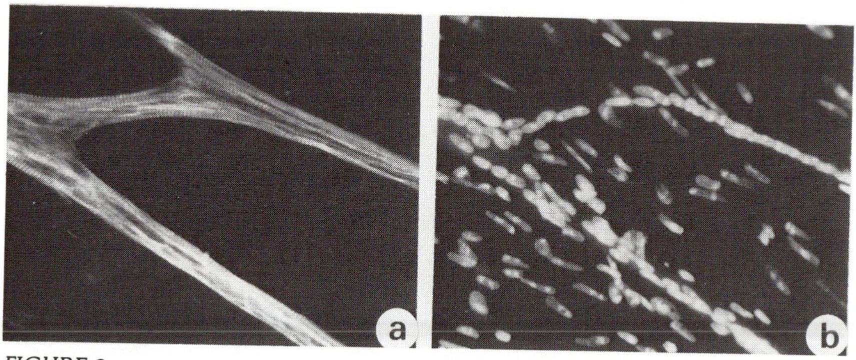

FIGURE 2
Fluorescent micrographs of the identical microscopic field of a day-4 control untreated muscle culture. The cells were stained with fluorescein-labeled anti-LMM (Croop et al. 1982) to localize the muscle-specific myosin (*a*), and with propidium iodide to localize the nuclei (*b*). The density and alignment of the nuclei within the two myotubes is indicated in *b*.

Effects of Taxol on Postmitotic Myoblasts and Immature Myotubes

Previous experiments suggest that 10–15% of the primary inoculum consists of postmitotic myoblasts (Holtzer et al. 1973; Dienstman and Holtzer 1975, 1977). Figure 1 shows that the number of LMM-positive cells remains stable over 4 days of taxol treatment. Since anti-LMM and antidesmin specifically stain postmitotic myoblasts and myotubes but do not stain replicating presumptive myoblasts or fibroblasts, they can be used to distinguish these two forms of replicating cells from definitive postmitotic myogenic cells. [^{3}H]thymidine labeling experiments of taxol-treated myogenic cultures reveal that after 4 days of exposure, fewer than 2% of the postmitotic myoblasts become labeled and/or form micronuclei (data to be presented elsewhere). Therefore, stability in the number of LMM-positive mononucleated cells over the course of 4 days of taxol treatment indicates that few definitive myoblasts in the original inoculum continue through the cell cycle.

Taxol-treated postmitotic myoblasts undergo a unique transformation to stubby star-shaped myoblasts, which first appear on day 2 of treatment. These cells develop four to six tapering arms and become striated over the course of 4 days in taxol (Fig. 3a–d). Distinctiveness of striations varies within the cell bodies due to overlapping of myofibrils or imperfect lateral alignment of individual thick filaments. However, striated processes less than 1 μm in diameter but over 200 μm in length can be observed in some cells (Fig. 3e). Labeled antibody to muscle phosphokinase and M-band protein (kindly supplied by S. Lowey, Rosenstiel Basic Medical Sciences Research Center, Waltham, Massachusetts) binds to a fine line bisecting each A band. Myofibrils are most conspicuous immediately beneath the cell

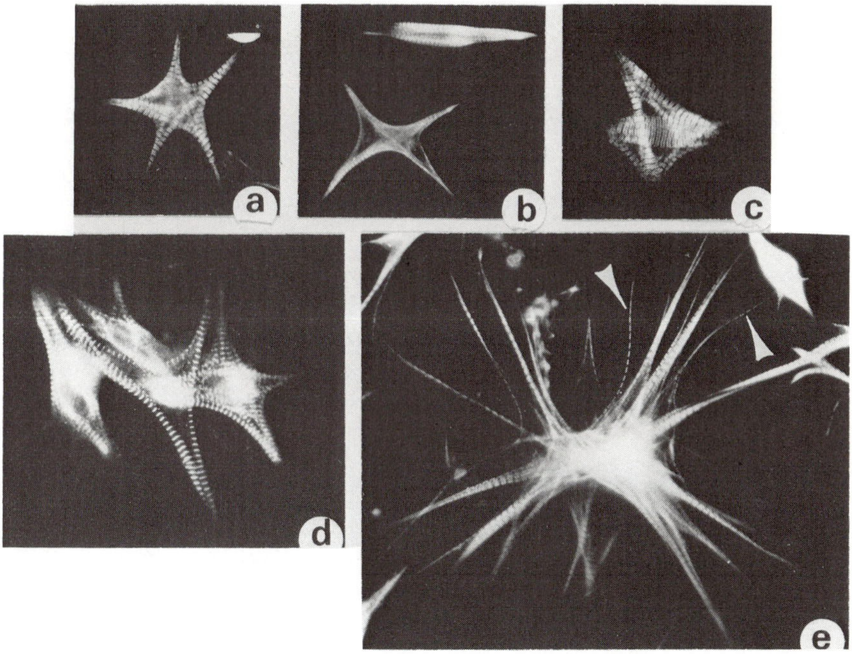

FIGURE 3

(*a–c*) Fluorescent micrographs of mononucleated, postmitotic, star-shaped myoblasts reared in taxol for 4 days and stained with labeled anti-LMM. The A bands of these myofibrils have the same dimensions as those in normal myoblasts and/or myotubes. (*d*) Fluorescent micrograph of four taxol-treated myoblasts given the same treatment as *a–c* but stained with anti-M-CPK. Note the lack of fusion. (*e*) Micrograph of a multinucleated, giant star myotube. This type of myotube is produced by adding taxol to day-3 cultures. Note the exceedingly fine, normally striated processes indicated by arrowheads. These processes would be difficult to detect under the phase microscope.

membrane, outlining the contours of the cell (Holtzer et al. 1957); other striated myofibrils encircle the nucleus or course from the upper to the lower cell surface.

Taxol blocks fusion of postmitotic, mononucleated myoblasts. As seen in Figure 3d, two to four star-shaped myoblasts can remain in contact for 3 to 4 days without fusing. The reason for formation of star-shaped myoblasts is not known. Assembly and elongation of myofibrils cannot be the sole impetus for this process as evidenced by cardiac cells, which assemble elongation myofibrils but do not become star-shaped when grown in taxol. Cartilage and nerve cells do not become star-shaped in taxol.

Both the star-shaped configuration and the blocking of fusion are reversible effects. If myogenic cultures are shifted to normal medium after 4 days in taxol, they lose their multiple arms, elongate, and fuse into exceptionally

thin myotubes. It may be noted that fusion occurs even though the cell density is approximately 10^5 cells/35-mm dish (1.5 ml of medium). The presence of fusion in these low-density embryonic chick cell cultures (see also Yeoh and Holtzer 1977) may be contrasted to the reported requirement of confluence for fusion in myogenic cell lines.

Colcemid, a microtubule-destabilizing agent, fragments immature myotubes into rounded, multinucleated "myosacs" (Bischoff and Holtzer 1967). Paradoxically, taxol also fragments immature myotubes into myosacs over a 24-hour period. The collapse is associated with a slow, peristalticlike retraction that can sever a single myotube into two to four irregularly shaped multinucleated sacs. However, unlike colcemid-induced myosacs, the taxol-induced multinucleated complexes develop numerous, sizable, tapering projections. These aggregates are best described as "giant stars" (Fig. 3e). In contrast to most myofibrils, those assembled within the tapered projection are also tapered. The interior of these giant stars is full of crisscrossing myofibrils, oriented at different angles to one another. Larger complexes of 20–30 nuclei remain formless in outline and are filled with randomly oriented striated myofibrils. Control 3–6-day myotubes contract regularly, whereas day-3 myotubes in taxol for 3–4 days rarely twitch.

Analysis by Electron Microscopy of Taxol-treated Myoblasts and/or Myotubes

Figures 4 and 5 are longitudinal and cross sections through myofibrils assembled in star-shaped, mononucleated myoblasts. The thick filaments are laterally aligned, approximately 1.6 μm in length, and display distinct M bands. Most striking is the vast number of microtubules and their interdigitation with the thick filaments comprising the early A bands. Normally, primary myofibrils are surrounded by longitudinally oriented microtubules that are rigorously excluded from the domain of interdigitating thick and thin filaments (e.g., Auber 1969). In taxol-treated myoblasts and/or myotubes, this exclusion breaks down. Approximately half of the striated myofibrils assembled in star-shaped myoblasts consist of interdigitating thick filaments and microtubules. Thin filaments were not detected in these structures, and Z-band material occurred in irregular patches or was totally absent. The remaining myofibrils vary from those with a normal ratio of thick to thin filaments, to those with various ratios of interdigitating thick to thin to microtubules. In many sections, 10-nm filaments appear to have been trapped into these unusual structures. The scanning of many sections has led to the estimate that per unit area, taxol-treated myoblasts and/or myotubes display 100–500 times the number of microtubules observed in equal areas of control cells.

In other regions of the star-shaped cells, bundles of interdigitating thick filaments and microtubules are less rigorously organized. However, within

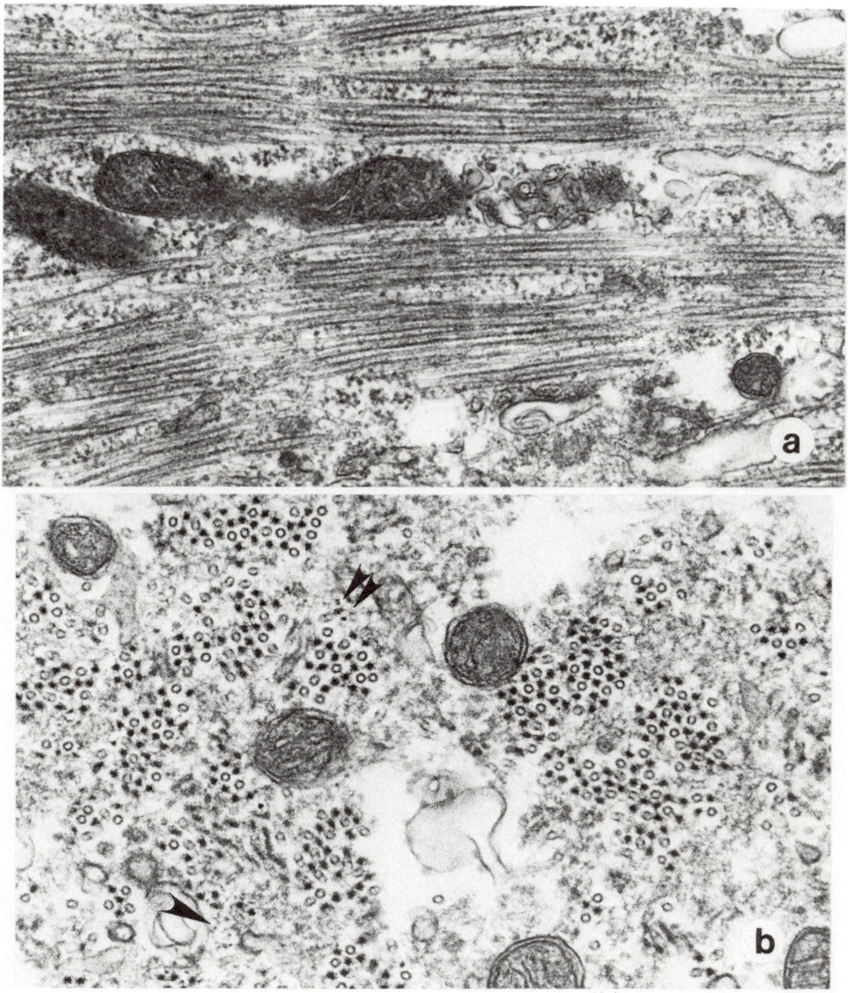

FIGURE 4
(*a*, *b*) Longitudinal and cross section of star-shaped myoblasts in taxol for 4 days. The density of microtubules in both *a* and *b* is vastly greater than that observed in untreated myoblasts. Note the interdigitation of microtubules and thick filaments. The only fibrillar component yet identified in these I bands is microtubules. Single microtubules can span more than three tandem sarcomeres. The single arrowhead points to what may be a thin filament; the double arrowhead points to what may be 10-nm filaments. The electron-opaque material characteristic of Z bands is not yet present in taxol-treated cells. For further details, see Antin et al. (1981).

a given bundle, the center-to-center spacing of thick filaments and microtubules remains remarkably constant. Individual bundles made up of varying ratios of thick filaments and microtubules can be many microns in

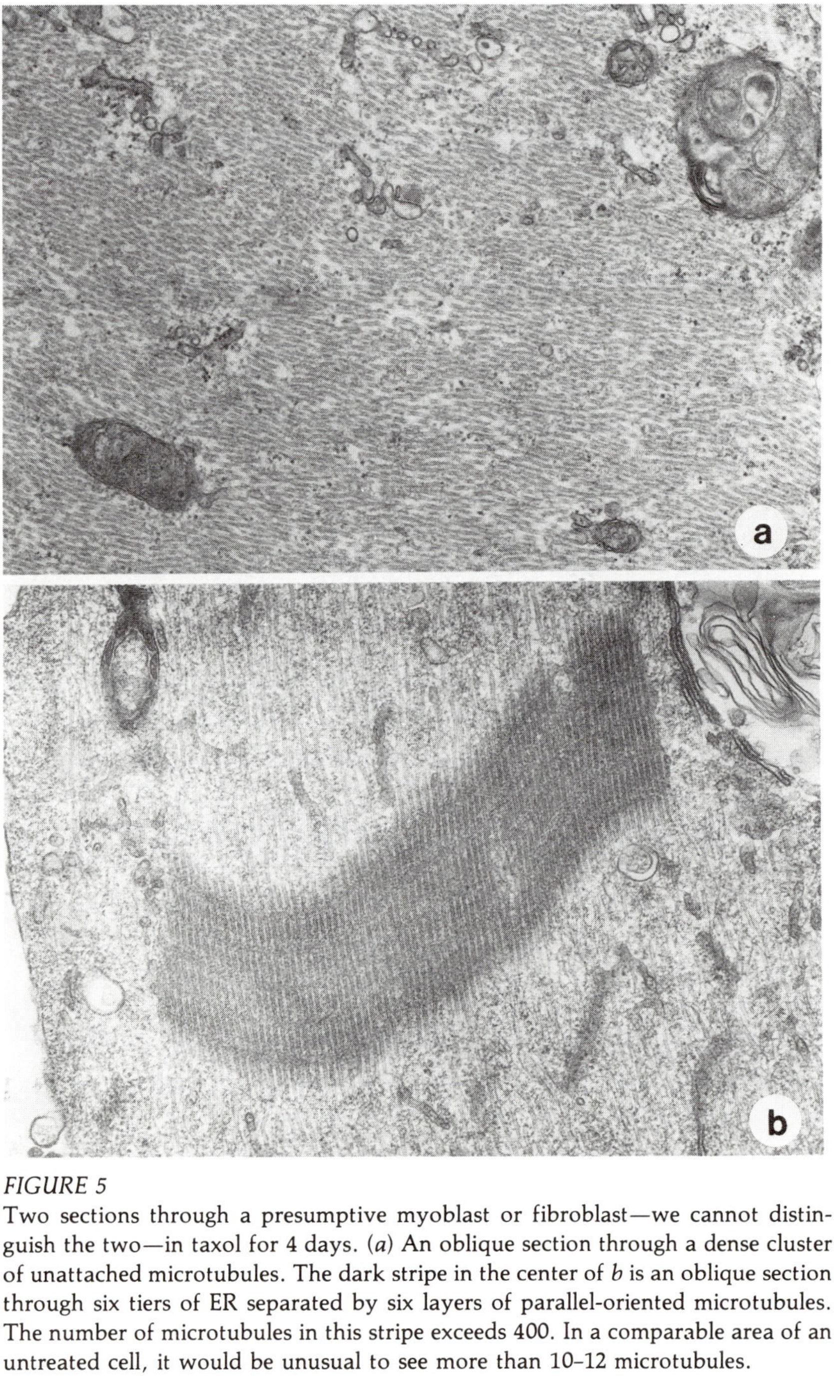

FIGURE 5
Two sections through a presumptive myoblast or fibroblast—we cannot distinguish the two—in taxol for 4 days. (*a*) An oblique section through a dense cluster of unattached microtubules. The dark stripe in the center of *b* is an oblique section through six tiers of ER separated by six layers of parallel-oriented microtubules. The number of microtubules in this stripe exceeds 400. In a comparable area of an untreated cell, it would be unusual to see more than 10–12 microtubules.

length. Single thick filaments are often randomly oriented but are intimately associated with one or more parallel-oriented microtubules. This affinity of thick filaments for microtubules is striking (see also Hayashi et al. 1980; Shima-Oka et al. 1980).

Analysis by Electron Microscopy of Taxol-treated Presumptive Myoblasts and Fibroblasts

The augmentation in numbers of microtubules induced by taxol in presumptive myoblasts and fibroblasts may be even greater than that induced in definitive myoblasts. More intriguing to the endogenous regulation of a cell's cytoarchitecture are the different distributions of the excess microtubules in these cells as compared with the definitive myoblasts. In replicating cells, the arrangement of microtubles falls into four overlapping categories. (1) Solid cables of over 300 microtubules with center-to-center spacings of 40–60 nm and lengths of over 10 μm. The cable itself and each microtubule in the cable is surrounded by an electron-lucent halo. Fine projections link adjacent microtubules. These cables are prominent in moribund cells and are even found extracellularly. Cell organelles are excluded from the domain of such cables. (2) Sheets, consisting of layers of microtubules coursing parallel to one another, and parallel to and connected with the cytoplasmic face of parts of the endoplasmic reticulum (ER) (see also Schiff and Horwitz 1980; Masurovsky et al. 1981). Stacks of two to six ER membranes plus associated microtubules are not uncommon (Fig. 4b) and may exceed 10 μm in length. (3) Lesser aggregates of 10–100 parallel-oriented microtubules separated by distances of over 100 nm. These microtubules intermingle with mitochondria, lysosomes, ER, and so on. Morphologically, they do not appear to interact with one another. (4) Long, curving, isolated microtubules that intersect with other long, curving, isolated microtubules. The electron-lucent halo surrounding most microtubules is particularly prominent when they are distributed in this fashion. Of the above four arrangements of microtubules found in presumptive myoblasts and fibroblasts, neither cables nor sheets of microtubules associated with ER have been observed in myoblasts and/or myotubes.

DISCUSSION

Stabilization of microtubules by taxol is manifested at various levels in myogenic cultures. The finding that taxol arrests and/or kills cells in mitosis (Schiff and Horwitz 1980; De Brabander et al. 1981; Holtzer et al. 1982b) has been substantiated in myogenic cultures. That replicating cells in chick myogenic cultures are selectively killed allows a separation of those cells that replicate during the culture period from those that have already

undergone their final mitosis prior to being cultured. Postmitotic, mononucleated myoblasts do not reenter the cell cycle and therefore are not killed by taxol, demonstrating that these cells are a stable phenotype. Their inability to fuse, along with the observation that TPA blocks fusion in confluent cultures (Cohen et al. 1977), indicates that more than contiguity is required for fusion to occur. Even postmitotic myoblasts that remain contiguous over 4 days in taxol do not fuse.

The morphological changes accompanying taxol treatment of postmitotic, mononucleated myoblasts and myotubes are unique to these cells. That stabilization of microtubules by taxol disturbs the normal bipolarity of postmitotic myoblasts and myotubes is an unanticipated phenomenon and is not readily reconcilable with the role of microtubules in elongating cells (Bischoff and Holtzer 1967). Although such cells become star-shaped in taxol, they also become strikingly striated. Ultrastructurally, taxol-treated postmitotic myoblasts display microtubule-myosin arrays from which thin filaments are absent. Such arrays appear striated, having normal M and H bands, but they have I bands consisting largely of microtubules. More recent work involving TPA (Croop et al. 1982) followed by taxol has revealed more perfectly aligned microtubule-myosin arrays, having normal M and H bands but no Z-band material, sarcoplasmic reticulum (SR), or T-system elements (Toyama et al. 1982). The I bands of such cells consist of laterally aligned microtubules. This unprecedented tandem array of microtubules and myosin filaments that simulate striated myofibrils raises many questions regarding normal myofibrillogenesis. Clearly, thin filaments and Z-band material are dispensible with respect to formation of striated myofibrils. It will be interesting to learn more about the turnover and intracellular localization of α-actin, tropomyosin, troponins, α-actinin, and so on, in these taxol-treated myoblasts and myotubes.

The differences in arrangement of excess microtubules in fibroblasts and presumptive myoblasts versus postmitotic myoblasts and/or myotubes again emphasize the distinctiveness of these cell types (Holtzer and Sanger 1972; Holtzer 1978). Replicating fibroblasts and presumptive myoblasts do not form microtubule-myosin arrays but instead organize their microtubules into immense aggregates including cables and sheets. Microtubules align laterally as well as along the ER in some cases. Thus, microtubules clearly form close associations with themselves, with the ER, and with myosin filaments in the case of postmitotic myoblasts.

ACKNOWLEDGMENTS

This research was supported in part by National Institutes of Health grants HL-15835 (to the Pennsylvania Muscle Institute), HL-18708, CA-18194, and 5-T32-HD07152 and by the Muscular Dystrophy Association of America.

REFERENCES

Abbott, J., J. Schlitz, and H. Holtzer. 1974. The phenotypic complexity of myogenic clones. *Proc. Natl. Acad. Sci.* **71:** 1506.

Antin, P.B., S. Forry-Schaudies, T.M. Friedman, S.J. Tapscott, and H. Holtzer. 1981. Taxol induces postmitotic myoblasts to assemble interdigitating microtubule-myosin arrays that exclude actin filaments. *J. Cell Biol.* **90:** 300.

Auber, J. 1969. La myofibrillogenese du muscle strie. I. Insectes. *J. Microsc.* **8:** 197.

Bennett, G.S., S. Fellini, and H. Holtzer. 1978. Immunofluorescent visualization of 100-Å filaments in different cultured chick embryo cell types. *Differentiation* **12:** 71.

Bennett, G.S., S. Fellini, Y. Toyama, and H. Holtzer. 1979. Redistribution of intermediate filament subunits during skeletal myogenesis and maturation *in vitro*. *J. Cell Biol.* **82:** 577.

Bischoff, R. and H. Holtzer. 1967. The effect of mitotic inhibitors on myogenesis. *J. Cell Biol.* **36:** 111.

——. 1970. Inhibition of myoblast fusion after one round of DNA synthesis in 5-bromodeoxyuridine. *J. Cell Biol.* **44:** 134.

Chi, J.C., A. Fellini, and H. Holtzer. 1975. Differences among myosins synthesized in non-myogenic cells, presumptive myoblasts, and myoblasts. *Proc. Natl. Acad. Sci.* **72:** 4999.

Cohen, R., M. Pacifici, N. Rubinstein, J. Biehl, and H. Holtzer. 1977. Effects of a tumor promotor (PMA) on myogenesis. *Nature* **266:** 583.

Croop, J., B. Dubyak, Y. Toyama, A. Dlugosz, A. Scarpa, and H. Holtzer. 1982. Effects of 12-O-tetradecanoyl-phorbol-13-acetate on myofibril integrity and Ca^{2+} content in developing myotubes. *Dev. Biol.* **89:** 460.

De Brabander, M., G. Geuens, R. Nuydens, R. Willebrords, and J. DeMey. 1982. Microtubule stability and assembly in living cells. The influence of metabolic inhibitors, taxol and pH. *Cold Spring Harbor Symp. Quant. Biol.* **46:** 227.

Dienstman, S. and H. Holtzer. 1975. Myogenesis: A cell lineage interpretation. In *The cell cycle and cell differentiation* (ed. J. Feinert and H. Holtzer), vol. 7, p. 1. Springer-Verlag, Berlin.

——. 1977. Skeletal myogenesis: Control of proliferation in a normal cell lineage. *Exp. Cell Res.* **107:** 355.

Hayashi, M., K. Ohnishi, and K. Hayashi. 1980. Dense precipitates of brain tubulin with skeletal muscle myosin. *J. Biochem.* **87:** 1347.

Holtzer, H. 1978. Cell lineage, stem cells, and the quantal cell concept. In *Stem cells and tissue homeostasis* (ed. B.I. Lord et al.), p. 1. Cambridge University Press, England.

Holtzer, H. and J.W. Sanger. 1972. Myogenesis: Old views rethought. In *Research in muscle development and the muscle spindle* (ed. B. Bander et al.), p. 122. Excerpta Medica, Amsterdam.

Holtzer, H., J. Marshall, and H. Fink. 1957. An analysis of myogenesis by the use of fluorescent antimyosin. *J. Cell Biol.* **3:** 705.

Holtzer, H., J. Sanger, H. Ishikawa, and K. Strahs. 1973. Selected topics in skeletal myogenesis. *Cold Spring Harbor Symp. Quant. Biol.* **37:** 549.

Holtzer, H., G.S. Bennett, S. J. Tapscott, J.M. Croop, and Y. Toyama, 1982a. Intermediate-sized filaments: Changes in synthesis and distribution in cells of

myogenesis. *Cold Spring Harbor Symp. Quant. Biol.* **37:** 549.

Holtzer, H., G.S. Bennett, S. J. Tapscott, J.M. Croop, and Y. Toyama, 1982a. Intermediate-sized filaments: Changes in synthesis and distribution in cells of myogenic and neurogenic lineages. *Cold Spring Harbor Symp. Quant. Biol.* **46:** 317.

Holtzer, H., G.S. Bennett, S.J. Tapscott, J.M. Croop, A. Dlugosz, and Y. Toyama. 1981. Changes in intermediate-sized filaments during myogenesis and neurogenesis. In *2nd International Symposium of Cell Biology,* Berlin (ed. H.G. Schweiger), p. 293. Springer-Verlag, Berlin.

Holtzer, H., P.B. Antin, S. Forry-Schaudies, S. Xue, H. Chang, D.S. Wang, and K.J. Wang. 1982b. Differential effects of taxol on postmitotic myoblasts, fibroblasts and CHO cells. In *8th International Symposium on Pharmacology,* Tokyo (ed. S. Ebashi), p. 173. Pergamon Press, Oxford.

Masurovsky, E.B., E.R. Peterson, S.M. Crain, and S.B. Horwitz. 1981. Microtubule arrays in taxol-treated mouse dorsal root ganglion-spinal cord cultures. *Brain Res.* **217:** 392.

Schiff, P.B. and S.B. Horwitz. 1980. Taxol stabilizes microtubules in mouse fibroblast cells. *Proc. Natl. Acad. Sci.* **77:** 1561.

Shimo-Oka, T., M. Hayashi, and Y. Satanabe. 1980. Tubulin-myosin interaction. Some properties of binding between tubulin and myosin. *Biochemistry* **21:** 4921.

Toyama, Y., C. West, and H. Holtzer. 1979. Differential response of myofibrils and 10 mm filaments to a co-carcinogen. *Am. J. Anat.* **156:** 131.

Toyama, Y., S. Forry-Schaudies, B. Hoffman, and H. Holtzer. 1982. Effects of taxol and colcemid on myofibrillogenesis. *Proc. Natl. Acad. Sci.* (in press).

Yeoh, G.C. and H. Holtzer. 1977. The effect of cell density, conditioned medium and cytosine arabinoside on myogenesis in primary and secondary cultures. *Exp. Cell Res.* **104:** 63.

MORPHOGENESIS OF THE CYTOSKELETON

Introduction: Myofibrillar Assembly

*DONALD A. FISCHMAN**
Department of Anatomy and Cell Biology
State University of New York
Downstate Medical Center
Brooklyn, New York 11203

The structure, composition, and functional properties of the myofibril in cross-striated muscle are well understood, but our knowledge of its assembly and turnover is inadequate. The precise double hexagonal lattice of interdigitating thick and thin myofilaments, the square array of the Z disc, and the transverse lattice bridges of the M band (Squire 1981) have been postulated to form via self-assembly processes underlying myofibrillogenesis (Fischman 1970). Compatible with such considerations are the well-recognized self-assembly properties of myosin to form bipolar thick filaments and the polymerization of G- to F-actin (Huxley 1963), which have provoked interest in models akin to those previously demonstrated for the morphogenesis of simple viruses and bacterial flagellae.

However, attractive as these models are, it should be recognized that almost 20 years have elapsed since Huxley's landmark study (1963) on the structural polymers formed by myofibrillar proteins. Since that publication, no confirmed reports have appeared that demonstrate the successful reconstruction of myofibrils or of length-regulated myofilaments in vitro from pure or impure components. It is likely that many biochemical processes and structural features of the cortical cytoplasm, inner face of the plasma membrane, and cytoskeleton will need to be replicated before successful myofibril assembly can be achieved in a cell-free system.

*Present address: Department of Cell Biology and Anatomy, Cornell University Medical College, New York, New York 10021.

It is the purpose of this brief introduction to emphasize some of the major gaps in our understanding of myofibrillogenesis with the hope that these can be addressed and filled in over the next few years.

I have divided the problem arbitrarily into the following categories: (1) differential accumulation, distribution, and functional properties of protein isoforms; (2) linkage of translation to macromolecular assembly; (3) roles of plasma membrane, cortical filaments, microtrabecular lattice, and cytoskeletal elements in assembly and subsequent morphogenesis of the myofibril; (4) elucidation of minor components of the myofibril and quantitation of their molar ratios; and (5) mechanisms underlying the remodeling and turnover of myofibrils.

PROTEIN ISOFORMS

One of the most striking developments of the past 10 years, largely from the introduction of two-dimensional gel electrophoresis, gene cloning, and monoclonal antibody procedures, has been the remarkable proliferation of contractile protein isoforms. It is apparent that actin (Engel et al. 1981; Firtel 1981; Shani et al. 1981), myosin heavy chains (MHCs) (Epstein et al. 1974; Masaki and Yoshizaki 1974; Gauthier et al. 1978; Rushbrook and Stracher 1979; Whalen et al. 1979; Umeda et al. 1981; Mahdavi et al. 1982) and myosin light chains (MLCs) (Hoh and Yeoh 1979; Lowey et al. 1979; Whalen and Sell 1980), C protein (Callaway and Bechtel 1981; Yamamoto and Moos 1981; Reinach et al. 1982), troponin (Cummins and Perry 1978; Dhoot et al. 1979; Toyota and Shimada 1981), and tropomyosin (Roy et al. 1979) are all products of multigene families. The functional properties of these protein isoforms must be established before their roles in myogenesis can be understood. Although it has been feasible to correlate the activity of different myosin ATPases with the speeds of shortening in muscles containing these respective isoforms (Barany 1967), it is less obvious what functions these different MHCs or MLCs might play during myogenesis. No clear-cut differences in the catalytic properties of fetal and adult myosin ATPases have been demonstrated (Dow and Stracher 1971; Sreter et al. 1975; Syrovy and Gutmann 1977; Benfield et al. 1981). Putative functions in myofilament assembly must be considered, and interesting differences in the aggregative properties of fetal myosin have been reported (Whalen et al. 1981). However, improved methods for quantifying myosin filamentogenesis need to be established before the self-assembly properties of fetal myosin can be rigorously compared with adult isoforms. It will then be feasible to compare and evaluate their interactive properties. Although much attention has been given to the isolation of factors that regulate the assembly and stability of actin polymers (Korn 1982), no corresponding molecules have been demonstrated that might trigger or regulate myosin polymerization in vertebrates. Paramyosin appears to be necessary for

proper in vivo assembly of myosin in invertebrates (Mackenzie and Epstein 1980). Recent reports detailing possible roles of myosin phosphorylation in smooth and nonmuscle myosin assembly (Suzuki et al. 1978; Kuczmarski and Spudich 1980; Scholey et al. 1980) should be extended to embryonic material. Amino acid sequences in the rod fragment of MHC, presumably derived from nucleotide sequencing procedures, will soon be available for most of the MHC isoforms (Mahdavi et al. 1982).

TRANSLATION AND ASSEMBLY

Although most subunits of the myofibrillar proteins have been synthesized in cell-free translation systems, very little is known of their subunit interactions or multimeric associations at physiological ionic strength. For example, each myosin molecule contains a dimer of heavy chains and two pairs of light chains. The MHCs are insoluble at 0.15 M salt, and rather drastic conditions are required to dissociate or reassociate these heavy chains in vitro. Do adjacent MHCs on sequential ribosomes of a polysome interact while undergoing translation? Are these nascent peptides kept in an extended, hydrophobic environment, perhaps on or within intermediate filaments (Lenk et al. 1977; Fulton et al. 1980), during synthesis and subunit association? No one has reported the isolation of ribosome dimers held together by nascent MHCs bridging adjacent ribosomes. It remains to be proved whether MHCs are released from the ribosomes into a soluble pool prior to their dimeric assembly.

Many of these questions will require a combination of protein biochemistry, high-resolution immunocytochemistry (Tokuyasu et al. 1981), and in situ hybridization with recombinant probes (Langer et al. 1981; Wu and Davidson 1981). Although technically demanding, it should nevertheless be feasible to identify which polysomes are synthesizing specific peptides and localize such to individual compartments of the cytoplasm. The recombinant probes may permit the localization of unique mRNA species to specific cytoplasmic domains and hopefully identify the polysomes and nontranslating messenger ribonucleoproteins that contain them. With monoclonal antibodies specific for MHC or actin isoforms, one can address questions such as, Do fetal MHCs codistribute with neonatal or adult isoforms in common molecules, thick myofilaments, or sarcomeres? Does α-actin form copolymers in vivo with β- or γ-actins and do these colocalize in the same myofibrils? Are the polysomes that synthesize one isoform of a protein spatially segregated from those synthesizing another? Recent studies using electron microscopy (Peng et al. 1981) emphasize the need for such analyses. As Porter and colleagues have emphasized for a number of years (Wolosewick and Porter 1979), the cytoplasmic ground substance is a complex array of filamentous material, and any comprehensive model of myofibrillar assembly must take this into account.

MEMBRANE, CORTEX, CYTOSKELETON, AND MICROTRABECULAR LATTICE

Provocative hypotheses have been proposed for the role of cytoskeletal elements in the assembly of myofibrils (Fulton 1981). It has been suggested that the cytoskeleton is essential for the vectorial movements and deposit of nascent myofibrillar proteins to sites of sarcomere assembly. It has been appreciated for some time that nascent myofibrils have an intimate association with the inner face of the myotube plasma membrane, and electron-dense material that might serve an adhesive function can be resolved at that location (Fischman 1967). Although Z-band attachments have been stressed, there may be additional points of myofibril insertion at the cell surface that should not be overlooked. The inner face of the plasma membrane may contain high-affinity binding sites for certain nascent peptides, and these arranged with the polarity required for the anisotropy of myofibrils (Shimada and Obinata 1977). With improved isolation of inside-out, plasma-membrane vesicles, it should be feasible to address this problem. In like manner, one might query the potential template function of cortical cytoplasmic components during myofibrillogenesis. Filamentous elements at the myotube cortex might provide both the attachment and assembly sites required for the initial steps in lattice assembly. Later growth of the lattice might then follow rules for quasi-equivalent structures proposed for regular viruses (Caspar and Klug 1963).

An interesting new approach has been reported (Taniguchi and Ishikawa 1982) in which partial reconstitution of A bands has been achieved in permeabilized myotubes following myosin extraction.

QUANTITATION OF MINOR MYOFIBRILLAR PROTEINS

A number of proteins associated with myofibrils, e.g., C protein (Offer 1973), M protein (Masaki et al. 1968; Morimoto and Harrington 1972; Trinick and Lowey 1977; Strehler et al. 1980), α-, β-, and eu-actinins (Goll et al. 1972; Maruyama 1976; Kuroda et al. 1981), end-filament protein (Trinick 1981), titin (Wang et al. 1979), nebulin (Wang and Williamson 198), desmin (Lazarides and Hubbard 1976), vimentin (Bennett et al. 1978; Franke et al. 1978), and synemin (Granger and Lazarides 1980), are found in relatively low concentrations, but their presence may be essential for myofibrillogenesis. Very few of these proteins are well understood; most are relatively insoluble in physiological buffers and are difficult to work with. This is particularly true for proteins of the intermediate filaments (desmin, vimentin, and synemin), which may be important in lateral alignment and linkage of adjacent myofibrils and their attachment to the cell surface (Lazarides 1980). It remains to be demonstrated what role, if any, phosphorylation of desmin plays in this process (Gard and Lazarides 1980).

TURNOVER OF MYOFIBRILS

Extensive turnover of myofibrillar proteins occurs throughout embryonic (Walker and Strohman 1978) and adult (Zak et al. 1979) life. That such turnover occurs in functional myofibrils is a remarkable phenomenon without adequate explanation. Turnover in all cases seems to follow first-order kinetics, implying that any molecule of a particular protein is as likely to be degraded, at one moment in time, as any other molecule of the same protein. Insertion and deletion mechanisms required for such turnover have not been uncovered. For example, are myosin molecules added or removed as single entities or as block myofilaments from existing sarcomeres? Is the longitudinal integrity of a myofilament, sarcomere, or myofibril disrupted during the turnover process? Since different isoforms of MHC are sequentially replaced during muscle development, a unique system may be available for following the turnover process with antibodies specific for the different isoforms.

Finally, although space does not permit significant discussion, attention should be drawn to the probable role of the extracellular matrix in establishing the basic anisotropy of the cell surface which, as discussed above, may provide the template for myofibril alignment and deposition in the myoplasm.

ACKNOWLEDGMENTS

The author gratefully acknowledges the criticisms of Fernando Reinach and David Bader during the preparation of this manuscript. Research support has been provided by the National Institutes of Health (AM-32147) and the Muscular Dystrophy Association of America.

REFERENCES

Barany, M. 1967. ATPase activity of myosin correlated with speed of shortening. *J. Gen. Physiol.* **50:** 197.

Benfield, P.A., S. Lowey, and D. Le Blane. 1981. Fractionation and characterization of myosins from embryonic chicken pectoralis muscle. *Biophys. J.* **33:** 243a.

Bennett, G.S., S.A. Fellini, J.M. Croop, J.J. Otto, J. Bryant, and H. Holtzer. 1978. Differences among 100 Å-filament subunits from different cell types. *Proc. Natl. Acad. Sci.* **75:** 4364.

Callaway, J.E. and P.J. Bechtel. 1981. C-protein from rabbit soleus (red) muscle. *Biochem. J.* **195:** 463.

Caspar, D.L.D. and A. Klug. 1963. Physical principles in the construction of regular viruses. *Cold Spring Harbor Symp. Quant. Biol.* **27:** 1.

Cummins, P. and S.V. Perry. 1978. Troponin I from human skeletal and cardiac muscles. *Biochem. J.* **171:** 251.

Dhoot, G.K., N. Frearson, and S.V. Perry. 1979. Polymorphic forms of troponin T

and troponin C and their localization in striated muscle cell types. *Exp. Cell Res.* **122:** 339.

Dow, J. and A. Stracher. 1971. Changes in the properties of myosin associated with muscle development. *Biochemistry* **10:** 1316.

Engel, J.N., P.W. Gunning, and L. Kedes. 1981. Isolation and characterization of human actin genes. *Proc. Natl. Acad. Sci.* **78:** 4674.

Epstein, H.F., R.H. Waterston, and S. Brenner, 1974. A mutant affecting the heavy chain of myosin *Caenorhabditis elegans. J. Mol. Biol.* **90:** 291.

Firtel, R.A. 1981. Multigene families encoding actin and tubulin. *Cell* **24:** 6.

Fischman, D.A. 1967. An electron microscope study of myofibril formation in embryonic chick skeletal muscle. *J. Cell Biol.* **32:** 557.

———. 1970. The synthesis and assembly of myofibrils in embryonic muscle. *Curr. Top. Dev. Biol.* **5:** 235.

Franke, W.W., E. Schmid, M. Osborn, and K. Weber. 1978. Different intermediate sized filaments distinguished by immunofluorescence microscopy. *Proc. Natl. Acad. Sci.* **75:** 5034.

Fulton, A.B. 1981. How do eucaryotic cells construct their cytoarchitecture? *Cell* **24:** 4.

Fulton, A.B., K.M. Wan, and S. Penman. 1980. The spatial distribution of polyribosomes in 3T3 cells and the associated assembly of proteins in the skeletal framework. *Cell* **20:** 849.

Gard, D.L. and E. Lazarides. 1980. The synthesis and distribution of desmin and vimentin during myogenesis *in vitro*. *Cell* **19:** 263.

Gauthier, G.F., S. Lowey, and A.W. Hobbs. 1978. Fast and slow myosin in developing muscle fibers. *Nature* **274:** 25.

Goll, D.E., A. Suzuki, J. Temple, and G.R. Holmes. 1972. Studies on purified α-actinin. I. Effect of temperature and tropomyosin on α-actinin/F-actin interaction. *J. Mol. Biol.* **67:** 469.

Granger, B.L. and E. Lazarides. 1980. Synemin: Two new high molecular weight proteins associated with desmin and vimentin filaments in muscle. *Cell* **22:** 727.

Hoh, J.F.Y. and G.P.S. Yeoh. 1979. Rabbit skeletal myosin isoenzymes from fetal, fast-twitch and slow twitch muscles. *Nature* **280:** 321.

Huxley, H.E. 1963. Electron microscope studies of the structure of natural and synthetic protein filaments from striated muscle. *J. Mol. Biol.* **7:** 281.

Korn, E.D. 1982. Actin polymerization and its regulation by proteins from nonmuscle cells. *Physiol. Rev.* **62:** 672.

Kuczmarski, C.R. and J.A. Spudich. 1980. Regulation of myosin self-assembly: Phosphorylation of *Dictyostelium* heavy chain inhibits formation of thick filaments. *Proc. Natl. Acad. Sci.* **77:** 7292.

Kuroda, M., T. Tanaka, and T. Masaki. 1981. Eu-actinin, a new structural protein of the Z-line of striated muscles. *J. Biochem.* **89:** 279.

Langer, P.R., A.A. Waldrop, and D.C. Ward. 1981. Enzymatic synthesis of biotin-labeled polynucleotides: Novel nucleic acid affinity probes. *Proc. Natl. Acad. Sci.* **78:** 6633.

Lazarides, E. 1980. Intermediate filaments as mechanical integrators of cellular space. *Nature* **283:** 249.

Lazarides, E. and B.D. Hubbard. 1976. Immunological characterization of the subunit 100 Å filaments from muscle cells. *Proc. Natl. Acad. Sci.* **73:** 4344.

Lenk, R., L. Ransom, Y. Kaufman, and S. Penman. 1977. A cytoskeletal structure with associated polyribosomes obtained from HeLa cells. *Cell* **10:** 67.

Lowey, S., P.A. Benfield, L. Silberstein, and L.M. Lang. 1979. Distribution of light chains in fast skeletal myosin. *Nature* **282:** 522.

Mackenzie, J.M., Jr. and H.F. Epstein. 1980. Paramyosin is necessary for determination of nematode thick filament lenght *in vivo*. *Cell* **22:** 747.

Mahdavi, V., M. Periasamy, and B. Nadal-Ginard. 1982. Molecular characterization of two myosin heavy chain genes expressed in the adult heart. *Nature* **297:** 659.

Maruyama, K. 1976. Actinins, regulatory proteins of muscle. *Adv. Biophys.* **9:** 157.

Masaki, T. and C.J. Yoshizaki. 1974. Differentiation of myosin in chick embryos. *J. Biochem.* **76:** 123.

Masaki, T., S. Takaiti, and S. Ebashi. 1968. "M-substance," a new protein constituting the M-line of myofibrils. *J. Biochem.* **71:** 355.

Morimoto, K. and W.F. Harrington. 1972. Isolation and physical chemical properties of an M-line protein from skeletal muscle. *J. Biol. Chem.* **247:** 3052.

Offer, G. 1973. C-protein and the periodicity in the thick filaments of vertebrate skeletal muscle. *Cold Spring Harbor Symp. Quant. Biol.* **37:** 87.

Peng, H.B., J.J. Wolosewick, and P.-C. Cheng. 1981. The development of myofibrils in cultured muscle cells: A whole-mount and thin section electron microscopic study. *Dev. Biol.* **88:** 121.

Reinach, F.C., T. Masaki, S. Shafiq, T. Obinata, and D.A. Fischman. 1982. Isoforms of C-protein in adult chicken skeletal muscle: Detection with monoclonal antibodies. *J. Cell Biol.* **95:** (in press.)

Roy, R.K., F.A. Sreter, and S. Sarkar. 1979. Changes in tropomyosin subunits and myosin light chains during development of chicken and rabbit striated muscles. *Dev. Biol.* **69:** 15.

Rushbrook, J.I. and A. Stracher. 1979. Comparison of adult, embryonic and dystrophic myosin heavy chains from chicken muscle by sodium dodecyl sulfate/polyacrylamide gel electrophoresis and peptide mapping. *Proc. Natl. Acad. Sci.* **76:** 4331.

Scholey, J.M., K.A. Taylor, and J. Kendrick-Jones. 1980. Regulation of nonmuscle myosin assembly by calmodulin-dependent light chain kinase. *Nature* **287:** 2331.

Shani, M., D. Zevin-Sonkin, O. Saxel, Y. Carmon, D. Katcoff, U. Nudel, and D. Yaffe. 1981. The correlation between the synthesis of skeletal muscle actin, myosin heavy chain and myosin light chain and the accumulation of corresponding mRNA sequences during myogenesis. *Dev. Biol.* **86:** 483.

Shimada, Y. and T. Obinata. 1977. Polarity of actin filaments at the initial stage of myofibril assembly in myogenic cells *in vitro*. *J. Cell Biol.* **72:** 777.

Squire, J. 1981. *The structural basis of muscular contraction.* Plenum Press, New York.

Sreter, F., M. Balint, and J. Gergeley. 1975. Structural and functional changes of myosin during development. Comparison with adult fast, slow and cardiac myosin. *Dev. Biol.* **46:** 317.

Strehler, E.E., G. Pelloni, C.W. Heizmann, and H.M. Eppenberger. 1980. Biochemical and ultrastructural aspects of M_r 165,000 M-protein in cross-striated chicken muscle. *J. Cell Biol.* **86:** 775.

Suzuki, H., H. Onishi, K. Takahashi, and S. Watanabe. 1978. Structure and function of chicken gizzard myosin. *J. Biochem.* **84:** 1529.

Syrovy, I. and E. Gutmann. 1977. Differentiation of myosin in soleus and extensor digitorum longus muscle in different species and during development. *Plfügers Arch.* **369:** 85.

Taniguchi, M. and H. Ishikawa. 1982. *In situ* reconsitution of myosin filaments within the myosin-extracted myofibril in cultured skeletal muscle cells. *J. Cell Biol.* **92:** 324.

Tokuyasu, K.T., A.H. Dutton, B. Geiger, and S.J. Singer. 1981. Ultrastructure of chicken cardiac muscle as studied by double immunolabeling in electron microscopy. *Proc. Natl. Acad. Sci.* **78:** 7619.

Toyota, N. and Y. Shimada. 1981. Differentiation of troponin in cardiac and skeletal muscles in chicken embryos as studied by immunofluorescence microscopy. *J. Cell Biol.* **91:** 497.

Trinick, J.A. 1981. End-filaments: A new structural element of vertebrate skeletal muscle thick filaments. *J. Mol. Biol.* **151:** 309.

Trinick, J. and S. Lowey. 1977. M-protein from chicken pectoralis muscle: Isolation and characterization. *J. Mol. Biol.* **113:** 343.

Umeda, P.K., A.M. Sinha, S. Jakovcic, S. Merten, H.J. Hsu, K.N. Subramanian, R. Zak, and M. Rabinowitz. 1981. Molecular cloning of two fast myosin heavy chain cDNAs from chicken embryo skeletal muscle. *Proc. Natl. Acad. Sci.* **78:** 2843.

Walker, C. and R. Strohman. 1978. Myosin turnover in cultured muscle fibers relaxed by tetrodotoxin. *Exp. Cell Res.* **116:** 341.

Wang, K. and C.L. Williamson. 1980. Identification of an N_2 line protein of striated muscle. *Proc. Natl. Acad. Sci.* **77:** 3254.

Wang, K., J. McClure, and A. Tu. 1979. Titin: Major myofibrillar components of striated muscle. *Proc. Natl. Acad. Sci.* **76:** 3698.

Whalen, R.G. and S.M. Sell. 1980. Myosin from fetal hearts contains the skeletal muscle embryonic light chain. *Nature* **286:** 731.

Whalen, R.G., K. Schwartz, P. Bouveret, S.M. Sell, and F. Gros. 1979. Contractile protein isozymes in muscle development. Identification of an embryonic form of myosin heavy chain. *Proc. Natl. Acad. Sci.* **76:** 5197.

Whalen, R.G., S.M. Sell, G.S. Butler-Browne, K. Schwartz, P. Bouveret, and I. Pinset-Härström. 1981. Three myosin heavy-chain isozymes appear sequentially in rat muscle development. *Nature* **292:** 805.

Wolosewick, J.J. and K.R. Porter. 1979. Microtrabecular lattice of the cytoplasmic ground substance artifact or reality. *J. Cell Biol.* **82:** 114.

Wu, M. and N. Davidson. 1981. Transmission electron microscopic method for gene mapping on polytene chromosomes by *in situ* hybridization. *Proc. Natl. Acad. Sci.* **78:** 7059.

Yamamoto, K. and C. Moos. 1981. C-proteins of vertebrate white, red and cardiac muscles. *J. Gen. Physiol.* **78:** 31a.

Zak, R., A.F. Martin, and R. Blough. 1979. Assessment of protein turnover by use of radioisotropic tracers. *Physiol. Rev.* **59:** 407.

Monoclonal Antibody Analysis of Myosin Heavy-chain and C-protein Isoforms during Myogenesis

TOMOH MASAKI[*†]
DAVID M. BADER[*]
FERNANDO C. REINACH[*]
TERUO SHIMIZU[*]
TAKASHI OBINATA[*‡]
SAIYID A. SHAFIQ[§]
DONALD A. FISCHMAN[*]
[*]*Department of Anatomy and Cell Biology*
[§]*Departments of Neurology and Biophysics*
State University of New York
Downstate Medical Center
Brooklyn, New York 11203

Several myofibrillar proteins, of both thick and thin myofilaments, undergo isoform transitions during myogenesis. For example, α-actin transcription and accumulation occurs at the time of myofibrillogenesis, whereas β- and γ-isoforms of actin are synthesized in prefusion myoblasts (Garrels and Gibson 1976; Whalen et al. 1976; Schwartz and Rothblum 1981). Two other thin-filament proteins, troponin and tropomyosin, exhibit isoform variants (Dhoot and Perry 1979; Obinata et al. 1979; Roy et al. 1979; Matsuda et al. 1981). There is reason to suspect that these protein changes are regulated coordinately during development (Devlin and Emerson 1979), but neural, hormonal, and other environmental influences on gene expression remain unresolved.

Biochemical and immunochemical studies have demonstrated that the myosin heavy chains (MHCs) in adult fast skeletal muscles differ from those

Permanent address: [†]Tokyo Metropolitan Institute of Medical Sciences, Tokyo, Japan; [‡] Department of Biology, Chiba University, Chiba, Japan.

in the identical embryonic muscle (Trayer et al. 1968; Huszar 1972; Masaki and Yoshizaki 1974; Sreter et al. 1975; Gauthier et al. 1978; Rushbrook and Stracher 1979; Whalen et al. 1979, 1981). Nevertheless, questions remain unresolved concerning the number of MHC isoforms, the size of this gene family, their transcriptive and translational regulation, and possible function of the variant proteins during morphogenesis. The size of the MHC, its associative properties, and antigenic complexity have complicated the analysis of myosin isoforms in developing muscle. Specific, high-affinity probes are needed to detect and quantitate what may be only slight differences in MHC structure. To this end, monoclonal antibodies (McAbs) were generated against myosin preparations from the pectoralis muscle of adult chickens to compare the antigenic relatedness between myosin isoforms during development. An unexpected by-product of the study was the production of several McAbs specific for C protein, and these have proved to be of value in analyzing isoforms of this protein.

MATERIALS AND METHODS

Proteins

Myosin was prepared from adult, neonatal, and embryonic pectoralis muscle (see below for specific ages) of White Leghorn chickens. Minced muscle was extracted in a modified Guba-Straub solution (0.3 M KCl, 0.1 M KH_2-PO_4, 0.05 M K_2HPO_4, 5 mM Mg-ATP) plus 0.1 mM phenylmethylsulfonyl fluoride (PMSF) for 15 minutes on ice, gauze-filtered, and precipitated at 0.05 M ionic strength. Mg-ATP was not included in this buffer during extraction of myosin from adult pectoralis muscle. The pellet was dissolved in 0.5 M KCl, 5 mM $MgCl_2$, and 5 mM ATP and cleared for 2 hours at 40,000 rpm in a Ti-45 rotor. The pellet obtained after one further precipitation of low ionic strength was labeled non-column-purified (NCP)-myosin to distinguish it from column-purified (CP)-myosin, which had been further column-purified by the method of Offer et al. (1973) using DEAE-Sephadex A-50. Adult pectoralis C protein was isolated from the DEAE-Sephadex A-50 effluent of myosin and further purified by hydroxylapatite chromatography (Starr et al. 1980; C. Moos, pers. comm.).

Radioimmunoassay

Protein samples (usually 1 μg) applied in 25 μl of 0.6 M NaCl to wells of flexible, polyvinyl-chloride microtiter plates were air-dried and stored at 4°C. After thorough immersion in phosphate-buffered saline (PBS) (0.12 M NaCl, 10 mM PO_4 buffer at pH 7.4), the wells were washed once in PBS containing 1% bovine serum albumin (BSA), and then reacted with 30 μl of McAb supernatant for 30 minutes at room temperature. After washing in

PBS, the samples were incubated in 40 μl (50,000 cpm) of ^{125}I-labeled goat anti-mouse (GAM) immunoglobulin (IgG) for 30 minutes, and bound ^{125}I was measured by γ spectrometry. The GAM IgG was purchased from Cappel Laboratories and iodinated while bound to a mouse IgG-Sepharose 4B column using the chloramine-T procedure of Hunter and Greenwood (1962). The labeled, affinity-purified GAM was diluted to 10^7 cpm/ml and stored frozen at −20°C in PBS-BSA for a maximum of 60 days.

McAbs

Spleen cells of mice injected with NCP-myosin were fused with myeloma cells by the procedures of Galfre et al. (1977), as modified by Gefter et al. (1977), with polyethylene glycol. Antigen injections contained 100 μg of myosin—the first injection in complete Freund's adjuvant, the subsequent injections in incomplete Freund's adjuvant. The myeloma cells used for the pectoralis myosin fusion are labeled P3U-1 (obtained from J. Unkeless, Rockefeller University, New York); P3NP cells (obtained from M. Scharff, Albert Einstein College of Medicine, Bronx, New York) were used for the cell fusion after mouse immunization with anterior latissimus dorsi (ALD) myosin. Both myeloma cell lines are hypoxanthine, aminopterin, thymidine (HAT)-sensitive and do not secrete IgG. Hybridomas were screened by solid-phase radioimmunoassay (RIA) with the dried immunogen, cloned twice in soft agar, and expanded in Dulbecco's modified Eagle's medium (DME) supplemented with glucose, glutamine, pyruvate, and 20% fetal calf serum (GIBCO).

Electrophoresis and Immunoblots

SDS-polyacrylamide gel electrophoresis (PAGE) was performed according to Laemmli (1970), and gel-displayed proteins were transferred electrophoretically in the presence of 40 mM Tris, 240 mM glycine (pH 8.7), and 20% ethanol for 2 hours with a constant current of 100 mÅ (Towbin et al. 1979; Burnette 1981). To improve transfer of MHC, 0.1% SDS was added to the back sponge. After protein transfer, the nitrocellulose sheets were reacted with first and second antibodies as described for the solid-phase RIA, but antibody incubations were shortened to 5 minutes each. Exposure of X-ray film was usually less than 24 hours.

Immunofluorescence

Glycerinated muscle was prepared essentially by the method of Huxley (1963) but with the addition of 10^{-4} M PMSF. For separation of myofibrils, the muscle strips were equilibrated in 0.1 M NaCl, 50 mM Tris HCl, and 1 mM EDTA (pH 7.5) and then Dounce-homogenized. Myofibrils were re-

acted with McAb and affinity-purified, fluorescein-conjugated GAM (15 min each), while adherent to glass microscope slides; they were then washed, mounted in the same buffer, sealed with nail polish, and photographed with phase and epifluorescent optics using a Zeiss microscope. Frozen sections were stained in the same manner, but fixed in 4% formaldehyde after the fluorescein-GAM, and mounted in 90% glycerol prior to photography.

RESULTS AND DISCUSSION

Characterization of McAb-binding Sites

Since NCP-myosin contains a minor but significant number of contaminants (Starr and Offer 1971) with prominent antigenicity (Pepe and Drucker 1975; Craig and Offer 1976), it was important to establish the binding specificity for each McAb. Three different approaches were used: (1) solid-phase RIA, (2) immunoblots, and (3) immunofluorescence.

NCP-myosin was fractionated by DEAE-Sephadex chromatography (Offer et al. 1973) and assayed for McAb binding by solid-phase RIA (Fig. 1a). Two groups of antibodies could be distinguished by this procedure. One group (e.g., MF-21) bound strongly to proteins not retained by the DEAE column, whereas the other McAbs (e.g., MF-20) reacted with proteins eluting in the main myosin peak. Immunoblots were then used to identify which peptide in the contaminant fractions was recognized by MF-21 (Fig. 1b). We observed MF-21 binding to a protein band of approximately 150 kD in both NCP-myosin and the flow-through fractions of the DEAE column. No binding of MF-21 to the 200-kD MHC band was detected (Fig. 1b). Other McAbs (MF-14, MF-20, and MF-30) bound to the 200-kD band (Fig. 1c), although binding was sometimes observed with peptides of lower molecular weight, which we believe to be proteolytic subfragments of the MHC.

FIGURE 1 *(see facing page)*
(*a*) DEAE-Sephadex A-50 purification of adult pectoralis myosin. The first small peak contains unretained contaminants (mainly C protein); the second peak consists of CP-myosin. 1 μg of selected fractions was reacted in RIA with MF-20 and MF-21. MF-21 (○) recognizes a determinant present in the first peak, whereas MF-20 (●) binds only to the second peak. (*b*) NCP-myosin (*1*), crude C protein (*2*), and molecular-weight markers (*3*) were run on a 7.5% SDS-PAGE, transferred to nitrocellulose paper, and reacted with MF-21. C protein (►) present in NCP-myosin (*1′*) and the crude C-protein (*2′*) preparation bound MF-21. No binding to myosin or molecular-weight markers (*3′*) was observed. (*c*) NCP-myosin (*1*) was processed for immunoblotting as in *b* and reacted with MF-14 (*2*), MF-30 (*3*), and MF-20 (*4*). All McAbs bound to the 200-kD MHC (►). (*d*) Phase and indirect immunofluorescence microscopy of adult pectoralis myofibrils with MF-30, MF-14, MF-20, and MF-21. All micrographs are printed at a final magnification of 1300×.

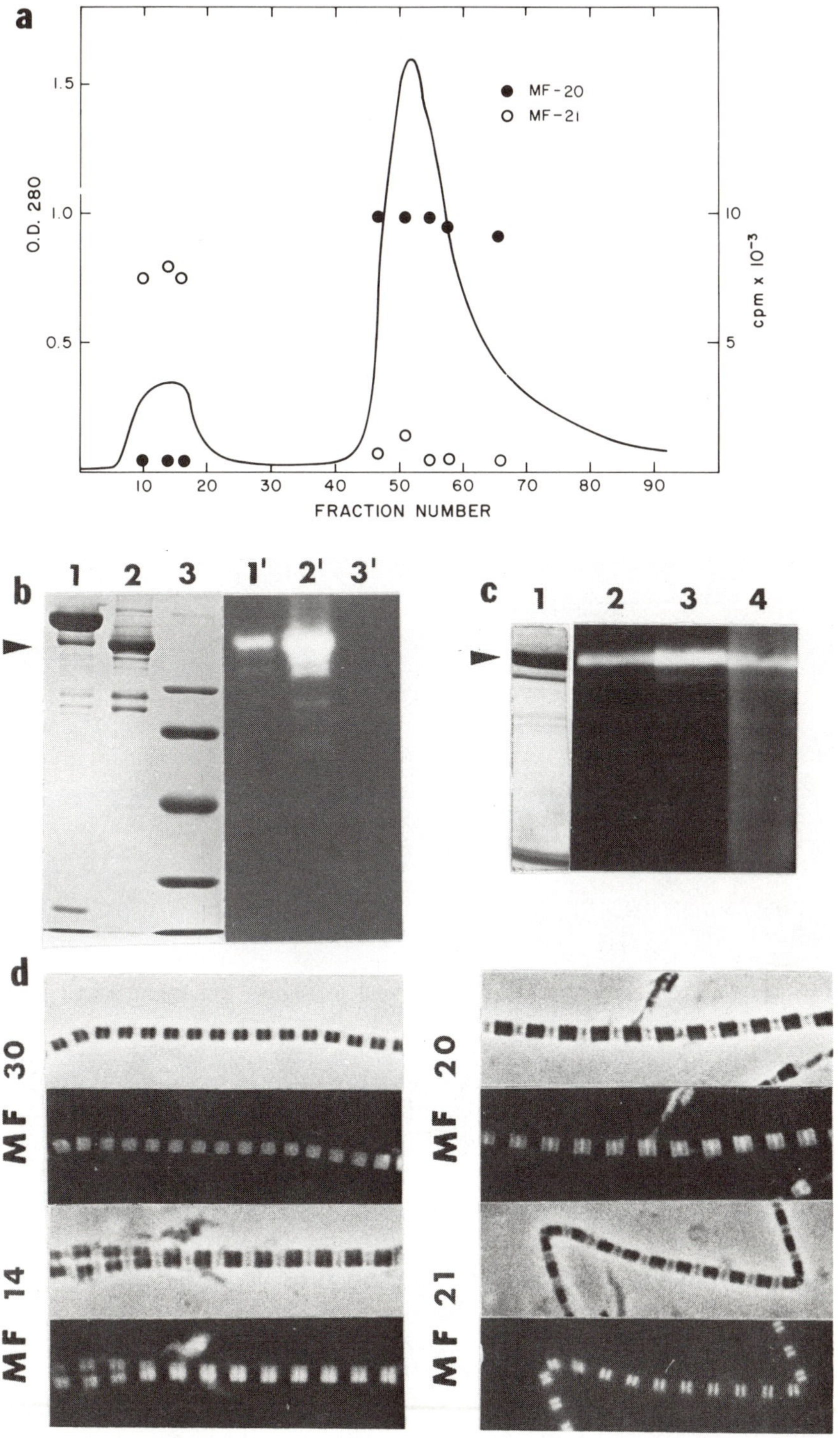
a
1.5
1.0
0.5
O.D. 280
10
5
cpm x 10⁻³
MF-20
MF-21
10
20
30
40
50
60
70
80
90
FRACTION NUMBER
b
1
2
3
1'
2'
3'
c
1
2
3
4
d
MF 30
MF 14
MF 20
MF 21

To establish the morphological distribution of specific epitopes within myofibrils, indirect immunofluorescent experiments were performed with glycerinated myofibrils from pectoralis muscle. Two transverse fluorescent stripes were observed in the A bands following MF-21 staining (Fig. 1d). The longitudinal distance between the outer edge of one stripe to the outer edge of the other is approximately 1 μm, and this measurement did not vary in sarcomeres of different length. This fluorescent pattern is identical to that previously reported by Pepe and Drucker (1975) and Moos (1981), with polyclonal antibodies to C protein. Furthermore, this fluorescent labeling pattern differed significantly from that observed with three McAbs specific for the MHC (Fig. 1d). MF-20 and MF-14, both of which bind to the light meromyosin (LMM) portion of the MHC (Fischman and Masaki 1982), stain the entire A band with a significant enhancement in its central region. McAb MF-30, which binds an epitope in chymotryptic subfragment 2 (S-2), stains the entire A band except for a thin transverse stripe coincident with the thick filament bare zone (Huxley 1963). The precise disposition of these epitopes must await analysis by electron microscopy.

Four pieces of evidence suggest that MF-21 binds to C protein and not myosin. First, the antigen is not retained by DEAE-Sephadex (Offer et al. 1973). Second, the antibody binds to a protein band of approximately 150 kD, significantly smaller than the MHC. Third, the immunofluorescent labeling pattern is identical to that previously reported with affinity-purified polyclonal antibodies to C protein. Finally, C protein has been purified by hydroxylapatite column chromatography (Starr et al. 1980), and MF-21, but not MF-14 or MF-20, bound to it.

Immunochemical Analysis of MHC Heterogeneity during Muscle Development

NCP-myosin prepared from pectoralis muscles of embryonic, neonatal, and adult stages of development was tested by solid-phase RIA for reactivity with serial dilutions of McAbs MF-14, MF-20, and MF-30 (Fig. 2A). No significant differences in the binding of MF-20 were observed with myosin from all stages tested. Immunoblots with these myosin preparations indicated that MF-20 bound exclusively to a 200-kD peptide. Thus, the epitope recognized by MF-20 exhibited no developmental transition. In contrast, MF-30 only bound slightly to NCP-myosin from 12-day-old embryos, and MF-14 did not bind at all, even at vast excess of McAb (Fig. 2A). By 19 days of embryonic development, the binding of MF-30 at antibody saturation was approximately 45% of that seen with adult pectoralis myosin, whereas MF-14 binding remained negative. Not until 2 days after hatching could MF-14 binding to pectoralis myosin be detected, and only by 28 days posthatching were adult values reached (Figs. 2A and 4B).

Confirmatory results were obtained with immunofluorescent analysis of frozen sectioned muscle (Fig. 3). At 12 days of incubation, only MF-20 ex-

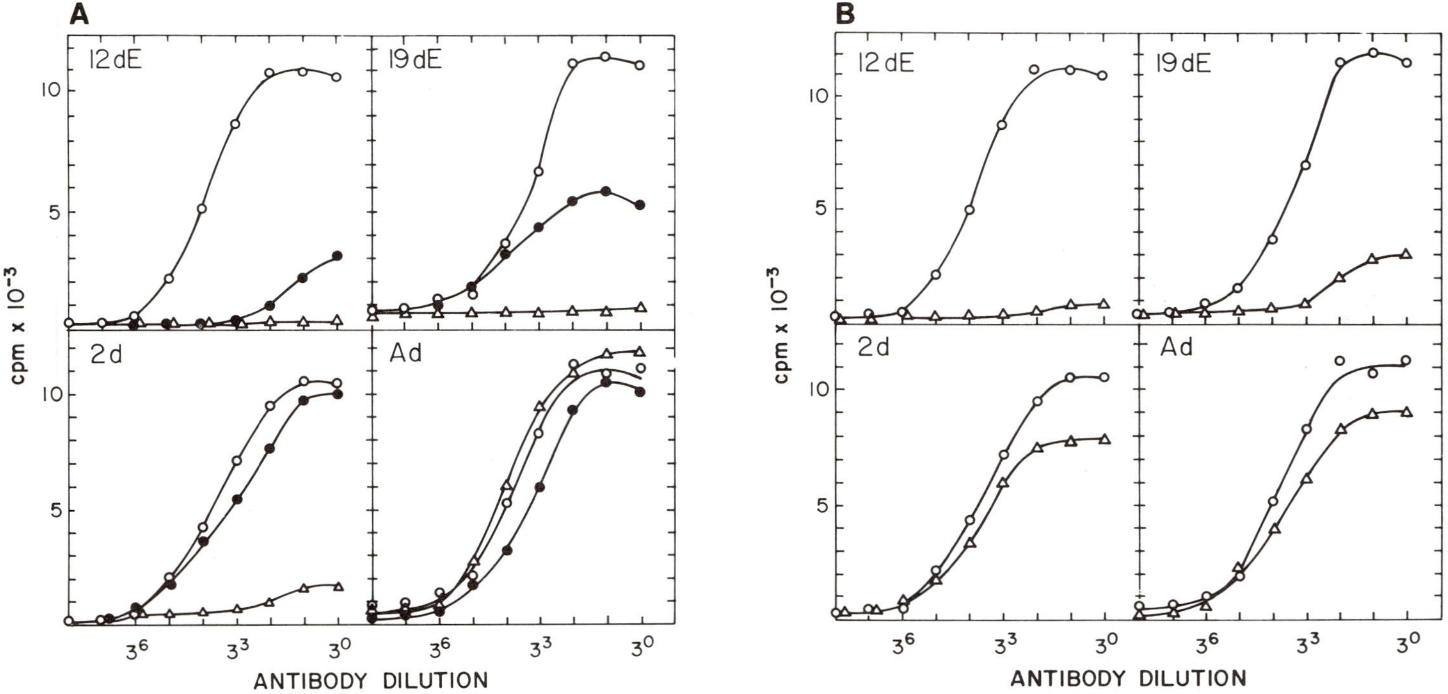

FIGURE 2
(*A*) NCP-myosin prepared from 12-day and 19-day embryonic, 2-day posthatch, and adult pectoralis muscles was reacted with serially diluted MF-20 (○), MF-30 (●), and MF-14 (Δ) by RIA. (*B*) The same NCP-myosin preparations used in *A* were reacted with serially diluted MF-20 (○) and MF-21 (Δ) by RIA.

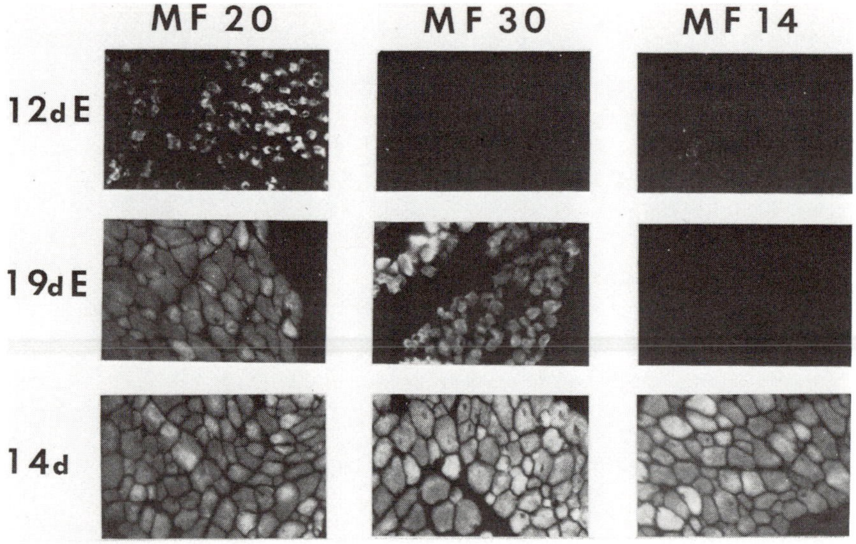

FIGURE 3
Cryostat sections of 12-day embryonic (12dE), 19-day embryonic (19dE), and 14-day posthatch (14d) pectoralis muscles were reacted with MF-20, MF-30, and MF-14 by indirect immunofluorescence. All exposures were 15 sec; with prolonged exposures, weak binding of MF-30 could be observed in 12dE.

hibited positive staining of pectoralis myocytes, and this McAb appeared to stain all of the muscle cells and myofibrils in this muscle. Binding of MF-30 was first detected by fluorescent staining at 15 days of incubation, whereas MF-14 reactivity first became apparent during the first week of posthatching. Since all of the myofibers in the adult pectoralis muscle stained positively with the three McAbs, these results suggest that the sequential expression of the epitopes specific for MF-20, MF-30, and MF-14 occurs in the same muscle fibers, and these myosin isoforms probably occupy the same myofibrils and sarcomeres. Our results are consistent with the concept that three MHC isoforms appear sequentially during myogenesis (Whalen et al. 1981). The absence of MF-30 and MF-14 binding to myosin from early muscle indicates that little or no adult-type MHC is present in those tissues.

Although previous studies have shown that embryonic fast muscle myosin has extensive immunological similarity to its adult counterpart (Masaki and Yoshizaki 1974; Gauthier et al. 1978; Rubenstein and Kelly 1981), the present data with McAbs indicate that epitopes present in the adult MHC are missing in embryonic MHC. This observation is consistent with the partial proteolytic peptide maps of rat myosin (Whalen et al. 1979) and chicken MHC (Rushbrook and Stracher 1979), which indicate that little if any adult MHC is present in embryonic muscle.

Several explanations might be considered for the present immunological results. First, it is conceivable that embryonic, neonatal, and adult MHCs are distinct gene products that accumulate sequentially during development. In this model, MF-20 would bind equivalently to each of the three MHC isoforms, MF-14 would have high affinity only for adult myosin, and MF-30 would recognize the neonatal and adult isoforms but not the embryonic MHC. Alternatively, the antigenic transitions might reflect sequential, posttranslational modifications of a single gene product. Such modifications are known to occur (Trayer et al. 1968; Huszar 1972) yet have not been documented to modify the immunochemical properties of MHC. Finally, it is now apparent that different myosin isoforms can exist within the same myofibers (Gauthier and Lowey 1979; Lutz et al. 1979). If adult pectoralis muscle contains two distinct MHC isoforms, one might entertain the possibility that the age-dependent binding by these three McAbs represents the noncoordinate accumulation of different adult MHC isoforms.

Immunochemical Analysis of C-protein Heterogeneity in Developing Muscle

McAb MF-21, which binds to C protein from adult pectoralis muscle, was tested for reactivity with embryonic pectoralis NCP-myosin (12-day embryos), and no binding could be detected (Fig. 2B). Similar results were obtained with several other McAbs to C protein from pectoralis (MF-1 and MF-3) and ALD (ALD-66) muscles. Another McAb (E101-D6), which was raised against NCP-embryonic myosin, bound to a 150-kD peptide on gels and exhibited a similar A-band fluorescent staining pattern to that of MF-21. Thus, we think it likely that a variant form of C protein is present in embryonic pectoralis muscle, and it must have distinct immunochemical properties (possibly a different primary amino acid sequence) from that present in adult pectoralis or ALD muscles.

By solid-phase RIA, the binding of MF-21 to NCP-myosin could be detected first at 17 days of egg incubation; by 2 days of posthatching, the saturation binding values had reached 90% of the adult values (Figs. 2B and 4B). Immunofluorescent analysis with frozen sections by double direct labeling with fluorescein isothiocyanate (FITC)-conjugated MF-20 and rhodamine-labeled MF-21 revealed that myofibers of 12-day-old pectoralis muscle were stained positively with MF-20, but none reacted with MF-21, with or without acetone fixation or Triton X-100 permeabilization (Fig. 4A). By 19 days of embryonic development, however, all fibers stained strongly with both McAbs. Thus, pectoralis muscle cells contain striated-muscle myosin at 12 days of incubation, but we could not detect any adult C-protein antigens recognized by our McAbs. Presumably, C protein, like MHC, undergoes isoform transitions during muscle development. The

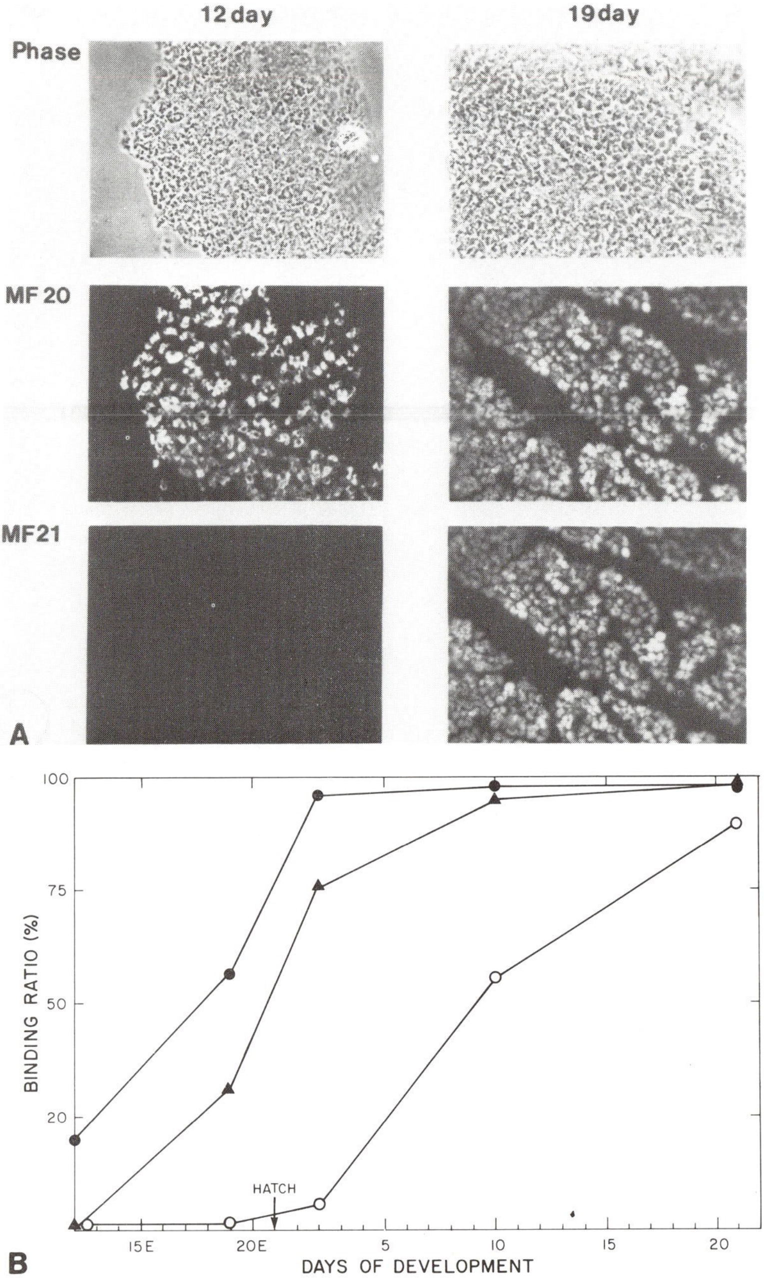

FIGURE 4
(*A*) Cryostat sections of 12-day and 19-day embryonic pectoralis muscles were reacted with directly labeled FITC-conjugated MF-20 and rhodamine-labeled MF-21. The sections were viewed for phase contrast and immunofluorescence with appropriate filters. (*B*) Ratios of binding of MF-30 (●), MF-21 (▲), and MF-14 (○) to MF-20 at saturation were calculated from data in Fig. 2a and b and plotted as a function of development.

changes in antigenic properties of C protein roughly parallel the changes in MHC antigenicity during myogenesis (Fig. 4B). It remains to be seen whether these transition differences reflect problems in assay sensitivity with the different McAbs or indicate noncoordinate accumulation of these contractile-protein isoforms. It should be noted that early in myogenesis (e.g., 12-day-old embryos) the pectoralis myofibrils must contain embryonic-type MHC and C-protein isoforms of unknown function. It remains to be seen what role, if any, these proteins play in myofibrillogenesis.

SUMMARY

The epitopes specified by three McAbs (MF-14, MF-20, and MF-30) have been mapped to distinct regions of the MHC. MF-14 and MF-20 bound to LMM, and MF-30 bound to S-2. When tested for reactivity with pectoralis myosin extracted from chickens at different stages of development, we observed that MF-20 exhibited equivalent binding to myosin from all developmental stages examined. However, MF-30 binding could first be detected 5–6 days before hatching, and MF-14 bound only to myosin extracted from posthatch chickens. Another McAb (MF-21) reacted with adult pectoralis C protein but failed to bind 12-day-old embryonic pectoralis muscle or its C protein. Evidence is presented that C protein, like MHC, exhibits antigenic changes during development, which may reflect isoform transitions for both of these myofibrillar proteins.

ACKNOWLEDGMENTS

We express our sincere gratitude for the excellent technical assistance of Donald Morgenstern, Carol Ramos, Lori Antich, and Ken Tanaka. Dr. Paul Dreizen was of great assistance in critically evaluating our myosin protein biochemistry. This study was supported by research grants from the National Institutes of Health (AM-28095), the Muscular Dystrophy Association of America, and the New York Heart Association. F.C.R. was supported by CNPq grant 200-096/80 from the Brazilian government.

REFERENCES

Burnette, W.N. 1981. "Western blotting": Electrophoretic transfer of proteins from sodium dodecyl sulfate-polyacrylamide gels to unmodified nitrocellulose and radiographic detection with antibody and radioiodinated protein A. *Anal. Biochem.* **112:** 195.

Craig, R. and G. Offer. 1976. The location of C-protein in rabbit skeletal muscle. *Proc. R. Soc. Lond. B* **192:** 451.

Devlin, R.B. and C.P. Emerson, Jr. 1979. Coordinate accumulation of contractile protein mRNAs during myoblast differentiation. *Dev. Biol.* **69:** 202.

Dhoot, G.K. and S.V. Perry. 1979. Distribution of polymorphic forms of troponin components and tropomyosin in skeletal muscle. *Nature* **278:** 714.

Fischman, D.A. and T. Masaki. 1982. Immunochemical analysis of myosin with monoclonal antibodies. In *Perspectives in differentiation and hypertrophy* (ed. W. Anderson), p. 279. Elsevier, Amsterdam.

Galfre, G., S.C. Howe, C. Milstein, G.W. Butcher, and J.C. Howard. 1977. Antibodies to major histocompatability antigens produced by hybrid cell lines. *Nature* **266:** 550.

Garrels, J.I. and W. Gibson. 1976. Identification and characterization of multiple forms of actin. *Cell* **9:** 793.

Gauthier, G.F. and S. Lowey. 1979. Distribution of myosin isoenzymes among skeletal muscle fiber types. *J. Cell Biol.* **81:** 10.

Gauthier, G.F., S. Lowey, and A.W. Hobbs. 1978. Fast and slow myosin in developing muscle fibers. *Nature* **274:** 25.

Gefter, M.L., H. Margulies, and M. Scharff. 1977. A simple method for polyethylene glycol-promoted hybridization of mouse myeloma cells. *Somat. Cell Genet.* **3:** 231.

Hunter, W.M. and F.C. Greenwood. 1962. Preparation of iodine-131 labeled human growth hormone of high specific activity. *Nature* **194:** 495.

Huszar, G. 1972. Developmental changes of the primary structures and histidine methylation in rabbit skeletal muscle myosin. *Nat. New Biol.* **240:** 260.

Huxley, H.E. 1963. Electron microscope studies on the structure of natural and synthetic protein filaments from striated muscle. *J. Mol. Biol.* **7:** 281.

Laemmli, U.K. 1970. Cleavage of structural proteins during the assembly of the head of bacteriophage T4. *Nature* **227:** 680.

Lutz, H., H. Weber, R. Billeter, and E. Jenny. 1979. Fast and slow myosin within single skeletal muscle fibers of adult rabbits. *Nature* **281:** 142.

Masaki, T. and C. Yoshizaki. 1974. Differentiation of myosin in chick embryos. *J. Biochem.* **76:** 123.

Matsuda, R., T. Obinata, and Y. Shimada. 1981. Types of troponin components during development of chicken skeletal muscle. *Dev. Biol.* **82:** 11.

Moos, C. 1981. Fluorescence microscope study of the binding of added C-protein to skeletal muscle myofibrils. *J. Cell Biol.* **90:** 25.

Obinata, T., Y. Shimada, and R. Matsuda. 1979. Troponin in embryonic chick skeletal muscle cells *in vitro*. An immunoelectron microscope study. *J. Cell Biol.* **81:** 59.

Offer, G., C. Moos, and R. Starr. 1973. A new protein of the thick filaments of vertebrate skeletal myofibrils. *J. Mol. Biol.* **74:** 653.

Pepe, F.A. and B. Drucker. 1975. The myosin filament III. C-protein. *J. Mol. Biol.* **99:** 609.

Roy, R.K., F.A. Sreter, and S. Sarkar. 1979. Changes in tropomyosin subunits and myosin light chains during development of chicken and rabbit striated muscles. *Dev. Biol.* **69:** 15.

Rubenstein, N.A. and A.M. Kelly. 1981. Development of muscle fiber specialization in the rat hindlimb. *J. Cell Biol.* **90:** 128.

Rushbrook, J.I. and A. Stracher. 1979. Comparison of adult, embryonic and dystrophic myosin heavy chains from chicken muscle by sodium dodecyl sulfate/polyacrylamide gel electrophoresis and peptide mapping. *Proc. Natl. Acad. Sci.* **76:** 4331.

Schwartz, R.J. and K.N. Rothblum. 1981. Gene switching in myogenesis: Differential expression of the chicken actin multigene family. *Biochemistry* **20:** 4122.

Sreter, F., M. Balint, and J. Gergeley. 1975. Structural and functional changes of myosin during development. Comparison with adult fast, slow and cardiac myosin. *Dev. Biol.* **46:** 317.

Starr, R. and G. Offer. 1971. Polypeptide chains of intermediate molecular weight in myosin preparations. *FEBS Lett.* **15:** 40.

Starr, R., P. Bennett, and G. Offer. 1980. X-protein and its polymer. *J. Muscle Res. Cell Motil.* **1:** 205.

Towbin, H., T. Staehelin, and J. Gordon. 1979. Electrophoretic transfer of proteins from polyacrylamide gels to nitrocellulose sheets: Procedure and some applications. *Proc. Natl. Acad. Sci.* **76:** 4350.

Trayer, I.P., C.I. Harris, and S.V. Perry. 1968. 3-Methylhistidine in adult and fetal forms of skeletal muscle myosin. *Nature* **217:** 452.

Whalen, R.G., G.S. Butler-Browne, and F. Gros. 1976. Protein synthesis and actin heterogeneity in calf muscle cells in culture. *Proc. Natl. Acad. Sci.* **73:** 2018.

Whalen, R.G., K. Schwartz, P. Bouveret, S.M. Sell, and F. Gros. 1979. Contractile protein isozymes in muscle development. Identification of an embryonic form of myosin heavy chain. *Proc. Natl. Acad. Sci.* **76:** 5197.

Whalen, R.G., S.M. Sell, G.S. Butler-Browne, K. Schwartz, P. Bouveret, and I. Pinset-Härström. 1981. Three myosin heavy-chain isozymes appear sequentially in rat muscle development. *Nature* **292:** 805.

Myosin Synthesis and Assembly in Nematode Body-wall Muscle

HENRY F. EPSTEIN
STEPHEN A. BERMAN
DAVID M. MILLER III
Program in Neurosciences
Department of Neurology
Baylor College of Medicine
Houston, Texas 77030

Myosin is the central piece in the protein machinery that produces muscle contraction. In this role, the myosin molecule must serve as an energy-transducing enzyme and as the major building block of the thick filament. In differentiated muscle cells, myosin is constantly being synthesized and assembled into thick filaments. Many issues regarding the processes of myosin synthesis and assembly in muscle remain unresolved. Is the synthesis and assembly of myosin coupled? Are additional protein components required for the assembly of myosin into native thick filaments? Do different myosin isoforms play distinct structural or functional roles within thick filaments?

The combined approaches of protein biochemistry, immunology, and electron microscopy have proved useful in establishing our present knowledge concerning the roles of myosin within muscle. The nematode *Caenorhabditis elegans* offers several experimental advantages in addition to these time-honored methods that may provide further insights concerning the process of myosin synthesis and assembly. This animal has an elegantly simple and well-defined development and anatomy of body-wall muscle cells (White et al. 1976; J.E. Sulston and E. Schierenberg, pers. comm.). Genetic analysis has characterized hundreds of specific mutants in over 20 genes affecting the development, structure, and function of body-wall muscle cells (Waterston et al. 1980; Zengel and Epstein 1980a). This

genetic approach, in combination with biochemical, immunological, and morphological methods, promises to offer significant insights into muscle development.

RESULTS AND DISCUSSION

Myosin Synthesis and Assembly Are Coupled

Previous work in our laboratory has provided both biochemical and morphological descriptions of muscle development during the free-living stages of the nematode. Quantitative measurements of the amounts of specific isoforms and of the total myosin heavy chains were obtained throughout the larval and early adult stages (Garcea et al. 1978). Total myosin heavy chain and the specific body-wall isoform, myosin B, increase exponentially during this period. During the same stages, the number of thick filaments per sarcomere increase significantly (Mackenzie et al. 1978b). In these previous reports, we commented upon the apparent correlation of the observed increase in body-wall myosin content and the increase in organized filament structure. This implies that myosin synthesis and assembly may be closely coupled processes.

To analyze this question further, we have reexamined the data of these previous reports and have made a more quantitative comparison between the biochemical measurements of myosin heavy chain (Garcea et al. 1978) and the morphological measurements of body-wall muscle sarcomeres and thick filaments (Mackenzie et al. 1978b), since some of the morphological measurements were only implicit in the previously published micrographs. Table 1 presents this comparison of the results of the two independent studies. From 24 to 60 hours of posthatching, the amount of body-wall myosin heavy chain increases 18-fold. The absolute values were calculated from Garcea et al. (1978, Fig. 2) using a constant factor of 0.85 for the fraction of total myosin in body-wall muscle. This factor is based on measurements in adult nematodes and may be high for 24-hour larvae. The numbers of myosin heavy chains per nematode were calculated from the determined weights by using a molecular weight of 210,000 (Epstein et al. 1974). The numbers of thick filaments per sarcomere cross section, the mean length of the sarcomeres, and the mean numbers of sarcomeres per body-wall muscle cell were measured from micrographs like those of Mackenzie et al. (1978b, Figs. 3 and 4). From these measurements (Table 1)—a body-wall thick filament length of 9.7 μm (Mackenzie and Epstein 1980) and the reasonable approximations that the thick filaments are effectively parallel to the long axis of each sarcomere and that the mean values of morphological parameters could be applied to all body-wall muscle—the numbers of thick filaments in the 95 body-wall muscle cells of nematodes at each stage were calculated.

Using Squire's model for the arrangement of myosins in thick filaments (Squire 1971, 1973, 1974) and the dimensions of nematode body-wall mus-

TABLE 1
Quantification of Increases in Myosin and Thick Filaments during Nematode Development

	Biochemical		Morphological				
Age post-hatching (hr)	amount of body-wall myosin heavy chains (nematode) (ng)	body-wall myosin heavy chains (nematode)	thick filaments (sarcomere cross section)	sarcomere length (μm)	sarcomeres (body-wall muscle cell)	thick filaments (nematode)	calculated thick filament myosin heavy chains (nematode)
24	1	2.5×10^9	189	60.4	2	2.1×10^5	2.4×10^9
44	6	1.5×10^{10}	257	102	4–5	1.1×10^6	1.1×10^{10}
60	18	4.4×10^{10}	313	166	7–8	3.8×10^6	4.0×10^{10}

cle thick filaments—diameter, 25 nm and length, 9.7 μm (Mackenzie and Epstein 1980)—the numbers of myosin molecules per 14.4-nm length and per individual thick filament were calculated to be 7.85 and 5400, respectively. From this value and the fact that each myosin molecule contains two heavy chains and the number of thick filaments per nematode body wall, we can calculate a total number of myosin heavy chains required to be within body-wall thick filaments for each stage. A comparison of the calculated total number of myosin heavy chains required to be within body-wall thick filaments and the biochemically determined number of body-wall myosin heavy chains for each stage is given in Figure 1. The plots show both a remarkable congruence that is model independent and numerical agreement that is dependent upon several assumptions.

With these considerations and potential qualifications in mind, the congruence and close numerical agreements between the total number of myosin molecules and the number calculated to be found within filaments could have several significant implications. The amount of myosin in pools outside of organized structure appears to be small throughout the developmental period. The similar rates of increase in total myosin and in filament myosin as a function of development suggest that the synthesis and assembly of myosin are coupled. The control of this coupling and the involvement of additional proteins are the subjects of current investigation.

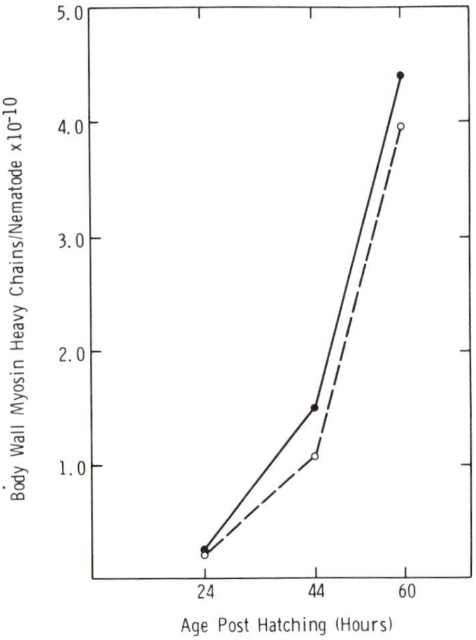

FIGURE 1
Comparison of total myosin heavy chains determined biochemically and myosin heavy-chain content of detectable thick filaments within nematode body-wall muscle cells during larval and early adult development. (●) Biochemical measurements; (○) structural calculations.

Nonmyosin Proteins in Thick Filaments

The thick filaments of nematode body-wall muscle cells contain paramyosin as a major protein component, in addition to myosin isoforms A and B (Epstein et al. 1974; Waterston et al. 1974; Harris and Epstein 1977). Mutants in the *unc-15* gene are associated with defective thick filaments and altered or absent paramyosin (Waterston et al. 1977). In the E1214 mutant, the absence of paramyosin leads to the assembly of abnormal thick filaments that are 1.53 μm in length and 32 nm in diameter (Mackenzie and Epstein 1980). This result suggests that the presence of myosins A and B and any other body-wall myofibrillar proteins are not sufficient to assemble in vivo normal thick filaments of 9.7 μm in length and 25 nm in diameter without paramyosin. Paramyosin is the first case of a nonmyosin protein required for the normal assembly of myosin.

It is likely that other myofibrillar proteins may also be necessary for the proper assembly of thick filaments. Figure 2 shows an autoradiograph of a two-dimensional gel separation (O'Farrell 1975) of the proteins from a ^{35}S-labeled nematode thick-filament preparation (Mackenzie and Epstein 1980, 1981). Other control experiments show that four of the spots represent the proteins of contaminating thin filaments. The rest copurify with myosin and paramyosin through Triton X-100 extraction, homogenization in a French press, and three sedimentations in buffer of low ionic strength. We assign them operationally as constituents of a nematode thick-filament preparation, but further experiments are required to establish these proteins as true components of nematode body-wall thick filaments. The location of such proteins may be determined by the reaction of specific antibodies with isolated nematode thick filaments and examination of the products by electron microscopy.

Location of Myosin B within Nematode Thick Filaments

Two myosin isoforms, A and B, exist within the same body-wall muscle cells and sarcomeres of *C. elegans* (Mackenzie et al. 1978a). One of these isoforms, myosin B, is absent in *unc-54* mutants such as E190 (Schachat et al. 1977), but thick filaments are still present in the body-wall muscle (Epstein et al. 1974). Such structures presumably contain the A isoform as the only myosin component. This result raises the issue as to whether the two isoforms reside within different thick filaments normally or whether a normal thick filament contains both myosins A and B. A third possibility is that some thick filaments contain both isoforms, and other filaments, one or the other myosin. The availability of monoclonal antibodies that preferentially react with specific nematode myosin isoforms (Epstein et al., this volume) has permitted us to begin investigation of the distribution of myosin isoforms in isolated nematode thick filaments. Clearly, this ap-

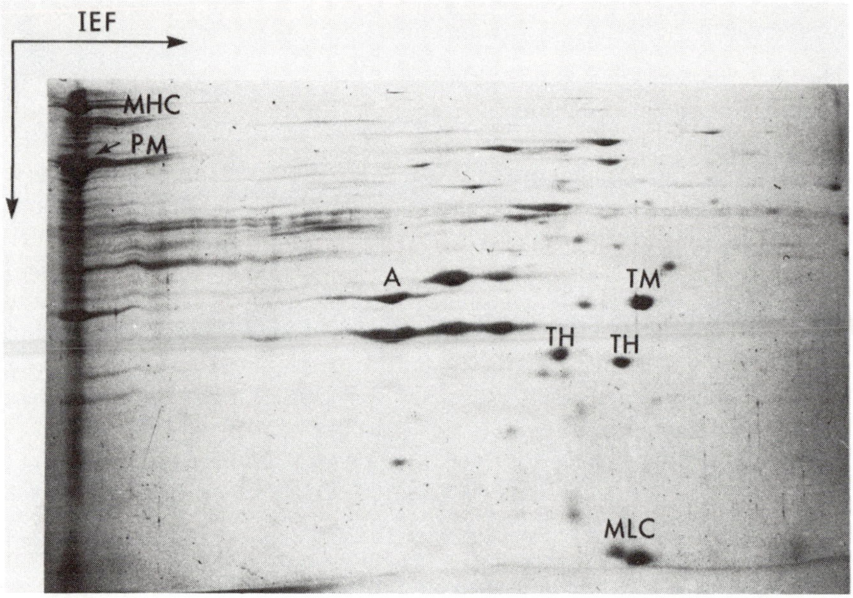

FIGURE 2
Two-dimensional gel separation of proteins of nematode thick-filament preparation. (MHC) Myosin heavy chain; (PM) paramyosin; (A) actin; (TM) tropomyosin; (TH) unidentified thin-filament-associated constituents; (MLC) myosin light chains.

proach will also be useful in establishing the locations of nonmyosin components.

Figure 3 shows the electron microscopy of a negatively stained nematode thick filament, a similarly prepared filament reacted with clone 28.2 antibody that preferentially reacts with myosin B (Epstein et al., this volume), and a filament that has been reacted with both clone 28.2 antibody and goat anti-mouse antibody. Close examination of the direct reaction of clone 28.2 antibody with nematode filaments reveals "bumps" along the entire filament. Reaction with secondary antibody leads to a dense, homogeneous coating of the entire lengths of all filaments in the preparation. We tentatively conclude that all thick filaments from normal body-wall muscle contain myosin B. This hypothesis has been confirmed by additional experimental tests performed under more stringent conditions (Epstein et al., this volume; D.M. Miller et al., unpubl.). An implication of these results that can now be experimentally verified is that the thick filaments of mutants such as E190 that do not contain myosin B must be abnormal.

SUMMARY

A combination of experimental approaches has provided several important insights with respect to myosin synthesis and assembly in body-wall muscle

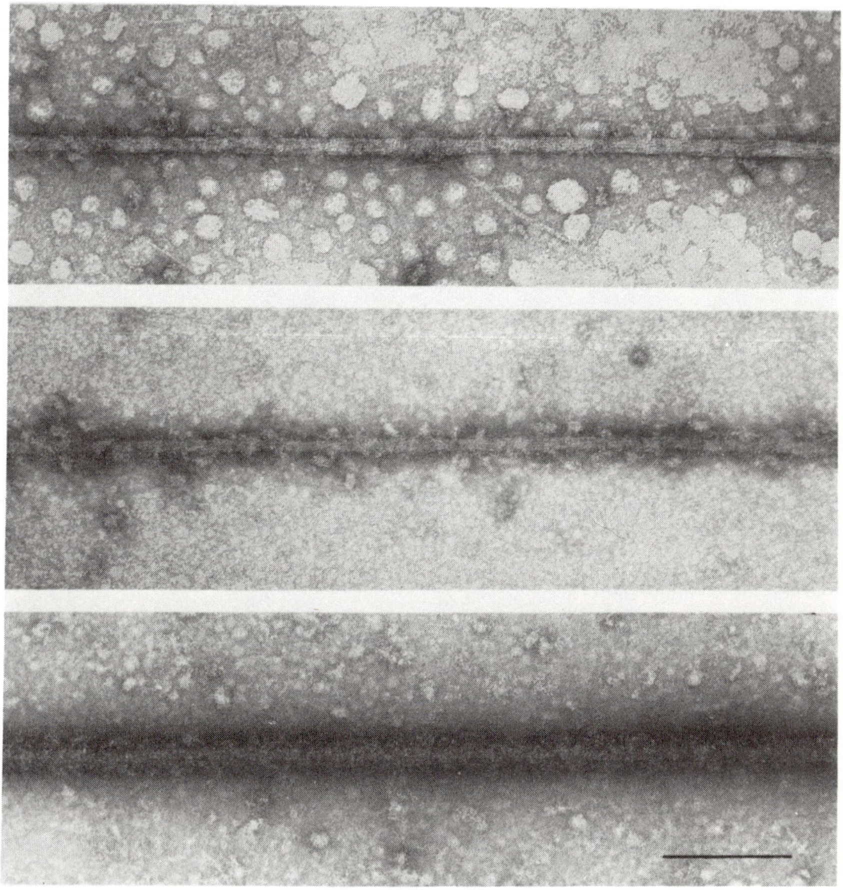

FIGURE 3
Electron microscopy of clone 28.2 antibody reaction with nematode thick filaments. (*Top*) Unreacted thick filament; (*middle*) thick filament reacted with 1 mg/ml of monoclonal antibody alone; (*bottom*) thick filament reacted with 1 mg/ml of monoclonal antibody followed by 1 mg/ml of rabbit anti-mouse immunoglobulin. All filaments were glutaraldehyde fixed, after reacting with antibody, and then washed extensively. All filaments were negatively stained with 1% uranyl acetate. Bar represents 0.2 μm. The magnification was identical in all three micrographs.

of the nematode *C. elegans.* The synthesis of myosin, primarily the B isoform, is coupled to thick-filament assembly within body-wall muscle cells during larval and early adult development. There are no measurable pools of myosin outside of organized filaments within an experimental error of 10% or less during this period of life cycle. This conclusion may not apply to early embryonic cells when they first produce myosin and paramyosin (Epstein et al., this volume). Mutants of the *unc-52* gene appear de-

fective concomitantly in the synthesis of myosin B and in the construction of myofibrillar structure during larval development (Mackenzie et al. 1978b; Zengel and Epstein 1980b). A nonmyosin protein, paramyosin, is clearly necessary for the proper assembly of myosin into thick filaments (Mackenzie and Epstein 1980); E1214 mutants deficient in paramyosin produce thick filaments with abnormal lengths and diameters. Additional proteins are isolated with myosin and paramyosin and thick filaments but are not found with actin and tropomyosin and thin filaments. The functional significance of some of these proteins may be assigned by examination of mutants in over 20 genes affecting muscle in *C. elegans* (Zengel and Epstein 1980a). The location of such proteins, as well as that of the A and B isoforms of myosin, within the substructure of body-wall thick filaments is being approached by electron microscopy of specific monoclonal antibodies bound to isolated filaments. Various hypotheses regarding the role of the different myosins, paramyosin, and other protein components in thick-filament assembly may be tested in appropriate mutants of *C. elegans*.

ACKNOWLEDGMENTS

We thank Irving Ortiz for outstanding assistance with electron microscopy and Dr. John M. Mackenzie, Jr., for valuable discussions. S.A.B. and D.M.M. are fellows of the Muscular Dystrophy Association of America, and H.F.E. holds a Research Career Development Award from the National Institute of Child Health and Human Development (NICHHD). Research was supported by grants from the National Institute of Aging, NICHHD, and the Jerry Lewis Neuromuscular Disease Research Center of the Muscular Dystrophy Association of America.

REFERENCES

Epstein, H.F., R.H. Waterston, and S. Brenner. 1974. A mutant affecting the heavy chain of myosin in *Caenorhabditis elegans*. *J. Mol. Biol.* **90:** 291.

Garcea, R.L., F. Schachat, and H.F. Epstein. 1978. Coordinate synthesis of two myosins in wild-type and mutant nematode muscle during larval development. *Cell* **15:** 421.

Harris, H.E. and H.F. Epstein. 1977. Myosin and paramyosin of *Caenorhabditis elegans*: Biochemical and structural properties of wild-type and mutant proteins. *Cell* **10:** 709.

Mackenzie, J.M., Jr. and H.F. Epstein. 1980. Paramyosin is necessary for determination of nematode thick filament length in vivo. *Cell* **22:** 247.

Mackenzie, J.M., Jr., F. Schachat, and H.F. Epstein. 1978a. Immunocytochemical localization of two myosins within the same muscle cells in *Caenorhabditis elegans*. *Cell* **15:** 413.

Mackenzie, J.M., Jr., R.L. Garcea, J.M. Zengel, and H.F. Epstein. 1978b. Muscle development in *Caenorhabditis elegans:* Mutants exhibiting retarded sarcomere construction. *Cell* **15:** 751.

O'Farrell, P.H. 1975. High resolution two-dimensional electrophoresis of proteins. *J. Biol. Chem.* **250:** 4007.

Schachat, F.H., H.E. Harris, and H.F. Epstein. 1977. Two homogeneous myosins in body-wall muscle of *Caenorhabditis elegans. Cell* **10:** 721.

Squire, J.M. 1971. General model for the structure of all myosin-containing filaments. *Nature* **233:** 457.

———. 1973. General model of myosin filament structure. III. Molecular packing arrangements in myosin filaments. *J. Mol. Biol.* **77:** 291.

———. 1974. Symmetry and three-dimensional arrangement of filaments in vertebrate striated muscle. *J. Mol. Biol.* **90:** 153.

Waterston, R.H., H.F. Epstein, and S. Brenner. 1974. Paramyosin of *Caenorhabditis elegans. J. Mol. Biol.* **90:** 285.

Waterston, R.H., R.M. Fishpool, and S. Brenner. 1977. Mutants affecting paramyosin in *Caenorhabditis elegans. J. Mol. Biol.* **117:** 825.

Waterston, R.H., J.N. Thomson, and S. Brenner. 1980. Mutants with altered muscle structure in *Caenorhabditis elegans.* **77:** 271.

White, J.G., E. Southgate, J.N. Thomson, and S. Brenner. 1976. The structure of the ventral nerve cord of *Caenorhabditis elegans. Phil. Trans. R. Soc. Lond. B* **275:** 327.

Zengel, J.M. and H.F. Epstein. 1980a. Identification of genetic elements associated with muscle structure in the nematode, *Caenorhabditis elegans. Cell Motil.* **1:** 73.

———. 1980b. Mutants altering coordinate synthesis of specific myosins during nematode muscle development. *Proc. Natl. Acad. Sci.* **77:** 852.

The M Protein Myomesin in Cross-striated Muscle Cells during Myofibrillogenesis

HANS M. EPPENBERGER
MARTIN BÄHLER
THOMAS C. DOETSCHMAN
MONIKA EPPENBERGER
JEAN-CLAUDE PERRIARD
DANIEL STUDER
THEO WALLIMANN
EMANUEL E. STREHLER
Institute for Cell Biology
Swiss Federal Institute of Technology
CH-8093 Zurich, Switzerland

In addition to the well-known and well-characterized major contractile proteins that form the paracrystalline myofibrils, the existence of "minor" protein components has become evident in recent years. It is assumed that some of these proteins have organizational functions and may serve as a scaffold for the assembly and alignment of myofibrils, whereas others may have different regulatory roles. Two characteristic domains of myofibrillar structure are the Z-line and M-line regions. Two protein components are known to contribute to the M line: the muscle-type isoenzyme of creatine kinase (MM-CK) (Turner et al. 1973; Wallimann et al. 1977) and the 165,000-molecular weight (M_r) M protein myomesin (Masaki et al. 1968; Masaki and Takaiti 1972, 1974; Etlinger et al. 1976; Strehler et al. 1980; Eppenberger et al. 1981).

In this paper we concentrate on myomesin or M protein, which is a tightly bound myofibrillar component and can only be extracted quantitatively under conditions where the myofibrils are virtually destroyed (Strehler et al. 1980). This is in contrast to MM-CK, which may form part of the so-called M bridges and accounts for the electron-dense material in the M line.

Upon the relatively easy removal of MM-CK, this electron-dense material disappears (Wallimann et al. 1978). The isolation and biochemical characterization of myomesin (165,000-M_r M protein) have been described (Masaki and Takaiti 1974; Trinick and Lowey 1977; Strehler et al. 1980) for chicken skeletal muscle, and the presence of myomesin in myofibrils lacking an electron-dense (visible) M line and MM-CK, e.g., those from chicken heart muscle, has also been demonstrated (Strehler et al. 1979). Myomesin synthesis in developing skeletal-muscle cell cultures has been measured, and it has been shown that significant synthesis and accumulation was not detected in replicating myogenic cells, whereas a significant increase in synthesis in cultures of postmitotic cells was observed concomitantly with the first appearance of the antigen in a cross-striated pattern, as revealed by immunofluorescence (Puri et al. 1980; Eppenberger et al. 1981).

To study the exact structural location and the functional role of myomesin, we have examined its localization, temporal appearance, tissue and species specificity, and molecular interaction with contractile proteins of myofibrils.

RESULTS AND DISCUSSION

Localization of Myomesin

The antibodies employed in this work were obtained using chicken skeletal-muscle myomesin (isolated according to Strehler et al. 1979) as an immunogen in rabbits (Strehler et al. 1980). The antisera were further purified by antigen-affinity chromatography. Incubation of either freshly prepared isolated myofibrils or myogenic cell cultures with antibody, employing the indirect immunofluorescence technique, reveals a very distinct location of the antigen within the M-line region of the sarcomere (Fig. 1). A similar pattern is revealed in suspension cultures of differentiated muscle cells, where M-line-containing myofibrils wound about the nucleus can be found (Puri et al. 1980). The antimyomesin antibody has also been used to stain the myotome of somites of early chick embryos as well as premuscle masses in developing limb buds, thus indicating the presence of already differentiated myogenic cells at these stages (Chiquet et al. 1981). Close inspection again reveals a distinct pattern of cross-striations. Obviously, the antigen is highly concentrated in the M-line region.

This has been further demonstrated by injecting fluorescence-labeled antibody into living myotubes (Fig. 2), in which the sole appearance of label at the sites of the M lines clearly shows that all myomesin appears to be bound to the M lines with high affinity. It is remarkable how precisely the small amounts of myomesin synthesized (contributing only $\sim 0.04\%$ of the total protein synthesized at the time of maximal synthesis in cell culture) are accumulated and structurally incorporated within the M-line region.

Ultrastructural studies revealed a decoration pattern of antimyomesin im-

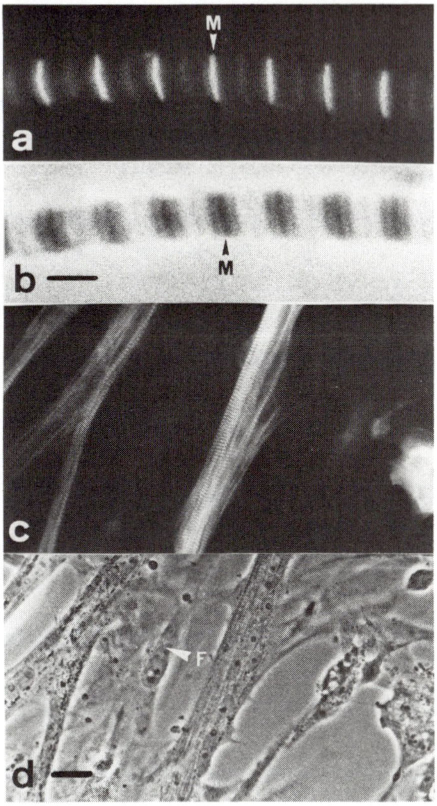

FIGURE 1
Indirect immunofluorescence using affinity-purified antibodies against myomesin. (*a*) Localization of myomesin in isolated chicken breast muscle myofibril; (*b*) phase-contrast picture of same myofibril (note phase-dense deposit of antibody in M-line region); (*c*) 6-day-old cell culture of chicken breast muscle stained for myomesin; (*d*) phase contrast (note presence of numerous unstained fibroblasts [F]). (M) M-line region. Bar represents 2 μm (*a,b*) and 20 μm (*c,d*).

munoglobulin (IgG) that stretches out over the whole area of the H zone (see Fig. 10 in Strehler et al. 1980), thus suggesting an arrangement of myomesin parallel to the thick filaments. The assumption of such an arrangement is corroborated by preliminary experiments done on ultrathin frozen sections incubated with antimyomesin IgG, where high-resolution electron microscopy indeed reveals a longitudinal decoration running parallel to the myosin filaments and spanning the H zone (E. Strehler and L. Thornell, pers. comm.). The arrangement of myomesin parallel to thick filaments leads us to believe that myomesin could contribute to the so-called M filaments described earlier (Knappeis and Carlsen 1968; Thornell and Sjöström 1975).

Appearance of Myomesin during Development

There is no indication for the presence of material cross-reacting with antimyomesin antibody in proliferating primary skeletal-muscle cell cultures (presumptive myoblasts), as shown in Figure 3a and b. A similar result was

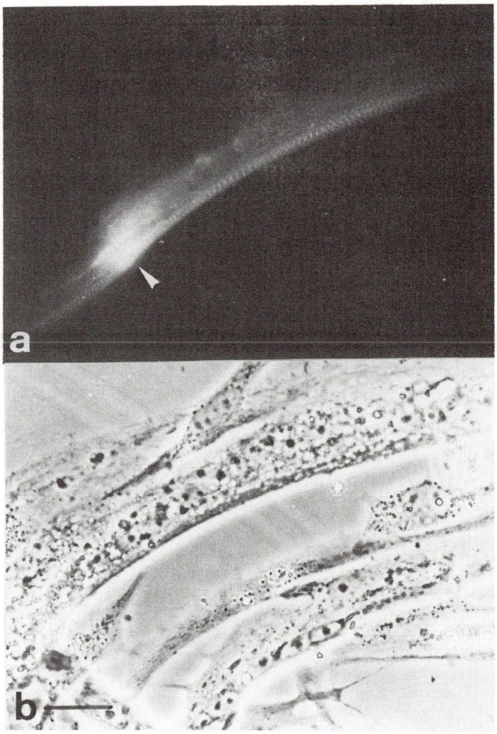

FIGURE 2
Direct immunofluorescence of microinjected rhodamine-labeled antimyomesin IgG in a live myotube. At 3 days of culture, affinity-purified, rhodamine-labeled antimyomesin IgG was microinjected into a live myotube. Fluorescence (*a*) and phase-contrast (*b*) photographs of the myotube 30 min after injection. Arrowhead shows site of injection. Bar represents 20 μm.

obtained when myogenic cells were cultured in a medium containing bromodeoxyuridine (Fig. 3c,d), which reversibly prevents replicating presumptive myoblasts from becoming postmitotic and thus prevents synthesis and accumulation of proteins specific for terminally differentiated muscle (Eppenberger et al. 1981). After 18 hours of culturing in standard medium (or after removal of bromodeoxyuridine), many cells have left the cell cycle and started aggregation and fusion into multinucleated myotubes. At this time, a faint pattern of cross-striation can be observed after staining for myomesin (Fig. 3e,f), whereas after 6 days myofibrils are very distinctly stained with antibody to myomesin (Fig. 3g,h). A similar developmental behavior has been observed for a number of muscle-specific polypeptides. It seems that the accumulation of myomesin in myoblasts and myotubes is coordinately (though not necessarily simultaneously) regulated with the accumulation of the major contractile proteins. Evidence for an early structural presence of myomesin in cells, where the alignment of thick filaments is in progress, has been shown by electron microscopy of very young myotubes (Eppenberger et al. 1981). There was no, or only traces of, electron-dense material detectable in the M-line region (hence, very little if any MM-CK), whereas incubation with antimyomesin IgG resulted in a

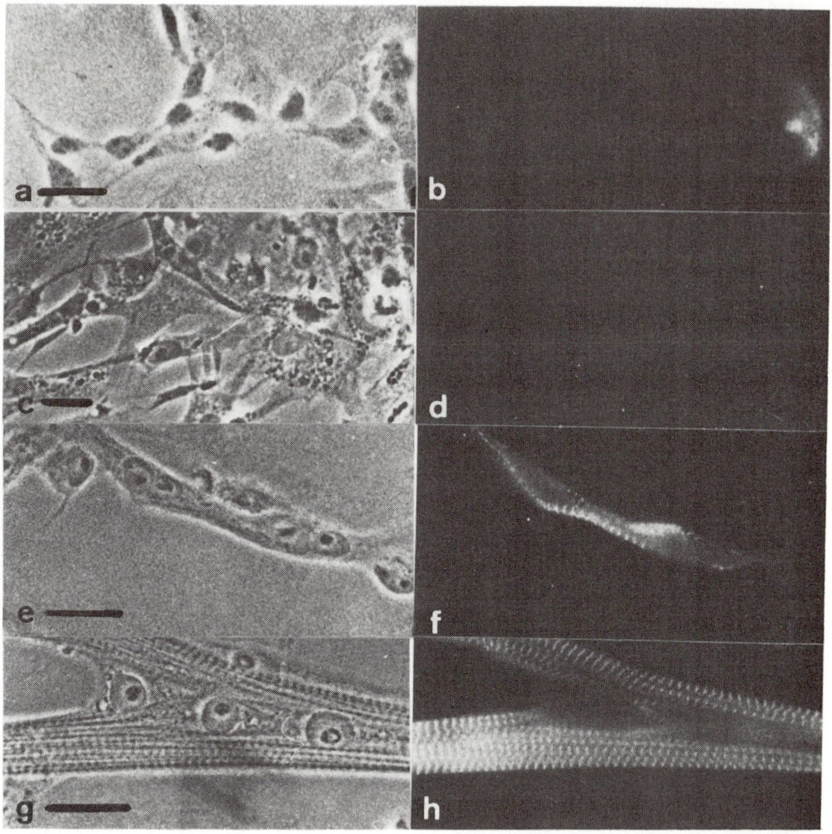

FIGURE 3
Indirect immunofluorescence using affinity-purified antibodies to myomesin. Cell cultures of chicken breast muscle 6 hr after plating, representing replicating presumptive myoblasts (*a,b*); grown for 96 hr in 2.5 μg/ml of bromodeoxyuridine-containing medium (*c,d*); 24 hr after plating, representing postmitotic myogenic cells (*e,f*); and 6 days after plating, representing a myotube (*g,h*). Bars represent 20 μm.

heavy decoration of the M-line region, indicating the presence of this antigen at this stage.

Tissue and Species Specificity of Myomesin

Extracts of several chicken tissues have been investigated for cross-reactivity with antimyomesin antibody using double immunodiffusion tests (Table 1). Out of 15 adult tissues, only breast and leg skeletal muscle, ventricular and atrial heart muscle, and thymus showed a cross-reaction. A reaction, however, with thymus extract had to be expected since thymus ap-

TABLE 1
Cross-reactivity of Rabbit Polyclonal Antibodies to Chicken Skeletal-muscle Myomesin (Immunofluorescence)

Species	Skeletal-muscle myofibrils	Heart-muscle myofibrils	Skeletal-muscle cells	Heart-muscle cells
Chick	+	+	+	+
Quail	+	n.d.[a]	+	n.d.
Rat	+	+	n.d.	n.d.
Mouse	+	n.d.	n.d.	n.d.
Hamster	+	+	+	+
Cow	n.d.	+	n.d.	n.d.
Cat	n.d.	+	n.d.	n.d.
Sheep	n.d.	+	n.d.	n.d.
Man	+	+	n.d.	n.d.
Xenopus	+	+	n.d.	n.d.
Trout	+	+	n.d.	n.d.
Crab	–	–	n.d.	n.d.
Drosophila flight muscle	–	n.d.	–	n.d.

[a]n.d., not determined.

pears to contain myogenic cells (Perriard et al., this volume). In addition, chicken gizzard (smooth muscle) and fibroblast cell cultures have been tested by the immunofluorescence technique (Eppenberger et al. 1981). Neither gave a positive reaction with antimyomesin antibody. These results point to a stringent distribution pattern for myomesin: Only cells and tissues containing myofibrillar organelles with very regular arrays of thick and thin filaments seem to synthesize and accumulate myomesin. Since many of the contractile proteins from different species are known to display a conservative behavior through evolution, it was also of interest to investigate myomesin in this respect. As is summarized in Table 1, antimyomesin antibody to chicken antigen reacts with isolated myofibrils and with muscle cells of a wide range of species, as judged by immunofluorescence. Interestingly, only myofibrils of vertebrate species cross-reacted with the antibody, whereas myofibrils of crab and of indirect flight muscle from *Drosophila melanogaster* did not show any reaction.

The question of whether there are no cross-reacting epitopes of myomesin or no myomesin at all in these invertebrate species cannot be answered at present.

Comparison of Myomesin and Myosin Molecules at the Electron Microscope Level

If the assumption is correct that myomesin has a role in either assembly and alignment of thick filaments during development or in keeping thick filaments in register during contraction, there may be a direct interaction of myomesin molecules with certain domains of myosin. In vitro binding experiments of myomesin to myosin and myosin fragments suggest a rather strong interaction (Mani and Kay 1978). To obtain information about the shape and the dimensions of myomesin relative to myosin, as well as possible interactions between these two molecules, experiments have been undertaken to simultaneously portray them at the electron microscope level (Fig. 4). A mixture of both molecules was dissolved in glycerol and ammonium acetate, sprayed on mica, and submitted to rotary shadowing. Myosin molecules show the well-known shape, whereas presumptive myomesin molecules appear as elongated, straight, or horseshoe-shaped specimens. Final identification of the elongated structures as the myomesin molecule is under way, using antibodies. Since our aim will be to demonstrate by this technique either a direct or an indirect interaction of myomesin with myosin and to determine the sites of such interactions, a number of different experimental conditions will have to be tested.

CONCLUSIONS AND SPECULATIONS

All cells where myomesin has been found have one specific feature in common: They contain myofibrils with very regular arrays of filaments. The

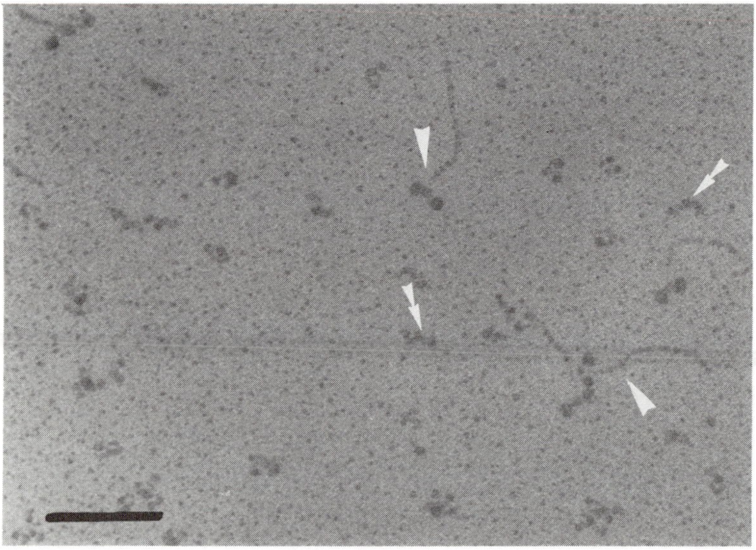

FIGURE 4
Electron micrograph of myomesin using myosin as a standard. Chicken myomesin was dissolved at a concentration of 15 μg/ml in a buffer containing 0.3 M ammonium acetate, 0.5 mM dithiothreitol, and 60% glycerol (pH 7.0). After the addition of a 1:10 molar ratio of chicken myosin to myomesin, the mixture was sprayed on mica, the glycerol evaporated in a Balzer's freeze-etching device, and the molecules rotary shadowed using Ta/W at an angle of 6°. Myomesin appeared as elongated molecules of 35 ± 5 nm in length and 4 ± 0.5 nm in width, respectively. Myosin molecules are indicated by arrowheads, and myomesin molecules, by arrows. Bar represents 0.1 μm.

protein is only synthesized and accumulated in cells that, at the same time, synthesize and accumulate the muscle-specific isoforms of contractile proteins. In cells that produce only the nonmuscle myosin or actin, myomesin is not expressed. This is a further indication for a coordinated, though not necessarily simultaneous, activation of specific muscle genes. The protein myomesin seems to be part of the "backbone" of myofibrils, and its role within the organelle seems to require that its structure remain conserved throughout vertebrate phylogeny from fish to man. We do not know whether there is an isoprotein of myomesin present in invertebrate myofibrils or whether another component substitutes myomesin in these species. We do not know at all what the actual function of myomesin is. We can only speculate that a protein that has been well preserved during evolution and that shows such a distinct specificity for myofibrils has an important and specific role in this organelle. It is quite clear that myomesin is a protein component located at a specific site—the M-line region—where it

could serve an organizational role during the assembly of myofibrils and also a role in the maintenance of the structural entity of the aligned thick filaments. Since myomesin is detectable in myogenic cells at a time when only rudimentary myofibrils can be seen, it may function as a kind of nucleation site for thick filaments. The specific components of the M-line region may be responsible for the proper polarity of the thick filaments during assembly. Myomesin may be an anchoring site for proteins like titin (Wang and Williamson 1980), which stretch throughout the sarcomere and may be part of a scaffold that keeps myofilaments in register.

New tools like monoclonal antibodies and new techniques like microinjection into living cells are now being used in our laboratory to find a clue for the function of myomesin.

ACKNOWLEDGMENTS

We are grateful to Hanni Moser for skillful technical assistance. This work was supported by grant 3.187-0.77 from the Swiss National Science Foundation, by a grant to H.M.E. from the Muscular Dystrophy Association, and by predoctoral training grants from the Swiss Federal Institute of Technology.

REFERENCES

Chiquet, M., H.M. Eppenberger, and D.C. Turner. 1981. Muscle morphogenesis: Evidence for an organizing function of exogenous fibronectin. *Dev. Biol.* **88:** 22.

Eppenberger, H.M., J.-C. Perriard, U.B. Rosenberg, and E.E. Strehler. 1981. The M_r 165,000 M-protein myomesin: A specific protein of cross-striated muscle cells. *J. Cell Biol.* **89:** 185.

Etlinger, J.D., R. Zak, and D.A. Fischman. 1976. Compositional studies of myofibrils from rabbit striated muscle. *J. Cell Biol.* **68:** 123.

Knappeis, G.G. and F. Carlsen. 1968. The ultrastructure of the M-line in skeletal muscle. *J. Cell Biol.* **38:** 202.

Mani, R.S. and C.M. Kay. 1978. Interaction studies of the 165,000 dalton protein component of the M-line with the S2 subfragment of myosin. *Biochim. Biophys. Acta* **536:** 134.

Masaki, T. and O. Takaiti. 1972. Purification of M-protein. *J. Biochem.* **71:** 355.

———. 1974. M-protein. *J. Biochem.* **75:** 367.

Masaki, T., O. Takaiti, and S. Ebashi. 1968. "M-substance," a new protein constituting the M-line of myofibrils. *J. Biochem.* **64:** 909.

Puri, E.C., M. Caravatti, J.-C. Perriard, D.C. Turner, and H.M. Eppenberger. 1980. Anchorage-independent muscle cell differentiation. *Proc. Natl. Acad. Sci.* **77:** 5297.

Strehler, E.E., G. Pelloni, C.W. Heizmann, and H.M. Eppenberger. 1979. M-protein in chicken cardiac muscle. *Exp. Cell Res.* **124:** 39.

———. 1980. Biochemical and ultrastructural aspects of M_r 165,000 M-protein in cross-striated chicken muscle. *J. Cell Biol.* **86:** 775.

Thornell, L.-E. and M. Sjöström. 1975. The myofibrillar M-band in cryosection-analysis of section thickness. *J. Microsc.* **104:** 263.

Trinick, J. and S. Lowey. 1977. M-protein from chicken pectoralis muscle: Isolation and characterization. *J. Mol. Biol.* **113:** 343.

Turner, D.C., T. Wallimann, and H.M. Eppenberger. 1973. A protein that binds specifically to the M-line of skeletal muscle is identified as the muscle form of creatine kinase. *Proc. Natl. Acad. Sci.* **70:** 702.

Wallimann, T., D.C. Turner, and H.M. Eppenberger. 1977. Localization of creatine kinase isoenzymes in myofibrils. I. Chicken skeletal muscle. *J. Cell Biol.* **75:** 297.

Wallimann, T., G. Pelloni, D.C. Turner, and H.M. Eppenberger. 1978. Monovalent antibodies against MM-creatine kinase remove the M-line from myofibrils. *Proc. Natl. Acad. Sci.* **75:** 4296.

Wang, K. and C.L. Williamson. 1980. Identification of an N_2 line protein of striated muscle. *Proc. Natl. Acad. Sci.* **77:** 3254.

Myofilamentous and Myofibrillar Connections: Role of Titin, Nebulin, and Intermediate Filaments

KUAN WANG
Clayton Foundation Biochemical Institute
Department of Chemistry and Cell Research Institute
University of Texas
Austin, Texas 78712

Titin and nebulin are two extremely large, major myofibrillar proteins found in a wide range of striated muscles. As is pointed out below, biochemical and biophysical analysis of these proteins is unexpectedly difficult; such analysis requires major technical modifications and adaptations of standard methods before it may be applied to these proteins. Perhaps more challenging are the conceptual difficulties of how and where to place these proteins within the framework of the two-filament sarcomere model, because they do not appear to be thick- or thin-filament-associated regulatory or anchoring proteins. These questions have in turn led us to reexamine several long-standing, paradoxical questions of muscle structure and behavior that cannot be accounted for satisfactorily by the current sarcomere model. Titin and nebulin may be crucial missing links in resolving some of these questions.

RESULTS AND DISCUSSION

Molecular Properties of Titin and Nebulin

Figure 1(a,b) shows an SDS gel pattern of myofibrillar proteins representative of a wide range of vertebrate striated muscles. Nearly all of the major protein bands below myosin heavy chain have been previously characterized. In contrast, the three major protein bands above myosin heavy

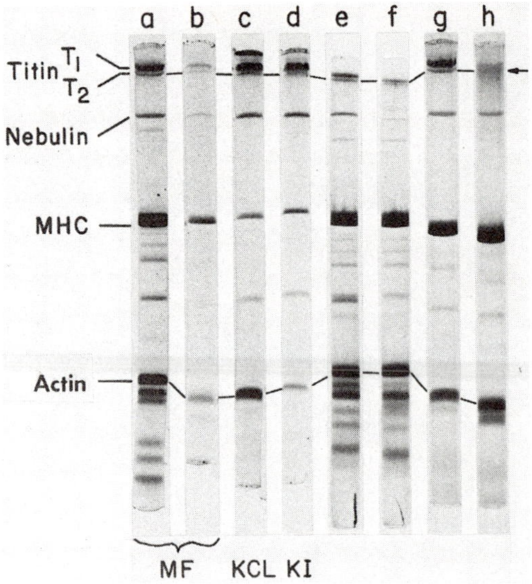

FIGURE 1
(*a–d*) Resistance of titin and nebulin to extraction. Rabbit skeletal myofibrils (*a,b,* two sample loadings) were sequentially extracted by 0.6 M KCl and then 0.6 M KI to remove the bulk of thick and thin filaments. The unextracted residues at each step were analyzed by gel electrophoresis. Note that titin and nebulin are the major residual proteins in the KCl (*c*) and KI (*d*) residues. (*e–f*) Degradation of titin and nebulin in situ. Rabbit myofibrils were prepared in the absence (*e*) and in the presence (*f*) of 1 mM Ca^{++}. Note the formation of large discrete fragments of similar mobility to titin and nebulin. Note also that in *f*, the upper band of the doublet is T_2 and not T_1 (established by co-electrophoresis experiments). This doublet is easily confused with the T_1-T_2 doublet of intact myofibrils. (*g–h*) Degradation of titin and nebulin in SDS. Rabbit myofibril SDS sample (*g*) was stored at room temperature for 5 days (*h*). Note the smearing of titin (→). Nebulin remained undegraded.

chain, which together constitute 12–15% of the total protein, have received little attention (Etlinger et al. 1976; Porzio and Pearson 1977; Wang and McClure 1978). We have designated the top doublet collectively as titin (derived from titan: anything of great size) because of its titanic size and because these two bands (T_1 and T_2) are immunologically related (Wang et al. 1979). The third band, distinct from titin, is referred to as nebulin because it is found associated with the N_2 line of the sarcomere—a nebulous striation within the I band (Franzini-Armstrong 1970; Wang and Williamson 1980; Wang 1981). The sizes of titin and nebulin were estimated as having approximate molecular weights of 1×10^6 and 0.5×10^6, respectively, because they comigrated under dissociating conditions on high-porosity

SDS gels with the hexamer and the trimer of cross-linked myosin heavy chains, respectively. Proteins of this size range do not enter SDS gels commonly used for normal-sized proteins and have easily been ignored. We have found, however, that even when a suitable high-porosity gel system was used, reproducible gel patterns were obtained only when strict precautions were taken to avoid degradation and aggregation of these proteins during preparation. The degradation of titin and nebulin presents a continuing problem that is not yet under complete control. So far, two types of degradation related to the physical states of these proteins have been recognized (K. Wang and D. Palter, in prep.). Within the myofibril, both titin and nebulin were rapidly cleaved by endogenous proteases (especially the calcium-activated neutral proteases) and by very low concentrations of exogeneously added proteases (such as trypsin and chymotrypsin) into a similar set of ladderlike large fragments. Under these conditions, no other major myofibrillar proteins showed any sign of degradation (see also Maruyama et al. 1981b). Since a rapid conversion of T_1 into T_2 was invariably observed as one of the initial events of in situ proteolysis, the T_2-to-T_1 weight ratio may be used as a sensitive indicator of the extent of proteolysis and/or perhaps protein turnover in the muscle. For many purposes, the inclusion of EDTA or EGTA to inhibit calcium-activated proteases was effective in minimizing (but not abolishing) their degradation during myofibril preparation. However, since in situ degradation yielded ladderlike, closely spaced fragments of similar mobilities to titin or nebulin, it has often been difficult to decide whether a protein band migrating closely to titin or nebulin is intact or degraded (see Fig. 1e,f). A different degradation pattern was observed after myofibrils were dissolved in SDS. Titin began to degrade when SDS samples were either left at room temperature for a few days or heated at 100°C for a few minutes. With time, titin appeared on SDS gels as an increasingly broad smear below its usual position (Fig. 1h). Nebulin, on the other hand, was stable under these conditions. Either the minute amount of residual proteolytic activity in SDS samples was sufficient to appreciably degrade titin at room temperature, or titin may contain heat-labile covalent bonds (see also King and Kurth 1980; Locker and Dianes 1980; Maruyama et al. 1981b).

Another nagging technical difficulty results from the relative insolubility and aggregation of titin and nebulin. Intact titin and nebulin are not extractable by a wide range of solutions commonly used to dissolve thick and thin filaments, such as 0.6 M KCl or KI, and therefore represent the major residual proteins of extracted myofibrils (Fig. 1a–d). To purify and analyze these proteins, it is necessary to use denaturants such as SDS, guanidinium chloride, and urea. However, we were surprised to discover that SDS did not consistently and quantitatively dissolve titin, unless the ionic strength of the myofibril suspension was very low (e.g., 5 mM Tris-Cl at pH 8.0). When SDS was added to the myofibril at higher (e.g., near physiological)

ionic strength, the majority of titin frequently remained as an elastic gel, whereas other myofibrillar proteins, including nebulin, were readily dissolved. Titin bands were either absent or greatly diminished in gel patterns of such SDS samples (K. Wang, in prep.). The SDS-insoluble titin gel undoubtedly must have been seen by many muscle biochemists and was probably discarded as uninteresting connective tissues.

Although all of the experimental factors that render titin insoluble in SDS are still unclear, we now know that purified titin easily aggregates after the removal of SDS. This titin aggregate, as viewed by light microscopy and electron microscopy, consists of entangled slender strands of filaments (2–10 nm) and amorphous material (Fig. 2) and is very difficult to redissolve in SDS (Wang and Ramirez-Mitchell 1979). It appears likely that lowering the ionic strength of myofibril suspension somehow decreases the tendency for titin to aggregate, perhaps by swelling the sarcomeric lattice to facilitate the entry of SDS.

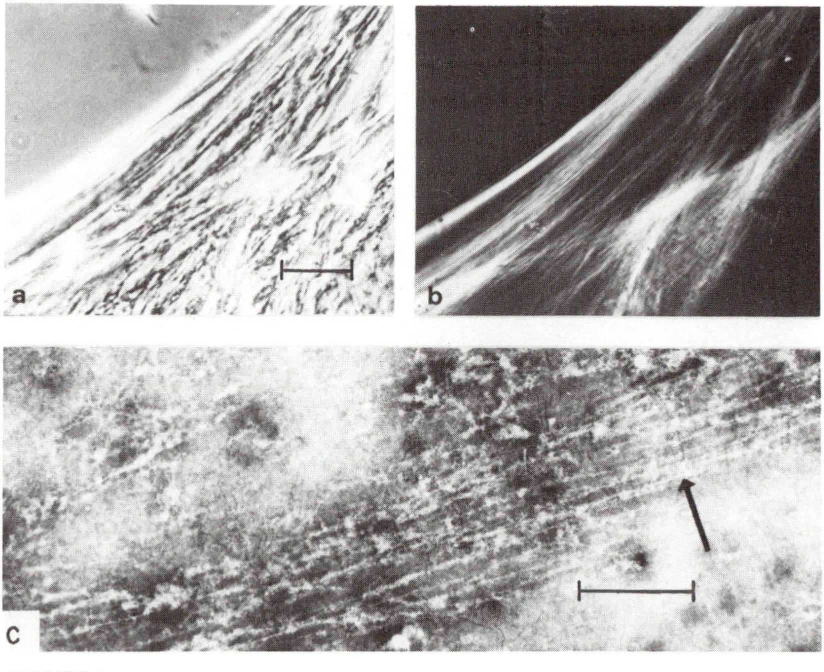

FIGURE 2

Aggregation of purified titin. Purified rabbit titin aggregated easily at physiological ionic strength into an elastic filamentous gel upon removal of SDS. Phase-contrast (*a*) and polarized-light (*b*) images of titin gel (bar represents 2 μm). Negative staining (*c*) of titin gel with uranyl acetate revealed the presence of entangled slender filaments (2–10 nm) and amorphous material (bar represents 0.1 μm).

To summarize, we have encountered unexpected technical difficulties in nearly every step of the seemingly straightforward task of obtaining reproducible SDS gel patterns: extremely rapid degradation of titin and nebulin during myofibril as well as SDS sample preparation and storage, inconsistent solubilization of titin by SDS, and the extremely low mobility of titin and nebulin in usual SDS gels. These technical difficulties are perhaps largely responsible for the evasiveness of these proteins and for the confusion and discrepancies of results and conclusions from several laboratories involved in the characterization of titin and nebulin. Maruyama and associates obtained an insoluble gel after extracting myofibrils with thick- and thin-filament solvents (0.6 M KCl and KI), followed by phenol, SDS, NaOH, or boiling urea (Maruyama et al. 1977). This harshly extracted elastic residue was designated as "connectin" (which is now known to be a heterogeneous mixture of predominately denatured, degraded, and aggregated titin and nebulin, plus variable amounts of myosin, actin, connective tissue, and intermediate-filament proteins (see Fig. 1d) (K. Wang, in prep.). As a consequence of this all-encompassing definition of connectin, various chemically distinct proteins or fragments and any structural features of the residues can be and indeed have been attributed to connectin in the literature. For example, King and Kurth (1980) isolated from sheep muscle a high-molecular-weight protein (~700K) that appeared on SDS gels as a band having a sharp upper edge with a trailing smear (degraded titin? See Fig. 1h). They designated it as connectin, on the basis of a similar amino acid composition with connectin. Locker and Dianes (1980) obtained a protein fraction appearing as a broad smear between 500K and 1000K (degraded titin?) and another band (nebulin?) on SDS gels and postulated that the smearing may represent a protein with variable degrees of cross-linking or degradation. Very recently, Maruyama et al. (1981a) isolated titin from extracted muscle residue, confirmed our immunofluorescent localization results, and then proceeded to "identify" connectin with titin. It should be pointed out, however, that the term connectin has undergone various stages of metamorphosis and has not been used consistently in the past. Most of the molecular and structural properties attributed to connectin were from ill-defined connectin fractions and not from purified titin.

After having overcome technical difficulties, we then proceeded to purify titin and nebulin from SDS-solubilized myofibrils using gel filtration chromatography, taking advantage of their large sizes and their differential solubility in SDS (Wang 1982). The amino acid compositions of titin and nebulin are clearly distinct from other myofibrillar proteins, yet not unusual in spite of their insolubility in aqueous solutions. T_1 and T_2 have very similar, if not identical, composition. It is worth noting that titin is enriched in proline (8–9%)—an α-helix-breaking residue, suggesting that native titin may not have as extensive an α-helical region as, e.g., myosin heavy chain (with ~2% proline). Although we consider it possible that

polypeptides of such sizes may be constructed out of (nondisulfide) covalently linked subunits, we have so far found no chemical evidence to support this view.

Distribution of Titin and Nebulin in the Sarcomere

The question of how titin and nebulin are arranged in the sarcomere has been approached using the fluorescent-antibody staining technique (Wang et al. 1979; Wang and Williamson 1980). Since these proteins are fairly antigenic, monospecific antibodies were successfully prepared in animals and then characterized by immunoblots and solid-phase immunoassays. It was found that titin and nebulin are immunologically distinct from other myofibrillar proteins, whereas T_1 and T_2 are related. When applied to isolated myofibrils, these antibodies gave highly reproducible staining patterns with standard fluorescent techniques. However, interpreting these patterns was far from straightforward, because a bewilderingly large number of patterns were obtained, which seem to be somehow related not only to the sarcomere length but also to the very manner in which the myofibrils were prepared. For example, monospecific titin antibody consistently labeled a wide zone centered on the edge of A bands (A-I junction). Yet in many but not all myofibrils it also stained one or more of the following loci: mid-A (M line?), Z lines, and throughout the entire A band. Furthermore, even for myofibrils with similar overall staining loci, the relative intensity, shape, and dimension of these fluorescent bands varied significantly from one myofibril to the next. For this reason, we were able to reach only some general conclusions. In particular, we concluded that titin is preferentially distributed in the A band. However, it was noted that the staining at the A-I junction, in fact, extended considerably into the I band, in sharp contrast to the labeling patterns of myosin, which are always confined within the boundaries of the phase-dense A band. This observation, as well as the resistance of intact titin to extraction by A-band solvents, raises major doubts whether titin is an A-band protein in the classical sense.

An important clue for interpreting titin staining came from the results of immunofluorescence labeling of nebulin: Antinebulin stained a subregion of the I band corresponding in location and in behavior to the N_2 line—the nebulous striation that flanks either side of the Z line (Franzini-Armstrong 1970; Wang and Williamson 1980). The N_2 line is an obscure yet highly interesting structure because it exhibits an unusual sarcomere length-dependent translocation: When the sarcomere is lengthened, the N_2 line translocates in such a way that it maintains the same proportional distance to the adjacent M line and Z line, as if it is associated with or is part of an "elastic longitudinal structure" anchored at M and Z lines. Furthermore, the N_2 line is located in the I-band region where thin filaments alter packing geometry from a square lattice at the Z line to a hexagonal array near the

A-I junction (Franzini-Armstrong 1970). Our work identified, for the first time, a major myofibrillar protein, nebulin, as an N_2-line-associated protein. Perhaps more significant is the fact that this finding encouraged us to look seriously into the possibility of interpreting fluorescent labeling patterns of titin and nebulin on the basis of unorthodox sarcomere models differing from the current two-filament model by the inclusion of an additional set of elastic longitudinal filaments (see below).

To more precisely determine the distribution of titin and nebulin relative to other myofibrillar proteins, we have applied double fluorescence techniques to study simultaneously the distribution of two different proteins in the same sarcomere (see Fig. 3A). A composite summary diagram containing all observed banding patterns of titin and nebulin is also presented in Figure 3A.

Although several interpretations of these patterns are plausible, most of these patterns can be explained on the basis of a speculative sarcomere model depicting the presence of a continuous longitudinal elastic filament consisting of titin and nebulin. It is perhaps useful to digress here and briefly discuss the structural evidence for the possible existence of a third filament in the sacomere.

Putative Elastic Filaments: A Third Filament of the Sarcomere

The possible presence of a third filament, distinct from thick and thin filaments, in the sarcomere of vertebrate striated muscle has been a source of speculation for a long time. These proposals were prompted by the detection of distinct, very thin (2–6 nm) filaments in the electron micrographs of extracted sarcomeres, where the A band had been removed, or of overly stretched sarcomeres, where gaps appeared between A bands and I bands (see Fig. 3B). Because these filaments were not observed in unextracted or unstretched sarcomeres, various suggestions have been put forward to predict their organization within the sarcomere. These include S filaments (connecting the ends of adjacent thin filaments) (Huxley and Hanson 1954), C or gap filaments (connecting the ends of adjacent thick filaments to Z lines) (Carlsen et al. 1961; Huxley and Peachey 1961; Sjostrand 1962; Page and Huxley 1963; Locker and Leet 1976), T filaments (connecting adjacent Z lines in between thin and thick filaments) (McNeil and Hoyle 1967; Walcott and Ridgeway 1967; Hoyle 1968), or core filaments (forming the core of thick filaments) (Guba et al. 1968; Locker and Leet 1976).

A continuity of the thick filaments with adjacent Z lines is well established structurally and biochemically in insect flight muscle (for review, see Pringle 1978; Saide 1981). This connecting filament (C filament) has been proposed as a structural basis for the high resting tension and for the phenomena of stretch activation (Thorson and White 1969). Although it has been generally assumed that such connecting filaments are limited to this

highly specialized muscle, more recent ultrastructural data are highly suggestive of their existence in vertebrate skeletal muscle as well (Locker and Dianes 1980). Locker and Leet (1976) proposed that each connecting "gap" filament served as a core to two thick filaments in adjacent sarcomeres, linking them through the Z line and passing between the thin filaments, such that only *one* end of each thick filament is connected to the adjacent Z line. Locker's data, however, are not inconsistent with a C-filament model analogous to that proposed for insect flight muscle in which *both* ends of

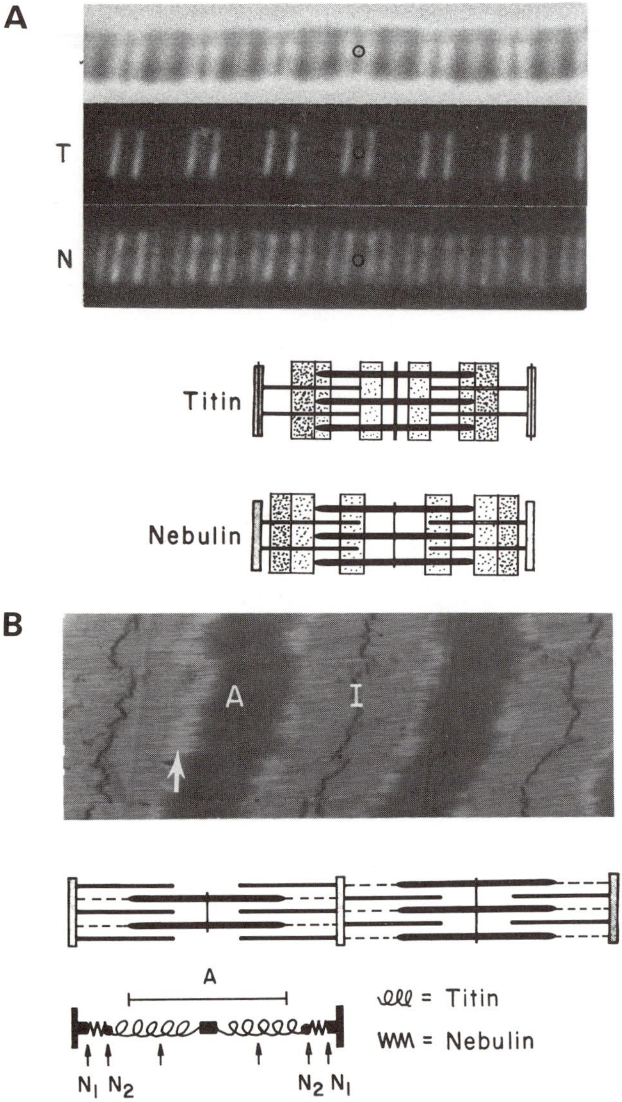

each thick filament are connected to the Z lines (see Fig. 3B). Although the details of this and other earlier models remain to be critically evaluated by further experimentation, the ultrastructural evidence, taken together, strongly suggests that a distinct elastic filament exists in vertebrate striated muscles.

The introduction of an elastic filament to the sarcomere may explain some puzzling muscle properties and behavior. For example, the presence of these filaments may provide a structural basis for the resting tension of muscle, which has been shown to reside to a large extent in the myofibril; the connection of both ends of thick filaments to Z lines through elastic filaments (such as that depicted in Fig. 3B) would explain why the A band remains centrally located between regularly spaced Z discs when relaxed muscle (where thin and thick filaments are disengaged) is stretched even to lengths beyond filament overlap.

Titin and Nebulin: Components of the Putative Third Filament

In view of the fact that the solubility properties and the distribution of titin and nebulin cannot easily be explained by the properties of thick or thin filaments, we also proposed the existence of a third type of filament in the sarcomere (Wang and Ramirez-Mitchell 1979, 1981; Wang and Williamson 1980). We now envision that titin and nebulin are the major, if not the exclusive, components of a set of longitudinal *continuous* filaments that connect Z lines from within the sarcomere. These filaments are intrinsically elastic and extensible along their length, except where they interact with extensible structures, such as thick or thin filaments or M and Z lines. Further-

FIGURE 3 (see facing page)
(*A*) Distribution of titin and nebulin in the sarcomere. (*Top*) An example of a myofibril doubly stained with fluorescein antititin (T) and rhodamine antinebulin (N) is shown with a phase-contrast image. (O) The Z line. This pair is selected to demonstrate the occasional codistribution of titin and nebulin on N_2 lines. Note also the staining of sub-A-band region by antinebulin. (*Bottom*) Composite diagrams incorporating all observed staining zones of titin and nebulin are superimposed on the two-filament sarcomere model. The density of dotted zones approximates relative staining intensity. Note that not all staining zones appeared on all myofibrils at all times (see text for details). (*B*) Putative elastic filaments. A third filament of the sarcomere. (*Top*) The appearance of slender filaments (→) in the gaps between separated A band and I bands of an overly stretched rabbit psoas muscle. (*Center*) A hypothetical connecting filament model that can explain the appearance of filaments such as those shown in the micrograph above. It also explains why the A bands remain centrally located during muscle contraction and relaxation. (---) The location of elastic filaments that connect thick filaments to Z lines. (*Bottom*) A proposed arrangement of titin and nebulin on the elastic filament. (→) Additional structural loci (besides Z lines, A-I junctions, and M lines), which may serve as anchoring sites of the elastic filaments (see text for details).

more, these filaments are prone to mechanical or proteolytic degradation resulting in discontinuous segments.

On the basis of this model, our fluorescent staining data can be interpreted as showing that titin is localized in the region between two N_2 lines across the A band and that nebulin constitutes the short stretch between the N_2 line and the N_1 line, a striation about 0.15 μm away from the center of the Z disc (Franzini-Armstrong 1970) (Fig. 3B). Because of the limited resolution of light microscopy, the assignment of the boundaries of titin and nebulin domains to the N lines, which are clearly defined only by electron microscopy, should be considered tentative. Some possible overlap of titin and nebulin near the N_2-line region cannot be ruled out.

The appearance of titin and nebulin staining at locations other than that specified above could be explained if, during myofibril preparation, breakage of the elastic filament had occurred. The elastic material would then retract and accumulate around the nearest locus where the filament is anchored to an inextensible sarcomeric structure such as thick or thin filaments. As indicated in Figure 3B, such loci, inferred from staining patterns, include Z line, N_1 line, N_2 line, A-I junction, M line, as well as a position about 0.4 μm from the M line. If some of the filaments were severed near the N_1 line, then nebulin would appear concentrated on the N_2 line, as observed previously (Wang and Williamson 1980). If, in addition, the N_2-line locus is also dislodged for some reason, then nebulin would also appear on the A-I junction, as occasionally observed. Similarly, if the filament is severed between the A-I junction and the N_2 line, then titin would appear on the N_2 line or even on the N_1 line (which is indistinguishable from the Z line by light microscopy) if the locus near the N_2 line is dislodged. Similar reasoning can be used to explain the location of nearly all of the observed staining zones of titin and nebulin of isolated rigor myofibrils. It also follows that, depending on the extent and location of filament damage, the relative intensity and dimension of the staining bands could vary accordingly.

It is important to emphasize that the interpretation described above is only one of several schemes that can be devised. However, it is one of the simplest ones in terms of the necessary structural and chemical assumptions. These structural assumptions are consistent with the general characteristics reported of the third filament (elastic, parallel, and longitudinal), without specifying its detailed arrangement. The chemical assumption that titin and nebulin are components of the third filaments is justified on the basis of their abundance, their solubility properties, their distribution in the sarcomere and, at least for titin, the ability to form elastic filamentous gels (Fig. 2; Wang and Ramirez-Mitchell 1979). In this respect, their great sizes could be particularly significant: They may be long enough that only one or a few molecules are needed to span the entire domain specified above.

Our interpetation, on the basis of these simple assumptions, thus reveals

the existence of several specific structural loci along the length of the elastic filaments, which appear to serve as sites of interaction with other sarcomeric components or with other third filaments. It is tempting to speculate that structural features of the third filament, especially these special loci, may contribute to X-ray diffraction patterns of muscle (Huxley and Brown 1967).

Myofibrillar Connections: Sarcomere-associated Transverse and Longitudinal Intermediate Filaments

When isolated myofibrils were extracted with 0.6–1.0 M KI to dissolve the majority of thick and thin filaments, residual longitudinal filaments were frequently seen connecting successive Z lines. These filaments were interpreted by dos Remedios and Gilmour (1978) as evidence of the third filaments consisting of cross-linked connectin (see also Locker and Dianes 1980; Maruyama et al. 1981a). We were intrigued by this possibility and were prompted to investigate the fine structure and spatial organization of these residual filaments by high-resolution transmission and scanning electron microscopy (Fig. 4; Wang and Ramirez-Mitchell 1983). We were led to a completely different conclusion from that of dos Remedios and Gilmour: These longitudinal residual filaments have the characteristic fine structure of the intermediate filaments (10 nm) (see also Price and Sanger 1979). Furthermore, they appear to connect the peripheries of successive Z discs by forming a sleeve ensheathing the myofibril. These longitudinal filaments are too thick and are in the wrong place to be the sought-after third filaments that are 2–6 nm thick and exist *within* the sarcomere.

These longitudinal intermediate filaments appear to be an integral part of a three-dimensional network of intermediate filaments that connect Z lines and M lines longitudinally as well as transversely within striated muscle. Although the transverse myofibrillar connections have been known for a long time and were proposed to play important roles in maintaining the remarkable transverse alignment of sarcomeric striation of adjacent myofibrils (see Lazarides 1980), the existence of longitudinal myofibrillar connections was an unexpected finding that immediately raised questions about their functional roles.

SUMMARY

Recent studies of these novel myofibrillar proteins have led us and others to seriously consider the probable existence of continuous myofilaments in the sarcomere of striated muscles. It appears likely that, in addition to the classical two sets of filaments (thick and thin filaments that exhibit discrete lengths and locations), two types of longitudinal filaments that span the sarcomere may also be present: a set of slender, elastic filaments consisting of

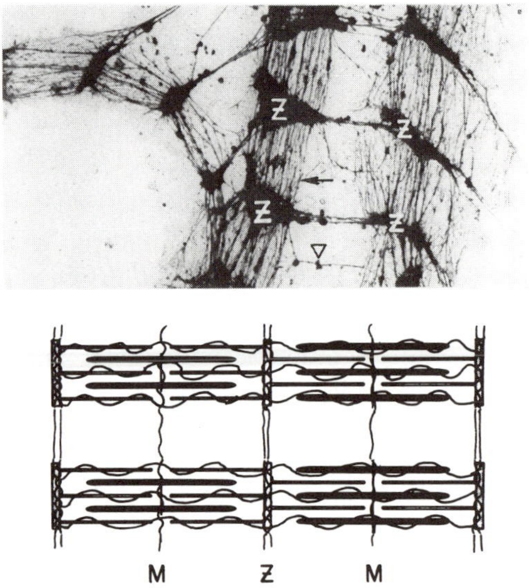

FIGURE 4
Myofibrillar connections: sarcomere-associated transverse and longitudinal intermediate filaments. (*Top*) Electron micrograph of residual filaments of a small myofibril bundle extracted with 0.6 M KI to remove the bulk of thick and thin filaments. Note the extensive transverse and longitudinal filamentous connections between adjacent Z lines and M lines. (→) Longitudinal filaments; (∇) transverse connection in the M-line position. These filaments are identified as intermediate filaments. (*Bottom*) A schematic diagram showing the arrangement of the transverse and longitudinal intermediate filaments relative to the sarcomeric striations. See text for details.

titin and nebulin, which exist within the sarcomere, and a set of parallel intermediate filaments, which surround each sarcomere.

What are the functions of these two sets of spanning filaments? They might have cytoskeletal roles in stabilizing the myofilament lattice and anchoring-associated membranous organelles; they might have mechanical roles of bearing at least part of the resting tension of the myofibrils; they might have regulatory roles in the assembly and turnover of the myofibril; and the third elastic filament might even be involved in active contraction, e.g., by modulating the interaction of thick and thin filaments by facilitating the smooth transition of the tetragonally arranged thin filaments to encounter hexagonally packed thick filaments.

ACKNOWLEDGMENTS

I thank the following co-workers who have shared with me the joys and sorrows in the pursuit of these elusive proteins: P. Louro, J. McClure, D. Palter, R. Ramirez-Mitchell, L. Somerville, A. Tu, and C.L. Williamson. This work was supported in part by grants from the U.S. Public Health Service (AM-20270 and CA-09182) and a grant from the American Heart Association, Texas affiliate.

REFERENCES

Carlsen, F., G.G. Knappeis, and F. Buchtal. 1961. Ultrastructure of the resting and contracted striated muscle fiber at different degrees of stretch. *J. Biophys. Biochem. Cytol.* **11:** 95.

dos Remedios, C. and D. Gilmour. 1978. Is there a third type of filament in striated muscle? *J. Biochem.* **84:** 235.

Etlinger, J.D., R. Zak, and D.A. Fischman. 1976. Compositional studies of myofibrils from rabbit striated muscle. *J. Cell Biol.* **68:** 123.

Franzini-Armstrong, C. 1970. Details of the I-band structure as revealed by the localization of ferritin. *Tissue Cell* **2:** 327.

Guba, F., V. Harsanyi, and E. Vajda. 1968. The muscle protein fibrillin. *Acta Biochim. Biophys. Acad. Sci. Hung.* **3:** 433.

Hoyle, G. 1968. *Symposium on Muscle* (ed. E. Ernst and F.B. Straub), p. 34. Akademiai Kiado, Budapest.

Huxley, H.E. and W. Brown. 1967. The low angle X-ray diagram of vertebrate striated muscle and its behavior during contraction and rigor. *J. Mol. Biol.* **30:** 383.

Huxley, H.E. and J. Hanson. 1954. Changes in the cross striations of muscle during contraction and stretch and their structural interpretation. *Nature* **173:** 973.

Huxley, A.F. and L.D. Peachey. 1961. The maximum length for contraction in vertebrate striated muscle. *J. Physiol.* **156:** 150.

King, N.L. and L. Kurth. 1980. SDS gel electrophoresis studies of connectin. In *Fibrous proteins: Scientific, industrial and medical aspects* (ed. D.A.D. Parry and L.K. Creamer), vol. 2, p. 57. Academic Press, London.

Lazarides, E. 1980. Intermediate filaments as mechanical integrators of cellular space. *Nature* **283:** 249.

Locker, R.H. and G.J. Dianes. 1980. Gap filaments—The third set in the myofibril. In *Fibrous proteins: Scientific, industrial and medical aspects* (ed. D.A.D. Parry and L.K. Creamer), vol. 2, p. 43. Academic Press, London.

Locker, R.H. and N.G. Leet. 1976. Histology of highly stretched beef muscle. II. Further evidence on the location and nature of gap filaments. *J. Ultrastruct. Res.* **55:** 157.

Maruyama, K., S. Kimura, K. Ohashi, and Y. Juwano. 1981a. Connectin, an elastic protein of muscle. Identification of titin with connectin. *J. Biochem.* **89:** 701.

Maruyama, K., M. Kimura, S. Kimura, K. Ohashi, K. Suzuki, and N. Katunuma. 1981b. Connectin: An elastic protein of muscle. Effects of proteolytic enzymes *in situ. J. Biochem.* **89:** 711.

Maruyama, K., S. Matsubara, S.R. Natori, Y. Nonomura, S. Kimura, K. Ohashi, F. Murakami, S. Handa, and G. Eguchi. 1977. Connectin, an elastic protein in muscle. Characterization and function. *J. Biochem.* **82:** 317.

McNeil, P.A. and G. Hoyle. 1967. Evidence for superthin filaments. *Am. Zool.* **7:** 483.

Page, S.G. and H.E. Huxley. 1963. Filament lengths in striated muscle. *J. Cell Biol.* **19:** 369.

Porzio, M.A. and A.M. Pearson. 1977. Improved resolution of myofibrillar proteins with sodium dodecyl sulfate-polyacrylamide gel electrophoresis. *Biochim. Biophys. Acta* **490:** 21.

Price M. and J.W. Sanger. 1979. Intermediate filaments connect Z discs in adult chicken muscle. *J. Exp. Zool.* **208:** 263.

Pringle, J.W.S. 1978. Stretch activation of muscle: Function and mechanism. *Proc. R. Soc. Lond. B* **201:** 107.

Saide, J.D. 1981. Identification of a connecting filament protein in insect fibrillar flight muscle. *J. Mol. Biol.* **153:** 661.

Sjostrand, F.S. 1962. The connection between A- and I-band filaments in striated frog muscle. *J. Ultrastruct. Res.* **7:** 225.

Thorson, J. and D.C.S. White. 1969. Distributed representations for actin-myosin interaction in the oscillatory contraction of muscle. *Biophys. J.* **9:** 360.

Walcott, B. and E.B. Ridgeway. 1967. The ultrastructure of myosin-extracted striated muscle fibers. *Am. Zool.* **7:** 499.

Wang, K. 1981. Nebulin, a giant protein component of N_2-line striated muscle. *J. Cell Biol.* **91:** 355a.

———. 1982. Purification of rabbit titin and nebulin. *Methods Enzymol.* **85:** 264.

Wang, K. and J. McClure. 1978. Extremely large proteins of vertebrate striated muscle myofibrils. *J. Cell Biol.* **79:** 334a.

Wang, K. and R. Ramirez-Mitchell. 1979. Titin: Possible candidate as components of putative longitudinal filaments in striated muscle. *J. Cell Biol.* **83:** 389a.

———. 1981. Myofibrillar connections. The role of titin, N_2-line protein and intermediate filaments. *Biophys. J.* **33:** 21a.

———. 1983. A network of transverse and longitudinal intermediate filaments is associated with sarcomeres of adult skeletal muscle. *J. Cell Biol.* (in press).

Wang, K. and C.L. Williamson. 1980. Identification of an N_2-line protein of striated muscle. *Proc. Natl. Acad. Sci.* **77:** 3254.

Wang, K., J. McClure, and A. Tu. 1979. Titin: Major myofibrillar components of striated muscle. *Proc. Natl. Acad. Sci.* **76:** 3698.

NEUROMUSCULAR JUNCTION FORMATION

Introduction: Neuromuscular Junction Formation

ZACH W. HALL
Department of Physiology
University of California
San Francisco, California 94143

Muscle cells are not only specialized to contract and generate force but also to receive the chemical signals from nerves that activate contraction. At the neuromuscular junction, the point of contact between nerve and muscle, both cells are specialized in the sense that structures occur here that are not found elsewhere in the cell. These are illustrated by the diagram in Figure 1.

Motor nerve terminals lie on the surface of the muscle in a shallow groove where they are capped by overlying Schwann cells. The terminals are filled with synaptic vesicles containing the neurotransmitter acetylcholine. These are not distributed uniformly but are concentrated near the synaptic surface, where they are clustered in bands that stretch across the terminal. These bands, called "active zones" (Couteaux and Pecot-Dechavissine 1970), are thought to be the sites at which acetylcholine is released into the synaptic cleft by exocytosis. Underneath the nerve, the muscle cell membrane is convoluted into folds that are in exact register with the active zones in the presynaptic terminals. The acetylcholine receptor (AChR), the protein that binds the released neurotransmitter, is concentrated in the postsynaptic membrane on the crests of these folds (Fertuck and Salpeter 1974). The density of the receptor in the subsynaptic membrane is over three orders of magnitude higher than that in the surrounding membrane. The postsynaptic membrane contains on its cytoplasmic surface a coating of amorphous material that lines the sides and tops of the membrane folds; bundles of intermediate filaments underlie the folds (Couteaux and Pecot-Dechavissine 1968; Heuser 1980). Lying in the synaptic cleft between presynaptic and postsynaptic membranes is the basal lamina, which envelops the entire muscle fiber. The basal lamina at the synapse has two distinctive features. First, it extends projections into the folds of the postsynaptic membrane. Second, it has attached to it a high concentration

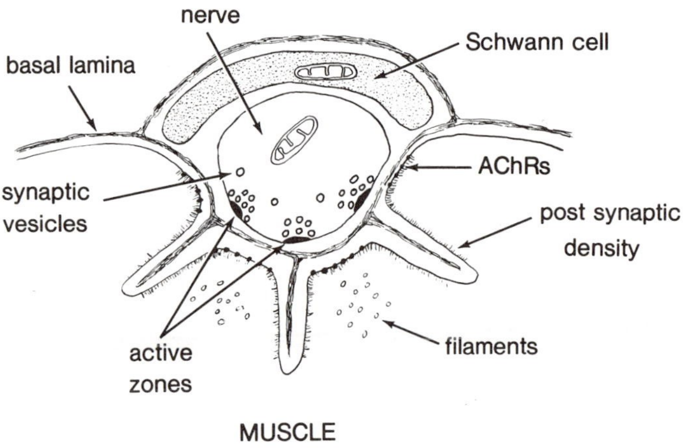

FIGURE 1
Diagrammatic representation of neuromuscular junction.

of acetylcholinesterase (AChE) (McMahan et al. 1978). This enzyme cleaves the released acetylcholine. Thus, the extracellular matrix, as well as presynaptic and postsynaptic cells, has unique features at the synapse.

One difficulty in studying how this complex structure arises during muscle development is that, unlike the contractile apparatus to which most of the muscle mass is devoted, the neuromuscular junction constitutes only a small fraction of the total tissue. Thus, only two of the macromolecules that are found in high concentration at the junction, the AChR and AChE, have been purified and characterized in detail. Although we know little about the biochemical structure of the neuromuscular junction, studies of the properties of these two proteins during development, particularly the AChR, have given us enough insight into the processes that occur to echo a theme reiterated in other papers in this volume, namely, that at the molecular level the process of differentiation is a complex, multistage process that occurs over several weeks.

This can be illustrated by considering the changes in properties of the AChR that occur during formation and maturation of the neuromuscular junction (Table 1). During development, AChRs first appear prior to innervation, when they are diffusely distributed over the entire surface of myotubes prior to innervation. The properties of the AChRs at this stage differ in several ways from those that are found at mature synapses. First, they normally occur at much lower density. In uninnervated myotubes the density of AChRs is a few hundred per square micron as opposed to approximately $10^4/\mu m^2$ in the subsynaptic membrane of adult muscles. Second, the AChR, like other membrane proteins, is in a dynamic metabolic state, and there is a dramatic difference in the metabolic stability of AChRs in embryonic and adult muscle. The AChRs in uninnervated myotubes have a

TABLE 1
Developmental Changes in the AChR in Rat Muscle

	AChRs at adult endplates	AChRs in uninnervated myotubes	Time of change[a]
Density	$3 \times 10^4/\mu m^2$	$\sim 10^2/\mu m^2$	ED 15–16
Half-life	~ 11 days	~ 1 day	ED 18–20
Mean channel open time	~ 1 msec	~ 3–4 msec	PD 6–14

The references for these values can be found in the text and in Fambrough (1979).
[a]ED, embryonic day; PD, postnatal day. Rats are born at 21 days of gestation.

metabolic half-life of about 1 day, whereas those at adult endplates turn over with a half-time of approximately 10 days.

Finally, the mean channel open time of AChRs, i.e., the average time that the transmembrane ionic channel remains open after binding acetylcholine, differs in the two cases, being several times shorter in adult than in embryonic muscle (for review, see Fambrough 1979).

For our purposes, the important point is that these three changes, in density, in metabolic turnover time, and in channel open time, each of which must reflect changes in the structure or the molecular associations of the AChR, occur at three distinct times over a period of about 3 weeks during muscle development. In the rat, receptors cluster at embryonic day 16 (Bevan and Steinbach 1971; Braithwaite and Harris 1979) but initially retain their rapid turnover time and prolonged channel open time. About 3–5 days later, their metabolic stability changes (Reiness and Weinberg 1981), and about 2 weeks later, during the second postnatal week, endplates whose AChRs have short channel open times first appear (Sakmann and Brenner 1978; Fischbach and Schnetze 1980). Several changes in the molecular properties of the AChR also occur over this time period, but the exact time at which these occur and their correspondence to the physiological changes is not known.

It is apparent that for an orderly sequence of events such as these to occur, not to mention the many other, mostly unknown, changes that are required for presynaptic and postsynaptic differentiation, nerves and muscles must have mechanisms for exchanging signals. Much of the work in the field at this time and, correspondingly, much of the work reported in this volume on neuromuscular differentiation, both in this section and in that on Muscle Development and Human Disease, concerns the nature of these signals. Most of the work has focused on the induction of AChR clustering, since that is the earliest sign of synaptic differentiation and is easily detected. In this section, Salpeter and Podleski report on the effects of a soluble factor from nervous tissue that causes clustering of AChRs in

myotubes cultured in the absence of nerves. Wallace et al. (also this section) report on the solubilization and partial purification of a factor from an insoluble fraction of *Torpedo* that has a similar effect. An alternate approach (Hall et al., this volume) has been to look for molecules that are associated with the extracellular matrix at the neuromuscular junction that might play a role in synaptic development. Finally, Kidokoro et al. (this volume) report on a soluble, nerve-derived factor that affects the channel open time of AChRs in *Xenopus* myotubes.

REFERENCES

Bevan, S. and J.H. Steinbach. 1977. The distribution of alpha-bungarotoxin binding sites on mammalian skeletal muscle developing *in vivo*. *J. Physiol.* **267:** 195.

Braithwaite, A.W. and A.J. Harris. 1979. Neural influence on acetylcholine receptor clusters in embryonic development of skeletal muscles. *Nature* **279:** 549.

Couteaux, R. and M. Pecot-Dechavassine. 1968. Particularities structurales du sarcoplasma sous-neural. *C.R. Acad. Sci. D* **266:** 8.

———. 1970. Vesicules synaptiques et posches au niveau de 'zone actives' de las jonction neuromusculaire. *C.R. Acad. Sci. D* **271:** 2346.

Fambrough, D.M. 1979. Control of acetylcholine receptors in skeletal muscle. *Physiol. Rev.* **59:** 165.

Fertuck, H.C. and M.M. Salpeter. 1974. Localization of acetylcholine receptor by ^{125}I-labelled alpha-bungarotoxin at mouse endplates. *Proc. Natl. Acad. Sci.* **71:** 1376.

Fischbach, G.D. and S.M. Schnetze. 1980. A postnatal decrease in acetylcholine channel open time at rat endplates. *J. Physiol.* **303:** 125.

Heuser, J.E. 1980. 3-D visualization of membrane and cytoplasmic specializations of the neuromuscular junction. In *Ontogenesis and functional mechanisms of peripheral synapses,* INSERM Symposium no. 13 (ed. J. Taxi), p. 139. Elsevier/North Holland, Amsterdam.

McMahan, U.J., J.R. Sanes, and L.M. Marshall. 1978. Cholinesterase is associated with the basal lamina at the neuromuscular junction. *Nature* **271:** 172.

Reiness, C.G. and C.B. Weinberg. 1981. Metabolic stabilization of acetylcholine receptors at newly formed neuromuscular junctions in rat. *Dev. Biol.* **84:** 247.

Sakmann, B. and H.R. Brenner. 1978. Changes in synaptic channel gating during neuromuscular development. *Nature* **274:** 68.

Synaptic Basal Lamina Components Made by a Muscle Cell Line

ZACH W. HALL
LAURA SILBERSTEIN
*NIBALDO C. INESTROSA**
Division of Neurobiology
Department of Physiology
University of California
San Francisco, California 94143

The assembly of a complex structure such as the vertebrate neuromuscular junction requires that signals be exchanged in an orderly manner between presynaptic and postsynaptic cells. One mechanism for such exchange is the activation of the postsynaptic cell produced by normal synaptic transmission (Lømo and Westgaard 1976); a second is the release and reception of soluble factors by the two cells (Christian et al. 1978; Podleski et al. 1978; Jessell et al. 1979; Henderson et al. 1981; see also Appel, this volume); a third is through the insoluble extracellular matrix (ECM) (Grobstein 1975; Sanes et al. 1978; Burden et al. 1979). The importance of the last possibility has been emphasized by the elegant work of McMahan and his colleagues (see Wallace et al., this volume), who have shown that the basal lamina at the synapse can direct both presynaptic and postsynaptic differentiation in regenerating synapses in adult frog muscle.

We have chosen to investigate the molecular properties of the synaptic basal lamina by using immunological methods. We have raised antisera to various basement-membrane extracts or collagen preparations and have examined these for antibodies that bind to components that are selectively concentrated in the basal lamina at the neuromuscular junction (Sanes and Hall 1979). We have found three such sera, termed JS (junction-specific)-I, JS-II, and JS-III, which recognize three different determinants in the synap-

*Present address: Laboratory of Neurophysiology, Department of Cell Biology, Catholic University of Chile, Santiago, Chile.

tic basal lamina (Table 1). A fourth component of the synaptic basal lamina is the enzyme acetylcholinesterase (AChE), which is highly concentrated at the neuromuscular junction, where it can been seen by histochemical methods (Couteaux 1955). In rat and mouse muscle, a particular form of AChE is concentrated in regions of muscle with endplates (Hall 1973; Vigny et al. 1976; Inestrosa et al. 1982). This form is thought to have a long, collagenlike tail (Bon et al. 1979), and it has been assumed, but not directly demonstrated, that the tailed form corresponds to the endplate enzyme seen by histochemistry. At least part of the AChE at the endplate is attached to the basal lamina (McMahan et al. 1978).

We have used these immunologically and enzymatically defined markers to investigate the synthesis of synapse-specific basal lamina. We undertook these experiments with the expectation of finding an example of intercellular cooperation in the biosynthesis of the ECM. We anticipated that the nerve might contribute, or specifically stimulate the muscle to contribute, those components of the synaptic basal lamina that are uniquely concentrated there. To our surprise, we found that myotubes derived from a mouse muscle cell line, cultured in the absence of nerves, are able to accumulate on their surface antigens recognized by each of the three antisera. Furthermore, these myotubes synthesize the endplate form of AChE, which occurs in focal patches associated with the ECM that partially surrounds cells.

RESULTS AND DISCUSSION

Synapse-specific Antigens

To assure cellular homogeneity, we chose for our experiments a muscle cell line, C2, derived from adult C3H mice by Yaffe and Saxel (1977). These

TABLE 1
Synaptic Basal Lamina Antisera

Anti-serum	Immunizing antigen	Adsorption required for endplate-specific staining	Staining collagenase-sensitive	Antigen extends beyond endplate
JS-I	anterior lens capsule	–	–	–
JS-II	collagen fraction from muscle	+	+	+
JS-III	collagen fraction from anterior lens capsule	+	+	–

See Sanes and Hall (1979) for further details.

cells proliferate in high-serum medium and fuse into myotubes following transfer into low-serum medium. Within several days after transfer into fusion medium, the cultures contain long, cylindrical myotubes that often undergo spontaneous and vigorous contractions. These cultures produce an extensive ECM, which can be visualized by staining with a serum that binds to both synaptic and extrasynaptic basal lamina in adult muscle sections (Fig. 1a). The pattern of staining shows that the ECM partially and incompletely ensheathes the myotubes and also extends between fibers as a fibrillar network.

A more limited pattern of staining was seen when the synapse-specific sera were used. JS-I and JS-II antigens occur most often as patches on the surfaces or sides of the myotubes (Fig. 1b,c). The JS-III antigen often had a distinctive pattern consisting of many speckles (Fig. 1d).

To test whether the antigens recognized by these sera in C2 cultures were similar to those at adult endplates, we examined the ability of C2 cells to bind the antibodies responsible for endplate staining. After adsorption with C2 cultures, all three sera failed to stain the endplates of adult mice.

AChE

Because aneural cultures of C2 cells contain each of the immunologically defined components of the synaptic basal lamina, we also investigated whether they produce the tailed form of AChE, which is recognized by its sedimentation coefficient (16S). Sucrose gradient sedimentation of C2 cell extracts of high ionic strength showed that this form can first be detected 2–3 days after transfer into fusion medium, and by 5 days constitutes approximately 30% of the total enzyme activity. Small amounts of 16S AChE have also been seen in another mouse muscle cell line (Sugiyama 1977). We found that the properties of the 16S AChE extracted from C2 myotubes are similar to those of the tailed forms of eel and *Torpedo* enzymes (Bon et al. 1979). Thus, the 16S mouse enzyme aggregates in solutions of low ionic strength, a property that is lost after treatment of the AChE with collagenase to remove the tail (data not shown).

To see whether the 16S enzyme was present on the cell surfaces of C2 myotubes, we examined the effect on AChE activity of enzyme inhibitors that do not cross the surface membrane. We found that BW284c51, a reversible, membrane-impermeable inhibitor, protected 16S AChE activity against subsequent irreversible inactivation by the permeable inhibitor diisopropyl fluorophosphonate (DFP) (Fig. 2). We conclude that the active site of the 16S enzyme is extracellular.

Further information about the localization of the enzyme was obtained by differential solubilization experiments. Over 90% of the 16S activity could be extracted by a solution of high ionic strength (1 M NaCl) lacking detergent. Less than 3% of the 16S activity was extracted at low ionic

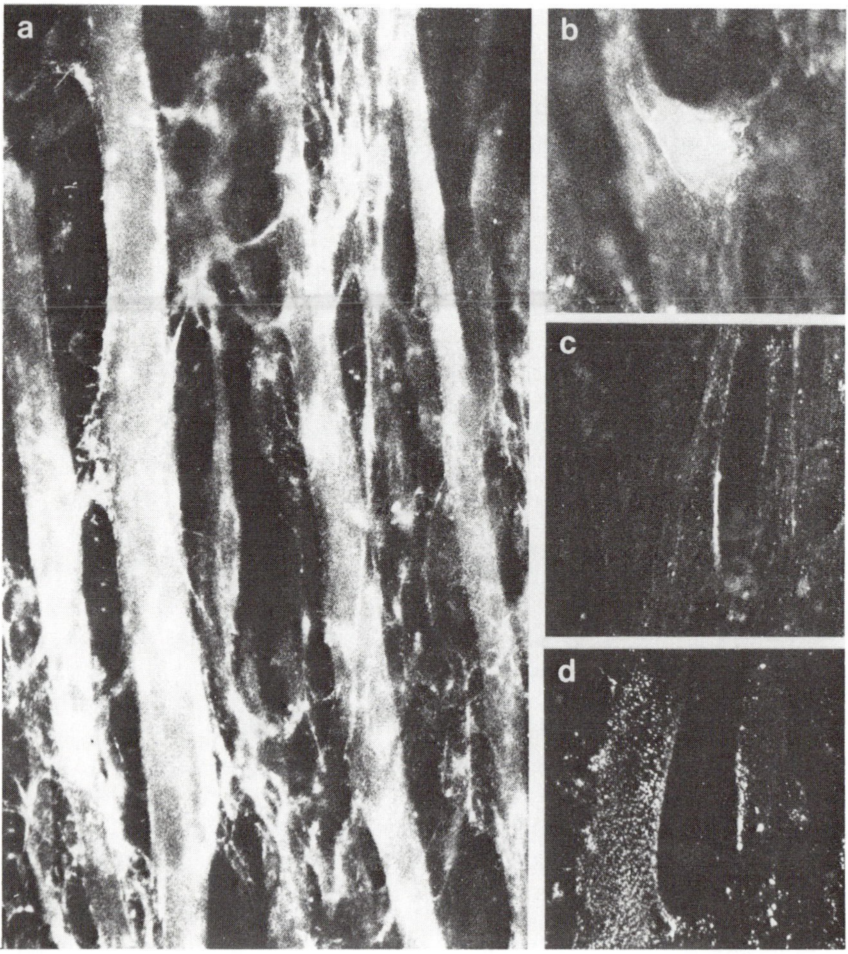

FIGURE 1
Immunological localization of extracellular matrix in C2 myotube cultures. Cells were plated on tissue-culture plastic in growth medium (20% fetal calf serum and 0.5% chick embryo extract in Dulbecco's modified Eagle's medium [DME]) and grown for 5 days in fusion medium (10% horse serum in DME); they were then incubated for 60 min at 37°C with a 1/100 dilution of rabbit antisera to muscle basement-membrane components as described in Sanes and Hall (1979). The cells were then incubated with a fluorescein-conjugated second antibody (Cappel Laboratories) at 1/100 dilution for 60 min at 37°C. The cells were washed in phosphate-buffered saline (PBS) containing 1 mM $CaCl_2$, mounted under 10% Gelvatol 20/30 (Monsanto) and 20% glycerol in PBS, and examined with a Leitz Ortholux II miscroscope under fluorescein optics. Reaction with antimuscle basement-membrane collagen (*a*), JS-I (*b*), JS-II (*c*), and JS-III (*d*). Magnification, 276×. (Reprinted, with permission, from Silberstein et al. 1982.)

strength, with or without 1% Trition X-100. Because detergent is not required for its extraction, the 16S protein is presumably not an integral membrane protein. It appears to be attached to the ECM by ionic interactions.

Histochemical staining of C2 cultures for AChE shows that the enzyme is not uniformly distributed but occurs in patches on the cell surface of some myotubes. These patches appear 3–5 days after transfer into fusion medium, shortly after the appearance of 16S AChE. The focal staining is abolished by 10 μM BW284c51 (Fig. 2). To determine whether the patches of AChE activity are related to the 16S enzyme, we treated intact C2 myotubes with highly purified bacterial collagenase. This treatment selectively removed the 16S AChE activity (Fig. 3), providing further evidence that this enzyme is extracellular, and abolished the patchy staining. We conclude that the patches consist of the 16S enzyme.

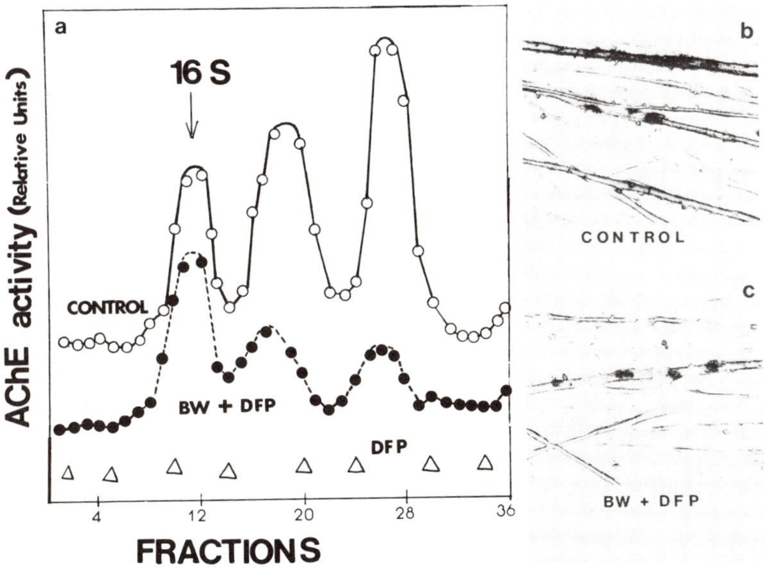

FIGURE 2

16S AChE and the patches of AChE activity are extracellular. (*a*) Sucrose gradient profile of AChE from C2 myotubes grown for 4 days in fusion medium. (●) Myotubes were preincubated for 30 sec in 750 μM BW284c51, a reversible inhibitor of AChE that penetrates the membrane, and then incubated with 2.5×10^{-5} DFP, an irreversible lipid-soluble inhibitor. (○) Control myotubes. Histochemical staining of myotubes in control cultures (*b*) and in cultures treated with BW284c51 and DFP as above (*c*). The cultures were stained by the method of Karnovsky and Roots (1964). Staining was for 3 hr at 37°C and was completely inhibited by 10 μM BW284c51. Because BW284c51 protects enzyme in the patches, we conclude that the active site of the enzyme is exposed to the extracellular medium.

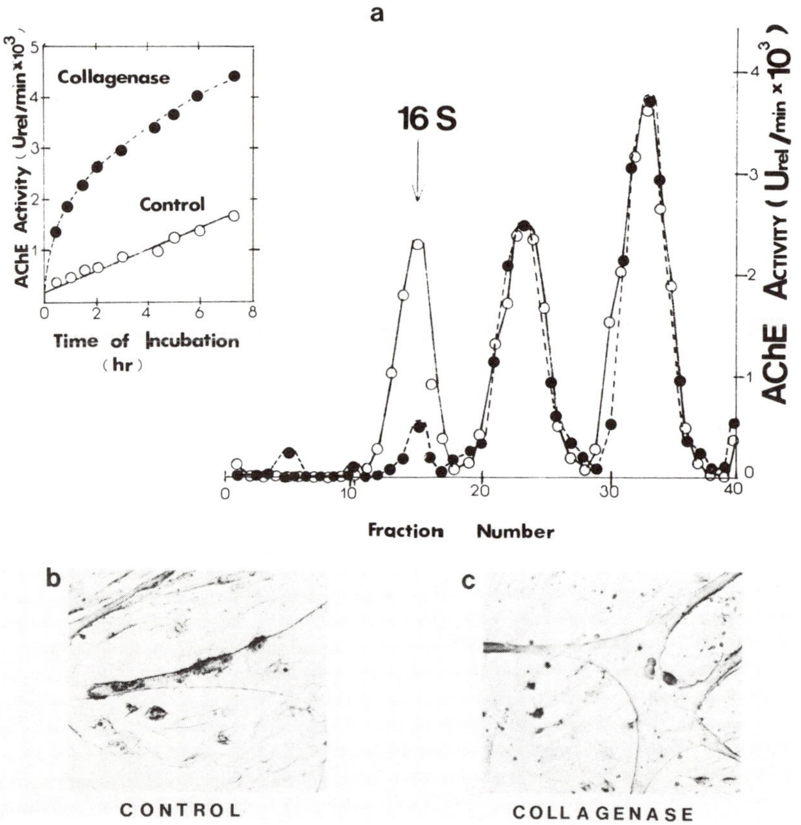

FIGURE 3
Collagenase treatment abolishes 16S AChE and the AChE patches seen by histochemistry. C2 cells were grown for 4 days in fusion medium, washed to remove serum, and incubated either in the presence (●) or absence (○) of 0.1 mg/ml of collagenase (form III, Advanced Biofactured, Lynbrook, New York), in 0.1 M Tris (pH 7.4) and 10 μM $CaCl_2$. (*a*) Sucrose gradient profile of AChE forms. (*Inset*) Time course of AChE release. Histochemical staining of cultures incubated without (*b*) and with (*c*) collagenase.

These experiments show that aneural C2 myotube cultures can make 16S AChE as well as the synapse-specific BL antigens. Moreover, the 16S enzyme, like the endplate enzyme, is located extracellularly in a focal patch, where it appears to be associated with an insoluble matrix.

Codistribution of Basal Lamina Components with Acetylcholine Receptors

Because C2 cultures contain each of the four synapse-specific basal lamina components, we asked whether a fully organized and mature synaptic basal lamina is formed. Are the components colocalized in the ECM? We could

not determine simultaneously the distribution of all four components, so we chose to examine the relationship of each to the acetylcholine receptor (AChR) patches that occur in C2 myotubes. AChR patches were visualized with rhodamine-labeled α-bungarotoxin; antibody binding was detected with a fluorescent second antibody, and AChE staining was done according to Moody-Corbett and Cohen (1981). We found that patches of one antigen, JS-I, often coincided with AChR clusters, whereas the other components were rarely related to AChR clusters.

In cultures examined 2–3 days after fusion, 30–50% of all AChR clusters were associated with patches of JS-I; patches of JS-I without AChR clusters were also seen. In older cultures, however, the coincidence of JS-I patches and AChR clusters was almost complete. In myotubes examined 5–6 days after fusion, over 95% of AChR clusters were associated with JS-I patches. The correspondence between the pattern of JS-I staining and the AChR distribution was often striking (Fig. 4). JS-II patches that codistributed with

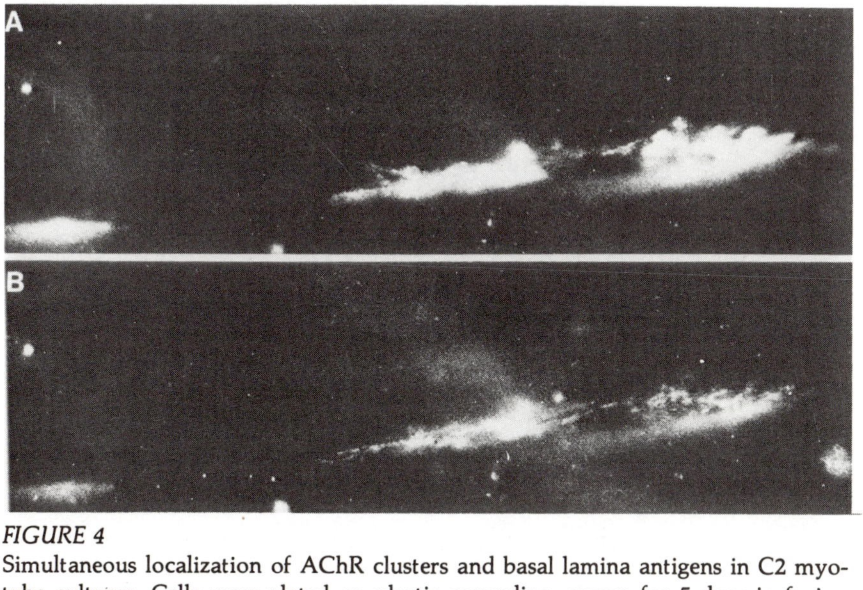

FIGURE 4
Simultaneous localization of AChR clusters and basal lamina antigens in C2 myotube cultures. Cells were plated on plastic coverslips, grown for 5 days in fusion medium, and incubated with fluorescent reagents as before. (*A*) AChR clusters are observed after incubation with 50 nM rhodamine-conjugated α-bungarotoxin; AChR clusters in control cultures, incubated with nonimmune serum and photographed and printed in identical conditions, could be seen with rhodamine but not with fluorescein filters. (*B*) Localization of JS-I antigen(s) in the same field by labeling with fluorescein-conjugated goat anti-rabbit serum as before. No receptor clusters were seen in cultures preincubated with unlabeled α-bungarotoxin (1 μM), followed by rhodamine-conjugated α-bungarotoxin. Magnification, 660×.

the AChR were occasionally seen; the pattern of AChE and of JS-III staining, however, appeared distinct from that of the AChR.

Since, in C2 cultures, some of the synaptic basal lamina components codistribute with AChR clusters and some do not, the distributions of each of the synaptic basal lamina antigens cannot be the same. These results suggest that although the C2 myotubes are able to make each of the synapse-specific basal lamina components that we have sought, they do not make a coherent and fully mature synaptic basal lamina. We are currently investigating the role of the nerve in assembly of the basal lamina at synapses.

SUMMARY

Myotubes of a mouse muscle cell line (C2) synthesize in culture the synaptic form (16S) of AChE and each of three immunologically defined components of the synaptic basal lamina in adult muscle. The 16S AChE has the properties expected of AChE forms with a collagenlike tail and is found in patches on the external cell surface where it appears to be associated with the ECM by ionic interactions. Surface patches of one of the synaptic antigens often occur in close association with AChR clusters, whereas the other antigens and the AChE patches are rarely, if ever, coincident with AChR clusters. We conclude that the C2 cells in the absence of nerves make and accumulate on their surface several components that are known to be concentrated in the basal lamina at adult synapses but that these are not assembled into a coherent whole as in the adult.

ACKNOWLEDGMENTS

We thank Dr. David Yaffe for the C2 cell line and Ms. Nicole Robitaille-Giguere for expert technical assistance. N.C.I. is a fellow of the Muscular Dystrophy Association of America. L.S. is a fellow of the National Institutes of Health. This research was supported by grants from the National Institutes of Health, the National Science Foundation, and the Muscular Dystrophy Association of America to Z.W.H.

REFERENCES

Bon, S., M. Vigny, and J. Massoulie. 1979. Asymmetric and globular forms of acetylcholinesterase in mammals and birds. *Proc. Natl. Acad. Sci.* **76:** 2546.

Burden, S.J., P.B. Sargent, and U.J. McMahan. 1979. Acetylcholine receptors in regenerating muscle accumulate at original synaptic sites in the absence of the nerve. *J. Cell Biol.* **82:** 412.

Christian, C.N., M.P. Daniels, H. Sugiyama, Z. Vogel, L. Jacques, and D.G. Nelson. 1978. A factor from neurones that increases the number of acetylcholine receptor aggregates on cultured muscle cells. *Proc. Natl. Acad. Sci.* **75:** 4011.

Couteaux, R. 1955. Localization of cholinesterases of neuromuscular junctions. *Rev. Cytol.* **4:** 355.

Grobstein, C.J. 1975. Developmental role of intercellular matrix: Retrospective and prospective. In *Extracellular matrix influences on gene expression* (ed. H.C. Slavkin and R.C. Grenlich), p. 9. Academic Press, New York.

Hall, Z.W. 1973. Multiple forms of acetylcholinesterase and their distribution in endplate and non-endplate regions of rat diaphragm muscle. *J. Neurobiol.* **4:** 343.

Henderson, C.E., M. Huchet, and J.-P. Changeaux. 1981. Neurite outgrowth from embryonic chicken spinal neurons is promoted by media conditioned by muscle cells. *Proc. Natl. Acad. Sci.* **78:** 2625.

Inestrosa, N.C., L. Silberstein, and Z.W. Hall. 1982. Association of the synaptic form of acetylcholinesterase with extracellular matrix in cultured mouse muscle cells. *Cell* **29:** 71.

Jessell, T., R. Siegel, and G.D. Fischbach. 1979. Induction of acetylcholine receptors on cultured skeletal muscle by a factor extracted from brain and spinal cord. *Proc. Natl. Acad. Sci.* **76:** 5397.

Karnovsky, M.J. and L. Roots. 1969. A "direct-coloring" thiocholine method for cholinesterases. *J. Histochem. Cytochem.* **12:** 219.

Lømo, T. and R.H. Westgaard. 1976. Control of ACh sensitivity in rat muscle fibers. *Cold Spring Harbor Symp. Quant. Biol.* **40:** 263.

McMahan, U.J., J.R. Sanes, and L.M. Marshall. 1978. Cholinesterase is associated with basal lamina at the neuromuscular junction. *Nature* **271:** 172.

Moody-Corbett, F. and M.W. Cohen. 1981. Localization of cholinesterase at sites of high acetylcholine receptor density on embryonic amphibian muscle cells cultured without nerve. *J. Neurosci.* **1:** 596.

Podleski, T.R., D. Axelrod, P. Ravdin, I. Greenberg, M.M. Johnson, and M.M. Salpeter. 1978. Nerve extract induces increase and redistribution of acetylcholine receptors on cloned muscle cells. *Proc. Natl. Acad. Sci.* **75:** 2035.

Sanes, J.R. and Z.W. Hall. 1979. Antibodies that bind specifically synaptic sites on muscle fiber basal lamina. *J. Cell Biol.* **83:** 357.

Sanes, J.R., L.M. Marshall, and U.J. McMahan. 1978. Reinnervation of muscle fiber basal lamina after removal of muscle fibers. *J. Cell Biol.* **78:** 176.

Silberstein, L., N.C. Inestrosa, and Z.W. Hall. 1982. Aneural muscle cells make synaptic basal lamina components. *Nature* **295:** 143.

Sugiyama, H. 1977. Multiple forms of acetylcholinesterase in clonal muscle cells. *FEBS Lett.* **84:** 257.

Vigny, M., J. Koenig, and F. Rieger. 1976. The motor endplate specific form of acetylcholinesterase: Appearance during embryogenesis and reinnervation of rat muscle. *J. Neurochem.* **27:** 1347.

Yaffe, D. and O. Saxel. 1977. Serial passaging and differentiation of myogenic cells isolated from dystrophic mouse muscle. *Nature* **270:** 725.

An Extract of Extracellular Matrix Fraction That Organizes Acetylcholine Receptors

BRUCE G. WALLACE
EARL W. GODFREY
RALPH M. NITKIN
*LEE L. RUBIN**
U. JACK McMAHAN
Department of Neurobiology
Stanford University School of Medicine
Stanford, California 94305

After damage to muscles, both myofibers and motor axons degenerate and are phagocytized, but the extracellular matrix, including the basal lamina sheaths of the myofibers, survives. New myofibers develop within the basal lamina sheaths of the original myofibers, axons grow to original synaptic sites on the basal lamina, and new neuromuscular junctions are formed (Marshall et al. 1977). The regenerated neuromuscular junctions perform and look much like the original ones: Nerve stimulation results in muscle contractions, active zones are distributed along the nerve terminal, and the portion of the myofiber plasma membrane directly beneath the nerve terminal has junctional folds and a high concentration of acetylcholine receptors (AChRs). Studies on structures within muscle that influence regeneration of the neuromuscular junction have provided evidence that the differentiation of regenerating nerve terminals and of the subsynaptic apparatus in myofibers is directed by factors attached to or a part of the synaptic portion of the basal lamina sheath of the myofiber (Sanes et al. 1978; Burden et al. 1979; McMahan et al. 1980; Bader 1981). Thus, regenerating axons that grow to the original synaptic sites on basal lamina sheaths in the absence of myofibers form active zones, and myofibers that regener-

*Present address: Rockefeller University, New York, New York 10021.

ate in the absence of nerve terminals form junctional folds and high concentrations of AChRs where they contact the synaptic portion of the basal lamina.

We have undertaken a project aimed at identifying the components of the synaptic basal lamina that influence the organization of AChRs on regenerating myofibers. Our plan is to make extracts of extracellular matrix, screen for those that induce aggregation of AChRs on cultured myotubes, purify the active components, and determine, by immunological techniques, whether or not the active components are similar to those in the synaptic basal lamina. We found that extracellular matrix fractions prepared from junctional regions of muscle had no discernible effect on AChR organization. Therefore, we turned to the electric organ of *Torpedo californica*, which has a much greater density of cholinergic synapses than muscle and has been used as a source for purification of other synaptic components. The extracellular matrix fraction from *Torpedo* was rich in a proteinaceous factor or factors that organized the AChRs into distinct clusters (Rubin et al. 1980; Rubin and McMahan 1982). Here, we report our progress in purifying and characterizing the active components from the *Torpedo* electric organ. In brief, the activity of our most pure extract is nearly 400-fold greater than that of crude electric-organ homogenate, and the molecular weights of the active components are between 50,000 and 100,000. We also show that at least some components of our extracellular-matrix extracts are antigenically similar to those in basal lamina at the neuromuscular junction.

RESULTS AND DISCUSSION

Assay of AChR-organizing Activity

The ability of extracts of electric organ to organize AChRs was assayed in 5–7-day-old cultures of myotubes prepared from 11–12-day chick embryo hind limbs (Fischbach 1972). Extracts were added, the cultures were incubated overnight, and the distribution of AChRs was determined by marking them with rhodamine-labeled α-bungarotoxin and examining the cultures with a fluorescence microscope. Figure 1 shows that the number of AChR clusters on myotubes was greatly increased by exposing the cultures to extracts of *Torpedo* extracellular matrix. The increase in number of AChR clusters began as early as 3 hours after adding extracts and continued up to 16 hours; prolonged exposure to extracts produced no further change in receptor distribution (Fig. 2A). Similar results were obtained when the AChRs were labeled before adding factor, indicating that the appearance of clusters was produced, at least in part, by the lateral migration of AChRs present in the plasma membrane prior to the addition of factor. Experiments using ^{125}I-labeled α-bungarotoxin showed little or no change in the number or half-life of AChRs on myotubes incubated with factor. Thus, exposure of

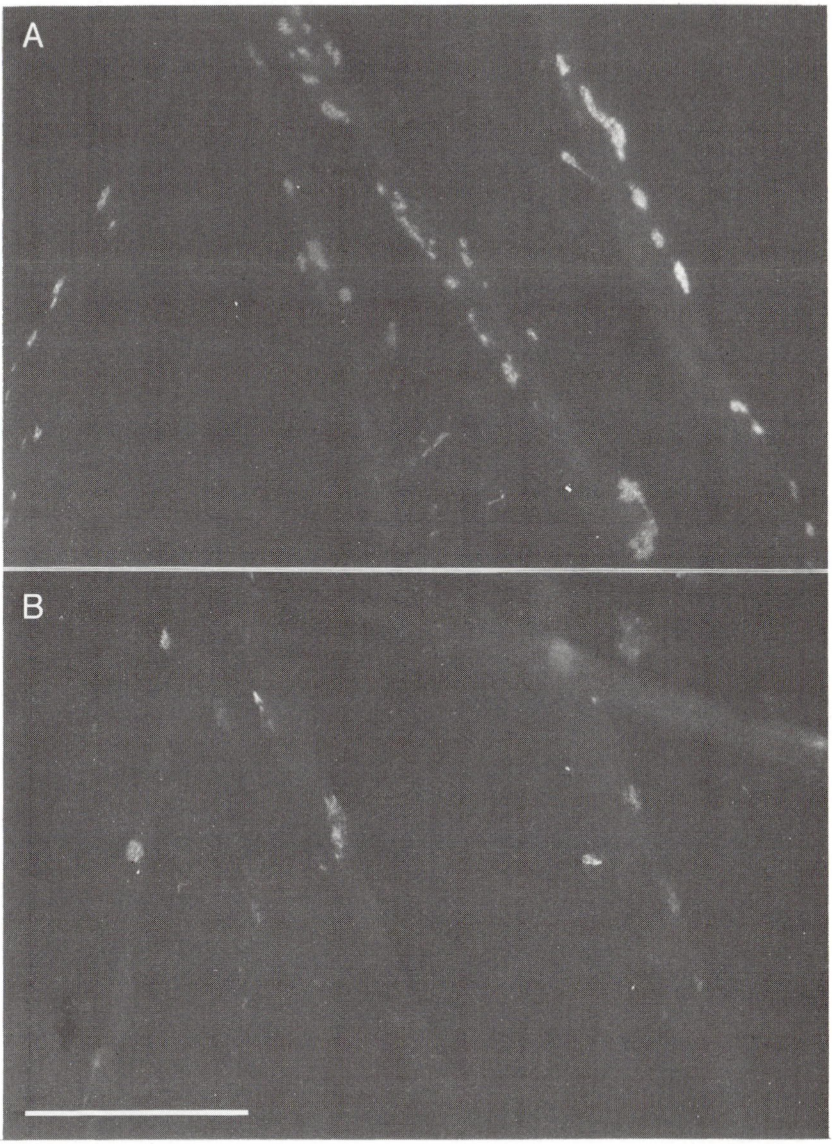

FIGURE 1

Extracts from *Torpedo* electric-organ extracellular matrix organize AChRs on cultured chick myotubes. Fluorescence micrographs of muscle cultures labeled with rhodamine-conjugated α-bungarotoxin and then fixed in cold 95% ethanol. (*A*) $MgCl_2$ extract was added to the culture 18 hr before labeling receptors. (*B*) Culture processed in parallel with that in *A* but without extract. Bar represents 100 μm.

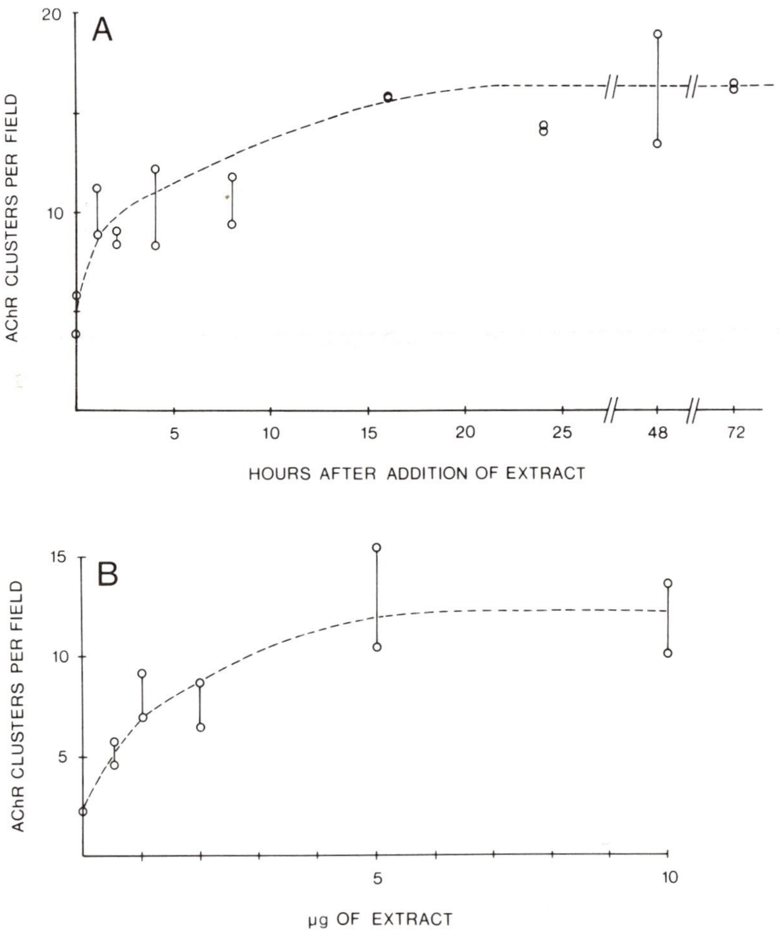

FIGURE 2

Response of chick myotubes to *Torpedo* AChR-organizing factor is rapid and dose-dependent. (*A*) Time course of receptor organization. Cell cultures of chick muscle were incubated with a saturating amount of the $MgCl_2$ extract for the times indicated; the AChRs were then marked with rhodamine-labeled α-bungarotoxin, and the number of receptor clusters was determined. The time of addition of extract was varied so that all cultures were the same age when receptor distribution was measured. (*B*) Dose dependence of receptor organization. Cell cultures of chick muscle were incubated for 18 hr with various amounts of the $MgCl_2$ extract. AChRs were then labeled, and the number of receptor clusters was determined. In both *A* and *B* the cultures were viewed at a total magnification of 313× so that several myofibers were included in each field (as in Fig. 1). Each point represents the mean of 20 fields from a single culture. The number of myofibers was constant from dish to dish.

myofibers to extracts of electric organ produces a reorganization of AChRs without significantly affecting their synthesis or degradation.

As the amount of extract added to the muscle cultures was increased, the number of receptor clusters per myotube increased, linearly at first, and then it reached a plateau (Fig. 2B). The magnitude of the response to a saturating amount of extract was similar from dish to dish in cultures prepared at the same time. However, it varied from plating to plating; the increase over control levels ranged from 3–20-fold. By defining one unit of activity as the amount of extract needed to cause a half-maximal response, we could measure the level of factor activity reproducibly from experiment to experiment. Thus, the rapid and dose-dependent response of chick myotubes provides a convenient quantitative assay for AChR-organizing activity.

Partial Purification and Characterization

The first step in our purification procedure was to make an extracellular-matrix fraction of the *Torpedo* electric organ on the basis of the method of Meezan et al. (1975) for isolating basal lamina. The electric organ was homogenized in 0.4 M NaCl. Particulate material was isolated and extracted with 3% Triton X-100. The remaining insoluble extracellular matrix fraction contained approximately one third of the AChR-organizing activity of the original homogenate (see Table 1). Most of this activity was solubilized by extracting the extracellular-matrix fraction with 2 M $MgCl_2$. This procedure has been used to solubilize acetylcholinesterase (Lwebuga-Mukasa et al. 1976), which is thought to be associated with the extracellular matrix

TABLE 1
Purification of AChR-organizing Activity

Fraction	Protein (mg)	AChR-organizing activity (units)	Purification (fold)	Yield (% activity)
Crude *Torpedo* electric-organ homogenate (1 kg wet weight)	11,500	958,000	1	100
Extracellular-matrix fraction	1,000	326,000	4	34
$MgCl_2$ extract	169	225,000	16	23
Gel-filtration pool	32.3	150,000	56	16
Ion-exchange pool	1.4	43,000	371	4

(Hall and Kelly 1971; Betz and Sakmann 1973; McMahan et al. 1978). When the $MgCl_2$ extract was dialyzed against 150 mM NaCl, most of the AChR-organizing activity remained in solution. We fractionated the soluble material on a Bio-Gel P-300 gel filtration column. The AChR-organizing activity eluted as a broad peak separated from most of the protein, which ran in the void volume. When compared to a set of molecular-weight standards, the *Torpedo* AChR-organizing factor had an apparent molecular weight in the range of 50,000–100,000. Glycerol (5%v/v) was included in this and all subsequent steps to prevent loss of activity due to low protein concentration. The active fractions were pooled, dialyzed, and fractionated further by chromatography on a DEAE-cellulose anion-exchange column. AChR-organizing activity bound to the resin at pH 8 in 10 mM buffer and eluted as a broad peak near the center of a 0–200-mM gradient of NaCl. The broad distribution of activity on gel filtration and ion exchange chromatography indicates some size and charge heterogeneity in the active components. The partially purified material was still active after several weeks of storage at −80°C. The purification, nearly 400-fold over crude electric-organ homogenate, is summarized in Table 1.

The fractions with the highest specific activity gave a half-maximal response at a protein concentration of approximately 10^{-9} M, assuming a molecular weight of 50,000–100,000 on the basis of our gel filtration data. This is near the concentration at which many proteinaceous growth factors and peptide hormones are active (Kahn 1976). Thus, the AChR-organizing factor may comprise an appreciable fraction of the total protein in our most purified material.

Extracts from brain, spinal cord, and conditioned media also influence the organization of AChRs on cultured myotubes (Christian et al. 1978; Podleski et al. 1978; Jessell et al. 1979; Bauer et al. 1981). The active factors are proteinaceous and, in nearly every case, have an apparent molecular weight greater than 50,000. Differences in methods of preparation and assay and incomplete characterization prevent detailed comparison of the active components in extracts from different sources. A feature that clearly distinguishes the *Torpedo* factor from the others is that it is initially isolated in an extracellular-matrix fraction.

Immunological Cross-reactivity with Components of Synaptic Basal Lamina

One of our aims is to determine, using immunological techniques, whether the AChR-organizing factor extracted from the electric organ is similar to active components of the synaptic basal lamina. To assess the antigenicity of the factor, rabbits were immunized with several injections of 1–3 mg of $MgCl_2$-solubilized *Torpedo* AChR-organizing factor (see Table 1) in Freund's adjuvant. Antiserum was obtained that blocked the effect of half-saturating

amounts of factor at a dilution of 1:500. Control rabbit serum was without effect. Several lines of evidence indicate that the rabbit antiserum blocked AChR organization by binding to the *Torpedo* factor and not to sites on the myotubes. (1) Factor activity could be removed from electric-organ extracts by immunoprecipitation with antiserum and fixed *Staphylococcus aureus* bacteria. (2) Inhibition of the AChR-organizing activity of *Torpedo* extract by antiserum could be reversed by the addition of excess factor. (3) The ability of antiserum to block the AChR-organizing activity of *Torpedo* extracts was not reduced by absorption with muscle cultures. (4) The response of muscle cultures to *Torpedo* AChR-organizing factor was unaffected by prior exposure of the cultures to antiserum. Thus, antibodies to the active component in extracts of *Torpedo* electric organ can be produced. At least some of these antibodies block activity and so are likely to be directed against determinants at or near the active site of the *Torpedo* AChR-organizing factor.

We demonstrated that antibodies to the *Torpedo* AChR-organizing extract cross-react with molecules in the basal lamina at the neuromuscular junction by incubating frog cutaneous pectoris muscles with the rabbit antiserum. Bound antibody was detected using horseradish-peroxidase-coupled goat anti-rabbit immunoglobulin (IgG). Figure 3A, a light micrograph, shows that horseradish-peroxidase stain was concentrated at the neuromuscular junction; a similar pattern of staining has been found with antibodies to extracellular material from other sources (Sanes and Hall 1979; Burden 1981). To establish that antibodies were bound to the extracellular matrix, muscles were frozen in situ so that all cellular components of the synapse (nerve terminals, Schwann cells, and myofibers) would degenerate and be phagocytized (McMahan et al. 1980). The nerves were evulsed, and the ends of the muscles were cut so that they would not regenerate. After 1 month, the only components of the synaptic sites that remained were the basal lamina sheaths. Antibodies stained the empty sheaths in a pattern similar to that seen by light microscopy in intact muscle (Fig. 3). Synaptic sites on the empty sheaths were identified in the electron microscope by the projections of basal lamina that had extended into the junctional folds (Fig. 4). Both the synaptic and extrasynaptic basal lamina and other extracellular structures in the vicinity of the synaptic site were more heavily labeled in preparations treated with anti-*Torpedo* antiserum than in preparations treated with control serum.

Because the *Torpedo* extract used as the antigen was impure, we do not know whether or not the antibodies that bound to basal lamina in frog muscle were the same as those that blocked the AChR-organizing activity of the extract. However, the finding that antiserum to partially purified *Torpedo* AChR-organizing factor cross-reacted with material at the neuromuscular junction means that at least some components of frog synaptic basal lamina are antigenically similar to molecules in our extract of *Torpedo* extracellular

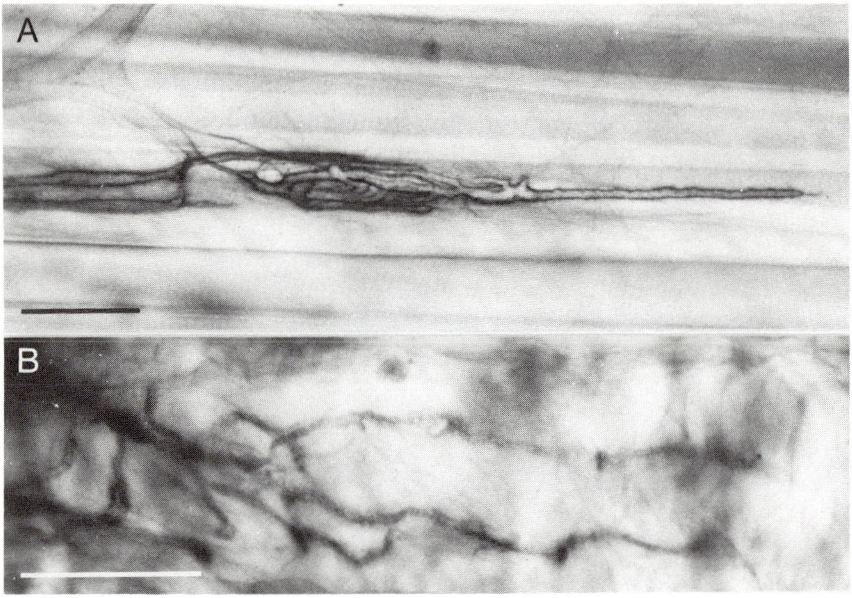

FIGURE 3
Rabbit antiserum to *Torpedo* AChR-organizing extract binds in high concentration to extracellular matrix of neuromuscular junctions. (*A*) Normal frog neuromuscular junction in whole mount. Muscle was incubated in rabbit antiserum for 1 hr, rinsed in Ringer's solution for 1 hr, incubated in horseradish-peroxidase-labeled goat anti-rabbit IgG for 1 hr, rinsed in Ringer's solution for 1 hr, fixed in glutaraldehyde for ½ hr, treated with 3,3-diaminobenzidine and H_2O_2 for 1 hr, treated with osmium tetroxide for 3 hr, and embedded in Epon. The stain outlines branches of an axon terminal arborization. (*B*) Synaptic site on empty basal lamina sheath. The muscle was damaged in situ to remove all cellular components (as described in text) and then prepared 30 days later for antibody labeling as in *A*. An arborization similar to that at normal neuromuscular junctions is stained. Bars represent 30 μm.

FIGURE 4 (see facing page)
Rabbit antiserum to *Torpedo* AChR-organizing extract binds to myofiber basal lamina. (*A*) Cellular components were removed from synaptic sites by freezing the tissue in situ, leaving only basal lamina and other extracellular structures after degeneration (as described in text). Synaptic sites (▭) on myofiber basal lamina (MBL) were recognized in cross sections by basal lamina that had projected into junctional folds (JBL) and by basal lamina (SBL) of Schwann cells (S), which cap nerve terminals (T). (*B*) In preparations treated with rabbit antiserum and processed as described in the legend to Fig. 3, horseradish-peroxidase reaction product was bound to synaptic and extrasynaptic basal lamina, to the basal lamina of Schwann cells, and to extracellular material associated with the Schwann cell basal lamina. (*C*) Control preparations had no discernible horseradish-peroxidase labeling. Bar represents 1 μm.

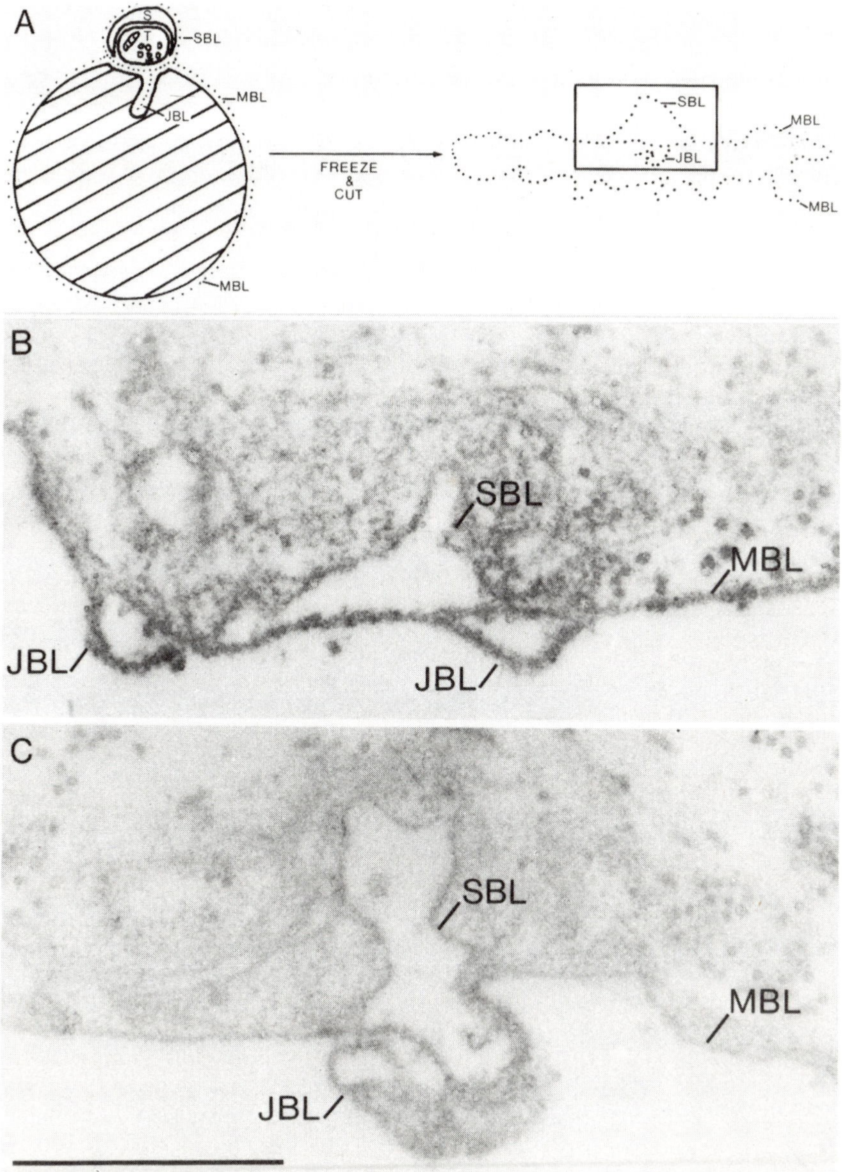
A
S
T
SBL
MBL
JBL
MBL
FREEZE
&
CUT
SBL
JBL
MBL
MBL
B
SBL
MBL
JBL
JBL
C
SBL
MBL
JBL

matrix. Experiments aimed at production of specific immunological reagents against the *Torpedo* AChR-organizing factor are being pursued.

CONCLUDING REMARKS

The results reported here demonstrate the feasibility of isolating AChR-organizing factors from the synapse-rich *Torpedo* electric organ and comparing them with those at the neuromuscular junction by immunological techniques. We have purified an AChR-organizing factor of several orders of magnitude. We have shown that antibodies bind at or near the active site of the factor and that antiserum against a *Torpedo* AChR-organizing extract cross-reacts with components of synaptic basal lamina of frog muscles. Identification of the factors in the synaptic basal lamina that organize AChRs is an important step toward understanding the molecular basis of regeneration at the neuromuscular junction.

ACKNOWLEDGMENTS

We thank Bob Marshall, Lyn Lazar, and Deborah Stairs for their excellent technical assistance. This research was carried out during the tenure of Muscular Dystrophy Association postdoctoral fellowships to R.M.N. and E.W.G. It was supported by grants from the Muscular Dystrophy Association of America and the U.S. Public Health Service (NS-14506 and NS-16440).

REFERENCES

Bader, D. 1981. Density and distribution of α-bungarotoxin-binding sites in postsynaptic structures of regenerated rat skeletal muscle. *J. Cell Biol.* **88:** 338.

Bauer, H.C., M.P. Daniels, P.A. Pudimat, L. Jacques, H. Sugiyama, and C.N. Christian. 1981. Characterization and partial purification of a neuronal factor which increases acetylcholine receptor aggregation on cultured muscle cells. *Brain Res.* **209:** 395.

Betz, W. and B.J. Sakmann. 1973. Effects of proteolytic enzymes on function and structure of frog neuromuscular junction. *J. Physiol.* **230:** 673.

Burden, S.J. 1981. Monoclonal antibodies to the frog nerve-muscle synapse. In *Monoclonal antibodies to neural antigens* (ed. R. McKay et al.), p. 247. Cold Spring Harbor Laboratory, Cold Spring Harbor, New York.

Burden, S.J., P.B. Sargent, and U.J. McMahan. 1979. Acetylcholine receptors in regenerating muscle accumulate at original synaptic sites in the absence of the nerve. *J. Cell Biol.* **82:** 412.

Christian, C.N., M.P. Daniels, H. Sugiyama, Z. Vogel, L. Jacques, and P.G. Nelson. 1978. A factor from neurons increases the number of acetylcholine receptor aggregates on cultured muscle cells. *Proc. Natl. Acad. Sci.* **75:** 4011.

Fischbach, G.D. 1972. Synapse formation between dissociated nerve and muscle cells in low density cell cultures. *Dev. Biol.* **28:** 407.

Hall, Z.W. and R.B. Kelly. 1971. Enzymatic detachment of endplate acetylcholinesterase from muscle. *Nat. New Biol.* **232:** 62.

Jessell, T.M., R.E. Siegel, and G.D. Fischbach. 1979. Induction of acetylcholine receptors on cultured skeletal muscle by a factor extracted from brain and spinal cord. *Proc. Natl. Acad. Sci.* **76:** 5397.

Kahn, C.R. 1976. Membrane receptors for hormones and neurotransmitters. *J. Cell Biol.* **70:** 261.

Lwebuga-Mukasa, J.S., S. Lappi, and P. Taylor. 1976. Molecular forms of acetylcholinesterase from *Torpedo californica*: Their relationship to synaptic membranes. *Biochemistry* **15:** 1425.

Marshall, L.M., J.R. Sanes, and U.J. McMahan. 1977. Reinnervation of original synaptic sites on muscle fiber basement membrane after disruption of muscle cells. *Proc. Natl. Acad. Sci.* **74:** 3073.

McMahan, U.J., D.R. Edgington, and D.P. Kuffler. 1980. Factors that influence regeneration of the neuromuscular junction. *J. Exp. Biol.* **89:** 31.

McMahan, U.J., J.R. Sanes, and L.M. Marshall. 1978. Cholinesterase is associated with the basal lamina at the neuromuscular junction. *Nature* **271:** 172.

Meezan, E., J.J. Hjelle, K. Brendel, and E.C. Carlson. 1975. A simple versatile nondisruptive method for the isolation of morphologically and chemically pure basement membranes from several tissues. *Life Sci.* **17:** 1721.

Podleski, T.R., D. Axelrod, P. Ravdin, I. Greenberg, M.M. Johnson, and M.M. Salpeter. 1978. Nerve extract induces increases and redistribution of acetylcholine receptors on cloned muscle cells. *Proc. Natl. Acad. Sci.* **75:** 2035.

Rubin, L.L. and U.J. McMahan. 1982. Regeneration of the neuromuscular junction: Steps toward defining the molecular basis of the interaction between nerve and muscle. In *Disorders of the motor unit* (ed. D.L. Schotland), p.187. Wiley, New York.

Rubin, L.L., A.S. Gordon, and U.J. McMahan. 1980. Basal lamina fraction from the electric organ of *Torpedo* organizes acetylcholine receptors on cultured myotubes. *Soc. Neurosci. Abstr.* **6:** 330.

Sanes, J.R. and Z.W. Hall. 1979. Antibodies that bind specifically to synaptic sites on muscle fiber basal lamina. *J. Cell Biol.* **83:** 357.

Sanes, J.R., L.M. Marshall, and U.J. McMahan. 1978. Reinnervation of muscle fiber basal lamina after removal of muscle fibers. *J. Cell Biol.* **78:** 176.

Acetylcholine Receptors on Primary Rat Muscle Cells Redistribute to Reach Junctional Site Densities after Exposure to Soluble Neuronal Extracts

MIRIAM M. SALPETER
THOMAS R. PODLESKI
Section of Neurobiology and Behavior
Division of Biology
Cornell University
Ithaca, New York 14853

Numerous studies have been performed to establish the role of innervation in determining the molecular organization and function of skeletal muscle. The debate regarding the relative roles of muscle activity and chemical substances released from nerve has given way to an appreciation that a whole set of complex phenomena, including activity, various chemicals, and a combination of these, may be involved. Activity itself is not a unitary event and may consist of voltage changes or ion fluxes at the junction or elsewhere. Intermediate steps involving second messengers may also play a role (Betz and Changeux 1979), and different muscles may have distinct responses. Thus, the many changes seen in muscle after innervation may be differently induced and differently maintained. For example, it has recently been shown that in the steps leading to the formation of the neuromuscular junction, accumulation of endplate-specific acetylcholinesterases (AChEs) needs activity and/or cyclic GMP (Betz et al. 1980; Cangiano et al. 1980; Lømo and Slater 1980; O'Brien et al. 1980; Rubin et al. 1980b), but the ac-

cumulation of junctional acetylcholine receptors (AChRs) does not (Steinbach et al. 1973; Frank and Fischbach 1979; Betz et al. 1980; Cohen and Weldon 1980).

We must also reexamine the classic distinction between junctional and extrajunctional AChRs (for reviews, see Edwards 1979; Fambrough 1979). Extrajunctional receptors are usually described as sparsely distributed ($\sim$300 sites/μm^2) (Fambrough 1974; Burden 1977a; Land et al. 1977), metabolically unstable (half-life of 1 day) (Berg and Hall 1975; Devreotes and Fambrough 1975; Merlie et al. 1976; Linden and Fambrough 1979; Betz et al. 1980; Reiness and Weinberg 1981), laterally mobile (Axelrod et al. 1976), and having a slow gating time (Katz and Miledi 1972; Dreyer et al. 1976; Neher and Sakmann 1976). Junctional AChRs, on the other hand, are described as immobile (Fertuck et al. 1975; Axelrod et al. 1976; Fambrough and Pagano 1978), concentrated at high density ($\sim$15,000–25,000 sites/μm^2) (Porter and Barnard 1975; Fertuck and Salpeter 1976; Matthews-Bellinger and Salpeter 1978), metabolically stable (half-life of $\sim$10 days) (Heinemann et al. 1977; Reiness et al. 1978; Linden and Fambrough 1979; Loring and Salpeter 1980; Reiness and Weinberg 1981), and having a fast gating time (Katz and Miledi 1972; Dreyer et al. 1976; Neher and Sakmann 1976). Other differences involving pharmacology (Chiu et al. 1974; Almon and Appel 1976), immunology (Almon and Appel 1975; Weinberg and Hall 1979), and isoelectric point (Brockes and Hall 1975; Teichberg and Changeux 1976) have also been described. However, it is now known that during development or after denervation, receptors at the junction may have properties often associated with extrajunctional receptors, such as rapid turnover rate (Burden 1977b; Levitt and Salpeter 1981) and slow gating time (Sakmann and Brenner 1978; Fischbach and Schuetze 1980; Michler and Sakmann 1980). Therefore, until it is known whether the different properties of the receptors are determined by different gene products or whether they represent posttranslational modifications, it may be most useful to reserve the terms junctional and extrajunctional for the anatomical localization of the receptors and not their properties.

Little is known about the mechanisms that cause these properties to develop and be maintained. The possibility that basal lamina plays a role in maintaining AChR clusters at adult neuromuscular junctions has been suggested by studies showing that AChR clusters can become localized in denervated regenerating muscle at the site of the old junctions, marked only by the residual basal lamina (Burden et al. 1979). Yet clustered and immobile receptors inserted at denervated neuromuscular junctions turn over with a half-life of 1 day (Levitt and Salpeter 1981). Thus, metabolic stabilization and AChR clustering require alternate and, as yet undetermined, neurally controlled events. The individual properties controlled by the nerve must therefore be identified, and the mechanisms by which the nerve exerts its control during development and in the adult must be determined separately for each.

One event of interest in this study is the clustering of AChRs. Since, in developing muscle, AChR clustering can occur in the absence of nerve (Fischbach and Cohen 1973; Sytkowski et al. 1973; Prives et al. 1976; Braithwaite and Harris 1979) or in response to a variety of nonphysiological stimuli (e.g., Jones and Vrbova 1974; Vyskocil and Syrovy 1979; Peng et al. 1981), it is not clear what the specific role of the nerve is in inducing clustering at the neuromuscular junction. Recently, several studies have investigated the effect of soluble factors from nervous tissue on the distribution and site density of AChRs of muscle developing in vitro. These studies led to some interesting results. We found that on the cloned rat muscle cells (L6) where AChRs are normally at low concentration and uniformly distributed (Steinbach 1975; Land et al. 1977), nerve extract causes not only a major (about fivefold) increase in AChR site density but also the formation of high-density AChR clusters (Podleski et al. 1978). Under the influence of nerve extracts, the AChRs on L6 cells thus acquire the nonuniform distribution seen on primary rat muscle cells. Nerve extract also causes formation of AChR clusters on cultured chick and rat myotubes and an increase in AChR site density in the chicken muscle but not in the rat (Christian et al. 1978; Jessel et al. 1979; Bauer et al. 1981). The increase in AChR number and formation of AChR clusters may be controlled by different factors, and different cells may have a different sensitivity to these factors. Since an increase in AChR number occurs after denervation, the possibility exists that stimulation of AChR synthesis may mimic the denervation effects of degenerating nerve, whereas AChR clustering may mimic innervation. The possible role of nerve factors in controlling the normal junctional development is of major interest. We therefore extended our studies on the effect of brain extract and asked specifically whether extract-induced clustering resembles events that occur during development of the neuromuscular junction.

RESULTS AND DISCUSSION

We used primary rat muscle cells since, in these cells, only clustering and no increase in AChR number has been reported. We found (Salpeter et al. 1982) that nerve extract did mimic specific events reported at developing neuromuscular junctions (see, e.g., Anderson and Cohen 1977), in that it caused the site density within AChR clusters to reach junctional values ($>10{,}000$ sites/μm^2) by a redistribution of preexisting receptors without the need for neuromuscular activity or a shift in metabolic stability.

These studies used quantitative light and electron microscope autoradiography for that purpose, after labeling AChRs with ^{125}I-labeled α-bungarotoxin (α-Bgt) (Fig. 1) (Salpeter et al. 1982). We have confirmed earlier studies that AChR clusters form on rat myotubes in vitro in the absence of any nerve or nerve factors. However, the calculated AChR concentrations in these clusters average ~ 3000 sites/μm^2 (Land et al. 1977) and usually do

FIGURE 1

(*A,B*) Extract effect on overall distribution of ^{125}I-labeled α-Bgt binding. Light microscope autoradiographs show nonextract-treated cells (*A*) and cells after 1 day in extract (*B*). Note increased receptor clustering after extract treatment. Autoradiographs are overexposed for illustration and thus cannot demonstrate the increased receptor site density seen in clusters after extract treatment. Magnification, 512×. (*C*) Histogram of AChR site densities by light microscope autoradiography, representing the top surface of myotubes, on the basis of seven experiments, 47 myotubes, and 10,600 developed grains. These cells were treated with cytosine arabinoside on day 4. For cells in extract for 1 day, the extract was added either on day 4 or day 6, and the cells were labeled and fixed 1 day later. For cells in extract for 3 days, extract was added on day 4 and again on day 5, and the cells were labeled and fixed on day 7. The percentage of the area occupied by clusters was ~3–6% in the absence of extract and ~7–10% after extract. Grains were counted within grid squares superimposed over the entire length of individual myotubes. The unit grid size was 39 μm^2. AChR site density was calculated as described previously (Land et al. 1977) and averaged over the entire tube (average). Any unit grid square with more than three times the average value was then designated as cluster, and all others were then averaged separately as intercluster. The tabulated square with the highest site density per tube was designated as peak density.

A

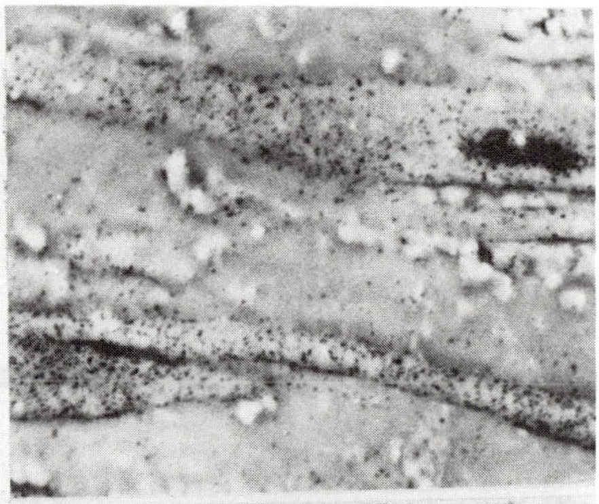

B

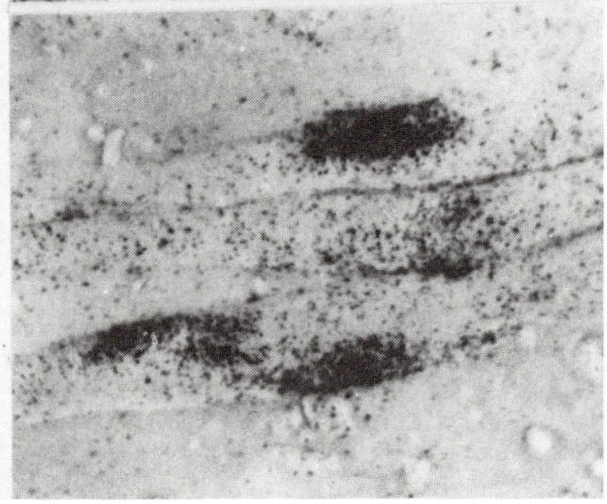

C

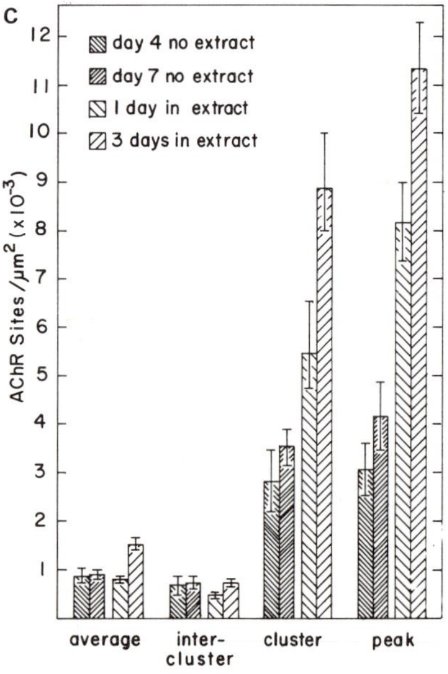

not exceed ~5000 sites/μm^2. After adding solubilized extract of central nervous tissue from fetal rat brain to rat myotubes in culture, we found that the receptor concentration within AChR clusters can reach peak site densities approaching those at the neuromuscular junction (>10,000 sites/μm^2 (Porter and Barnard 1975; Fertuck and Salpeter 1976; Matthews-Bellinger and Salpeter 1978, 1979). (See legend to Fig. 1 for definition of peak density.) After 3 days in extract, the average peak site density from all of the myotubes was ~11,000 ± 1000, but values of ~20,000 were encountered in individual myotubes, and 55% of myotubes had at least one peak density greater than 10,000 sites/μm^2. On the other hand, no myotubes had any peak density over 10,000 sites/μm^2 in the absence of extract.

To test whether clusters can form by a redistribution of existing receptors, we labeled AChRs with ^{125}I-labeled α-Bgt prior to exposure to extract and then kept the cells (with or without extract) for 1 day or 3 days before fixing and preparing for light autoradiography. We found that within 1 day in extract (Table 1) there was a redistribution of prelabeled receptors into clusters, giving a ratio of cluster to intercluster site density similar to that seen when receptors were labeled after exposure to extract (Table 1; Fig. 1c). In one experiment, the redistribution of receptors was seen as early as 3 hours in extract, but more often it occurred by 8 hours.

An interesting observation by light autoradiography in these studies was that after 3 days in extract, there was an increase in AChR site density by about a factor of 2, even in cases where receptors were labeled prior to the addition of extract (Table 1). Thus, an increase in receptor synthesis could not explain this higher site density. The increase in site density could result if extract caused (1) a slowing in degradation rate of AChR or (2) an introduction of labeled receptors from a source not detected by the emulsion in the light autoradiographs. Such a source could be the bottom of the myotubes, since the Ilford L4 emulsion used in this study is insensitive to gamma radiation and since the range of the low-energy electrons from ^{125}I is only about 2000 Å (Fertuck and Salpeter 1974). Thus, our light autoradiograms "see" only labeled receptors from the top of the myotube.

Possibility 1 was tested by calculating (from data in Table 1) the approximate half-life of degradation of AChRs, both in clusters and in intercluster regions. These are approximately 13 hours with and without extract and, hence, are compatible with other reports for the turnover rate of embryonic receptors (e.g., Berg and Hall 1975; Chang and Huang 1975; Devreotes and Fambrough 1975; Merlie et al. 1976; Burden 1977b; Linden and Fambrough 1979; Michler and Sakmann 1980). No obvious decrease in the degradation rate occurred after 3 days in extract.

To test possibility 2, i.e., whether receptors could be coming from the bottom of the tube, we examined electron microscope autoradiographs of myotubes sectioned longitudinally, at right angles to the dish (Fig. 2). We found that after 3 days in extract, there was indeed a shift in relative AChR

TABLE 1
Effect of Extract on AChR Site Density on Primary Rat Muscle Cells by Light Autoradiography

		AChR site density (sites/μm^2)[a]				
Time after labeling (days)	treatment	average	cluster[b]	intercluster	ratio Cl/ICl	area occupied by cluster (%)
0[c]	no extract	895 ± 105	3519 ± 360	724 ± 128	5.7 ± 1.3	5 ± 0.9
	extract	1534 ± 149	8860 ± 797	726 ± 95	16.2 ± 2.9	10.1 ± 0.9
	ratio X/NX	1.7	2.5	1.0	2.8	2.0
1[d]	no extract	238 ± 56	1033 ± 191	204 ± 51	6.9 ± 1.6	4.5 ± 1.4
	extract	304 ± 48	1631 ± 238	186 ± 37	11.4 ± 1.7	8.6 ± 0.6
	ratio X/NX	1.3	1.6	0.9	1.7	1.9
3[e]	no extract	16 ± 6	82 ± 6	13 ± 5	6.7 ± 1.6	4.2 ± 1.1
	extract	45 ± 4	360 ± 27	18 ± 6	19.0 ± 9	7.7 ± 0.9
	ratio X/NX	2.8	4.4	1.4	3.4	1.7

[a] AChR site densities decrease with time after labeling with a half-life of <1 day; this was calculated using the equation $t_{1/2} = -\ln 2(t/\ln [R/I_0])$, where R is the residual label after time t, and I_0 is the initial label. Values are mean ± S.E. for about five myotubes from each of three to five dishes per condition. X, extract-treated; NX, nonextract-treated; Cl, cluster; ICl, intercluster region.

[b] A cluster is defined as any tabulation grid square (39 μm^2) in which the site density is more than three times the average value. This definition was chosen since it correlated well with visually identifiable "hot spots," which ranged in size from one to three contiguous grid squares. Since the grain density was often higher within a fraction of such a grid square, all of the "cluster" values are slight underestimates. Occasionally, nonextract-treated myotubes had no clusters by our definition. Such myotubes were included in the average value but not in the cluster or intercluster values.

[c] Extract treatment involved adding extract on day 4 and day 5; cells were labeled and fixed on day 7 (data as in the legend to Fig. 2).

[d] Cells were labeled on day 4 or day 6 (with or without a simultaneous single addition of extract) and fixed 1 day later.

[e] Cells were labeled on day 4 (with or without addition of extract on day 4 and day 5) and fixed on day 7.

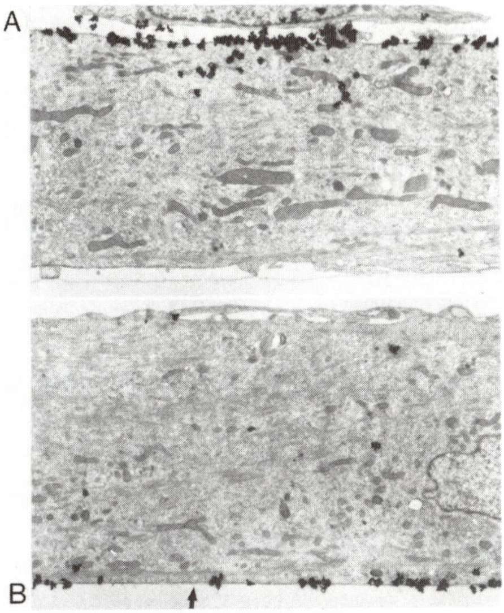

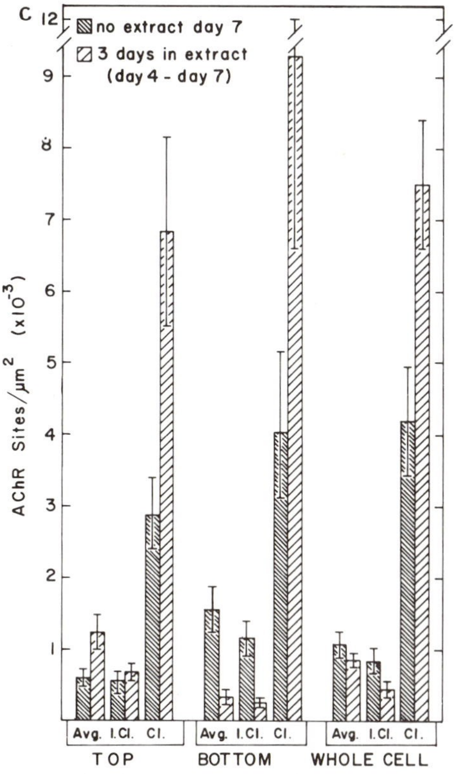

FIGURE 2
(*A,B*) Overexposed electron microscope autoradiographs show typical localization of the receptor cluster on the top surface after extract treatment (*A*) and at the bottom surface in nonextract-treated cells (*B*). The bottom of the cell can be identified by the characteristic line (→) derived from the tissue-culture dish. Magnification, 2450×. (*C*) Histogram of AChR site densities by electron microscope autoradiography, on the basis of eight myotubes, 1800 developed grains, and > 7000-μm length of myotubes examined. The extract was added on day 4 and day 5; cells were labeled and fixed on day 7. For each surface (TOP, BOTTOM, or WHOLE CELL), AChR site densities were either averaged (Avg.) or tabulated separately for cluster (Cl.) or intercluster (I.Cl.) regions. Since it was not easy to tabulate an entire myotube by electron microscope autoradiography, a cluster was defined locally as any membrane region with a grain density five times higher than that on equal lengths of membrane on either side of that cluster. By such a definition, a cluster could be less than five times the average over the whole tube. The site densities of AChR at the top surface seen by electron microscope autoradiography are similar to that seen by light autoradiography; this proves that the smaller sample used for the electron microscope autoradiography was statistically valid. Note that although the average and intercluster AChR site densities at the bottom of the tube decreased in response to extract, the density within clusters increased as it did at the top. We found, however, that the number of clusters at the top increased about threefold, whereas those at the bottom decreased about sevenfold.

concentration from the bottom to the top of the cell. Earlier studies have reported a tendency for receptors and receptor clusters to favor the bottom of the myotubes (Axelrod et al. 1976; Land et al. 1977; Bloch and Geiger 1980) and that extract causes clusters to appear at the top surface of myotubes (Podleski et al. 1978; Axelrod et al. 1981). In this study we analyzed such observations quantitatively and found that in the absence of extract, there is an average receptor concentration twofold to threefold higher (Fig. 2c) and an area occupied by clusters fivefold to sixfold larger at the bottom of the myotube than at the top. After 3 days in extract, the top surface has an average receptor site density twofold to threefold higher and an area occupied by clusters threefold larger than at the bottom. It is of interest that even though there were fewer clusters at the bottom of the myotube after extract, the site density within clusters was elevated to the same extent at the bottom as at the top. The overall AChR site density for the whole myotube (top plus bottom) did not, however, increase in response to extract, and the overall number of receptor clusters decreased somewhat. We suggested (Salpeter et al. 1982) that a redistribution of receptors from the bottom of the cells to the top, in response to extract, could quantitatively account for the increase in average receptor density observed with light autoradiography.

Our examination of the electron microscope autoradiographs was extended to include fine-structural specializations, which are characteristic of the neuromuscular junction: electron density of plasma membrane and the presence of basal lamina. Dense membrane was defined as plasma membrane that appeared thicker, more rigid, and often contained a cytoplasmic fuzzy layer. Basal lamina was defined as a continuous surface layer separated from the plasma membrane by a zone that was less electron dense. Since our methods do not permit preservation and observation of all of the basal lamina, no exact values can be given. Within these limitations, we found that membrane regions containing clustered receptors were enriched (more than fourfold) with both electron-dense membrane and with basal lamina (Fig. 3). This was true whether or not cells were treated with extract. Extract caused about a twofold to threefold increase in both dense membrane and basal lamina at the top surface of the cells, paralleling the increase in AChRs and AChR clusters. Dense-membrane specialization and basal lamina in myotubes appear morphologically similar to those seen in cells cocultured with nerve (Cohen and Weldon 1980; Michler and Sakmann 1980) and in neonatal endplates (J. Matthews-Bellinger and M.M. Salpeter, in prep.). There is somewhat less intense cytoplasmic "fuzz" than that seen at the postsynaptic membrane of adult neuromuscular junctions (Fertuck and Salpeter 1976). Extract did not notably change myofibrillar organization (see Fig. 3), nor did it affect the cytoplasmic label that has been reported when intact embryonic or denervated muscle cells are labeled with α-Bgt (Loring and Salpeter 1978; Podleski et al. 1979; and J. Matthews-Bellinger and M.M. Salpeter, unpubl.). We thus reported that

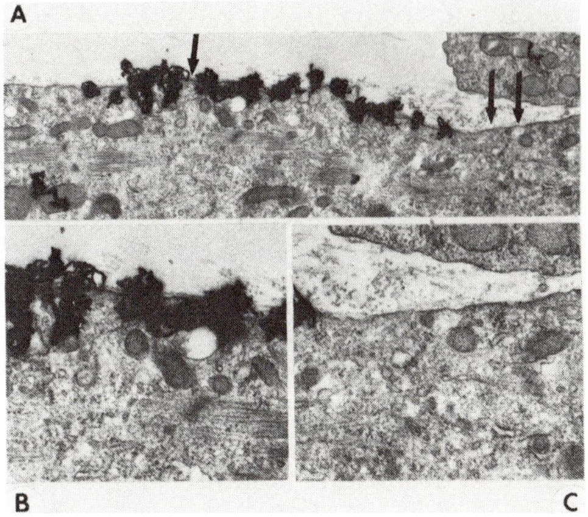

FIGURE 3
Overexposed electron microscope autoradiographs of extract-treated cells showing top surface. (*A*) Characteristic AChR cluster on slight membrane bulge. The area within the cluster (→) is enlarged in *B*, and that from the adjacent intercluster region (⇉) is enlarged in *C*. Note the basal lamina overlying the electron-dense membrane in the cluster region. Basal lamina, by definition, was a continuous line separated from the plasma membrane by a less dense zone. Dense membrane was characterized by being thickened, more electron dense, and more "rigid" in appearance than the rest of the membrane and was often associated with a cytoplasmic "fuzz." Magnifications: (*A*) 6944×; (*B*,*C*) 13,888×.

brain extract can induce (1) a redistribution of preexisting receptors within 1 day to form high-density AChR clusters with peak site densities approaching junctional values, (2) a shift of receptors from the bottom to the top of the cell within 3 days in extract, and (3) an increase in the relative amount of dense membrane and basal lamina on the top of the cell.

One possible mechanism that could produce a shift in receptor concentration into clusters, or from the bottom to the top of the cell, could be a change in the relative number of mobile versus immobile AChRs. For example, if, in the absence of extract, there were a higher fraction of immobile receptors at the bottom of the cell and, after extract treatment, there were a higher fraction at the top, then once the mobile receptors have equilibrated, one would see a shift in receptor site density from the bottom to the top, as we have reported, With a diffusion constant of 10^{-10} cm^2/sec (Axelrod et al. 1976) or faster and a myotube of 30 μm in diameter, mobile receptors would equilibrate well within the 3 days of exposure to extract used in our study. An increase in the fraction of immobile receptors at the top of myotubes has been reported in rat muscle cells exposed to conditioned medium from neuroblastoma/glioma hybrids (Axelrod et al. 1981). It remains to be established whether there is a larger number of immobile receptors at the bottom of the myotube in the absence of neuronal factors. In addition, since the shift of receptors from bottom to top is accompanied by a net loss of clusters, but not of AChRs, there may be a translocation and coalescence of

whole clusters. Alternatively, there may be a dispersal of AChRs from clusters into the mobile pool followed by a redistribution.

Immobilization of receptors or receptor clusters may involve specializations such as are reflected in membrane electron density with its associated cytoplasmic fuzz and/or basal lamina, both of which are affected by extract and are normally enriched in clustered regions. The approximate threefold increase in basal lamina in response to extract and the reported role of basal lamina in localizing AChR clusters in denervated regenerating muscle (Burden et al. 1979) and in causing AChR clustering (Rubin et al. 1980a) raise the question of whether or not basal lamina may serve as an "intermediate step" in the extract-induced clustering phenomenon, possibly by immobilizing receptors. At present we must emphasize that we have no evidence that would relate any morphological specialization to the extract-induced effect on AChR, and the work of Weinberg et al. (1981) indicates that some of the antigenic determinants associated with synaptic basal lamina develop after AChR clustering. However, these antigens may not be relevant to the clustering process, and our results on basal lamina and dense membrane encourage further investigation.

Several metabolic, immunologic, and physiologic characteristics of AChRs have been linked to neural influence and to different stages of development of the neuromuscular junctions (for reviews, see Edwards 1979; Fambrough 1979; Fischbach et al. 1979; and more recent reports by Brett and Younkin 1979; Betz et al. 1980; Fischbach and Schuetze 1980; Levitt et al. 1980; Loring and Salpeter 1980; Michler and Sakmann 1980; Levitt and Salpeter 1981). Current evidence indicates that the different properties of the AChR in the development and maintenance of the neuromuscular junction are independently controlled by the nerve or nerve-activity-related events.

SUMMARY

It is often assumed that AChR clustering does not require the nerve. However, since receptor clusters seen in the absence of nerve or nerve factors do not reach junctional concentrations, it now appears that the acquisition of the high AChR density within clusters at the developing neuromuscular junction is possibly under neural control. We showed that brain extract produces AChR clusters with junctional site densities and enhanced surface specializations. The extract-induced receptor clustering does not require a decrease in turnover rate and can be produced by a redistribution of existing receptors. It obviously occurs in the absence of nerve-muscle activity. In these respects it resembles events reported at developing neuromuscular junctions (Anderson and Cohen 1977; Burden 1977b; Schuetze et al. 1978; Frank and Fischbach 1979; Michler and Sakmann 1980). Our data thus suggest that a soluble factor(s) released from

the ingrowing nerve could cause the accumulation of AChR seen at the neuromuscular junction.

ACKNOWLEDGMENTS

We thank Mary Johnson, Chari Smith, Rose Harris, William Harris, Susan Spanton, Kristine Holley, and Karla Neugebauer for technical assistance and Stephen Jones for helpful discussions. This research was supported by grants GM-10422 (M.M.S.) and NS-14679 (T.R.P.).

REFERENCES

Almon, R.R. and S.H. Appel. 1975. Interaction of myasthenic serum globulin with the acetylcholine receptor. *Biochim. Biophys. Acta* **393:** 66.

———. 1976. Cholinergic sites in skeletal muscle. II. Interaction of an agonist and two antagonists with the acetylcholine site. *Biochemistry* **15:** 3667.

Anderson, M.J. and M.W. Cohen. 1977. Nerve-induced and spontaneous redistribution of acetylcholine receptors on cultured muscle cells. *J. Physiol.* **268:** 757.

Axelrod, D., H.C. Bauer, M. Stya, and C.N. Christian. 1981. A factor from neurons induces partial immobilization of nonclustered acetylcholine receptors on cultured muscle cells. *J. Cell Biol.* **88:** 459.

Axelrod, D., P. Ravdin, D.E. Koppel, J. Schlessinger, W.W. Webb, E.L. Elson, and T.R. Podleski. 1976. Lateral motion of fluorescently labeled acetylcholine receptors in membranes of developing muscle fibers. *Proc. Natl. Acad. Sci.* **73:** 4594.

Bauer, H.C., M.P. Daniels, P.A. Pudimat, L. Jacques, H. Sugiyama, and C.N. Christian. 1981. Characterization and partial purification of a neuronal factor which increases acetylcholine receptor aggregation on cultured muscle cells. *Brain Res.* **209:** 395.

Berg, D.K. and Z.W. Hall. 1975. Loss of α-bungarotoxin from junctional and extrajunctional acetylcholine receptors in rat diaphragm muscle in vivo and in organ culture. *J. Physiol.* **252:** 771.

Betz, H. and J.-P. Changeux. 1979. Regulation of muscle acetylcholine receptor synthesis *in vitro* by cyclic nucleotide derivatives. *Nature* **278:** 749.

Betz, H., J.-P. Bourgeois, and J.-P. Changeux. 1980. Evolution of cholingeric proteins in developing slow and fast skeletal muscles in chick embryo. *J. Physiol.* **302:** 197.

Bloch, R.J., and B. Geiger. 1980. The localization of acetylcholine receptor clusters in areas of cell-substrate contact in cultures of rat myotubes. *Cell* **21:** 25.

Braithwaite, A.W. and A.J. Harris. 1979. Neural influence on acetylcholine receptor clusters in embryonic development of skeletal muscles. *Nature* **279:** 549.

Brett, T.S. and S.G. Younkin. 1979. Accelerated degradation of junctional α-bungarotoxin acetylcholine receptor complexes in denervated rat diaphragm. *Neurosci. Abstr.* **5:** 765.

Brockes, J.P. and Z.W. Hall. 1975. Acetylcholine receptors in normal and denervated rat diaphragm muscle. II. Comparison of junctional and extrajunctional receptors. *Biochemistry* **14:** 2100.

Burden, S. 1977a. Development of the neuromuscular junction in the chick embryo: The number, distribution and stability of acetylcholine receptors. *Dev. Biol.* **57:** 317.

———. 1977b. Acetylcholine receptors at the neuromuscular junction: Developmental change in receptor turnover. *Dev. Biol.* **61:** 79.

Burden, S.J., P.B. Sargent, and U.J. McMahan. 1979. Acetylcholine receptors in regenerating muscle accumulate at original synaptic sites in the absence of nerve. *J. Cell Biol.* **82:** 412.

Cangiano, A., T. Lømo, L. Lutzemberger, and O. Sveen. 1980. Effects of chronic nerve conduction block on formation of neuromuscular junctions and junctional AChE in the rat. *Acta Physiol. Scand.* **109:** 283.

Chang, C.C. and M.C. Huang. 1975. Turnover of junctional and extrajunctional acetylcholine receptors of the rat diaphragm. *Nature* **253:** 643.

Chiu, T.H., A.J. Lapa, E.A. Barnard, and E.X. Albuquerque. 1974. Binding of d-tubocurarine and α-bungarotoxin in normal and denervated mouse muscles. *Exp. Neurol.* **43:** 399.

Christian, C.N., M.P. Daniels, H. Sugiyama, Z. Vogel, L. Jacques, and P.G. Nelson. 1978. A factor from neurons increases the number of acetylcholine receptor aggregates on cultured muscle cells. *Proc. Natl. Acad. Sci.* **75:** 4011.

Cohen, M.W. and P.R. Weldon. 1980. Localization of acetylcholine receptors and synaptic ultrastructure at nerve muscle contacts in culture: Dependence of nerve types. *J. Cell Biol.* **86:** 388.

Devreotes, P.N. and D.M. Fambrough. 1975. Acetylcholine receptor turnover in membranes of developing muscle fibers. *J. Cell Biol.* **65:** 335.

Dreyer, F., C. Walther, and K. Peper. 1976. Junctional and extrajunctional acetylcholine receptors in normal and denervated frog muscle fibers. *Pflugers Arch. Gesamte Physiol.* **366:** 1.

Edwards, C. 1979. Effects of innervation on the properties of acetylcholine receptors in muscle. *Neuroscience* **4:** 565.

Fambrough, D.M. 1974. Acetylcholine receptors. Revised estimates of extrajunctional receptor density in denervated rat diaphragm. *J. Gen. Physiol.* **64:** 468.

———. 1979. Control of acetylcholine receptors in skeletal muscle. *Physiol. Rev.* **59:** 165.

Fambrough, D.M. and R.E. Pagano. 1977. Positional stability of acetylcholine receptors at the neuromuscular junction. *Carnegie Inst. Washington Yrbk.* **76:** 28.

Fertuck, H.C. and M.M. Salpeter. 1974. Sensitivity in electron microscope autoradiography for ^{125}I. *J. Histochem. Cytochem.* **22:** 80.

———. 1976. Quantitation of junctional acetylcholine receptors by electron microscope autoradiography after ^{125}I-α-bungarotoxin binding at mouse neuromuscular junctions. *J. Cell Biol.* **69:** 144.

Fertuck, H.C., W.W. Woodward, and M.M. Salpeter. 1975. In vivo recovery of muscle contraction after α-bungarotoxin binding. *J. Cell Biol.* **66:** 209.

Fischbach, G.D. and S.A. Cohen. 1973. The distribution of acetylcholine sensitivity over uninnervated and innervated muscle fibers grown in cell culture. *Dev. Biol.* **31:** 147.

Fischbach, G.D. and S.M. Schuetze. 1980. A postnatal decrease in acetylcholine channel open time at rat endplates. *J. Physiol.* **303:** 125.

Fischbach, G.D., E. Frank, T.M. Jessell, L.L. Rubin, and S.M. Schuetze. 1979. Accumulation of acetylcholine receptors and acetylcholinesterase at newly formed nerve-muscle synapses. *Pharmacol. Rev.* **30:** 411.

Frank, E. and G.D. Fischbach. 1979. Early events in neuromuscular junction formation *in vitro*. *J. Cell Biol.* **83:** 143.

Heinemann, S., S. Bevan, R. Kullberg, J. Lindstrom, and J. Rice. 1977. Modulation of acetylcholine receptor by antibody against the receptor. *Proc. Natl. Acad. Sci.* **74:** 3090.

Jessell, T.M., R.E. Siegel, and G.D. Fischbach. 1979. Induction of acetylcholine receptors on cultured skeletal muscle by a factor extracted from brain and spinal cord. *Proc. Natl. Acad. Sci.* **76:** 5397.

Jones, R. and G. Vrbova. 1974. Two factors responsible for the development of denervation hypersensitivity. *J. Physiol.* **236:** 517.

Katz, B. and R. Miledi. 1972. The statistical nature of the acetylcholine potential and its molecular components. *J. Physiol.* **224:** 665.

Land, B.R., T.R. Podleski, E.E. Salpeter, and M.M. Salpeter. 1977. Acetylcholine receptor distribution on myotubes in culture correlated to acetylcholine sensitivity. *J. Physiol.* **269:** 155.

Levitt, T.A. and M.M. Salpeter. 1981. Denervated endplates have a dual population of junctional acetylcholine receptors. *Nature* **291:** 239.

Levitt, T.A., R.H. Loring, and M.M. Salpeter. 1980. Neuronal control of acetylcholine receptor turnover rate at a vertebrate neuromuscular junction. *Science* **210:** 550.

Linden, D.C. and D.M. Fambrough. 1979. Biosynthesis and degradation of acetylcholine receptors in rat skeletal muscle. Effects of electrical stimulation. *Neuroscience* **4:** 527.

Lømo, T. and C.R. Slater. 1980. Control of junctional acetylcholinesterase by neural and muscular influence in the rat. *J. Physiol.* **303:** 191.

Loring, R.H. and M.M. Salpeter. 1978. I-125-α bungarotoxin binding to denervated muscle: A survey study using light and EM autoradiography. *Neuorsci. Abstr.* **4:** 604.

———. 1980. Denervation increases turnover rate of junctional acetylcholine receptors. *Proc. Natl. Acad. Sci.* **77:** 2293.

Matthews-Bellinger, J. and M.M. Salpeter. 1978. Distribution of acetylcholine receptors at frog neuromuscular junctions with a discussion of some physiological implications. *J. Physiol.* **279:** 197.

———. 1979. Distribution of AChR in developing and dystrophic mouse neuromuscular junctions. *Neurosci. Abstr.* **5:** 485.

Merlie, J.P., J.-P. Changeux, and F. Gros. 1976. Acetylcholine receptor degradation measured by pulse chase labeling. *Nature* **264:** 74.

Michler, A. and B. Sakmann. 1980. Receptor stability and channel conversion in the subsynaptic membrane of the developing mammalian neuromuscular junction. *Dev. Biol.* **80:** 1.

Neher, E. and B. Sakmann. 1976. Noise analysis of drug induced voltage clamp currents in denervated frog muscle fibers. *J. Physiol.* **258:** 705.

O'Brien, R.A.D., A.J. Ostberg, and G. Vrbova. 1980. Effect of acetylcholine on the function and structure of the developing mammalian neuromuscular junction. *Neuroscience* **5:** 1376.

Peng, H.B., P.C. Cheng, and P.W. Luther. 1981. Formation of acetylcholine receptor clusters induced by positively charged latex beads. *Nature* **292:** 831.

Podleski, T.R., S. Nichols, P. Ravdin, and M.M. Salpeter. 1979. Cloned myogenic cells during differentiation: Membrane biochemistry and fine structural observations. *Dev. Biol.* **68:** 239.

Podleski, T.R., D. Axelrod, P. Ravdin, I Greenberg, M.M. Johnson, and M.M. Salpeter. 1978. Nerve extract induces increase and redistribution of acetylcholine receptors on cloned muscle cells. *Proc. Natl. Acad. Sci.* **75:** 2035.

Porter, C.W. and E.A. Barnard. 1975. The density of cholinergic receptors at the endplate postsynaptic membrane: Ultrastructural studies in two mammalian species. *J. Membr. Biol.* **20:** 31.

Prives, J., I. Silman, and A. Amsterdam. 1976. Appearance and disappearance of acetylcholine receptor during differentiation of chick skeletal muscle in vitro. *Cell* **7:** 543.

Reiness, C.G. and C.B. Weinberg. 1981. Metabolic stabilization of acetylcholine receptors at newly formed neuromuscular junctions in rat. *Dev. Biol.* **84:** 247.

Reiness, C.G., C.B. Weinberg, and Z.W. Hall. 1978. Antibody to acetylcholine receptor increases degradation of junctional and extrajunctional receptors in adult muscle. *Nature* **274:** 68.

Rubin, L.L., A.S. Gordon, and U.J. MacMahan. 1980a. Basal lamina fractions from electric organ of *Torpedo* organizes acetylcholine receptors in cultured myotubes. *Abstr. Soc. Neurosci.* **6:** 330.

Rubin, L.L., S.M. Schuetze, C.L. Weill, and G.D. Fischbach. 1980b. Regulation of acetylcholinesterase appearance at neuromuscular junctions *in vitro. Nature* **283:** 264.

Sakmann, B. and H.R. Brenner. 1978. Change in synaptic channel gating during neuromuscular development. *Nature* **276:** 401.

Salpeter, M.M., S. Spanton, K. Holley, and T.R. Podleski. 1982. Brain extract causes acetylcholine receptor redistribution which mimics some early events at developing neuromuscular junctions. *J. Cell Biol.* **93:** 417.

Schuetze, S.M., E.F. Frank, and G.D. Fischbach. 1978. Channel open time and metabolic stability of synaptic and extrasynaptic acetylcholine receptors on cultured chick myotubes. *Proc. Natl. Acad. Sci.* **75:** 520.

Steinbach, J.H. 1975. Acetylcholine responses on clonal myogenic cells *in vitro. J. Physiol.* **247:** 393.

Steinbach, J.H., A.J. Harris, J. Patrick, D. Schubert, and S. Heinemann. 1973. Nerve-muscle interaction *in vitro:* Role of acetylcholine. *J. Gen. Physiol.* **62:** 255.

Sytkowski, A.J., Z. Vogel, and M.W. Nirenberg. 1973. Development of acetylcholine receptor clusters on cultured muscle cells. *Proc. Natl. Acad. Sci.* **70:** 270.

Teichberg, V.I. and J.-P. Changeux. 1976. Presence of two forms of acetylcholine receptor with different isoelectric points in the electric organ of *Electrophorus electricus* and their catalytic interconversion *in vitro. FEBS Lett.* **67:** 264.

Vyskocil, F. and I. Syrovy. 1979. Do peripheral nerves contain a factor inducing acetylcholine sensitivity in skeletal muscle? *Experientia* **35:** 218.

Weinberg, C.B. and Z.W. Hall. 1979. Antibodies from patients with myasthenia gravis recognize determinants unique to extrajunctional acetylcholine receptors. *Proc. Natl. Acad. Sci.* **76:** 504.

Weinberg, C.B., J.R. Sanes, and Z.W. Hall. 1981. Formation of neuromuscular junctions in adult rats: Accumulation of acetylcholine receptors, acetylcholinesterase, and components of synaptic basal lamina. *Dev. Biol.* **84:** 255.

Developmental Changes in Acetylcholine Receptor Distribution and Channel Properties in Xenopus *Nerve-muscle Cultures*

YOSHI KIDOKORO
*PAUL BREHM**
RAPHAEL GRUENER[†]
The Salk Institute
San Diego, California 92138

During formation of the neuromuscular junction, the acetylcholine receptors (AChRs) undergo changes in density, distribution, and channel properties. Prior to contact by nerve, the AChRs are diffusely distributed over the entire muscle surface. Soon after the establishment of a functional synapse, however, receptors accumulate beneath the region of nerve contact. These early developmental phenomena can be conveniently reproduced in *Xenopus* nerve-muscle cultures, which offer an opportunity to examine the underlying molecular mechanisms. Cultures were prepared by dissecting nerve and muscle tissue from embryos of *Xenopus laevis* and by dissociating them in solution containing collagenase, as described previously (Kidokoro et al. 1980). We measured AChR distribution by fluorescence microscopy using tetramethyl-rhodamine-conjugated α-bungarotoxin (TMR-αBgt) or by autoradiography using ^{125}I-labeled α-Bgt. The extent of functional AChR accumulation was quantitated by iontophoretic application of acetylcholine. AChR channel properties were examined by noise analysis of acetylcholine-induced current fluctuations and also by recording single-channel currents (Neher and Sakmann 1976).

Present addresses: *Department of Physiology, Tufts University School of Medicine, Boston, Massachusetts 02111; [†]Department of Physiology, College of Medicine, University of Arizona, Tucson, Arizona 85724.

Neurons from embryonic neural tube mediate changes in both the density and the functional properties of AChRs in the *Xenopus* muscle cell. The neuronally mediated increase in the receptor density is restricted to the region closely associated to the nerve-muscle contact. In contrast to this regional effect of nerve, a nerve-dependent alteration in functional properties is observed for junctional and nonjunctional receptors alike. This paper is directed at summarizing the findings of recent experiments performed at The Salk Institute, which examined these neural influences on the spectral distribution and functional properties of AChRs.

NEURONS INDUCE AChR ACCUMULATION AT THE SITE OF NERVE-MUSCLE CONTACT

Embryonic *Xenopus* muscle cells in culture have AChRs diffusely organized over the entire surface with occasional spots of high receptor density, termed hot spots (Anderson et al. 1977). The growing neurites appear to contact the muscle surface randomly, without seeking out such hot spots. When the muscle cell is contacted by cholinergic neurons, functional synaptic transmission can be detected within 10 minutes following the initial contact (Kidokoro and Yeh 1981a). After less than 24 hours of nerve contact, AChRs accumulate to the nerve contact area in many muscle cells. Nerve-induced AChR clusters have a characteristic tractlike pattern and are readily distinguishable from aggregates that existed prior to nerve contact (Fig. 1). Accumulation of AChRs to such regions of nerve contact is primarily the result of migration of receptors that already existed (Anderson and Cohen 1977). However, the preferential insertion of newly synthesized AChRs to the nerve contact region may also contribute to such high-density aggregates.

Nerve-induced receptor aggregates are observed on only approximately 42% of nerve-contacted muscle cells in *Xenopus* cultures (Kidokoro and Yeh 1981b). The following data indicate that the absence of nerve-induced receptor aggregates on the remaining cells often results from contact by noncholinergic neurons. To identify cholinergic neurites among the variety of neural types, we tested for the presence of curare or α-Bgt blockable synaptic potentials in the neurite-contacted muscle cell. In 37 neurite-contacted muscle cells exhibiting characteristic receptor localizations, all were cholinergic by electrophysiological criteria. Conversely, no nerve-induced receptor localizations were observed in 48 cells in which synaptic activities were absent. Therefore, only cholinergic neurons appear effective in causing receptor localization. However, not all cholinergic neurons were capable of causing AChR accumulation in cell culture. Spinal cord explants from old tadpoles formed functional synaptic contact with muscle cells but did not cause receptor accumulation (Kidokoro et al. 1980).

The nature of the signal that is responsible for the accumulation of AChRs in *Xenopus* muscle cells is not known. Probably acetylcholine does

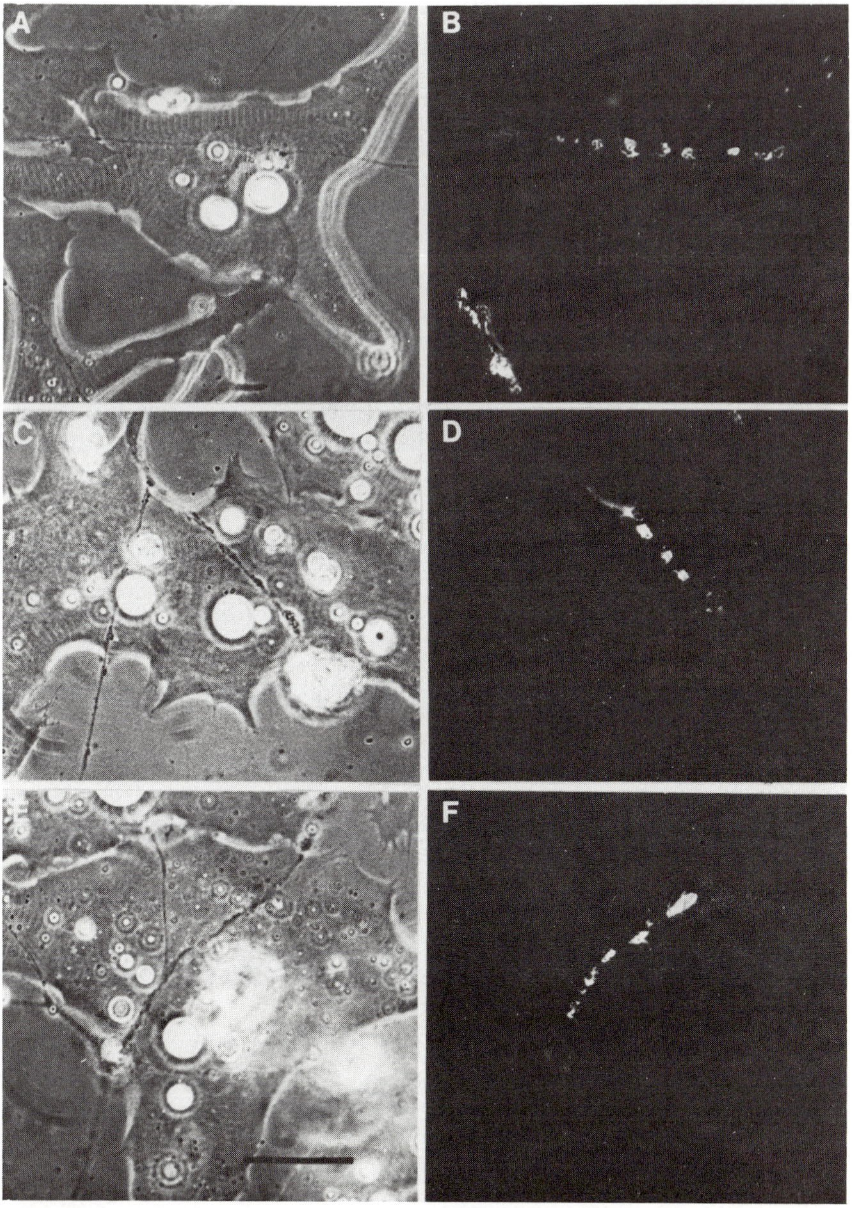

FIGURE 1
Phase (*A*, *C*, *E*) and fluorescence (*B*, *D*, *F*) micrographs of nerve-contacted muscle cells at advanced stages of AChR accumulation. The bar in *E* represents 30 μm in all micrographs. (Reprinted, with permission, from Kidokoro et al. 1980.)

not serve as the signal, since blockade of all functional receptors with α-Bgt or curare did not prevent AChR accumulation (Anderson and Cohen 1977), nor did suppression of spontaneous acetylcholine release by increasing magnesium concentration in culture medium affect receptor accumulation (Y. Kidokoro and E. Yeh, unpubl.).

During the time over which AChR accumulation occurred in culture, the amplitude of spontaneous synaptic potentials increased about fivefold (Fig. 2). This increase in amplitude was anticipated, since more acetylcholine molecules will bind to the receptors before diffusing away from the synaptic cleft when the receptor density is higher (Anderson et al. 1979; Kidokoro et al. 1980). The question remains, however, whether the extent of the change in receptor density was adequate to explain the fivefold increase in amplitude. Autoradiographic examination of α-Bgt binding sites at the junctional region yielded a mean density of $1000/\mu m^2$, which is about nine times higher than the density prior to nerve contact ($104/\mu m^2$) (Fig. 3; Kidokoro and Gruener 1982). Concomitant with the increase in the receptor density at the junctional region was a fivefold increase in acetylcholine sensitivity as measured by acetylcholine iontophoresis (Gruener and Kidokoro 1982). Therefore, both quantitative techniques indicate that the extent of receptor accumulation was sufficient to explain the increase in the amplitude of synaptic potentials.

COCULTURE WITH NEURAL TUBE CELLS OR CONDITIONED MEDIUM INCREASES ACETYLCHOLINE SENSITIVITY OF MUSCLE CELLS AT NONJUNCTIONAL REGIONS

Since receptor accumulation results primarily from migration of AChRs out of the nonjunctional region, a lowering in nonjunctional acetylcholine sensitivity might be anticipated in nerve-contacted cells. This possibility was

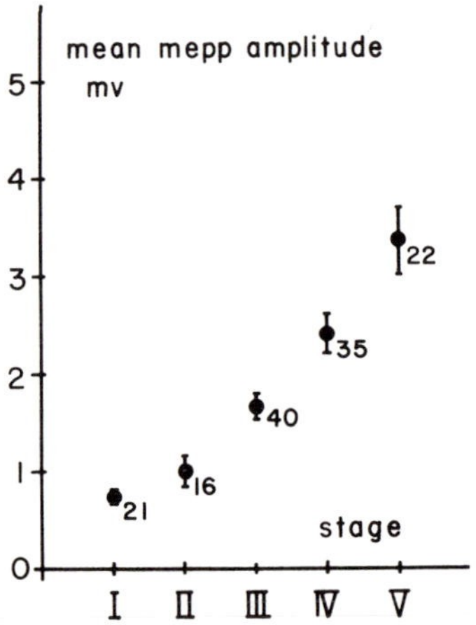

FIGURE 2
Mean miniature endplate potential (mepp) amplitude of cells in different stages. Half-lengths of the bar and number attached to each circle indicate the S.E.M. and the number of cells examined, respectively. The extent of AChR accumulation was classified as stages I–V. The criteria of classification are described in detail elsewhere. (Reprinted, with permission, from Kidokoro et al. 1980.)

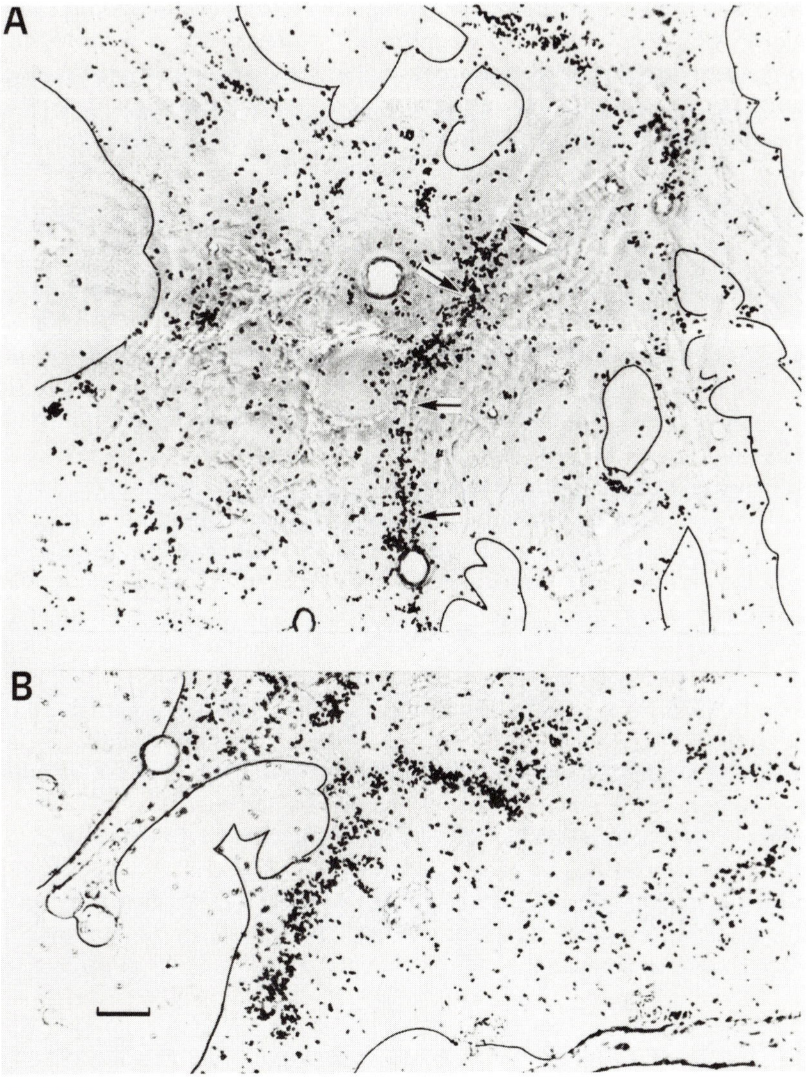

FIGURE 3
(*A*) Autoradiogram of a nerve-induced band of high-grain-density area with a width of about 5 μm. (→) The neurite path. (*B*) A similar-size band of high-grain-density area as *A* from a cell where no nerve was seen. The nerve that induced this grain aggregation was probably sheared off during repeated washing procedures for autoradiography. The bar in *B* represents 10 μm and applies to both *A* and *B*. (Reprinted, with permission, from Kidokoro et al. 1982.)

tested by comparing the acetylcholine sensitivity at nonjunctional regions of muscle cells showing receptor accumulation with that of muscle cells lacking nerve contact in the same culture. The acetylcholine sensitivity was

found to be the same in those two groups of muscle cells (Gruener and Kidokoro 1982). This result was confirmed by measuring α-Bgt binding sites in autoradiography (Kidokoro and Gruener 1982). Therefore, the migration of AChRs to the nerve contact region does not result in a significant decrease in the nonjunctional density.

Muscle activity has been shown to regulate the nonjunctional acetylcholine sensitivity on denervated skeletal muscle (Drachman and Witzke 1972; Lømo and Rosenthal 1972). It was of interest to determine whether muscle activity in embryonic muscle caused a similar decrease in acetylcholine sensitivity. Innervated *Xenopus* muscle in culture often twitches due to the large amplitude of the spontaneous synaptic potentials. Comparison of twitching muscle cells to noninnervated muscle cells in the same culture indicated no significant difference in nonjunctional acetylcholine sensitivity (Kidokoro and Gruener 1982). Therefore, the muscle activity caused by spontaneous synaptic potentials is apparently not adequate for decreasing AChR density in the nonjunctional region.

When the acetylcholine sensitivity at the nonjunctional region of muscle cells cocultured with neural cells was compared with that of muscle cells grown alone, there was a significant increase. Addition of neural tube cells to muscle cultures caused an average 50% increase in acetylcholine sensitivity over the entire muscle surface. Increased sensitivity to acetylcholine was also found in the muscle cells grown in conditioned medium, which was prepared by growing muscle cells with neural cells. This suggests that a diffusible factor released from neural cells or from contaminating glial cells affected acetylcholine sensitivity of muscle cells. This increase in the acetylcholine sensitivity was not accompanied by changes in passive electrical properties of muscle cells, such as the resting membrane potential or the input impedance (Kidokoro and Gruener 1982). This observation raised a question as to whether this increase in the acetylcholine sensitivity in nonjunctional regions was due to an increase in AChR density or due to changes in AChR channel properties.

COCULTURE WITH NEURAL TUBE CELLS DOES NOT INCREASE AChR DENSITY AT NONJUNCTIONAL REGION

The density of nonjunctional AChRs, as measured by the autoradiographic technique, was compared between muscle cells grown in the absence of neural tube cells and muscle cells cocultured with neural tube cells. There was no significant difference under these two conditions. Nor in the majority of cases were there any differences in the site density between muscle cells cultured in conditioned medium and those grown in standard culture medium (Kidokoro and Gruener 1982). We conclude that the neural effect on nonjunctional acetylcholine sensitivity was not mediated by a change in AChR density.

COCULTURE WITH NEURAL TUBE CELLS CHANGES INDIVIDUAL AChR CHANNEL PROPERTIES

The mean channel open time of the AChR channel was estimated by extracellularly recorded acetylcholine-induced noise. Over a 3-day coculture period, the spectral shape of acetylcholine-induced noise underwent a dramatic reduction in the high frequency power. Similar changes in spectral characteristics were observed from muscle cells cultured in neural tube conditioned medium. Spectra from muscle cells cultured alone, as well as those cocultured with neural tube cells, were best fit by a curve described by two Lorentzian functions: one corresponding to a short open time (0.7 msec), and the other, to a long open time (3.0 msec). This finding suggested that the change in the spectral shape following coculture with neural tube cells reflected an increase in channel open time and, further, that this change resulted from an increase in the relative proportion of a channel type with a long open time rather than a gradual shift in the open time value of all receptors on the muscle surface (Brehm et al. 1982). Prolongation of the mean open time results in a greater charge transfer during the activation of AChRs. As a consequence, activation of an equivalent number of AChRs would lead to a larger depolarization. An increased acetylcholine sensitivity in muscle cells cocultured with neural tube cells would result without any increase in AChR density.

TWO TYPES OF AChR CHANNELS. COCULTURE WITH NEURAL TUBE CELLS CHANGES THE RELATIVE PROPORTION OF THESE TWO TYPES

Single-channel current recording provided additional information that allowed characterization of individual channels on the basis of conductance. In muscle cells cultured alone, the amplitude histogram of the unitary current had two distinct peaks (Fig. 4). When these two populations were separated and the duration histograms were constructed separately, it was evident that the population with the large unitary current had a shorter mean open time when compared to the population with the smaller current. The above observation is in accord with that published by Clark and Adams (1981). In muscle cells cultured alone, the population of channels that have a larger unitary current and a shorter open time is evident (triangles in Fig. 4). However, in muscle cells cocultured with neural tube cells, the channel type with a smaller unitary current and longer open time becomes predominant (Fig. 4) (P. Brehm and Y. Kidokoro, unpubl.).

Preferential migration of AChR channels with a short open time to the nerve contact region could result in prolongation of the mean open time at the nonjunctional region. However, this was not the case. Prolongation of channel open time in muscle cells cocultured with neural cells occurred regardless of the presence of direct nerve contact. Moreover, prolongation

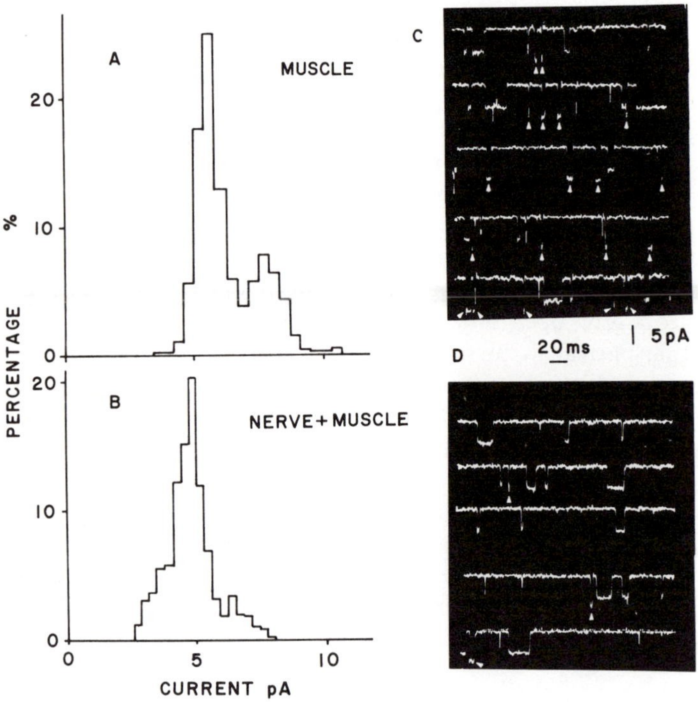

FIGURE 4
Single AChR channel currents from *Xenopus* muscle cells. (*A,B*) The amplitude histograms from a muscle cell cultured alone and cultured with neural tube cells, respectively. In both cases, the tip of the patch clamp electrode was maintained at +60 mV during the recording. This caused the membrane under the electrode to be effectively hyperpolarized 60 mV over the resting potential. The resting membrane potential of muscle cells under this recording condition was measured separately and was between −80 mV and −90 mV. (*C,D*) Oscilloscope traces corresponding to *A* and *B*, respectively. (▲) A class of larger current deflections, which have shorter duration in average. Note that triangles are more frequent in *C* than in *D*. Correspondingly, two peaks were clearly separated in *A* but not in *B*. The slight difference in the major peak in *A* and *B* is insignificant. The amplitudes of about 1000 single-channel currents were recorded on an FM tape recorder (Racal) and were analyzed off-line on a PDP-11/23 microcomputer (Digital). Due to the frequency limitation of our recording system, the amplitude of current deflections less than 0.5 msec in duration cannot be measured accurately. Therefore, these histograms include only currents lasting more than 0.5 msec. It should be noted that, due to this limitation, the proportion of the class of larger current deflections, which had a shorter average duration, is underrepresented in these histograms.

of channel open time occurred for both junctional and nonjunctional receptors alike (Brehm et al. 1982).

A PARADOX REMAINS

In the *Xenopus* tadpole tail muscle, Kullberg et al. (1981) showed that the mean channel open time shortened progressively as the animal developed. Shortening of the channel open time occurred both at the junctional and nonjunctional regions. Similar progressive shortening was observed in *Xenopus* muscle cells in culture. However, when neural tube cells were added to muscle cultures, the mean channel open time was *prolonged* (Brehm et al. 1982). This was unexpected since, in vivo, the channel open time *shortened* in the presence of nerve. Neither the specificity nor the possible functional significance of the effect is known. An attractive idea is that the effect is the result of a juvenilizing effect on muscle receptor synthesis by embryonic neural tube cells. In this manner the expression of junctional receptors could be suppressed until the cells in the spinal cord mature. An alternative possibility is that expression of nonjunctional receptors on embryonic muscle is somehow influenced by the presence of neural factors, as in the case of denervated muscle.

By examining various properties of AChRs on embryonic muscle, we hope to elucidate the fundamental mechanisms controlling developmental changes.

ACKNOWLEDGMENTS

This work was supported by grants from the National Institutes of Health (NS-11918 to Y.K.) and from the Muscular Dystrophy Association of America. P.B. was supported by a postdoctoral fellowship from the Muscular Dystrophy Association of America and by the Galster Foundation.

REFERENCES

Anderson, M.J. and M.W. Cohen. 1977. Nerve-induced and spontaneous redistribution of acetylcholine receptors on cultured muscle cells. *J. Physiol.* **268:** 757.

Anderson, M.J., M.W. Cohen, and E. Zorychta. 1977. Effects of innervation on the distribution of acetylcholine receptors on cultured muscle cells. *J. Physiol.* **268:** 731.

Anderson, M.J., Y. Kidokoro, and R. Gruener. 1979. Correlation between acetylcholine receptor localization and spontaneous synaptic potentials in cultures of nerve and muscle. *Brain Res.* **166:** 185.

Brehm, P., J.H. Steinbach, and Y. Kidokoro. 1982. Channel open time of acetylcholine receptor on *Xenopus* muscle cells in dissociated cell culture. *Dev. Biol.* **91:** 93.

Clark, R.B. and P.R. Adams. 1981. ACh receptor channel populations in cultured *Xenopus* myocyte membranes are non-homogeneous. *Soc. Neurosci. Abstr.* **7:** 838.

Drachman, D.B. and F. Witzke. 1972. Trophic regulation of acetylcholine sensitivity of muscle: Effect of electrical stimulation. *Science* **176:** 514.

Gruener, R. and Y. Kidokoro. 1982. Acetylcholine sensitivity of innervated and noninnervated *Xenopus* muscle cells in culture. *Dev. Biol.* **91:** 86.

Kidokoro, Y. and R. Gruener. 1982. Distribution and density of α-bungarotoxin binding sites on innervated and noninnervated *Xenopus* muscle cells in culture. *Dev. Biol.* **91:** 78.

Kidokoro, Y. and E. Yeh. 1981a. Initial synaptic transmission at the growth cone in *Xenopus* nerve-muscle cultures. *Soc. Neurosci. Abstr.* **7:** 6.

———. 1981b. Synaptic contacts between embryonic *Xenopus* neurons and myotubes formed from a rat skeletal muscle cell line. *Dev. Biol.* **86:** 12.

Kidokoro, Y., M.J. Anderson, and R. Gruener. 1980. Changes in synaptic potential properties during acetylcholine receptor accumulation and neurospecific interactions in *Xenopus* nerve-muscle cell culture. *Dev. Biol.* **78:** 464.

Kullberg, R.W., P. Brehm, and J.H. Steinbach. 1981. Nonjunctional acetylcholine receptor channel open time decreases during development of *Xenopus* muscle. *Nature* **289:** 411.

Lømo, T. and J. Rosenthal. 1972. Control of ACh sensitivity by muscle activity in the rat. *J. Physiol.* **221:** 493.

Neher, E. and B. Sakmann. 1976. Single channel currents recorded from membrane of denervated frog muscle fibers. *Nature* **260:** 799.

MUSCLE DEVELOPMENT AND HUMAN DISEASE

Introduction: Human Neuromuscular Disorders

STANLEY H. APPEL
Department of Neurology and Program in Neuroscience
Jerry Lewis Neuromuscular Disease Research Center
Baylor College of Medicine
Houston, Texas 77030

One of the major goals of basic research in muscle is to gain a greater understanding of human neuromuscular development and to translate these findings into effective therapy for patients with neuromuscular disease. Among the most common disorders of the neuromuscular system are the muscular dystrophies and, specifically, Duchenne muscular distrophy (DMD) and myotonic muscular dystrophy (MYD). Both of these conditions are inherited disorders. But despite the most intense investigative efforts, the inborn errors of metabolism remain unknown. Over the years, attention has been directed at deficiency states in catecholamines, amino acids, vitamins B6 and E, or steroids, including anabolic steroids and corticosteroids (Dubowitz and Heckmatt 1980), or at deficits of the vascular supply of muscle, neuronal innervation, cytoskeletal constituents, or widespread disturbances in surface membranes as the key problem in these disorders (Appel and Roses 1978). Unfortunately, none of these approaches has provided either reproducible explanations of the dysfunction or fruitful insights into therapy.

DMD AND MYD

There is still great interest in membranes as the locus of the fundamental defect in DMD. This enthusiasm stems largely from the early demonstration of elevated creatinine phosphokinase (CPK) levels in the blood of young males with DMD and evidence that the CPK is primarily derived from muscle. The demonstration of δ lesions in DMD muscles, the discontinuities in DMD muscle plasma membranes (Mokri and Engel 1975), the increase in

calcium within muscle myoplasm, and the changes in particle distribution by freeze-fracture analysis (Schotland et al. 1977) and in lectin binding to muscle surfaces (Bonvilla et al. 1978) all point to definite abnormalities in plasma membranes of muscle. However, such studies cannot distinguish abnormalities of a primary nature from those that are secondary, namely due to altered extracellular or intracellular metabolic processes.

The red blood cell has been extensively studied as a means of defining the membrane locus of the genetic abnormality. However, no consistent or reproducible abnormality in red blood cell membrane structure or function has helped define the inborn error of metabolism in DMD or MYD (Plishker and Appel 1980). In fact, in studies of membranes from diverse tissues, there is no convincing evidence that the link between the abnormal gene product and the membrane is direct. In red cells, as in muscle, even where reproducible alterations have been recorded, abnormalities of circulating metabolites or intracellular constitutents could readily explain the membrane perturbations.

The failure of membrane studies to define a specific defect has prompted investigators to refocus on muscle tissue as the primary target of the dystrophic process. In past studies of muscle, changes of degeneration, regeneration, fat, and connective tissue appeared to obscure the inborn error of metabolism in DMD. The problem with the direct analysis of muscle is that the many changes in intermediary metabolism more likely reflect the secondary process of damage and repair rather than the primary defect. A similar criticism can be leveled at almost all biochemical investigations of the heterogeneous tissue obtained at biopsy. In fact, the primary deficit may already have been defined and not be recognized because of the plethora of abnormalities of lipid, protein, nucleic acid, and carbohydrate already present in damaged tissue.

These failures have prompted a new strategy, namely the culture of muscle obtained at biopsy from patients with DMD and MYD. Normal skeletal-muscle fibers are known to contain a population of spindle-shaped cells that lie between the plasma membrane and the basal lamina and are termed satellite cells. Konigsberg (1963) demonstrated that these satellite cells gave rise to new muscle fibers in vitro. Thus, muscle obtained from patients at any age may be dissociated in vitro to yield myoblasts that subsequently fuse to form myotubes. Muscle-specific markers, such as CPK and nicotinic acetylcholine receptor, as well as myosin, actin, tropomyosin, and troponin can be demonstrated in such cultures and appear in increasing concentration concomitant with the process of fusion. Such cultures may be used to help define the primary metabolic error. However, in removing muscle cells from their environment in vivo, one may be removing them from the cues that are responsible for the dystrophic process, or one may be placing them in an artificial environment that prevents full maturation and complete expression of the genetic defect. In either instance, the defect would not be ap-

parent in vitro, unless the abnormal in vivo environment could be recreated or unless the appropriate degree of maturation could be fostered.

Several laboratories have now reported abnormalities in muscle tissue cultured from patients with DMD. Reports have appeared of altered morphology, including clumping of myoblasts (Thompson et al. 1977) and reduced adhesion of fibroblasts (Jones and Witkowski 1979), or altered biochemistry, such as altered protein synthesis (Ionasescu et al. 1971) or decreased CPK-specific activity. Not all laboratories have been able to confirm such morphological or biochemical findings. Our own laboratory has not been able to demonstrate any change in morphological or electrical properties in primary cultures of myotubes from patients with DMD.

HUMAN MUSCLE CELL CULTURES

In this volume, Blau and co-workers present data in which they have cloned myoblasts from tissue obtained from DMD patients and from normal controls. In these studies, they have examined detailed properties of growth, morphological appearance, and specific biochemical characteristics, including muscle structural proteins, CPK isozymes, and nicotinic acetylcholine receptor. No systematic differences were found in any parameter examined. Thus, neither in primary cultures nor in clonal lines can reproducible abnormalities be detected in muscle myotubes obtained from patients with DMD. In no sense does this rule out an intrinsic muscle genetic defect, since myotubes do not fully differentiate in vitro. Clearly, the next step would be to attempt to enhance maturation in vitro, possibly by coculturing with nerve cells, fibroblasts, or in the presence of growth or trophic factors, and then to examine differences in the properties of well-differentiated DMD or control myotubes in vitro.

The failure to detect abnormalities in DMD, using primary or cloned cultures, also highlights the need for a sensitive assay for the DMD muscle cells in vitro. Although the clinical presentation, muscle pathology, and elevated serum enzymes serve to define the DMD phenotype in vivo, it is not clear whether any of these parameters will suffice to define either the phenotype or the genotype in vitro. In addition, it is not even self-evident that a primary metabolic error of muscle be necessarily expressed in myotubes in culture.

Miranda et al. (this volume) present data that in certain myopathic disorders, muscle tissue cultures demonstrate no apparent alteration in the enzyme activity known to be deficient in biopsy tissue; whereas in other diseases, the enzyme activity known to be deficient in biopsy tissue is also deficient in myotube cultures. When discrepancies occur between in vitro and in vivo situations, it may be explained by the presence of several isozymes of the particular enzyme known to be present in normal tissue. For example, myotubes grown in vitro from biopsies of patients with Mc-

Ardle syndrome (myophosphorylase deficiency) had *normal* enzyme activity due to the presence of a "fetal" isozyme of phosphorylase. In acid-maltase deficiency, by way of contrast, myotubes grown from biopsy muscle showed the same deficiency in enzymatic activity as noted in the tissue extracts themselves.

Thus, these studies demonstrate the importance of a specific, as well as a sensitive, assay for defining genetic defects in vitro and emphasize that certain genetic defects may only express themselves later in development—possibly at a time beyond the stage of differentiation normally reached in culture.

MYD appears to differ from DMD, since a defect is expressed in tissue culture even at a relatively immature myotube stage. In studies of MYD, our laboratory employed electrophysiologic assays for the defect (Merickel et al. 1981). Morphologic and general biochemical parameters were normal, yet electrical studies demonstrated an abormal phenotype. Control myotubes demonstrated a hyperpolarization afterpotential (HAP), whereas MYD myotubes possessed depolarizing afterpotentials (DAPs) and repetitive bursts. Raising the extracellular or the intracellular calcium concentration in control myotubes caused their HAP to become depolarizing, but raising extracellular calcium in MYD myotubes caused their DAP to become hyperpolarizing. Increasing extracellular calcium depressed the slow outward current in MYD myotubes. These studies indicate that MYD myotubes express physiologic alterations in the absence of nerve cells and provide a convenient model for further exploration of the inborn error of metabolism.

The paper presented by Walsh et al. (this volume) describes several techniques, including monoclonal antibodies and animal-human cell hybridization, as approaches to the detection of specific muscle proteins. These approaches have been effectively used to characterize surface and intracellular constitutents in other systems and offer the potential for sensitive assays of in vitro maturation and even for assays of the altered constituents of dystrophic muscle.

MYASTHENIA GRAVIS

Advances in our understanding of the pathogenesis and treatment of myasthenia gravis have been far greater than advances in our understanding of the muscular dystrophies. Myasthenia gravis is now well accepted as an autoimmune disease in which antibodies directed against the muscle nicotinic acetylcholine receptor result in a loss of receptors, atrophy of the muscle endplate junctional folds, and widening of the intersynaptic cleft, leading to impairment of neuromuscular transmission and clinical weakness and fatigability. Why such antibodies arise is not clear, but the presence of thymic hyperplasia or thymoma in the overwhelming number of myasthe-

nia gravis patients and the beneficial clinical response of early thymectomy suggest a prominent, although as yet unknown, role for the thymus in this process. These research advances have paid handsome dividends in improved treatment of myasthenia gravis patients, extending from the use of anticholinesterases through thymectomy to the judicious use of corticosteroids, cytotoxic agents such as azathioprine, and plasmapheresis.

In this volume, Souroujon et al. review studies from their laboratory in which polyclonal and monoclonal antibodies are raised against structural or nonstructural antigenic determinants of the acetylcholine receptor. Such antibodies may not only help to further characterize the acetylcholine receptor, but may also help to further define the immunologic defect in myasthenia gravis.

REFERENCES

Appel, S.H. and A.D. Roses. 1978. The muscular dystrophies. In *The metabolic basis of inherited disease* (ed. J.B. Stanbury et al.), p. 1260. McGraw-Hill, New York.

Bonvilla, E., D.L. Schotland, and Y. Wakayama. 1978. Duchenne dystrophy: Focal alterations in the distribution of concanavalin A binding sites at the muscle cell surface. *Ann. Neurol.* **4:** 117.

Dubowitz, V., and J. Heckmatt. 1980. Management of muscular dystrophy: Pharmacological and physical aspects. *Br. Med. Bull.* **36:** 139.

Ionasescu, V., H. Zellweger, and T.W. Conway. 1971. Ribosomal protein synthesis in Duchenne muscular dystrophy. *Arch. Biochem. Biophys.* **144:** 51

Jones, G.E. and J.A. Witkowski. 1979. Reduced adhesiveness between skin fibroblasts from patients with Duchenne muscular dystrophy. *J. Neurol. Sci.* **43:** 465.

Konigsberg, I.R. 1963. Clonal analysis of myogenesis. *Science* **140:** 1273.

Merickel, M., R. Gray, P. Chauvin, and S.H. Appel. 1981. Cultured muscle from myotonic muscular dystrophy patients: Altered membrane electrical properties. *Proc. Natl. Acad. Sci.* **78:** 648

Mokri, B. and A.H. Engel. 1975. Duchenne dystrophy: Electron microscopic findings pointing to a basic or early abnormality in the plasma membrane of the muscle fiber. *Neurology* **25:** 1111.

Plishker, G.A. and S.H. Appel. 1980. Red blood cell alterations in muscular dystrophy: The role of lipids. *Muscle and Nerve* **3:** 70.

Schotland, D.L., F. Bonvilla, and Y. Wakayama. 1977. Duchenne dystrophy: Alterations in plasma membrane structure. *Science* **196:** 1005.

Thompson, E.J., R. Yasin, G. VanBeer, K. Nurse, and S. Al-Ani. 1977. Myogenic defect in human muscular dystrophy. *Nature* **268:** 241.

Developmentally Regulated Isozyme Transitions in Normal and Diseased Human Muscle

ARMAND F. MIRANDA[*]
SARA SHANSKE[†]
SALVATORE DiMAURO[†]
*Departments of *Pathology and †Neurology*
College of Physicians and Surgeons
Columbia University
New York, New York 10032

During differentiation of muscle in vivo and in culture, genetically regulated isozyme transitions take place. These transitions often involve a switch from one form of enzyme prevalent in many different tissues to a muscle-specific form characteristic of mature muscle. For instance, creatine kinase (CK) activity in myoblasts is due to a homodimer enzyme, BB-CK. During muscle differentiation, when myoblasts fuse to form multinucleated myotubes, the muscle-specific form MM-CK and a hybrid, MB-CK, begin to appear (for review, see Miranda et al. 1979a). As muscle matures, BB-CK activity declines and MM-CK activity increases manyfold. In mature skeletal muscle, virtually all creatine kinase activity is due to the muscle-specific form. In other tissues, however, such as brain and connective tissue (fibroblasts), BB-CK prevails throughout adult life. Creatine kinase isozymes can be identified by differences in electrophoretic mobility and antigenic properties. This enzyme system is therefore a convenient biochemical marker for analyzing stages of muscle development.

Since subunits of isozyme systems are under separate genetic control, genetic defects of either the "fetal" or "mature" forms could be expected to occur. Genetic metabolic myopathies involving isozyme subunits that are

prevalent in embryonic muscle and absent or catalytically inactive in mature muscle have not yet been described. On the other hand, several human genetic metabolic myopathies identified to date involve a deficiency of the muscle subunit of isozyme systems, which are developmentally regulated. For these metabolic myopathies, disease expression should coincide with the stage of muscle development at which the muscle-specific enzyme activity normally appears. In genetic metabolic myopathies affecting enzymes that are not tissue-specific and that do not show any change in isozyme composition during myogenesis, it is to be expected that the disease would be expressed in both immature and mature muscle. Systematic analysis of normal muscle cultures and cultures from patients with metabolic myopathies will permit us to better understand the regulatory mechanisms of enzyme transitions and the metabolic consequences of the enzyme defect when this is expressed in tissue culture.

RESULTS AND DISCUSSION

Isozyme Transitions

Isozymes or isoenzymes are multiple molecular forms of enzyme with the same catalytic activity, occurring in a single species and encoded by different genes. Many isozymes are tissue specific, and several metabolic diseases in humans are due to genetic defects of muscle-specific isozymes. Gene loci coding for muscle-specific isozyme subunits of some of these, such as muscle-type lactate dehydrogenase (M-LD) and muscle-type phosphofructokinase (M-PFK), have been assigned to chromosome 11 and chromosome 1, respectively (Boone et al. 1972; Vora et al. 1981). The loci coding for muscle-specific subunits of phosphorylase (M-PPL), phosphoglycerate mutase (M-PGAM), and adenylate deaminase (M-AMPD) are still unmapped, but based on indirect evidence, particularly the observation that nonmuscle isozymes are not affected in these hereditary metabolic myopathies, it can be assumed that these are also encoded at gene loci different from those of the nonmuscle forms. Not only are isozymes different in different tissues but, within the same tissue, the isozyme pattern may change during development and tissue regeneration in vivo and in culture, as reviewed by Schapira (1978) and Miranda et al. (1979a). Regression toward a less differentiated isozyme profile may also occur in some pathologic conditions (Sato et al. 1975; Schapira et al. 1975) or when differentiation is inhibited by viral transformation in muscle cultures (Miranda et al. 1981b).

Tissue culture is an ideal system for analyzing the temporal sequence of developmentally regulated isozyme transitions, because unlike the intact organism, it is not subject to interaction with multiple tissues. In this simplified in vitro system, it may therefore be easier to identify factors that regulate isozyme transitions.

We have begun to study muscle cultures of normal controls and patients with genetic metabolic myopathies involving developmentally regulated isozymes and to compare these disorders with muscle diseases involving enzymes that are not tissue specific and that do not change during myogenesis. In addition, we are trying to determine how the tissue culture data correlate with myogenesis in vivo by studying changes in isozyme compositions during normal fetal development. For all of these studies, creatine kinase is used as a marker of muscle differentiation, because transition from BB to MM isozyme is well documented in human embryonic development (Goto et al. 1969; Tzvetanova 1971; Foxall and Emery 1975), as well as in human muscle cultures (Miranda et al. 1979a; Samaha et al. 1981). In fetal muscle, the first electrophoretically detectable MM-CK activity occurs at about 7–8 weeks of gestation. This stage coincides with the initiation of myoblast fusion (Fidzianska 1980). By 16 weeks, MM-CK predominates, and BB-CK activity is less than 5% of the total. In the same samples, however, PGAM, another developmentally regulated dimer enzyme, still shows about 30% of total activity in the BB form. Even at 18 weeks and 20 weeks, when creatine kinase is present almost exclusively as MM-CK, approximately 30% of PGAM is still due to the nonmuscle BB form. Similar immature isozyme patterns are also observed for PPL (Fig. 1). These studies show that isozyme transitions do not occur synchronously for these enzymes and are probably differentially regulated. In agreement with these observations in fetal muscle, normal muscle cultures grown in the absence of nerve or embryo extract

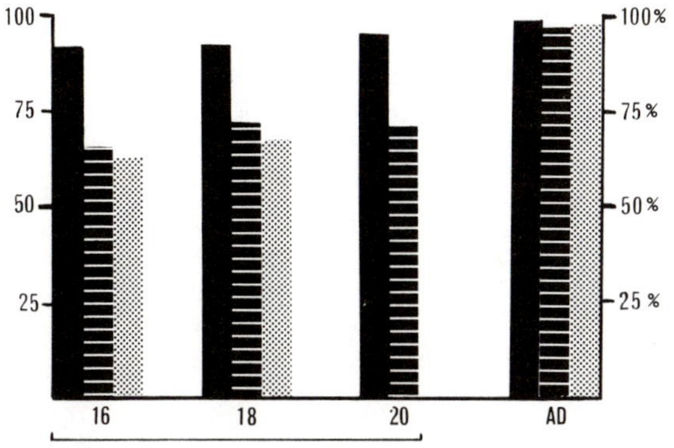

FIGURE 1
Percentage of inhibition of creatine kinase (■), PGAM (▤) and PPL (▨) by antibodies to muscle isozyme (creatine kinase and PPL) or by mercuric chloride (PGAM) in normal adult muscle compared with fetal muscle at 16 weeks, 18 weeks, and 20 weeks of gestation. PPL at 20 days of gestation was not studied.

show a similar lack of coordination in isozyme transition. In these cultures, MM-CK activity is already detectable at the time of myoblast fusion, and it represents 30% or more of total activity in myotube cultures 24–48 hours after optimal fusion. At this stage of muscle differentiation, neither MM-PGAM nor muscle-specific PPL activities are present in detectable amounts by electrophoresis or immunological studies (Fig. 2; Di Mauro et al. 1978, 1981b).

METABOLIC MYOPATHIES IN MUSCLE CULTURES

PPL

The first genetic metabolic myopathy to be studied in tissue culture was myophosphorylase deficiency (McArdle disease), and the findings appeared surprising because PPL activity was clearly detectable by histochemistry in cultured myotubes from patients (Roelofs et al. 1972). Following this initial observation, "reappearance" of PPL activity in muscle cultures from genetically deficient patients has been confirmed and studied in more detail (Meienhofer et al. 1977; Sato et al. 1977; DiMauro et al. 1978). Electrophoresis showed that the reappearance of PPL activity was due to a different isozyme, present in fetal as well as in cultured muscle. In immunochemical studies, antibodies to muscle-specific PPL, which completely inhibited activity of normal mature muscle, failed to significantly inhibit activity of normal or McArdle muscle cultures (Sato et al. 1977; DiMauro et al. 1978). Abnormal accumulation of glycogen, which is the pathologic hallmark of myophosphorylase-deficient muscles from patients, could not be demonstrated either by staining with periodic acid Schiff (PAS) reagent or by electron microscopy in cultured myotubes from these patients.

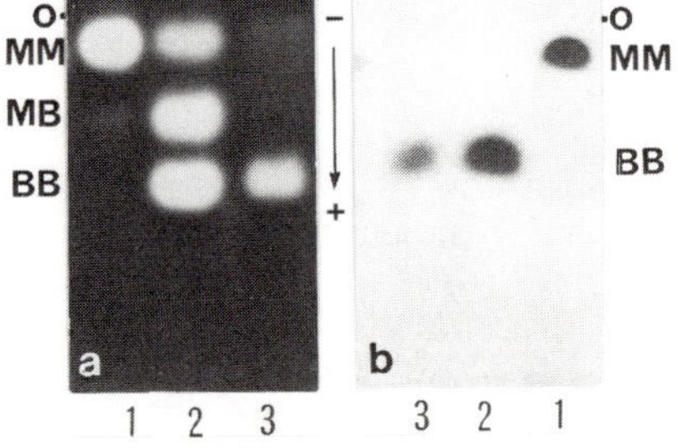

FIGURE 2
Cellulose acetate electrophoresis of creatine kinase (*a*) and PGAM (*b*). Normal adult muscle shows almost exclusively MM enzyme activity (*1*). Muscle cultures prior to fusion show only BB enzyme activity (*3*). After fusion (24 hr), creatine kinase isozyme transition is already advanced, but the same cultures still lack detectable MM-PGAM activity (*2*).

PFK

As is the case with PPL deficiency, we observed reappearing enzyme in muscle cultures from patients with PFK deficiency. Total enzyme activity was similar to that of the controls and of cultured myotubes from normal controls and patients stained histochemically with similar intensity for enzyme activity (Fig. 3). This activity in cultures was not significantly inhibited by antibodies to muscle-specific PFK. Electrophoretically, the only band with PFK activity in myotube cultures from controls and patients migrated somewhat faster than the band in normal mature muscle. There was no electron microscope evidence of increased glycogen in cultured myotubes from patients (Miranda et al. 1980). Even though only fetal PFK isozyme is observed electrophoretically in early myotube cultures, it may be possible to detect the early appearance of small amounts of M-PFK in these cultures using more sensitive techniques. This is not unlikely because even skin fibroblast cultures have recently been shown to contain small amounts of M-PFK (Kahn et al. 1981; Vora 1981). However, the absence of minute amounts of PFK-M in myotube cultures from patients should not result in a significant reduction of total PFK activity and should not affect normal glycogen metabolism. In fact, glycogen accumulation, characteristic of intact PFK-deficient muscle, does not occur in myotube cultures from patients.

PGAM, AMPD, and LD Deficiencies

These isozymes are also developmentally regulated and, as expected, reappearance of enzyme activity due to fetal isozymes was noted as well (DiMauro et al. 1980, 1981b; Miranda et al. 1980). We have not yet studied muscle cultures from patients with M-LD deficiency, a newly discovered genetic disorder of muscle (Kanno et al. 1980). Unlike PPL, PFK, PGAM,

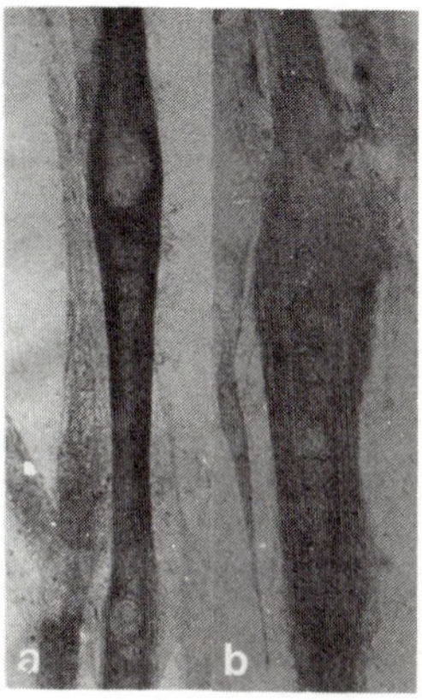

FIGURE 3
Histochemical staining for PFK is similar in cultured myotubes from normal control (*a*) and a PFK-deficient patient (*b*).

and AMPD deficiencies, M-LD deficiency should be detectable in muscle cultures because muscle-specific lactate dehydrogenase is normally present in large amounts even at early stages of muscle differentiation (Miranda et al. 1979a). A lack of M-LD should therefore result in a marked decrease of total lactate dehydrogenase activity and in the disappearance of electrophoretic bands containing M-LD subunits, as has been shown in mature muscle from patients (Kanno et al. 1980).

Acid Maltase, Debrancher, and Phosphoglycerate Kinase Deficiencies

Besides genetic metabolic myopathies affecting tissue-specific and developmentally regulated isozymes, there are several other muscle diseases affecting enzymes that appear to exist in a single molecular form, found in most tissues, including muscle, at all stages of development. In these disorders, the enzyme activity that is lacking in the mature muscle of patients is also lacking in muscle cultures from these individuals. This has been documented in acid maltase deficiency, a lysosomal disorder, in debrancher deficiency, and in a newly recognized variant of phosphoglycerate kinase (PGK) deficiency. For acid maltase and debrancher deficiencies, abnormal storage of glycogen, a characteristic feature in muscle biopsies, was also seen in muscle cultures (Fig. 4; Askanas et al. 1976; Miranda et al. 1980, 1981a). The morphology of PGK-deficient muscle cultures remains to be studied, but because the mutant enzyme retains some residual activity, glycogen accumulation in cell culture may not occur in this condition, as has been observed in the muscle biopsy (DiMauro et al. 1981a).

Carnitine Palmitoyltransferase Deficiency and Triglyceride Storage Myopathy

Two disorders of lipid utilization that we are studying are carnitine palmitoyltransferase (CPT) deficiency and triglyceride storage myopathy, and both of these disorders can be identified in culture. CPT activity was lacking in myotube cultures from one patient (Miranda et al. 1980). It is not known whether CPT exists in multimolecular forms, but these findings suggest that a single form of CPT is present in mature and immature muscle.

Triglyceride storage myopathy is characterized by accumulation of lipid in muscle, as well as in other tissues of patients. This disorder is expressed in fibroblast, myoblast, and myotube cultures, as demonstrated by a 20–40-fold increase of triglycerides (Miranda et al. 1979b). The deficient enzyme in this disorder has not yet been identified, but since abnormal triglyceride storage occurs in fibroblast as well as in muscle cultures at all stages of development, we can conclude that the deficient enzyme is neither tissue-specific nor developmentally regulated.

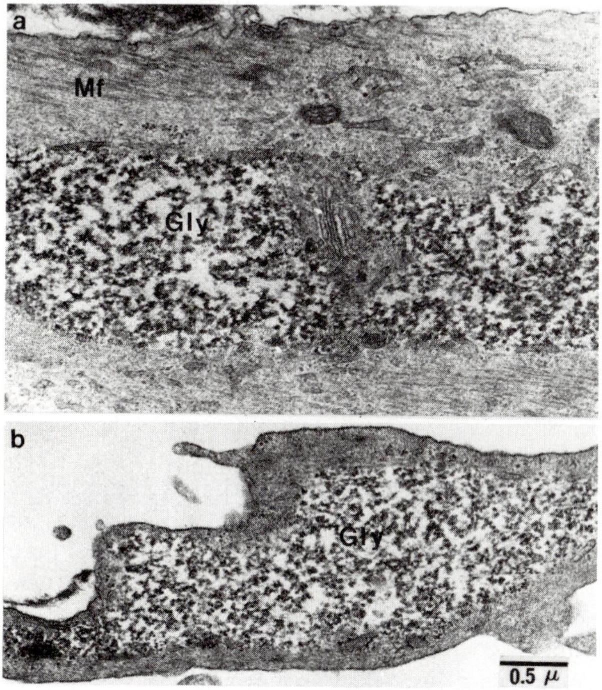

FIGURE 4
Transmission electron micrographs illustrating glycogen accumulation in myotube (*a*) and mononuclear cell (*b*) in muscle culture from a patient with debrancher enzyme deficiency. (Gly) Glycogen; (Mf) myofilaments.

In genetic disorders of unknown etiology that lack characteristic pathology in muscle culture, such as Duchenne muscular dystrophy, the genetic lesion may affect a developmentally regulated isozyme or it may be due to a "factor" that becomes increasingly important as the muscle matures. To study these diseases it will be necessary to obtain a greater degree of muscle differentiation in vitro.

Mucolipidosis Type II (I-cell Disease)

In all hereditary metabolic myopathies due to lack of a specific isozyme, the genetic defect appears to involve the isozyme characteristic of mature skeletal muscle. And yet, undoubtedly, there must be genetic errors affecting isozymes characteristic of embryonic muscle. These diseases may not be easily recognized, either because they are lethal in utero or because they are clinically benign and are no longer expressed in mature muscle. We may have identified one such disease in mucolipidosis type II, or I-cell disease, a

lysosomal storage disorder in which several acid hydrolases are decreased intracellularly and found in high concentrations in body fluids. Similarly, cultures of skin fibroblasts show abundant lysosomal inclusions and decreased activities of several acid hydrolases, whereas the same enzymes are present in excessive amounts in the medium of cultures. In one patient with I-cell disease, we studied both mass monolayers and clonal muscle cultures and found that the disease was expressed both morphologically and biochemically in myoblasts and in early myotubes and that myoblast fusion was much less efficient than in normal muscle cultures. However, whenever mature myotubes with distinct cross-striations did develop, the characteristic lysosomal inclusion bodies were either absent or much less in number (Shanske et al. 1981). In agreement with these data, electron microscope studies of a muscle biopsy from the same patient showed that satellite cells, the stem cells of regenerating muscle, were filled with lysosomal inclusions, but mature muscle fibers showed virtually no abnormality (S.A. Shafiq, unpubl.). Those lysosomal hydrolases that were decreased in both fibroblasts and immature muscle cultures were normal or even increased in the patient's muscle. The enzyme deficient in I-cell disease has recently been identified as the activity that regulates the normal packaging of acid hydrolases into lysosomes (Hasilik et al. 1981; Reitman et al. 1981). It will be interesting to determine whether this enzyme is developmentally regulated in muscle, in which case I-cell disease may be due to a genetic defect of fetal isozyme. Alternatively, the affected enzyme may have an important functional role in immature muscle but may not be of crucial importance in mature muscle.

CONCLUDING REMARKS

These studies illustrate several important points:

1. A comparison of the temporal sequence of isozyme transition of several developmentally regulated isozymes indicates that different enzymes "switch" at different times during development.
2. Primary myopathies are not necessarily expressed in differentiating muscle cultures. In the case of several hereditary metabolic myopathies, this is due to the fact that the mutant muscle-specific isozyme is subject to developmental control and is not expressed in detectable amount in immature muscle cultures. A different isozyme, under separate genetic control, is expressed at that stage of differentiation.
3. Genetic disorders of fetal, rather than mature, muscle isozymes may exist but remain to be documented.
4. For genetic metabolic muscle disorders of unknown etiology that are expressed in muscle cultures, the deficient enzyme is likely to be found in multiple tissues and may not change during muscle development.

5. For genetic muscle diseases of unknown etiology, such as Duchenne muscular dystrophy, the apparent lack of characteristic muscle pathology may indicate that the disorder is due to a developmentally regulated protein.

Future studies should include an analysis of muscle cultures at more advanced stages of differentiation, which may be obtained by manipulation of the medium or by innervation. In these culture systems one should be able to witness the precise stage of muscle development at which genetic myopathies become expressed and the early pathologic consequences of these disorders.

ACKNOWLEDGMENTS

This work was supported by center grant NS-11766-08 from the National Institute of Neurological and Communicative Disorders and Stroke and a center grant from the Muscular Dystrophy Association of America.

REFERENCES

Askanas, V., W.K. Engel, S. DiMauro, M. Mehler, and M.R. Brooks. 1976. Adult onset acid maltase deficiency: Morphologic and biochemical abnormalities reproduced in cultured muscle. *N. Engl. J. Med.* **294:** 573.

Boone, C.M., T.R. Chen, and F.H. Ruddle. 1972. Assignment of LDH-A locus in man to chromosome C-11 using somatic cell hybrids. *Proc. Natl. Acad. Sci.* **69:** 510.

DiMauro, S., M. Dalakas, and A.F. Miranda. 1981a. Phosphoglycerate kinase (PGK) deficiency: A new cause of myoglobinuria. *Ann. Neurol.* **10:** 90. (Abstr.).

DiMauro, S., S. Arnold, A.F. Miranda, and L.P. Rowland. 1978. McArdle disease: The mystery of reappearing phosphorylase activity in muscle culture: A fetal isozyme. *Ann. Neurol.* **3:** 60.

DiMauro, S., A.F. Miranda, S. Kahn, and K. Gitlin. 1981b. Human muscle phosphoglycerate mutase deficiency: Newly discovered metabolic myopathy. *Science* **212:** 1277.

DiMauro, S., A.F. Miranda, A.P. Hays, G.C. Hoffman, R.S. Schoenfeldt, and N. Singh. 1980. Myoadenylate deaminase deficiency: Muscle biopsy and muscle culture in a patient with gout. *J. Neurol. Sci.* **47:** 191.

Fidzianska, A. 1980. Human ontogenesis. I. Ultrastructural characteristics of developing human muscle. *J. Neuropathol. Exp. Neurol.* **39:** 476.

Foxall, D.D. and A.E.H. Emery. 1975. Changes in creatine kinase and its isoenzymes in human fetal muscle during development. *J. Neurol. Sci.* **24:** 483.

Goto, I., M. Nagamorie, and S. Katsuki. 1969. Creatine phosphokinase isozymes in muscles. Human fetus and patients. *Arch. Neurol.* **20:** 422.

Hasilik, A., A. Waheed, and K. von Figura. 1981. Enzymatic phosphorylation of lysosomal enzymes in the presence of UDP-*N*-acetylglucosamine. Absence of the activity in I-cell fibroblasts. *Biochem. Biophys. Res. Commun.* **98:** 761.

Kahn, A., D. Weil, D. Cottreau, and J.C. Dreyfus. 1981. Muscle phosphofructokinase deficiency in man: Expression of the defect in blood cells and cultured fibroblasts. *Annu. Hum. Genet.* **45:** 5.

Kanno, T., K. Sudo, I. Takeuchi, S. Kanda, N. Honda, Y. Nishimura, and K. Oyama. 1980. Hereditary deficiency of lactate dehydrogenase M-subunit. *Clin. Chim. Acta* **108:** 267.

Meienhofer, M.C., V. Askanas, D. Proux-Daegelen, J.C. Dreyfus, and W.K. Engel. 1977. Muscle-type phosphorylase activity present in muscle cells cultured from three patients with myophosphorylase deficiency. *Arch. Neurol.* **34:** 779.

Miranda, A.F., H. Somer, and S. DiMauro. 1979a. Isoenzymes as markers of differentiation. In *Muscle regeneration* (ed. A. Mauro), p. 453. Raven Press, New York.

Miranda, A.F., S. DiMauro, A. Antler, L.Z. Stern, and L.P. Rowland. 1981a. Glycogen debrancher deficiency is reproduced in muscle culture. *Ann. Neurol.* **9:** 283.

Miranda, A.F., P.B. Fisher, W.G. Johnson, S. Shanske, and S. Khan. 1981b. Characterization of SV40-transformed human muscle cells. *Neurology* **31:** 58. (Abstr.).

Miranda, A.F., C. Trevisan, S. Shanske, A. Antler, and S. DiMauro. 1980. The expression of genetic enzyme defects in muscle cultures. Now you see it, now you don't. *Neurology* **30:** 367. (Abstr.).

Miranda, A.F., S. DiMauro, A.B. Eastwood, A.P. Hays, W.G. Johnson, W.G. Olarte, R. Whitlock, R. Mayeux, and L.P. Rowland. 1979b. Lipid storage myopathy, ichthyosis and steatorrhea. *Muscle and Nerve* **2:** 1.

Reitman, M.L., A. Varki, and S. Kornfeld. 1981. Fibroblasts from patients with I-cell disease and pseudo-Hurler polydystrophy are deficient in uridine 5′-diphosphate *N*-acetylglucosamine: Glycoprotein *N*-acetylglucosaminyl-phosphotransferase activity. *J. Clin. Invest.* **67:** 1574.

Roelofs, R.I., W.K. Engel, and P.B. Chauvin. 1972. Histochemical phosphorylase activity in regnerating muscle fibers from myophosphorylase-deficient patients. *Science* **177:** 795.

Samaha, F.J., S.T. Iannaccone, and B. Nagy. 1981. Maturation of human skeletal muscle cells in culture. *Neurology* **31:** 47. (Abstr.).

Sato, K., F. Imai, F. Hatayama, and R.I. Roelofs. 1977. Characterization of glycogen phosphorylase isoenzymes present in cultured skeletal muscle from patients with McArdle disease. *Biochem. Biophys. Acta* **78:** 663.

Sato, K., T. Sato, H.P. Morris, and S. Weinhouse. 1975. Carcinofetal alterations in glycogen phosphorylase isoenzymes in rat hepatomas. *Ann. N.Y. Acad. Sci.* **259:** 273.

Schapira, F. 1978. Isoenzymes and differentiation. *Biomedicine* **28:** 1.

Schapira, F., A. Hatzfeld, and A. Weber. 1975. Resurgence of some fetal isozymes in hepatoma. In *Isozymes* (ed. C.L. Markert), vol. 3, p. 987. Academic Press, New York.

Shanske, S., A.F. Miranda, A.S. Penn, and S. DiMauro. 1981. Mucolipidosis II (I-cell disease): Studies of muscle biopsy and muscle cultures. *Pediatr. Res.* **15:** 1334.

Tzvetanova, E. 1971. Creatine kinase isoenzymes in muscle tissue of patients with neuromuscular diseases and human fetuses. *Enzyme* **12:** 279.

Vora, S. 1981. Isozymes of human phosphofructokinase in blood cells and cultured cell lines: Molecular evidence of a trigenic system. *Blood* **57:** 724.

Vora, S., S. Durham, B. de Martinville, D.L. George, and U. Francke. 1981. Assignment of the human gene for muscle-type phosphofructokinase (PFKM) to chromosome 1 (Region cen—q32) using somatic cell hybrids and monoclonal anti-M antibody. *Somat. Cell Genet.* **8:** 95.

Experimental Autoimmune Myasthenia Gravis: Specificity of Antibodies to Acetylcholine Receptor

MIRY C. SOUROUJON
DARIA MOCHLY-ROSEN
DANIEL BARTFELD
SARA FUCHS
Department of Chemical Immunology
The Weizmann Institute of Science
Rehovot 76100, Israel

In our laboratory we are studying several immunological aspects of the nicotinic acetylcholine receptor (AChR), with an emphasis on the involvement of AChR in myasthenia gravis (Fuchs 1979a). Here, we report mainly on our studies on monoclonal antibodies (McAbs) to the *Torpedo* AChR as probes for muscle AChR.

Since AChR is a high-molecular-weight antigen that expresses many antigenic determinants, we wanted to have monospecific antibodies to analyze individual antigenic determinants in the receptor molecule and to define antibodies that are specifically responsible for blocking α-bungarotoxin (α-Bgt) binding, accelerating receptor degradation, and inducing myasthenia, all of which have been detected previously with polyclonal antibodies.

Our initial approach to the structural and functional analysis of the AChR was the elicitation of polyclonal antibodies by immunization with AChR and its derivatives. Antibodies were fractionated by appropriate affinity chromatography. Such studies also led to the development of derivatives of AChR with a potential to suppress the immunologic response of experimental autoimmune myasthenia gravis (EAMG) (Bartfeld and Fuchs 1978). By using polyclonal antibodies, we have classified two major

antibody populations in anti-AChR antibodies: (1) antibodies to nonconformational (primary structure) determinants and (2) antibodies to conformational (three-dimensional) antigenic determinants in the AChR molecule. Since we have also applied this classification for screening our anti-AChR McAbs, we briefly summarize our studies with polyclonal antibodies and the properties of two AChR derivatives that led to the fractionation and characterization of these two antibody subpopulations.

RESULTS AND DISCUSSION

AChR Modification and Antibody Fractionation

The first derivative that we have studied in detail is a denatured AChR preparation that was prepared by reduction and carboxymethylation of AChR in 6 M guanidine hydrochloride and designated reduced and carboxymethylated AChR (RCM-AChR; Bartfeld and Fuchs 1977). RCM-AChR does not bind cholinergic ligands and does not induce EAMG. Nevertheless, it is a good immunogen and elicits antibodies that cross-react with the intact receptor. The immune response to RCM-AChR is directed only against nonstructural antigenic determinants that are shared by both the native and denatured receptors and are not involved in the myasthenic activity of the receptor (Bartfeld and Fuchs 1978; Fuchs 1979b). Anti-AChR antibodies exhibit only a partial cross-reactivity with RCM-AChR. We have prepared trypsin-treated AChR by a different strategy and have demonstrated that a controlled tryptic digestion of AChR did not change either the pharmacological specificity or the pathological myasthenic activity of the receptor. The product obtained after tryptic digestion of AChR and repurification by affinity chromatography was designated T-AChR and was shown to be a simpler and smaller molecule of AChR, representing mainly the regions in the molecule that are resistant to proteolytic digestion (Bartfeld and Fuchs 1979a).

Any antiserum of animals immunized with AChR contains a certain portion of antibodies reacting with either RCM-AChR or T-AChR. We have fractionated anti-AChR sera by an immunoadsorbtion on RCM-AChR-Sepharose (Bartfeld and Fuchs 1979b) and have isolated antibodies that bind T-AChR, but not RCM-AChR, and block the binding of α-Bgt to AChR in vitro. The remaining antibody fraction binds to RCM-AChR and only to a very limited extent to T-AChR and does not block α-Bgt binding to AChR (Bartfeld and Fuchs 1979a,b). We have demonstrated that this latter antibody fraction (anti-dAChR) is not involved in the myasthenic activity of anti-AChR antibodies. Thus, by appropriate fractionation of polyclonal anti-AChR antibodies, it is possible to obtain antibody subpopulations of defined specificities which, for the sake of convenience, we define as having conformational and nonstructural antigenic specificities, recognizing primary structure and three-dimensional structure, respectively.

McAbs

McAbs with anti-AChR activity were prepared by fusing spleen cells from mice immunized with *Torpedo* AChR with NS-1 nonsecreting myeloma cells. Hybridomas with anti-AChR antibody activity were cloned and propagated as ascitic fluids (Mochly-Rosen et al. 1979). The various clones were characterized by measuring their binding to various preparations and derivatives of *Torpedo* AChR and, later, to muscle AChR preparations. Whereas anti-AChR sera bind to all various antigens derived from AChR, as mentioned above, the individual McAbs possess more restricted specificities. Some McAbs are directed selectively against conformational antigenic determinants of AChR and bind T-AChR but not RCM-AChR. Other McAbs are directed against nonstructural antigenic determinants represented by RCM-AChR and do not bind T-AChR (Mochly-Rosen et al. 1979).

We then checked the reactivity of the McAbs with the individual subunits of *Torpedo* AChR. The procedure for isolating the individual subunits involves SDS denaturation; therefore, they were not expected to bind, and indeed they did not bind to any of the McAbs to conformational antigenic determinants. Of the McAbs to nonstructural determinants, only two (McAb 1.17 and McAb 1.22) bound exclusively to a single subunit. Other McAbs bound to more than one subunit, although in all cases the binding to the α-subunit was higher (Fig. 1). It seems that the α-subunit bears the most immunopotent determinants. The binding of a McAb to more than one subunit probably reflects the sequence homology between the subunits, as has been reported by Raftery et al. (1980).

Following the characterization of the McAbs with *Torpedo* AChR and its derivatives, we asked whether there were McAbs to the cholinergic binding site of the *Torpedo* AChR or to muscle AChR. Both those specificities may be involved in blocking AChR at the neuromuscular junction in myasthenia.

Antibodies to the Cholinergic Binding Site

When the binding of the McAbs to *Torpedo* AChR was compared with their binding to the α-Bgt-AChR complex, only one McAb, designated McAb 5.5, did not bind to the receptor when complexed with α-Bgt (Fig. 2). This particular McAb belongs to the group of antibodies specific to conformational sites; it binds T-AChR and does not bind RCM-AChR. We have demonstrated that not only α-Bgt but also other cholinergic agonists and antagonists inhibited the binding of AChR to McAb 5.5 in a pharmacological manner (Mochly-Rosen and Fuchs 1981). We have thus concluded that McAb 5.5 is directed against the cholinergic binding site of AChR. McAb 5.5 also blocked the carbamylcholine-induced sodium influx into chick primary muscle cell cultures (Table 1) (D. Mochly-Rosen et al., unpubl.) and reduced the acetylcholine sensitivity of these cells (Goldberg et al. 1981). None of the other anti-AChR McAbs tested had such effects on muscle cell functions.

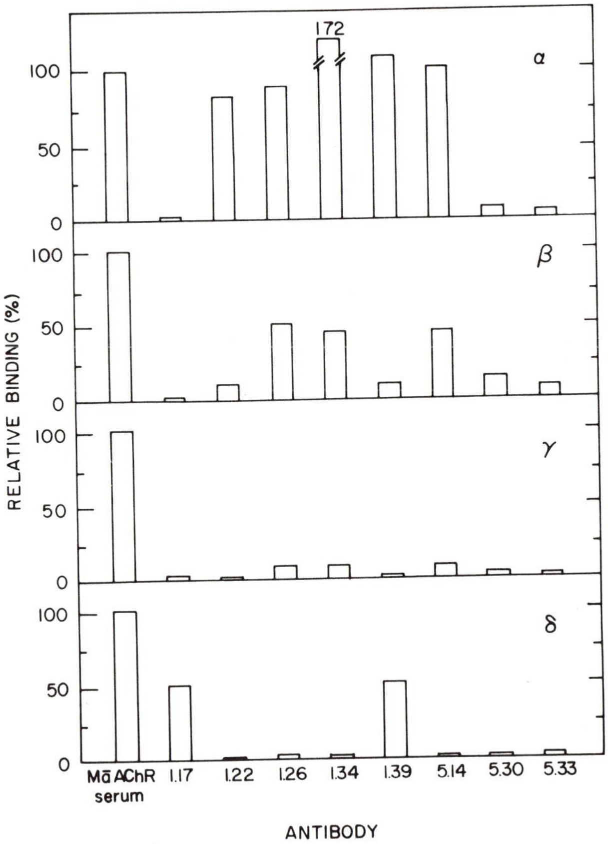

FIGURE 1
Subunit specificity of mouse anti-AChR serum (MāAChR serum) and of McAbs, determined by a radioimmunoassay of the various ascitic fluids with ^{125}I-labeled isolated subunits.

Antibodies to Muscle AChR

The cross-reactivity between *Torpedo* and muscle AChR is a prerequisite for the ability to induce EAMG with *Torpedo* AChR. The cross-reactivity of the McAbs with muscle AChR was first assessed by radioimmunoassay with Triton extracts from muscle homogenates, in which the receptor was radioactively labeled with ^{125}I-labeled α-Bgt. No significant differences between the reactivity of the McAbs with Triton-extracted AChR from innervated and denervated mouse leg muscle were observed (Fig. 2). These two receptor preparations represent junctional and extrajunctional AChR, respectively. Also, there seems to be no difference between the cross-reactivity of the McAbs with AChR extracted from primary muscle cultures of various species (Fig. 2).

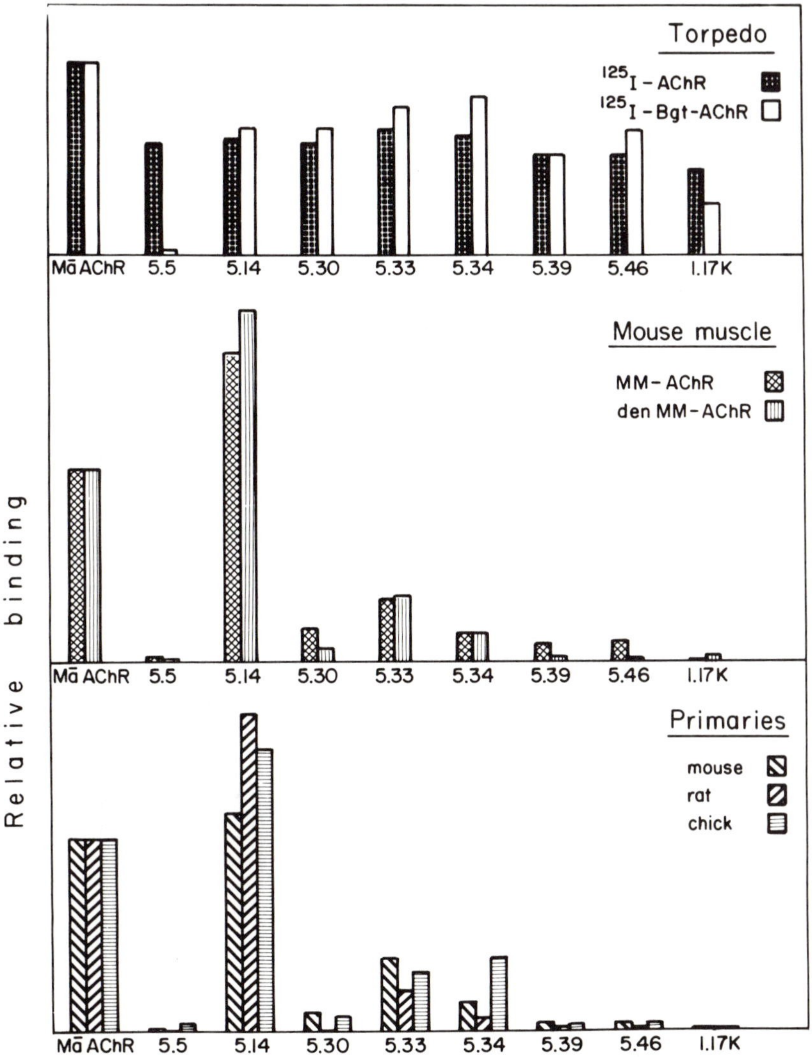

FIGURE 2
Relative binding of *Torpedo* AChR, AChR preparations from denervated and nondenervated mouse muscle, and from mouse, rat, and chicken primary muscle cultures to various anti-AChR McAbs.

Of the various antibodies analyzed with Triton-extracted AChR from muscles, McAb 5.14 reacted the best (Souroujon et al. 1981). This, however, was not the case when we tested the binding of the McAbs to membrane-bound AChR, as measured by their direct binding to muscle cell cultures and to crude membrane preparations of such cells, or by their ability to inhibit the binding of α-Bgt to muscle cell cultures. None of these assays required prelabeling with α-Bgt and therefore were applicable also to

TABLE 1
Effect of McAbs on $^{22}Na^+$ Influx to Chick Primary Muscle Cell Cultures

Antibody (dilution)	Agonist	Influx of $^{22}Na^+$ (cpm)	Inhibition (%)
—	—	4,000	
—	carbamylcholine	17,300	
—	McAb 5.5 (1/10)	3,500	
McAb 5.5 (1/10)	carbamylcholine	6,700	80
McAb 5.5 (1/100)	carbamylcholine	9,500	59
McAb 5.5 (1/1000)	carbamylcholine	15,600	13
McAb 5.14 (1/10)	carbamylcholine	14,700	20
McAb 5.34 (1/10)	carbamylcholine	14,300	23
McAb 5.46 (1/10)	carbamylcholine	15,100	17
McAb 1.17 (1/10)	carbamylcholine	14,000	25
McAb 1.39 (1/10)	carbamylcholine	15,300	15
Mouse anti-AChR serum (1/10)	carbamylcholine	16,000	10
Normal mouse serum (1/10)	carbamylcholine	15,600	13

The cell cultures were preincubated for 2 hr at 37° C with the tested antibody. The antibody solution was then removed and sodium influx was measured, as described by Catterall (1975), following an incubation for 30 sec at room temperature with 2 ml of transport medium containing 10^{-3} M carbamylcholine and 5 μCi $^{22}Na^+$.

anti-binding site antibodies. McAb 5.5 exhibited the highest binding to membranous AChR, followed by McAb 5.34 and then McAb 5.14 (M.C. Souroujon et al., in prep.). McAb 5.14, which bound very efficiently to Triton-extracted muscle AChR, did not inhibit α-Bgt binding to muscle cell cultures. Thus, there seem to be antigenic differences between Triton-solubilized and membranous muscle AChR.

One of the mechanisms proposed for the role of antibodies in myasthenia is the loss of AChR molecules from the muscle cell surface via accelerated degradation of the receptor. By using McAbs it is possible to find out which antibody specificity is responsible for this effect. We have thus tested the ability of our McAbs to accelerate the degradation of AChR in primary chick muscle cultures. Of the McAbs tested, only McAb 5.34 and McAb 5.5 accelerated the degradation significantly (Table 2). Since McAb 5.5 binds in AChR to the same site as α-Bgt does, the effect of McAb 5.5 on the degradation rate could be observed only when subsaturating amounts of ^{125}I-labeled α-Bgt were used to label the cultures before the application of the antibodies (Table 2) (M.C. Souroujon et al., in prep.).

The ability to bind to muscle cultures or specifically to the binding site of AChR is not necessarily associated with the ability to affect the AChR degradation rate, since McAb 5.14, which binds to muscle cultures to the

TABLE 2
Antibody-mediated AChR Degradation: Correlation with α-Bgt Concentration

Antibody (dilution)	α-Bgt concentration[a]	
	5×10^{-10} M	5×10^{-6} M
—	20.0	20.0
McAb 5.34 (1:25)	13.0	11.5
McAb 5.34 (1:100)	13.5	13.5
McAb 5.5 (1:25)	15	20.5
McAb 5.5 (1:100)	14	20.0

Degradation of AChR was measured as described by Prives et al. (1979). ^{125}I-labeled α-Bgt was added to primary muscle cultures (5-day-old cultures) in 2 ml of medium to a final concentration of 5×10^{-10} M or 5×10^{-6} M for 1 hr at 37°C. The cells were thoroughly washed with medium, and the antibodies were added in 2 ml of medium at different concentrations. Every 3 hr the medium was removed for counting and replaced by new medium. After 24 hr, the cells were solubilized, and residual radioactivity in the tissue culture plates was determined. For each plate the total originally bound radioactivity was calculated, and then the percentage of total radioactivity released to the medium as a function of time was depicted on a semilog scale. The time at which 50% of the originally bound radioactivity was released to the medium represents the half-life of the AChR molecules in the tissue culture.

[a]Half-life of AChR (hr).

same extent as McAb 5.34 does, did not have a significant effect on the rate of receptor degradation.

CONCLUDING REMARKS

We have not yet analyzed in detail the role of the individual McAbs in EAMG. Some neuromuscular dysfunction was observed using sensitive neurophysiological tests in mice, following a passive transfer of McAb 5.5 or McAb 5.34. However, the effects caused by these McAbs are much weaker than those observed by transfer of polyclonal antibodies or by AChR immunization (Pachner et al. 1981). Lennon and Lambert (1980) and Richman et al. (1980) have reported on the transfer of EAMG by McAbs, derived from AChR-immunized rats, which bind to muscle AChR but are not against the receptor binding site. These investigators did not test their McAbs for accelerating AChR turnover. It seems that more than one antibody specificity can damage the neuromuscular junction, thus leading to myasthenic symptoms. We are now eliciting antiidiotypes to various McAbs to use them for regulating anti-AChR response and, possibly, for suppressing the immunologic response of myasthenia. Antiidiotypes to polyclonal rabbit anti-AChR antibodies had a suppressive effect on EAMG in rabbits (Feingold and Fuchs 1980).

In addition to the importance of anti-AChR McAbs for elucidating the role of antibodies in myasthenia, such antibodies should provide a useful tool in analyzing the structure of AChR in various tissues and species.

REFERENCES

Bartfeld, D. and S. Fuchs. 1977. Immunological characterization of an irreversibly denatured acetylcholine receptor. *FEBS Lett.* **77:** 214.

———. 1978. Specific immunosuppression of experimental autoimmune myasthenia gravis by denatured acetylcholine receptor. *Proc. Natl. Acad. Sci.* **75:** 4006.

———. 1979a. Active acetylcholine receptor fragment obtained by tryptic digestion of acetylcholine receptor from *Torpedo californica. Biochem. Biophys. Res. Commun.* **89:** 512.

———. 1979b. Fractionation of antibodies to acetylcholine receptor according to antigenic specificity. *FEBS Lett.* **105:** 303.

Catterall, W.A. 1975. Sodium transport by the acetylcholine receptor of cultured muscle cells. *J. Biol. Chem.* **250:** 1776.

Feingold, C. and S. Fuchs. 1980. Regulation of experimental autoimmune myasthenia gravis in rabbits by anti-idiotypes. *Isr. J. Med. Sci.* **16:** 805.

Fuchs, S. 1979a. Immunology of the nicotinic acetylcholine receptor. *Curr. Top. Microbiol. Immunol.* **85:** 1.

———. 1979b. Immunosuppression of experimental myasthenia. In *Plasmapheresis and the immunobiology of Myasthenia gravis* (ed. P.C. Dau), p. 20. Houghton Mifflin, New York.

Goldberg, G., D. Mochly-Rosen, S. Fuchs, and Y. Lass. 1981. Blocking of acetylcholine induced channels by a monoclonal antibody to the cholinergic binding site in cultured chick myoblasts. *Soc. Neurosci. Abstr.* **7:** 702.

Lennon, V.A. and E.H. Lambert. 1980. Myasthenia gravis induced by monoclonal antibodies to acetylcholine receptor. *Nature* **285:** 238.

Mochly-Rosen, D. and S. Fuchs. 1981. Monoclonal anti-acetylcholine receptor antibodies directed against the cholinergic binding site. *Biochemistry* **20:** 5920.

Mochly-Rosen, D., S. Fuchs, and Z. Eshhar. 1979. Monoclonal antibodies against defined determinants of acetylcholine receptor. *FEBS Lett.* **106:** 389.

Pachner, A.R., F.S. Kantor, and S. Fuchs. 1981. The role of poly- and monoclonal antibodies in murine myasthenia. *Ann. Neurol.* **10:** 83.

Prives, J., L.H. Hoffman, R. Tarrab-Hazdai, S. Fuchs, and A. Amsterdam. 1979. Ligand induced changes in stability and distribution of acetylcholine receptors in surface membranes of muscle cells. *Life Sci.* **24:** 1713.

Raftery, M.A., M.W. Hunkapiller, C.D. Strader, and L.E. Hood. 1980. Acetylcholine receptor: Complex of homologous subunits. *Science* **208:** 1454.

Richman, D.B., C.M. Gomez, P.W. Berman, S.A. Burres, F.W. Fitch, and B.G.W. Arnason. 1980. Monoclonal anti-acetylcholine receptor antibodies can cause experimental myasthenia. *Nature* **286:** 738.

Souroujon, M.C., D. Mochly-Rosen, and S. Fuchs. 1981. Studies of muscle acetylcholine receptor by monoclonal antibodies. *Abstr. Proceedings of the Annual Meeting of the Israel Biochem. Soc.*, p. 5.

Human Muscle Antigens during Development

FRANK S. WALSH
CHRISTOPHER A. QUINN
ROSE YASIN
EDWARD J. THOMPSON
OREST HURKO
Muscular Dystrophy Research Laboratories
Institute of Neurology
London WC1N 3BG, England

Duchenne muscular dystrophy is a clinically distinctive human disorder with a highly specific pattern of skeletal-muscle necrosis, fibrosis, and resultant weakness (Gardner-Medwin 1980). Although it has long been known that this disease is caused by a mutation in the X chromosome, the precise location of this mutant gene is not known, in part, because of a paucity of suitably linked genetic markers (Emery 1980; Jacobs et al. 1981). Furthermore, it is not known whether this is a structural or regulatory gene, or what gene products result from the expression of the "Duchenne gene" or its normal wild-type allele. Finally, it is not known which cell types express this gene, whether the skeletal-muscle degeneration in Duchenne dystrophy is a direct consequence of defective gene expression in muscle cells, or whether it is a secondary effect resulting from the interaction of intrinsically normal muscle with nerve cells, fibroblasts, or distant tissues that are defective themselves, by virtue of expression of the Duchenne gene.

To approach such questions, it is useful to separate these various cell types so that they may be observed as pure populations or in simple, experimentally defined combinations. We profess, as an initial working hypothesis, that the Duchenne gene is expressed in myogenic cells. Therefore, we seek to identify those gene products of the human X chromosome that are expressed in pure cultures of human muscle cells or of rodent-human hybrid muscle cells that contain only a limited complement of human genetic material. To perform such an analysis, it is first necessary

to be able to routinely establish tissue cultures of muscle from human fetal and biopsy sources. Second, it is necessary to be able to identify the major cell types in such muscle cultures and to do so rapidly and reliably, as a guide to their isolation as pure populations. Third, techniques must be developed for the isolation and transfer of relevant portions of the human genome that continue to express identifiable human muscle-specific functions as well as constitutive "housekeeping" products. In this paper we briefly summarize work on the characterization of antigens in human muscle cultures recently performed in our laboratory.

RESULTS AND DISCUSSION

Characterization of Single Cell Clones from Adult Skeletal Muscle

In 1977, Yasin et al. reported a quantitative and reproducible method for establishing dispersed cell cultures from dissociated human biopsies, with the aim of studying the factors important to human myogenesis and the identification of cellular and molecular abnormalities associated with Duchenne muscular dystrophy. However, a major problem with such primary muscle cultures is that they contain a mixture of heterogenous cell types, thus making it difficult to relate biochemical and physiological properties specific to the different cell types, such as fibroblasts, myoblasts, and myotubes, in both the normal and the diseased state. To circumvent this problem we have developed a single cell-cloning technique to isolate pure cell populations in large numbers from adult muscle. Our procedure is based on plating cells derived from dissociated muscle biopsies on glass shards. A shard with a single cell is then isolated and grown up to clonal cell populations of 10^6–10^7 cells (Yasin et al. 1980). These cells maintain their phenotypes for up to 45–50 days in culture even after freezing and thawing.

Four clones prepared from the same dystrophic muscle sample have been analyzed in detail. Clones D2 and D4 are myogenic but differ markedly in their myogenic capacity, in that 70% of D2 myoblasts fused to form large branched multinucleated myotubes, whereas only 15% of D4 myoblasts fused into short myotubes (Fig. 1). Clones D1 and D3 exhibited no musclelike characteristics (Fig. 1). Myogenic clones D2 and D4 expressed the muscle-specific surface antigenic marker 5.1H11 (see below) at all stages of growth and synthesized fibronectin; at the myotube stage, both synthesized muscle-specific creatine kinase (MM-CK), but only clone D2 synthesized nicotinic acetylcholine receptors.

Identification of Cell-type-specific Surface Antigens in Human Skeletal Muscle

Immunological techniques have recently been applied to the study of muscle cell-surface components in a number of species. First, Friedlander and

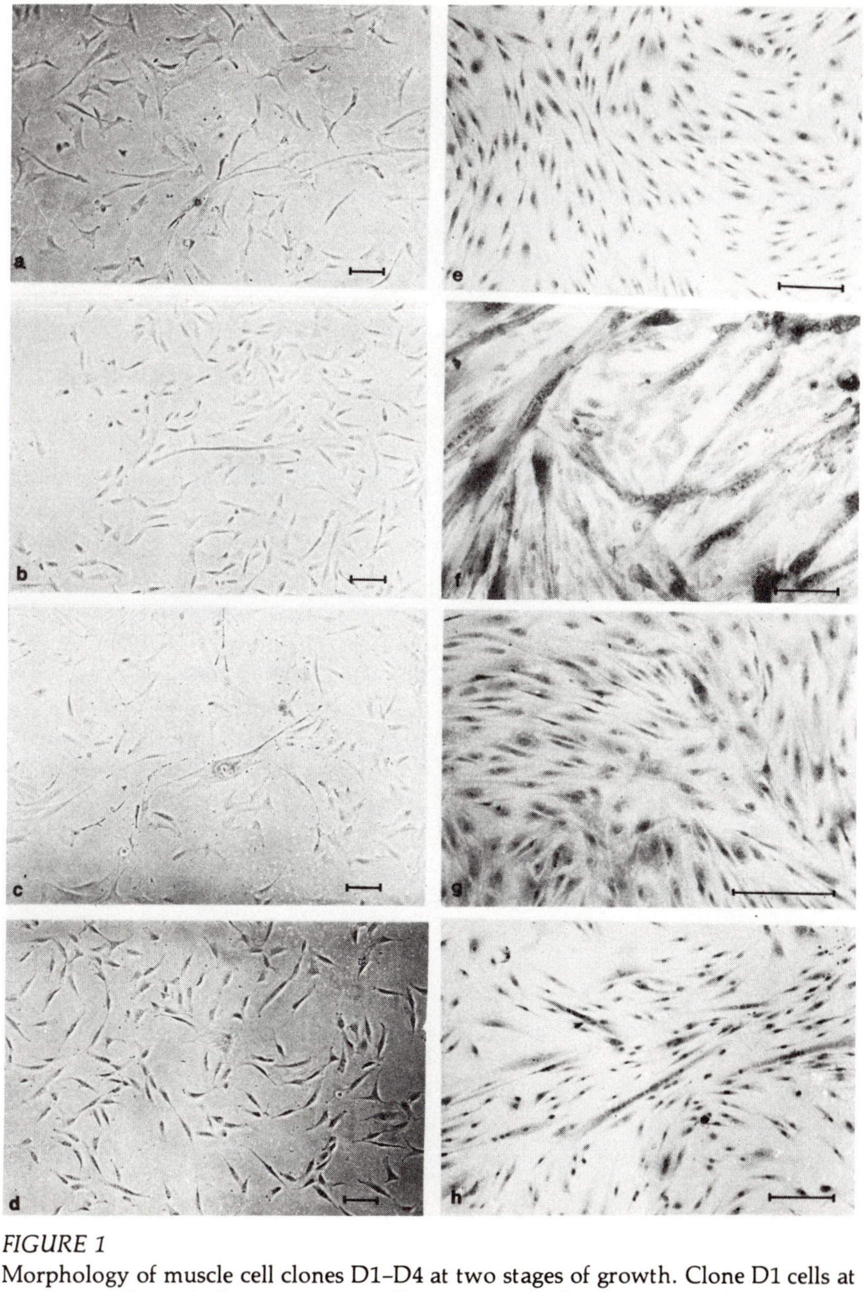

FIGURE 1

Morphology of muscle cell clones D1–D4 at two stages of growth. Clone D1 cells at low density (*a*) and after transfer to differentiation medium for 5 days (*e*). Clone D2 cells at low density (*b*) and 5 days after transfer to differentiation medium (*f*). Clone D3 cells at low density (*c*) and 4 days after transfer to differentiation medium (*g*). Cells from clone D4 at low density (*d*) and 6 days after transfer to differentiation medium (*h*). Bar represents 100 μm.

Fischman (1979) prepared rabbit antibodies that reacted with specific protein antigens on chick myoblasts. More recently, monoclonal antibody methodologies have been applied to rat L6 (Lee and Kaufman 1981) and chick muscle cells (Grove et al. 1981), resulting in the identification of a number of myoblast, myotube, and extracellular matrix specificities. We have attempted to define cell-type-specific antigens in clonal isolates of human muscle cultures and have found that the use of two antibody reagents is sufficient to positively identify all of the major muscle cell types. The two antibody reagents are 5.1H11 monoclonal antibody and antibody to Thy-1 antigen (Walsh and Ritter 1981).

5.1H11 monoclonal antibody was produced from a cell fusion involving spleen cells from a mouse immunized with human myotube cells and P3X63 Ag8 myeloma cells. 5.1H11 antibody reacts with a trypsin-sensitive, neuraminidase-insensitive, cell-surface antigen that is readily soluble in detergents such as sodium deoxycholate but not in Triton X-100. Indirect immunofluorescent staining of a variety of human cell types has shown that 5.1H11 antigen is restricted in expression to skeletal muscle cells, and 5.1H11 antibody reacts only with myoblasts and myotubes but not with fibroblasts in human muscle cultures. The antigen responsible appears to be developmentally regulated in vivo, in that it can be readily identified by indirect immunofluorescent staining in frozen sections of 15- and 18-week-old human fetal muscle, appears only faintly at 21 weeks, and is absent at 25 weeks and normal adult muscle sections. 5.1H11 antigen is also detected on the "immature" or regenerating small muscle fibers observed in biopsies of polymyositis and Duchenne muscular dystrophy but is absent from mature muscle fibers on these same sections.

Thy-1 antigen exhibits a different pattern of reactivity to 5.1H11 antigen in human muscle cultures. Here, all cultured skin or muscle fibroblast cells are positive for Thy-1 antigen but negative for 5.1H11 antigen. Thy-1 antigen is present on the surface of myoblasts, but after fusion, the antigen is not detectable on the surfaces of myotubes as was observed in rats by Lesley and Lennon (1977). Exactly when this antigen is lost during myotube formation and maturation is not known at present.

Because of the differing patterns of reactivity of 5.1H11 and Thy-1 antibodies with muscle cells, we have used both antibodies and double indirect immunofluorescence to assign a specific antigenic phenotype to the major cell types in culture. Cultures of fibroblasts were found to be Thy-1^+/5.1H11^-, in contrast to myoblasts that were double labeled (Thy-1^+/5.1H11^+) (Fig. 2). In postfusion cultures the myotubes were found to be Thy-1^-/5.1H11^+, whereas the residual cells were Thy-1^+/5.1H11^- (Fig. 2). The small number of nonproliferating myoblasts present in these cultures was found to have a similar staining profile to myotubes, i.e., Thy-1^-/5.1H11^+. Further confirmation of these antigenic phenotypes was derived by double immunofluorescent staining with antibody to human MM-CK and Thy-1 or 5.1H11 antibodies. MM-CK immunoreactivity was found in myotubes and in selected myoblasts only (data

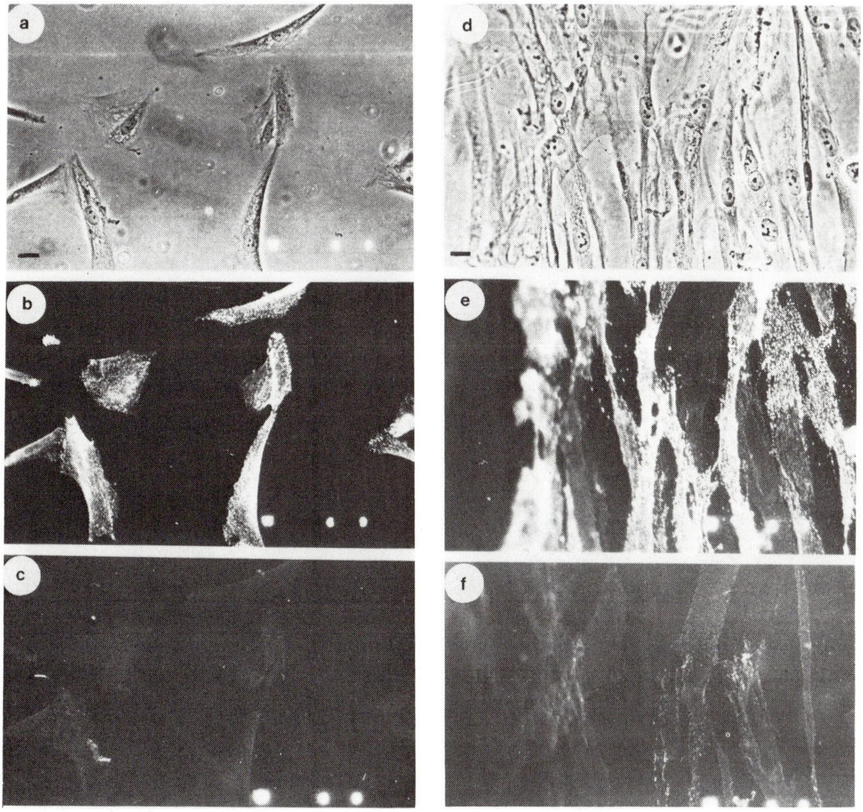

FIGURE 2
Double immunofluorescent staining of fetal human muscle cultures with rabbit anti-rat Thy-1 antibody and mouse 5.1H11 antibody. (*a*) Phase-contrast micrograph of myoblasts. (*b,c*) The same fields stained with Thy-1 and 5.1H11 antibodies, respectively. (*d*) Phase-contrast micrograph of a myotube culture. (*e,f*) The same fields stained with Thy-1 and 5.1H11 antibodies, respectively. Bar represents 20 μm.

not shown). These cells were always Thy-1^- and 5.1H11^+ and were considered to be nonproliferating myoblasts since they did not incorporate [^{3}H]thymidine. Thus, a combination of the Thy-1 and 5.1H11 antibody reagents has permitted the assignment of a specific antigenic phenotype to the major skeletal-muscle cell types. Whether these phenotypes are altered in any way in diseased muscle cultures is currently under investigation.

Expression of Human Muscle Gene Products in Rat-Human Muscle Cell Hybrids

The techniques of somatic cell hybridization have only recently been applied to the analysis of muscle cell development. Cell hybrids between rat

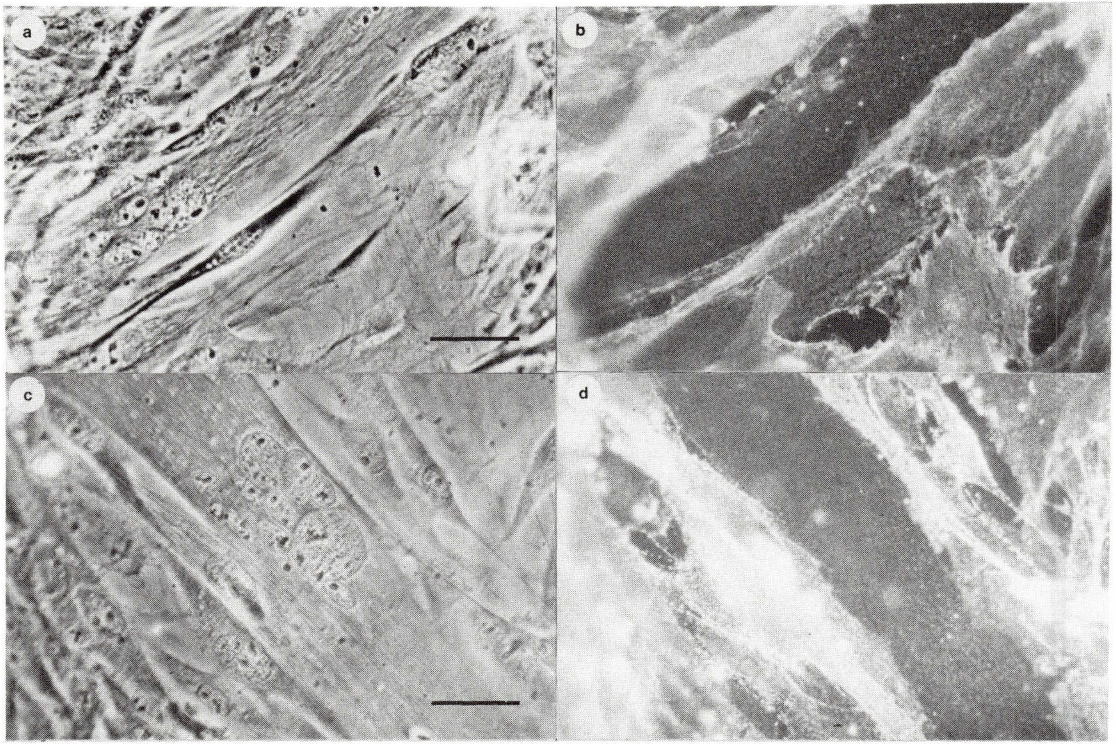

FIGURE 3
Indirect immunofluorescent staining of postfusion (myotube) 37-11 hybrid muscle cells with anti-human Thy-1-specific monoclonal antibody F15-42-1 (*a*,*b*) and with anti-rat Thy-1-specific monoclonal antibody OX7 (*c*,*d*). Bar represents 20 μm.

L6 cells and fibroblasts do not express muscle-specific gene products (Buckingham et al. 1974; Dufresne et al. 1976; Brzeski et al. 1980). Expression of muscle-specific functions can be achieved when both parental lines are myogenic (Quinn et al. 1981b; Wright and Gros 1981). We have observed that permanent cell hybrids between L6 cells and human skeletal-muscle cells can be prepared at high frequency, and some clones differentiate to form myotubes under appropriate conditions (Quinn et al. 1981b). Isozyme analysis demonstrated that these were true cell hybrids. Myogenic clone 37-11 expressed the X-chromosome-encoded enzyme glucose-6-phosphate dehydrogenase as an interspecific heterodimer, as well as native rat and human dimers. Indirect immunofluorescent staining of clone 37-11 cultures demonstrated 5.1H11 antigen on the surface of myoblasts and myotubes but Thy-1 on myoblasts only. Species-specific monoclonal anti-Thy-1 reagents demonstrated that both the rat and human forms of the antigen were coordinately regulated (Quinn et al. 1981a) (Fig. 3). Creatine kinase isozymes were also developmentally regulated in the 37-11 hybrid cells, with mononucleate cells synthesizing human immature brain creatine kinase (BB-CK) and myotubes initiating synthesis of the human M subunit. Such synthesis of the M subunit is a novel example of the activation of human muscle-specific genes in cell hybrids.

Such well-differentiated muscle hybrids are attractive antigen sources for the development of monoclonal antibody reagents specific to human muscle, particularly since segregation of human chromosomes in these hybrids affords the prospect of chromosomal assignment of the genes related to such antigens. Our initial immunizations of Wistar rats with 37.11 cells have resulted in a specific immune response to human muscle surface antigens and no discernible reactivities to the cell surface of L6, to which the rat is tolerant. Similar protocols with chromosomally restricted interspecific hybrids have successfully identified antigens coded by human chromosomes 11 and X (Buck and Bodmer 1976, 1977). We believe that the use of somatic genetic and immune procedures can be used in a search for muscle-specific gene products linked to the Duchenne locus on the X chromosome. We are therefore preparing human L6 muscle hybrids with limited human chromosomal complements including either normal or Duchenne X chromosomes for use in future immunizations.

ACKNOWLEDGMENTS

This work was supported by grants from the Muscular Dystrophy Group of Great Britain. C.A.Q. was supported by a postgraduate studentship from the Muscular Dystrophy Group of Great Britain. O.H. holds a William O. Mosely, Jr., traveling fellowship from Harvard Medical School.

REFERENCES

Brzeski, H., S. Linder, V. Krondahl, and N.R. Ringertz. 1980. Pattern of polypeptide synthesis in myoblast hybrids. *Exp. Cell Res.* **128:** 267.

Buck, D. and W.F. Bodmer. 1974. The human species antigen on chromosome 11. *Cytogenet. Cell Genet.* **14:** 257.

———. 1976. Serological identification of an X-linked human cell surface antigen SA-X. *Cytogenet. Cell Genet.* **16:** 376.

Buckingham, M.E., A. Cohen, F. Gros, D. Luzzata, D. Charmot, and G. Drugeon. 1974. Expression of the myosin gene in a hybrid cell derived from a rat myoblast and a mouse fibroblast. *Biochimie* **56:** 1571.

Dufresne, M.J.P., J. Rogers, B. Coutter, E. Ball, T. Lo, and B.D. Sanwal. 1976. Apparent dominance of serine auxotrophy and the absence of expression of muscle specific proteins in rat myoblast X mouse L-cell hybrids. *Somat. Cell Genet.* **2:** 521.

Emery, A.E.H. 1980. Duchenne muscular dystrophy. Genetic aspects, carrier detection and antenatal diagnosis. *Br. Med. Bull.* **36:** 117.

Friedlander, M. and D.A. Fischman. 1979. Immunological studies of the embryonic muscle cell surface: Antiserum to the prefusion myoblast. *J. Cell Biol.* **81:** 193.

Gardner-Medwin, D. 1980. Clinical features and classification of the muscular dystrophies. *Br. Med. Bull.* **36:** 109.

Grove, B.K., G. Schwartz, and F.E. Stockdale. 1981. Quantitation of changes in cell surface determinants during skeletal muscle cell differentiation using monospecific antibody. *J. Supramol. Cytol.* **17:** 147.

Jacobs, P.A., P.A. Hunt, M. Mayer, and R.D. Bart. 1981. Duchenne muscular dystrophy (DMD) in a female with an X/autosome translocation: Further evidence that the DMD locus is at Xp21. *Am. J. Hum. Genet.* **33:** 513.

Lee, H.U. and S.J. Kaufman. 1981. Use of monoclonal antibodies in the analysis of myoblast development. *Dev. Biol.* **81:** 81.

Lesley, J.F. and V.A. Lennon. 1977. Transitory expression of Thy-1 antigen in skeletal muscle development. *Nature.* **268:** 163.

Quinn, C.A., P.N. Goodfellow, and F.S. Walsh. 1981a. Regulations of Thy-1 antigen in rat-human muscle cell hybrids. *Cell Biol. Int. Rep.* **5:** 767.

Quinn, C.A., P.N. Goodfellow, S. Povey, and F.S. Walsh. 1981b. Human-rat muscle somatic cell hybrids form myotubes and express human muscle gene products. *Proc. Natl. Acad. Sci.* **78:** 5031.

Walsh, F.S. and M.A. Ritter. 1981. Surface antigen differentiation during human myogenesis in culture. *Nature* **289:** 60.

Wright, W.E. and F. Gros. 1981. Coexpression of myogenic functions in L6 rat × T984 mouse myoblast hybrids. *Dev. Biol.* **86:** 236.

Yasin, R., D. Kundu, and E.J. Thompson. 1980. Cell clones derived from adult human skeletal muscle. *Cell Biol. Int. Rep.* **4:** 783.

Yasin, R., G. Van Beers, K.C.E. Nurse, S. Al-Ani, D.N. Landon, and E.J. Thompson. 1977. A quantitative technique for growing human adult skeletal muscle in culture starting from mononucleate cells. *J. Neurol. Sci.* **32:** 347.

Isolation and Characterization of Pure Populations of Normal and Dystrophic Human Muscle Cells

HELEN M. BLAU*
CECELIA WEBSTER*
CHOY-PIK CHIU*
SUSAN GUTTMAN*
BRUCE ADORNATO†
FRANCES CHANDLER*
*Departments of *Pharmacology and †Neurology*
Stanford University School of Medicine
Stanford, California 94305

Human muscular dystrophies are heritable diseases characterized by progressive muscular weakness. Since the primary defects are not known, classification has relied on such characteristics as age of onset, mode of inheritance, muscle groups involved, and rate of progression. A particularly well-defined clinical entity is X-linked recessive Duchenne muscular dystrophy (DMD). However, even in this disease, the lack of information concerning the primary defect has made carrier detection and prenatal diagnosis problematic (Emery et al. 1979; Golbus et al. 1979; Emery 1980). At present, definitive counseling of female carriers relies largely on pedigree analysis and prenatal sex determination. DMD constitutes a problem of sufficient magnitude to warrant concern: It is a chronic, lethal disease that produces considerable psychological and physical suffering. The incidence in the general population is 2.1×10^{-4} (Emery 1977), and the postulated mutation rate of 1.0×10^{-4} is high (Gardner-Medwin 1970).

Many attempts to elucidate the primary defect in DMD have involved analyses of biochemical properties of biopsied muscle. A major difficulty in the biochemical analysis of dystrophic muscle, however, has been its heterogeneous composition. Tissue culture offers a means by which muscle cells

can be investigated under controlled conditions, free of neuronal and circulatory influences. Yet, studies using muscle explants or dissociated cells in culture have also been complicated by cellular heterogeneity. The proportion of fibroblasts is greatly increased in dystrophic muscle tissue, and the rate of proliferation of these contaminating fibroblasts frequently exceeds that of muscle upon plating in culture. Due to this variability, numerous experiments analyzing the properties of normal and dystrophic muscle in tissue culture have given contradictory results (Geiger and Garvin 1957; Bishop et al. 1971; Bateson et al. 1972; Gallup et al. 1972; Witkowski and Dubowitz 1975; Thompson et al. 1977; Thompson 1980; Cavanagh et al. 1981; Franklin et al. 1981).

Clearly, homogeneous cultures of muscle cells, free of other cell types, are required to test hypotheses concerning the etiology of DMD and other dystrophies in a systematic fashion. Toward this end, we have developed an in vitro system for the study of human muscle cells under standardized conditions (Blau and Webster 1980, 1981). Our approach extends the culture methods of Yaffe (1973) and Hauschka (Hauschka and Konigsberg 1966; Hauschka 1974a,b), so that pure populations of human muscle cells can now be obtained in sufficient quantity for biochemical and molecular analysis of differentiation. Muscle cells are routinely isolated by cloning, and several clones from the same individual pooled and plated at high density in mass cultures for studies of differentiation. Frozen stocks are maintained in liquid nitrogen for use in long-term comparative analyses. By using these procedures over the past 2 years, we have accumulated a bank of cloned human muscle cells from 38 different individuals. These cells are used in much the same manner as a continuous cell line, i.e., the cell number and stage of differentiation of a series of replicate plates are precisely manipulated for use in a given set of experiments. Yet, in contrast with continuous lines, these biopsied muscle cells are karyotypically diploid, not altered by adaptation to long-term culture, and have a finite lifespan.

CULTURE SYSTEM

We have used postnatal rather than fetal muscle as a source. A major advantage in the use of postnatal muscle for the study of DMD is that a definitive diagnosis of the disease is only possible after birth. The fetus of an obligate carrier of this X-linked disorder would have a 50% chance of being either an affected male or a carrier female. Thus, fetal muscle cells of male karyotype have at best the uncertainty of a 50% risk for the disease. In addition, since rat muscle cells isolated from animals at late stages of development are capable of producing synaptic cholinesterase when plated in culture —whereas cells from early stages are not (Koenig and Vigny 1978)—human postnatal satellite cells may be more mature than fetal cells in culture.

Our method is based on the proliferative capacity of satellite cells, the mononucleate muscle cells positioned between the basement membrane and the plasma membrane of multinucleate fibers (Mauro 1961; Bischoff 1975;

Konigsberg et al. 1975.) Although our methods were developed for the satellite cells of the vastus lateralis muscle of children, the muscle routinely biopsied for diagnosis of DMD, they appear to have more general applicability. We have obtained proliferative clones and differentiated mass cultures of skeletal muscle from individuals ranging in age from 1 to 80 years and material up to 16 hours postmortem. In addition, we have successfully applied our methods to the culture of muscle from the eye, a muscle whose development, function, and structure differ significantly from that of skeletal muscle (Wolff 1961).

Pure populations of muscle cells are obtained by protease digestion of tissues and plating dissociated cells at clonal density, as described previously (Blau and Webster 1981). Individual clones are separately tested for their myogenic potential by plating a few cells in 16-mm culture wells and scoring these for fusion. The remainder of the cells from three to four clones are pooled, grown to near confluence, and frozen for storage in liquid nitrogen. This approach ensures that the frozen cells are homogeneously muscle, have not initiated differentiation, and spend a minimum of time in culture prior to use. The cells derived from each individual are stored and analyzed separately.

Following freezing and thawing, the capacity of muscle cells to proliferate and to differentiate is maintained. The lag before the onset of cell proliferation in culture and the cell generation time are unaltered, and 99% of cells stored in liquid nitrogen are capable of giving rise to cultures that differentiate to form myotubes. Like rat, mouse, and chicken muscle culture systems (Paterson and Strohman 1972; Yaffe 1973; Jakob et al. 1978), not all human muscle cells of a given clone or mass culture fuse to form myotubes in vitro. In well-differentiated cultures, an average of 70% of total cell nuclei are contained in myotubes, possibly due to developmental asynchrony of the cells at the time of isolation from the tissue.

The composition of the growth and fusion media are of critical importance to the induction of either proliferative or differentiative behavior (Fig. 1). A mitogen and nutrient-rich medium containing 20% fetal calf serum (FCS) and 0.5% chick embryo extract in Ham's F-10 medium stimulates growth. On the other hand, a medium that is relatively nutrient-poor, but high in calcium, such as Dulbecco's modified Eagle's medium with 2% horse serum, promotes differentiation (Blau and Webster 1981). A combination of nutrient deprivation, known to inhibit cell proliferation and induce commitment (Yaffe and Saxel 1977; Nadal-Ginard 1978), and elevated calcium levels, which stimulate fusion in muscle of other species, (Paterson and Strohman 1972) also enhances differentiation of human muscle. Striations and slow rhythmic contraction of myotubes are frequently observed.

STUDIES REGARDING MYOGENIC ETIOLOGY OF DMD

Several hypotheses regarding the etiology of DMD have been based on a myogenic defect. These include abnormal turnover of muscle proteins, a de-

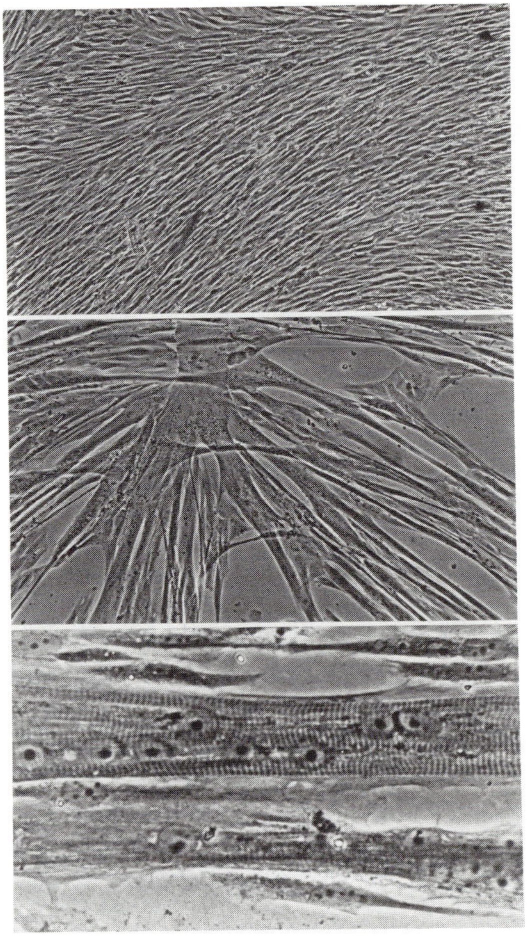

FIGURE 1
Differentiation of frozen-thawed cells in mass culture. Cells grown in medium containing 20% FCS (*top*) continue to proliferate, and cells grown in medium containing 2% horse serum (*center*) differentiate. Visualized using a Leitz inverted microscope with phase-contrast optics. Magnification, 48×. (*Bottom*) Striations are evident in myotubes fixed in 2% glutaraldehyde in Hank's balanced salt solution for 20 min at 37°C and stained with 2% orcein in 45% propionic acid. Magnification, 366×.

fective muscle membrane resulting in muscle enzyme leakage, and defective regulation of expression of differentiation-specific muscle protein isoforms. A number of findings are consistent with an abnormality in the plasma membrane of DMD muscle, which could result in an ineffective cellular barrier (for review, see Roland 1980). The most striking biochemical finding is a 10–100-fold elevation of serum levels of creatine kinase in DMD patients (Zweig et al. 1980). Other muscle enzymes such as aldolase, lactate dehydrogenase, and glutamic oxaloacetic transaminase are three to six times increased in serum (Munsat et al. 1973). The specific increase in serum creatine kinase might reflect a primary membrane lesion or more general damage of muscle fibers. Little is known about these properties of creatine kinase in DMD. Creatine kinase is a dimer composed of a combination of two subunits, B and M. Consequently, three different isozymes are possible.

The muscle and serum of DMD patients, like the muscle of a normal fetus, contains a large proportion of BB and MB in addition to MM creatine kinase isozymes (Zweig et al. 1980). In contrast, normal adult muscle contains primarily MM creatine kinase (Foxall and Emery 1975; Miranda et al. 1979). The presence of the embryonic or brain isozyme in adult skeletal muscle could reflect either an aberrant regulatory mechanism controlling the transition from the fetal to the adult state, as suggested by Kuby and coworkers (1977) or, alternatively, an increased proportion of undifferentiated myoblasts associated with a high degree of regenerative activity (Zweig et al. 1980).

Here, we report the results of experiments conducted in our laboratory to test a number of these hypotheses under well-controlled conditions in tissue culture. Using homogeneous populations of muscle cells from individuals with DMD and matched normal controls, we have systematically compared several independent parameters of muscle differentiation.

The growth properties of DMD and normal muscle cells were compared after muscle was dissociated with trypsin and cells were plated at clonal density. The number of viable cells obtained per gram of muscle biopsy was greatly decreased, and with increasing age of the patient the number of muscle relative to nonmuscle clones declined (Table 1) (H.M. Blau and C. Webster, in prep.). These differences could possibly reflect decreased viability or decreased plating efficiency of DMD muscle cells.

The extent of muscle cell differentiation was examined morphologically by observing the fusion of mononucleate cells into multinucleate syncytia or myotubes. The morphology of the myotubes formed, the kinetics of fusion, and the percentage of total cells fused in both normal and DMD cultures were determined by light microscopy. Neither the morphology of DMD cells at clonal density nor the morphology of DMD myotubes, which differentiated in mass cultures of pooled clones from each of the six DMD individuals, differed from normal (Fig. 2). Furthermore, the rate and extent of cell fusion paralleled normal cultures.

As an assay of possible defects in DMD muscle membranes, the properties of a functionally significant major constituent of mature muscle membranes, the acetylcholine receptor (AChR), were examined. The AChR is a membrane glycoprotein that undergoes posttranslational modification in the Golgi apparatus and follows the intracellular secretory pathway typical of the majority of protein components of biological membranes. The synthesis of AChRs is induced following muscle cell fusion to form myotubes. The accumulation of AChRs in DMD and normal muscle cultures was examined as a function of time in culture. The kinetics of receptor accumulation on the surface of DMD muscle in culture were similar to the values observed in normal cultured myotubes, and the receptors were distributed along the myotubes in an unclustered fashion (Figs. 2 and 3).

Since elevated serum creatine kinase in DMD patients is one of the major diagnostic criteria of the disease, intracellular and extracellular activities of

TABLE 1
Clonal Analysis of Cell Yield

Normal					Dystrophic				
sample	age (yr)[a]	muscle[b]	fb[b]	muscle (%)[c]	sample[d]	age (yr)[a]	muscle[b]	fb[b]	muscle (%)[c]
1	1.6	165	2	99	1	2.6	122	136	52
2	2.0	195	34	85	2	5.0	106	6	95
3	2.0	18	6	75	3	6.0	10	26	43
4	2.0	163	7	96	4	7.5	32	74	41
5	4.0	90	8	92	5	8.5	15	31	49
6	5.0	63	10	86	6	13.0	66	249	39
7	6.3	56	1	98					
Mean ± S.E.M.				90 ± 3	mean ± S.E.M.				53 ± 9

[a] Age of donor.
[b] Number of muscle and fibroblast (fb) clones observed.
[c] Percentage of total clones that were muscle.
[d] Two pairs of dystrophic samples, 3 and 5, 4 and 6, are brothers.

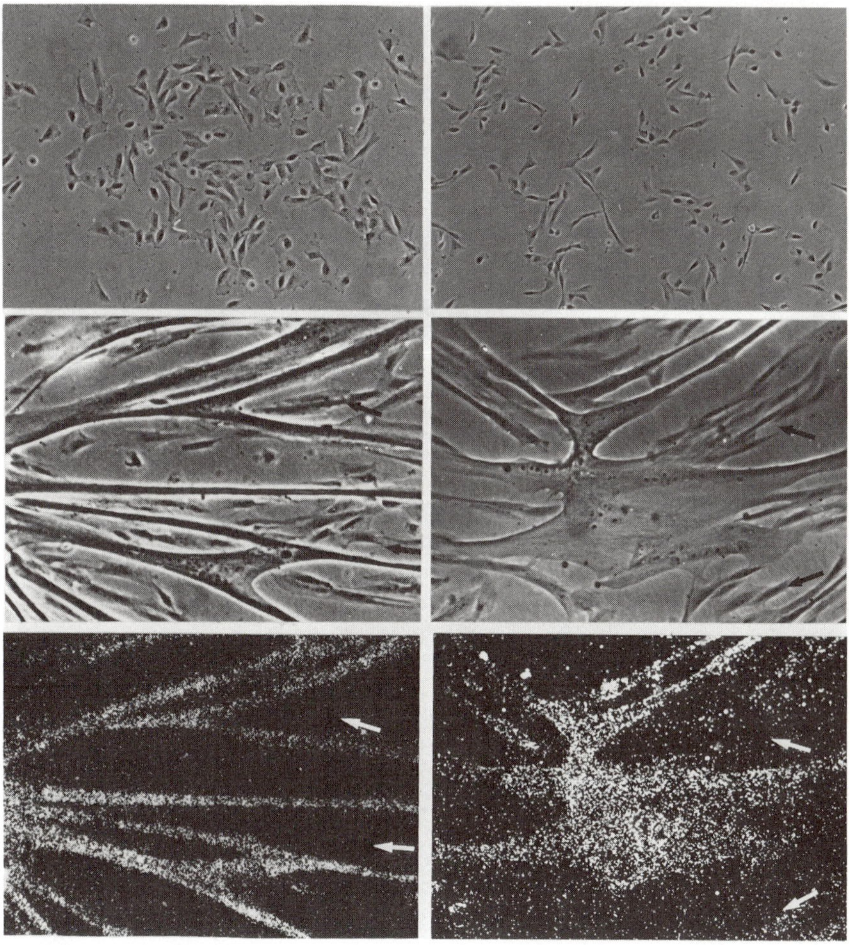

FIGURE 2

Morphology and distribution of AChRs in normal and DMD muscle cells in culture. Normal (*left*) and DMD (*right*). Proliferating clones (*top*) and differentiating myotubes (*center*) shown with phase-contrast optics. (*Bottom*) Same myotubes as above in dark-field illumination reveal the pattern of silver grains and the location of AChRs. Cells were labeled with ^{125}I-labeled α-bungarotoxin (sp. act. 95–140 Ci/mmole, New England Nuclear) for 45 min at 5×10^{-8} M and processed for autoradiography. Specificity of binding was controlled by incubating replicate dishes of cells with unlabeled α-bungarotoxin (3×10^{-6} M) for 20 min prior to and during labeling. Arrows indicate undifferentiated myoblasts that do not have AChRs. Magnification, 89×.

creatine kinase in normal and dystrophic muscle cultures were analyzed as a function of time during differentiation. In both DMD and normal cultures, intracellular creatine kinase activities increased in the course of myogenesis in vitro and frequently reached peak levels of 3000 milliunits/mg of protein

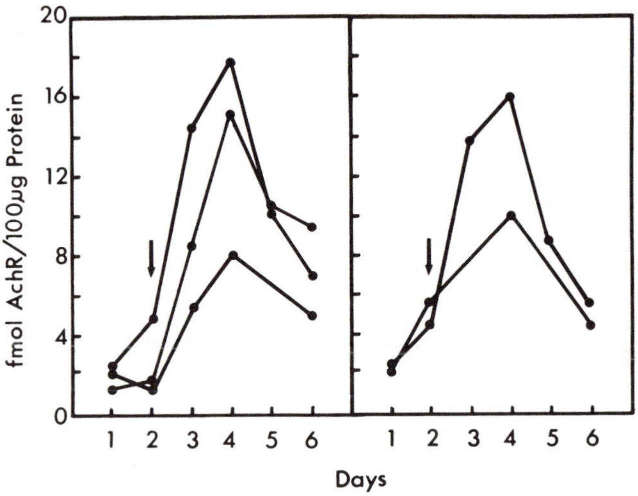

FIGURE 3
Accumulation of AChRs on normal and DMD muscle fibers. Normal (*left*) and DMD (*right*). Days indicate elapsed time after change to fusion medium; (→) onset of fusion. Each point represents the average of total AChR in duplicate dishes reacted with ^{125}I-labeled α-bungarotoxin, as described in the legend to Fig. 2. Cells were solubilized in 1 M NaOH and assayed in a gamma counter with 75% efficiency. Total protein was determined by the method of Lowry et al. (1951).

(Fig. 4). These values compare favorably with those observed in well-differentiated rat primary cells and lines (Shainberg et al. 1971; Turner et al. 1974). Increased intracellular creatine kinase activities due to enzyme accumulation were not observed in DMD cells.

Extracellular creatine kinase activity in the medium was assayed to determine whether creatine kinase leaked from DMD myotubes in accordance with a postulated defect in membrane permeability (Fig. 5). Creatine kinase activity was detectable in the medium of both normal and DMD cultures, but even when creatine kinase was at its peak in the cells, extracellular activity was negligible. The relative increase in extracellular creatine kinase at later times in culture was correlated with myotube death. Although a serum inhibitor of creatine kinase activity has been documented (Simpson et al. 1979), its effect on extracellular activity would presumably be similar in both normal and DMD cultures. Thus, we conclude that neither intracellular nor extracellular creatine kinase levels differ significantly between DMD and control cultures. Leakage of creatine kinase does not occur prematurely in differentiating DMD myotubes but accompanies the necrosis typical of normal end-stage myotubes in culture.

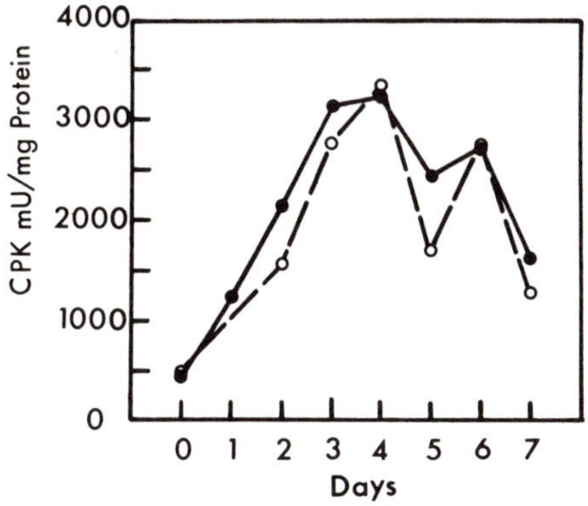

FIGURE 4
Accumulation of creatine kinase in normal and DMD cells in culture. Creatine kinase activity is shown as a function of differentiation in normal (o) and DMD (●) cultures. Days indicate time after change to fusion medium. Each point represents the average of duplicate dishes of cells harvested in 0.1% 2-mercaptoethanol and 1% Nonidet P-40 in 0.05 M glycylglycine (pH 7.5 at 4°C). Following sonication, creatine kinase activity was assayed spectrophotometrically by the method of Shainberg et al. (1971). All creatine kinase activities were corrected for the myokinase (adenylate kinase) present in samples by subtraction of enzyme activity measured in the absence of creatine phosphate. Fusion medium alone had no detectable activity. Specific activity is expressed in milliunits of creatine kinase activity (CPK) per milligram of cell protein. Intracellular activity reached a peak on day 4 in each case.

The creatine kinase isozyme composition was examined as a function of differentiation of normal and DMD myotubes in culture (Fig. 6). A transition from BB to MM isozymes was observed in both the normal and DMD cultures. The BB isozyme was observed in myoblasts, MB appeared upon fusion, and MM was evident with further differentiation. Thus, the shift from B- to M-subunit synthesis, which normally accompanies muscle differentiation, is observed in our DMD and normal muscle cultures. The persistence of the BB form in these cultures is likely to reflect the proportion of unfused myoblasts (~30%).

CONCLUSION

The methods we have developed for the study of human muscle permit replication of experiments with homogeneous muscle material and comparative studies of normal and dystrophic muscle cells under identical conditions (Blau and Webster 1981). By manipulating culture conditions, either proliferation or differentiation of muscle cells can be maximized. Furthermore, the muscle phenotype is stable and cells can be stored in liquid nitrogen for long-term use in numerous comparative experiments.

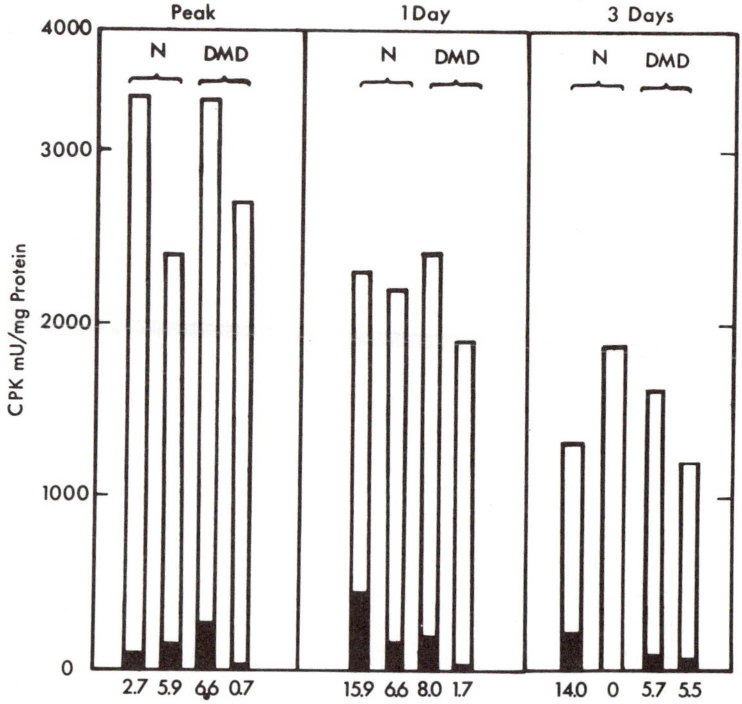

FIGURE 5
Intracellular and extracellular creatine kinase activities in normal (N) and DMD cells in culture. Creatine kinase activity in medium (solid bars) is shown relative to intracellular activity (open bars) at peak, 1 day, and 3 days thereafter. The abscissa indicates the percentage of total creatine kinase activity in medium for each sample. At each time point, results for two normal and two DMD samples are shown, each represented by its own bar graph. Creatine kinase activity (CPK) in cells and media was assayed as described in the legend to Fig. 4 and expressed in milliunits per milligram of total cell protein. Creatine kinase activity in media is similar for normal and DMD cells under identical conditions.

Although the cells isolated by our methods can be manipulated in much the same fashion as a continuous cell line, they do not exhibit a transformed phenotype. As a result, the cells have limited longevity (Hayflick 1965) and senesce after approximately 45 cell doublings in culture so that stocks must be renewed by cell isolation from tissue. However, these cells are karyotypically diploid and their differentiative potential appears not to have been influenced by long-term adaptation to a tissue culture milieu.

The ability to isolate and characterize postnatal muscle and nonmuscle cells makes it possible to obtain cells from individuals with diagnosed human genetic muscle disease. These cells constitute mutants in myogenesis

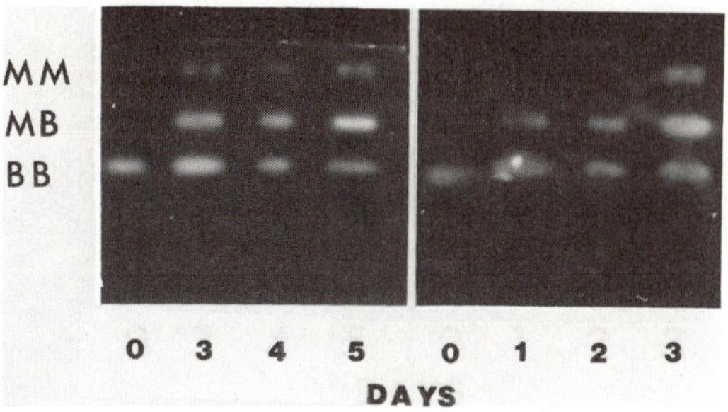

FIGURE 6
Creatine kinase isozymes during differentiation of normal and DMD cells in culture. Normal (*left*) and DMD (*right*). MM, MB, and BB creatine kinase isozymes were separated by electrophoresis on nondenaturing 12% polyacrylamide gels and visualized according to the method of Perriard et al. (1978). Cells were harvested as described in the legend to Fig. 4. Days indicate time after addition of fusion medium. The onset of synthesis of the M subunit is observed in each case.

due to single gene defects. Such mutants are notably lacking in mammalian muscle biology. Characterization of human muscle mutants under well-defined conditions in vitro should greatly contribute to our understanding of the prerequisites for normal muscle differentiation and development and aid in the prenatal detection and treatment of genetic muscle disease.

We have begun to test in a systematic manner several long-standing hypotheses concerning the myogenic etiology of DMD (see also H.M. Blau et al., in prep.). We have analyzed pure populations of normal and dystrophic muscle cells grown under well-controlled conditions and compared simultaneously. Our methods eliminate the problem of cell heterogeneity in muscle cultures, a problem of particular importance in the study of dystrophic muscle in which the proportion of fibroblasts is greatly increased and highly variable. No delay or inhibition of fusion was observed in DMD cultures, and striations and contractions were evident in differentiated myotubes. The induction of synthesis of membrane AChRs was similar between normal and DMD cultures, as was intracellular creatine kinase accumulation and leakage of creatine kinase into the medium. By these criteria, our experiments did not reveal the presence of a membrane defect in DMD muscle cells. Furthermore, the shift from the BB to the MM isozyme was observed. Therefore, the MB isozyme found in the serum of DMD patients is unlikely to be due to aberrant regulation of expression of the M subunit. Rather, it may be derived from the large proportion of regenerating muscle fibers that are developmentally in transition. In summary, pure populations of DMD muscle cells are very similar to normal cells in terms of several features of differentiation that are exhibited in culture.

ACKNOWLEDGMENTS

This work was supported by grants to H.B. from the Muscular Dystrophy Association of America, the March of Dimes Birth Defects Foundation, and the National Institutes of Health (GM-26717), and by a National Institutes of Health postdoctoral fellowship to S.G. (AM-06565).

REFERENCES

Bateson, R.G., D. Hindle, and J. Warren. 1972. Growth pattern *in vitro* of normal and diseased adult human skeletal muscle. *J. Neurol. Sci.* **15:** 183.

Bischoff, R. 1975. Enzymatic liberation of myogenic cells from adult rat muscle. *Anat. Rec.* **180:** 645.

Bishop, A., B. Gallup, Y. Skeate, and V. Dubowitz. 1971. Morphological studies on normal and diseased human muscle in culture. *J. Neurol. Sci.* **13:** 333.

Blau, H.M. and C. Webster. 1980. Isolation of myogenic clones from adult human muscle. *J. Cell Biol.* **87:** 31a.

———. 1981. Isolation and characterization of human muscle cells. *Proc. Natl. Acad. Sci.* **78:** 5623.

Cavanagh, N.P.C., G.I. Franklin, B.P. Hughes, R. Yasin, E. Phillips, G. van Beers, and E.J. Thompson. 1981. Creatine kinase isoenzymes in cultured human muscle cells. II. A study of carrier females for Duchenne muscular dystrophy by needle and open biopsy. *Clin. Chem. Acta* **115:** 191.

Emery A.E.H. 1977. Genetic considerations in the x-linked muscular dystrophies. In *Pathogenesis of human muscular dystrophies* (ed. L.P. Rowland), p. 42. Excerpta Medica, Amsterdam.

———. 1980. Duchenne muscular dystrophy: Genetic aspects, carrier detection, and antenatal diagnosis. *Br. Med. Bull.* **36:** 117.

Emery, A.E.H., D. Burt, V. Dubowitz, I. Rocker, D. Donnai, R. Harris, and P. Donnai. 1979. Antenatal diagnosis of Duchenne muscular dystrophy. *Lancet* **i:** 847.

Foxall, C.D. and A.E.H. Emery. 1975. Changes in creatine kinase and its isoenzymes in human fetal muscle during development. *J. Neurol. Sci.* **24:** 483.

Franklin, G.I., N.P.C. Cavanagh, B.P. Hughes, R. Yasin, and E.J. Thompson. 1981. Creatine kinase isoenzymes in cultured human muscle cells. I. Comparison of Duchenne muscular dystrophy with other myopathic and neurogenic diseases. *Clin. Chem. Acta* **115:** 179.

Gallup, B., H. Strugalska-Cynowska, and V. Dubowitz. 1972. Histochemical studies on normal and diseased human and chick muscle in tissue culture. *J. Neurol. Sci.* **17:** 109.

Gardner-Medwin, D. 1970. Mutation rate in Duchenne type of muscular dystrophy. *J. Med. Genet.* **7:** 334.

Geiger, R.S. and J.S. Garvin. 1957. Pattern of regeneration of muscle from progressive muscular dystrophy patients cultivated *in vitro* as compared to normal human skeletal muscle. *J. Neuropathol. Exp. Neurol.* **16:** 532.

Golbus, M.S., J.D. Stephens, M.J. Mahoney, J.C. Hobbins, F.P. Haseltine, C.T. Caskey, and B.Q. Banker. 1979. Failure of fetal creatine phosphokinase as a diagnostic indicator of Duchenne muscular dystrophy. *N. Engl. J. Med.* **300:** 860.

Hauschka, S.D. 1974a. Clonal analysis of vertebrate myogenesis. II. Environmental influences upon human muscle differentiation. *Dev. Biol.* **37:** 329.

———. 1974b. Clonal analysis of vertebrate myogenesis. III. Developmental changes in the muscle-colony-forming cells of the human fetal limb. *Dev. Biol.* **37:** 345.

Hauschka, S.D. and I.R. Konigsberg. 1966. The influence of collagen on the development of muscle clones. *Proc. Natl. Acad. Sci.* **55:** 119.

Hayflick, L. 1965. The limited *in vitro* lifetime of human diploid cell strains. *Exp. Cell Res.* **37:** 614.

Jakob, H., M.E. Buckingham, A. Cohen, L. Dupont, M. Fiszman, and F. Jacob. 1978. A skeletal muscle cell line isolated from a mouse teratocarcinoma undergoes apparently normal terminal differentiation *in vitro*. *Exp. Cell Res.* **114:** 403.

Koenig, J. and M. Vigny. 1978. Neural induction of the 16S acetylcholinesterase in muscle cell cultures. *Nature* **271:** 75.

Konigsberg, U.R., B.H. Lipton, and I.R. Konigsberg. 1975. The regenerative response of single mature muscle fibers isolated *in vitro*. *Dev. Biol.* **45:** 260.

Kuby, S.A., H.J. Keutel, K. Okabe, H.K. Jacobs, F. Ziter, D. Gerber, and F.H. Tyler. 1977. Isolation of the human ATP-creatine transphosphorylases (creatine phosphokinases) from tissues of patients with Duchenne muscular dystrophy. *J. Biol. Chem.* **252:** 8382.

Lowry, O.H., N.J. Rosebrough, A.L. Farr, and R.J. Randall. 1951. Protein measurement with the folin phenol reagent. *J. Biol. Chem.* **193:** 265.

Mauro, A. 1961. Satellite cell of skeletal muscle fibers. *J. Biophys. Biochem. Cytol.* **9:** 493.

Miranda, A.F., H.Somer, and S. DiMauro. 1979. Isoenzymes as markers of differentiation. In *Muscle regeneration* (ed. A. Mauro), p. 453. Raven Press, New York.

Munsat, T.L., R. Baloh, C.M. Pearson, and W. Fowler, Jr. 1973. Serum enzyme alterations in neuromuscular disorders. *J. Am. Med. Assoc.* **226:** 1536.

Nadal-Ginard, B. 1978. Commitment, fusion and biochemical differentiation of a myogenic cell line in the absence of DNA synthesis. *Cell* **15:** 855.

Paterson, B. and R.C. Strohman. 1972. Myosin synthesis in cultures of differentiating chicken embryo skeletal muscle. *Dev. Biol.* **29:** 113.

Perriard, J.-C., E.R. Perriard, and H.M. Eppenberger. 1978. Detection and relative quantitation of mRNA for creatine kinase isoenzymes in RNA from myogenic cell cultures and embryonic chicken tissues. *J. Biol. Chem.* **253:** 6529.

Rowland, L.P. 1980. Biochemistry of muscle membranes in Duchenne muscular dystrophy. *Muscle and Nerve* **3:** 3.

Shainberg, A., G. Yagil, and D. Yaffe. 1971. Alterations of enzymatic activities during muscle differentiation *in vitro*. *Dev. Biol.* **25:** 1.

Simpson, A.C., D. Holmes, and R.J.T. Pennington. 1979. Dilution effect on serum creatine kinase in carriers of Duchenne muscular dystrophy. *Ann. Clin. Biochem.* **16:** 54.

Thompson, E.J. 1980. Tissue culture of dystrophic muscle cells. *Br. Med. Bull.* **36:** 181.

Thompson, E.J., R. Yasin, G. van Beers, K. Nurse, and S. Al-Ani. 1977. Myogenic defect in human muscular dystrophy. *Nature* **268:** 241.

Turner, D.C., V. Maier, and H.M. Eppenberger. 1974. Creatine kinase and aldolase isoenzyme transitions in cultures of chick skeletal muscle cells. *Dev. Biol.* **37:** 63.

Witkowski, J.A. and V. Dubowitz. 1975. Growth of diseased human muscle in combined cultures with normal mouse embryonic spinal cord. *J. Neurol. Sci.* **26:** 203.

Wolff, E. 1961. *Wolff's anatomy of the eye and orbit*, fifth edition. (Revised by R.J. Last), p. 259. Saunders, Philadelphia.

Yaffe, D. 1973. Rat skeletal muscle cells. II. Preparation of primary cultures. In *Tissue culture methods and applications* (ed. P.F. Kruse, Jr. and M.K. Patterson, Jr.), p. 106. Academic Press, New York.

Yaffe, D. and O. Saxel. 1977. A myogenic cell line with altered serum requirements for differentiation. *Differentiation* **7:** 159.

Zweig, M.H., B. Adornato, A.C. Van Steirteghem, and W.K. Engel. 1980. Serum creatine kinase BB and MM concentrations determined by radiommunoassay in neuromuscular disorders. *Ann. Neurol.* **7:** 324.

SUMMARY

Regulatory Mechanisms in Muscle Development: A Perspective

MARK L. PEARSON
Basic Research Program-LBI
Cancer Biology Program
Frederick Cancer Research Facility
Frederick, Maryland 21701

HENRY F. EPSTEIN
Department of Neurology and
Program in Neurosciences
Baylor College of Medicine
Houston, Texas 77030

This volume summarizes much of our current knowledge of the molecular and cellular factors regulating muscle development. Clearly, we are learning how to identify the individual macromolecules important for the proper development of muscle, even though the regulatory mechanisms governing their biosynthesis and assembly are not understood as well as those for similar processes in prokaryotic systems. No specific strategy for the orderly temporal regulation of muscle gene expression or macromolecular assembly has yet been proposed. Nevertheless, we are now in a position, thanks largely to technological advances in molecular and cell biology, to characterize specific macromolecules and the patterns of their synthesis much more precisely than seemed possible just a few short years ago. Now we can formulate direct questions about how muscle genes are activated and how muscle proteins interact to form functional muscle fibers. In this paper we summarize some of the questions that we think are worth considering by those interested in the formation and maintenance of muscle tissue.

MUSCLE PROTEINS

Analysis of the detailed molecular properties of many of the muscle proteins has uncovered a bewildering diversity of molecular forms. What are the structural and, more important, the functional distinctions between the multiple forms of the muscle proteins? Why do cells contain more than one isozyme for a given functional activity? Is this simply an artifact of evolution reflecting the chromosomal context in which the structural gene for a given isoform finds itself in the genome? Or are blocks of the chromosomes activated at different times during development in such a way that the genes in these blocks are selectively expressed at that particular time? If the latter speculation is correct, it raises the question of what determines the activation of a particular chromosomal domain at a particular time in development.

Alternatively, structural subtleties distinguishing one isozyme from another may signal the presence of secondary functional differences that we have not yet been able to detect. Perhaps such differences are important to the synthesis and assembly of particular isoproteins in distinct subcellular structures. In this regard, it is worth noting that the physiological functions of some of the polypeptides that are commonly recognized in muscle remain unclear. For example, what is the real function of the myosin light chains? Clearly, there are cases of specific myosins in which the light chains are required for calcium regulation of ATPase activity and contraction. In other muscles, do the light chains possibly represent the vestiges of functions "lost" during evolution? In the case of creatine kinase, what is the function of the brain form? Could it play a role in motility in other cell types similar to that of the muscle form in contraction? The muscle form is localized in the M line, where it is presumably necessary for the generation of high local concentrations of ATP during contraction. Perhaps the more ubiquitous brain isozyme is required to generate ATP during cell division as a structural component of the fission apparatus. What is the functional significance of posttranslational modification for the structure and activity of many of the muscle proteins? Methylation, acetylation, phosphorylation, and proteolytic processing, especially at the amino terminus, are known to modify the structure and function of many muscle proteins. For instance, phosphorylation of myosin light chains appears to have a significant role in the regulation of myosin activity and assembly in smooth muscle and nonmuscle cells and may modulate production of tension in skeletal muscle.

GENE REGULATION

The mechanism by which synthesis of different proteins characteristic of muscle is turned on and off during myogenesis and during further maturation of muscle fibers is not yet known. What factors regulate the specific initiation, elongation, termination, and subsequent processing of muscle-

characteristic RNAs? Are there global mechanisms that control the activation of gene expression, perhaps activating different blocks of genes at different times during development? Do gross changes in chromatin structure as reflected in the nuclease sensitivity for transcribed genes reflect the causes or effects of enhanced expression? Is there any spatial relationship between activated domains in chromatin and parts of the nuclear matrix or the inner surface of the nuclear membrane? We need to know both the specific macromolecular composition and the dynamic properties of such structures during development. Similarly, we need to isolate and characterize the regulatory factors that influence the specificity of transcription. Do the same control mechanisms operate on genes activated in muscle at the same time during development?

Much of the work presented in this volume indicates that muscle gene activation during development results in the accumulation of specific mRNAs, suggesting that regulation of these genes occurs primarily at the level of transcription. In muscle, as in other eukaryotic systems, processing of the nascent transcripts is required for the production of active mRNAs. It is not clear whether mRNA processing and transport to the cytoplasm (reactions likely localized to the nuclear envelope) or intranuclear initiation or termination events are the primary sites of control determining the steady-state mRNA levels in muscle. Furthermore, it is important to recognize that alternative control points do exist subsequent to transcription. mRNA compartmentalization into messenger ribonucleoprotein (mRNP) particles and selective processing and turnover of mRNA in the cytoplasm may well affect the level of specific muscle proteins. Although transcriptional activation is necessary for the production of a particular gene product, it is not sufficient to guarantee its appearance in the cell cytoplasm.

The difficulty in determining the rate-limiting step in the control of the appearance of functionally active mRNA in the cytoplasm of a differentiating myoblast stems from technical limitations in labeling, isolating, and fractionating transcription intermediates in living cells. It also reflects the absence in eukaryotic cells, in general, and muscle, in particular, of faithful in vitro systems capable of supporting the proper initiation, elongation, termination, and processing of mRNA molecules.

Data reported in this volume show that gene amplification could provide a significant alternative mechanism for the rapid temporal regulation of gene expression during muscle development, at least in the case of the chicken actin gene. It is very important to see whether this observation can be duplicated in other systems or for other muscle genes. The data indicate that the amplified gene primary structure differs in restriction enzyme sensitivity from that of the unamplified chromosomal copy, a finding that raises the question of whether other primary sequence rearrangements are involved in the differential activation of muscle gene expression. There is no evidence at present that this is the case for muscle proteins other than actin, but the immunoglobulin gene rearrangements accompanying differentiation

in the immune system provide a precedent for the physiological significance of this mode of control. We can expect rapid progress in the analysis of gene primary structure in the next few years as more muscle-characteristic genes are isolated and their sequences are examined both for possible rearrangements and for structural similarities in their flanking regulatory regions.

More is known about the extrinsic factors regulating gene expression during myogenesis than about the intrinsic factors mentioned above. Based primarily on work in cell culture, various hormones and components of the basal lamina have been found that can affect muscle development. How do these factors interact with the cell surface and influence metabolic changes in the cytoplasm or affect gene expression in the nucleus? Future progress in this area will be facilitated by the development of completely defined culture media and the use of cloned cell lines.

Muscle differentiation is accompanied by a cessation of DNA synthesis in terminally differentiated myoblasts. The DNA polymerase-α activity in such cells appears to decrease dramatically. Why? We do not know if initiation, elongation, or termination factors required for DNA replication become depleted during differentiation. Since muscle-characteristic genes are still actively transcribed in mature myotubes, the template properties of the DNA itself appear to remain unaffected, and thus the DNA should retain its competence for DNA replication.

Are there specific DNA-binding proteins able to recognize sequences within the genome that are somehow important to muscle development? We do not understand what distinguishes active DNase-I-sensitive gene sets from inactive DNase-I-insensitive ones. More precise molecular probes of nucleosome and chromatin structure as well as mutants defective in their assembly and activity are required. Is specific DNA modification, especially methylation of specific sequences, required for proper gene expression? In the case of 10T1/2 mouse fibroblasts, 5-azacytidine treatment stimulates the formation of myotubes; this pyrimidine analog is known to interfere with proper DNA methylation in this and other cell types. Besides cytosine methylation, nucleosome phasing, chromosome supercoiling, and nuclear protein methylation, acetylation, phosphorylation and proteolysis may all play roles in the activation of genes during myogenesis. We need to define the specificity of any such modification and to demonstrate that they are a direct cause, not an effect, of selective gene activation. Finally, at least a few genes expressed in myoblasts before fusion do not remain active once myotubes are formed. The identification of these genes and their products and the study of the molecular factors involved in their selective repression need to be pursued.

MEMBRANES

The membranes of myotubes presumably are qualitatively different from the membranes of myoblasts as shown by physical, chemical, and im-

munological analyses. The identification and characterization of new surface antigens that may appear, perhaps transiently, on the surface of differentiating myoblasts will be greatly aided by the more widespread application of monoclonal antibody technology. Not only new proteins and glycolipids but also carbohydrate side chain modifications should be detectable with such probes for surface markers of differentiation. At the same time, antigens existing in a masked state in myoblasts may become unmasked in myotubes, as in the case of the acetylcholine receptor. Some of the developmentally regulated membrane proteins exposed on the myoblast surface may be integral membrane proteins that span the membrane and functionally connect the extracellular space and the cytoplasm proper. Work on the identification of any such proteins and the characterization of their roles in myogenesis remains to be performed.

Do surface antigens exist that play an active role in membrane fusion? What initiates the fusion process? Is myoblast fusion different in any way from the "spontaneous" fusion that occurs infrequently under the usual cell-culture conditions employed for fibroblasts or other cell types? These questions could be approached by examination of the ability of specific monoclonal antibodies to block fusion, although such studies might be compromised by the "capping" and internalization of such surface-bound antibodies. No one as yet has been able to identify and localize surface proteins that are specific to nucleation sites for membrane fusion. How do proliferating myoblasts recognize other myoblasts as distinct cell types different from fibroblasts or chondroblasts (which develop from the same cell lineage), or from quiescent (G_0) satellite cells that might be found in developing muscle?

Membrane fluidity of the lipid bilayer decreases transiently at the time of fusion. Are these gross changes that involve the entire membrane or are they regional changes that result only in local lipid disorder? Alterations of this kind in membrane fluidity may facilitate the diffusion and "rafting" together of membrane proteins in new functional associations that could be significant to membrane fusion. There may be enhanced functional association between the myoblast membrane and the cytoskeleton; perhaps this too has something to do with fusion, although with present methods identification of the nature and character of such associations is difficult.

During myogenesis, myoblasts appear to permit the passive transfer of small molecules (m.w. ~ 1000) via gap junctions. Is this developmentally significant? There is a temporal correlation between the transient appearance of gap junctions and the initiation of fusion; however, the identities of any small molecules that signal or promote myogenic differentiation by passing from cell to cell remain unknown. Does capping of membrane antigens and subsequent internalization occur during muscle development and does it play an important role in myoblast fusion? Endogenous lectinlike substances with hemagglutinating activity are present in many eukaryotic cells, including muscle, and these compounds may function to promote cell-cell interaction. However, their role in muscle development re-

mains unclear. Internalization of hormones—e.g., glucocorticoids, fibroblast growth factor, insulin—bound to their cognate receptors and their subsequent translocation to the nucleus could facilitate molecular signaling from the extracellular fluid through to the nucleus, thus influencing myogenesis. Alternatively, glycosyl transferases may promote membrane fusion by intercellular glycosylation of acceptor sites on adjacent cells, as has been proposed for the organotypic specificity of cell aggregation during embryonic development.

What is the physiological significance of secretion in muscle? There are differences in the electrophoretic patterns of the proteins secreted by myoblasts and myotubes, although the functional properties of any of these proteins are not known. The best-studied example of such a protein is fibronectin, a membrane glycoprotein known to promote fusion in some instances, although its mechanism of action remains unclear. The spectrum of secreted proteins seems complex, as judged by the polyacrylamide gel profile of labeled proteins recovered from culture fluids, but this finding may be caused in part by the membrane-associated protease digestion of a relatively simple set of secreted proteins. How do elements of the extracellular matrix affect myoblast fusion and muscle regeneration? Could the basal lamina play distinctly different roles in embryonic development and in the maintenance of adult fibers? These roles may not be the same as that of the extracellular matrix in promoting acetylcholine receptor localization at the neuromuscular junction. What is the role of proteolysis in membrane fusion? Neutral proteases do exist on the muscle cell membrane; although they appear not be under developmental control, their precise location is unknown and their endogenous substrates remain to be identified.

EMBRYOGENESIS

During embryonic development, myoblasts become committed to forming myotubes and synthesizing the spectrum of muscle-characteristic proteins. What are the molecular correlates of commitment? Is commitment reversible and, if so, when? What are the cell lineages leading to committed muscle cells and how are they related to other mesodermal lineages, especially those leading to chondrocyte and adipocyte differentiation? Is commitment regulated by extrinsic factors that can be identified and characterized? Morphogenetic gradients presumably are established early in embryonic differentiation of mesodermal lineages. If so, what are the chemical determinants in such gradients? What are the sources of such gradients; how are they maintained and subsequently remodeled during development? What are the interactions between the specific cell types involved? If such myogenic factors exist, are they specific regulatory proteins that interact with particular DNA sequences? What do the nearby genes code for? The

most direct approach to such questions in development would seem to lie in the genetic approaches being employed in *Caenorhabditis* and *Drosophila*. A concentration of effort on the isolation and characterization of mutants affecting muscle and nerve function in these organisms is likely to provide the biological material required to dissect the pathway of differentiation and to define the molecular interactions involved in commitment and embryonic development.

MORPHOGENESIS

We do not understand in molecular detail how the individual structural specializations characteristic of muscle, such as thick filaments, thin filaments, basal lamina, endplate, and sarcoplasmic reticulum, are assembled from their component parts. We need to know if their assemblies follow ordered pathways that can be genetically and biochemically analyzed in vitro, as has been done successfully in the case of bacteriophage morphogenesis. What is the role of the associated macromolecules, not only proteins but also nucleic acids or lipids, in catalyzing the assembly of such structures? The posttranslational modification of muscle proteins may be another significant factor in determining assembly or disassembly. The muscle plasma membrane and the cytoskeleton itself may serve as essential transitory nucleation points for the assembly of specialized structures not normally associated with either the membrane or the cytoskeleton. For instance, do desmin- and vimentin-containing filaments participate in organizing Z disks? Are the M-line and Z-line constituents critical to the organization of thick and thin filaments? Does the cytoskeleton organize acetylcholine receptors into endplate membranes? Does it participate in the assembly of the membranes of the sarcoplasmic reticulum or the T-tubule system? All of these problems are now more amenable to analysis with advances in the use of monoclonal antibodies, improvements in protein purification techniques, plus the development of more sophisticated analyses.

Such methods have been applied to muscle development in the past with emphasis on the static rather than the dynamic aspects of organelle structure. However, the assembly of such structural proteins may be closely coupled with their biosynthesis, as is the case in other morphogenetic systems. Can the assembly of the contractile apparatus proceed in vivo without protein synthesis or protein processing and modification and even degradation? The answers to such questions have critical implications for any attempts to assemble such structures in vitro. In this regard, coupling of synthesis and assembly may be required physiologically, although possibly not from a physical-chemical point of view. To try and dissect what are likely to be complicated interactions among many components in the pathway of assembly of such structures, we will probably require novel

mutants or a new generation of very specific drugs to perturb each of the step-by-step interactions along these pathways. In our view, the solution of such problems by chemistry alone seems unlikely.

HUMAN DISEASE

One of the major limitations in dealing with human muscle diseases is the difficulty of determining whether the disorder is really caused by a dysfunction of the muscle fiber itself. A primary defect expressed in nerve could be expressed as a secondary defect in muscle fibers because of impaired nerve-muscle interaction. For example, Duchenne muscular dystrophy may be caused by a nerve defect that subsequently affects commitment and subsequent myotube differentiation or it may be caused by an intrinsic defect of muscle cells themselves. Determinations of this kind require the culture of muscle cells obtained by biopsy from patients afflicted with neuromuscular diseases. Once in culture, these cells can be readily manipulated for biochemical and physiological assays. Not only can radiolabeling of newly synthesized metabolites be done, but also larger absolute numbers of cells can be obtained by proliferation of the biopsy samples in culture. Clearly, coculture of motor neurons with their cognate muscle cells in the normal and diseased state could offer significant advantages to analysis of the target tissue and the biochemical nature of the primary defect.

Culturing muscle biopsy material often results in the expression of genes and proteins that appear to be specific to embryonic muscle and not the neonatal or the adult forms characteristic of the tissue from which the sample was obtained. Since some muscle disorders, such as McArdle disease (glycogen phosphorylase deficiency), affect specifically proteins of the mature adult fibers, muscle biopsies may not be as useful as one would wish for analysis of diseases of this kind unless it becomes possible to stimulate the maturation of embryonic-type myotubes to form adult-type fibers in culture. In contrast, other muscle disorders, e.g., phosphofructose kinase deficiency, may appear to affect proteins expressed in both embryonic and adult muscle, making their analysis amenable to studies involving muscle biopsy cultures.

A major problem is determination of the molecular genetic basis of inherited human muscle disease at the DNA level. For disorders with biochemically defined lesions, such as McArdle disease, it is now theoretically possible to use recombinant DNA technology to clone cDNAs coding for the deficiency in question and to use them as probes for the isolation of genomic DNAs from normal and affected individuals. Analysis of the nucleotide sequences of the mutant and wild-type alleles can localize the mutations down to the nucleotide level, if they lie in or near the structural gene deficiency. Although tedious in practice, depending on the abundance of the mRNA coding for the defective enzyme and the size of the gene, this method is straightforward. However, for inherited disorders with unknown

biochemical defects or with regulatory gene mutations, such as Duchenne muscular dystrophy, there is no clear-cut way to approach this problem. An indirect method being pursued for this and other human genetic diseases relies upon the discovery of restriction site polymorphisms that can be linked to the diseased state by the analysis of DNA taken from the members of affected families. The principle employed is the same as that used to map actin genes in the nematode. In humans, in which the genetic locus being investigated is unknown, the choice of a suitable probe is difficult. As a consequence, success in such a search relies on a combination of luck and the judicious choice of a cloned DNA probe containing a sequence repeated many times in the genome. Many different sequences with varying repetition frequencies exist in the human genome making such searches less daunting than might otherwise be the case. Analysis of restriction site polymorphisms using such probes on Southern blots of DNA from patients with known chromosomal abnormalities and syndromes related to disease may make the identification of genetic lesions associated with that disease possible. One other approach to the genetic analysis of human muscle disease involves the use of fine-structure karyotyping to search for specific chromosomal abnormalities in patients with a high probability of such anomalies, such as boys who suffer from both mental retardation and from Duchenne muscular dystrophy.

In the more distant future, gene transfer technology may be used to assay and ultimately clone those genes that correct biochemical defects expressed in muscle cells in tissue culture. This type of analysis does not necessarily require foreknowledge of the nature of the biochemical defect if there is some selectable phenotype that distinguishes normal from diseased muscle cells. The isolation of genes from several different human tumors capable of conferring a neoplastic transformed phenotype on normal NIH-3T3 fibroblasts is one example of the successful use of this method to clone genes on the basis of their biological activity. Having such genes cloned, even as just a fragment of the coding sequence, would permit the preparation of antibodies to portions of the protein gene product which, in turn, might be used to identify the polypeptide that is responsible for the primary defect in the disease state. Current genetic engineering and immunological methods, coupled with advances in the growth of cells from muscle biopsies, make us optimistic that the next few years will see great advances in our understanding of the basic underlying defects in some human muscle diseases. The challenge will be to use this information to better disagnose and treat the clinical manifestations of such diseases in muscle and nerve as well as in other tissues.

ACKNOWLEDGMENTS

This research was sponsored by grants from the National Institute of Aging and a Research Career Development Award from the National Institute of

Child Health and Human Development, and support from the Jerry Lewis Neuromuscular Disease Research Center of the Muscular Dystrophy Association of America (to H.F.E.) and the National Cancer Institute, Department of Health and Human Services, under contract N01-C0-23909 with Litton Bionetics (M.L.P.). The contents of this publication do not necessarily reflect the views or policies of the Department of Health and Human Services, nor does mention of trade names, commercial products, or organizations imply endorsement by the U.S. Government.

Author Index

Subject Index